SPRINGER SERIES IN PHOTONICS   4

Springer
Berlin
Heidelberg
New York
Barcelona
Hong Kong
London
Milan
Paris
Singapore
Tokyo

Physics and Astronomy

ONLINE LIBRARY

http://www.springer.de/phys/

SPRINGER SERIES IN PHOTONICS

Series Editors: T. Kamiya   B. Monemar   H. Venghaus

The Springer Series in Photonics covers the entire field of photonics, including theory, experiment, and the technology of photonic devices. The books published in this series give a careful survey of the state-of-the-art in photonic science and technology for all the relevant classes of active and passive photonic components and materials. This series will appeal to researchers, engineers, and advanced students.

Series homepage – http://www.springer.de/phys/books/ssp/

Norbert Grote   Herbert Venghaus (Eds.)

Fibre Optic Communication Devices

With 281 Figures

 Springer

Dr. Norbert Grote
Dr. Herbert Venghaus
Heinrich-Hertz Institut für Nachrichtentechnik Berlin GmbH
Einsteinufer 37
10587 Berlin, Germany
e-mail: grote@hhi.de
e-mail: venghaus@hhi.de

Series Editors:

Professor Takeshi Kamiya
Dept. of Electronic Engineering, Faculty of Engineering
University of Tokyo
7-3-1 Hongo, Bunkyo-ku, Tokyo, 113, Japan

Professor Bo Monemar
Dept. of Physics and Measurement Technology, Materials Science Division
Linköping University
58183 Linköping, Sweden

Dr. Herbert Venghaus
Heinrich-Hertz-Institut für Nachrichtentechnik Berlin GmbH
Einsteinufer 37
10587 Berlin, Germany

ISSN 1437-0379
ISBN 3-540-66977-9 Springer-Verlag Berlin Heidelberg New York

Library of Congress Cataloging-in-Publication Data.

Fibre optic communication Devices/Norbert Grote, Herbert Venghaus, eds. p.cm. – (Springer series in photonics; v. 4). Includes bibliographical references and index. ISBN 3540669779 (alk. paper). 1. Optical communications – Equipment and supplies. 2. Optoelectronic devices. I. Grote, Norbert, 1950– . II. Venghaus, Herbert, 1943– . III. Series. TK5103.59 D52 2000   621.382'7–dc21   00-039465

Springer-Verlag Berlin Heidelberg New York
a member of BertelsmannSpringer Science+Business Media GmbH

© Springer-Verlag Berlin Heidelberg 2001
Printed in Germany

Typesetting: Dataconversion by LE-TEX, Leipzig
Cover concept: eStudio Calamar Steinen
Cover production: *design & production* GmbH, Heidelberg

Printed on acid-free paper   SPIN: 10574904   57/3141tr   5 4 3 2 1 0

Preface

Modern communication technologies have revolutionized society and economy on a global scale, and progress in this field is still breathtaking. Among the crucial features of these developments is the possibility to transmit/receive ever-increasing bit rates at lower and lower cost. This is due to some extent to improved electronics, but it is much more due to the capabilities of optical, fibre-based communication systems. Optical communication technologies were initially introduced in order to improve traditional information exchange, but they have recently become a key prerequisite for the tremendous growth in internet traffic, and optical technologies will be even more important in enabling and supporting the future expansion of internet traffic with annual growth rates forecast at 100%–200%.

It is the purpose of the present book to describe the enabling components of optical communication systems, in particular their fundamentals, principles and current status, as well as the potential for future developments. Following such a scheme, it is the editors' hope that this book will not only attract the attention of experts already working in the field, but be, at least, likewise directed to interested newcomers. The book is organized into eleven chapters addressing the essential optical and optoelectronic components which form the hardware basis of today's, and most likely of future, optical networks. It has not been the intention of the editors and the authors to present a consistently written book; instead it is comprised of 'stand-alone' articles to be readable in themselves. In some cases, therefore, minor overlaps will be encountered. The articles have been written by different authors and coauthors, all of them renowned experts contributing outstanding work to their specific field. A brief introduction to them is given at the end of this book. The editors owe them many thanks for their great efforts, being aware that there is always lack of time for doing such demanding work, which not only requires a comprehensive, in-depth technical knowledge but also didactic abilities.

The first chapter on the characteristics of optical communication systems illustrates basic principles, but also identifies the most important components and their functionality and thus sets the frame for the subsequent chapters, which focus on the different components in more detail.

An optical transmitter, a receiver and the optical fibre link in between are the most basic parts of any kind of optical communication system. Consequently, the first three components-related chapters focus on these elements. Chap. 2 presents an introduction to the optical fibre, the transport medium for the optical signals travelling in the optical network. Starting with a thorough theoretical treatment of the waveguiding properties of those fibres, from which various designs can be deduced, an overview will be given on fibre materials and basic optical properties. Another section has been devoted to nonlinear effects occurring in the fibre transmission at high intensities. Finally, the problem of pulse propagation and solitons will be addressed.

The following chapter deals with laser diodes used as optical transmitters. Following a theoretical description of the basic properties of lasers an introduction to the design of laser structures is given. Of utmost importance for optical transmission are single-mode lasers to cope with the impact of fiber dispersion. The theoretical and practical aspects of those devices will be treated, and the chapter will finish with an outline of wavelength-tunable laser structures.

In Chap. 4 photodiodes and related optical receivers will be treated. The various structures of photodiodes developed will be presented together with their optical and electrical characteristics. Thereafter, receivers incorporating electronic amplifier stages are discussed, and technological approaches for their monolithic integration are outlined.

Optical amplifiers have revolutionized optical networks, allowing signal transmission over hundreds of kilometers without optical/electronic conversion. Nowadays, amplifiers based on erbium doped fibres are predominantly employed in the 1.55-μm transmission window. This amplifier type will be comprehensively treated in part 1 of Chap. 5, whereas in part 2 semiconductor optical amplifiers are presented which in principle can be tailored to any desired wavelength band in the $1.3-1.6\,\mu m$ range. In particular at $1.3\,\mu m$ they have proven attractive because of the lack of an efficient fibre amplifier in that window.

Glass integrated-optics devices (Chap. 6) comprise passive and active devices as well. Passive devices – due to their extremely low loss and mature technological state – have already evolved into standard commercial products, while active devices are still in the laboratory state. In addition to being compact, active glass devices may eventually achieve high functionality, but so far a number of unresolved technological problems prevent a real competition between active glass components and their widely used fibre-based counterparts. Particular emphasis is given to the complex physics governing waveguide structures with multiple doping and to related technological issues, but it includes a number of examples of devices already fabricated.

Wavelength division multiplexing (WDM) has been initially introduced primarily in order to enhance the optical fibre transport capacity, and more than 100 channels can already be transmitted simultaneously over a single fibre. In addition, WDM will increase the functionality of optical networks significantly as it enables efficient routeing, in particular in combination with wavelength conversion. Key elements are wavelength-selective transmitters but also corresponding wavelength-selective devices such as demultiplexers and filters. These are of prime importance for WDM networks and they are covered in Chap. 7. Fabry-Perot and dielectric multilayer filters are already commercially available, the same holds for devices relying on gratings, e.g. Bragg gratings in optical fibres or bulk gratings for wavelength demultiplexers. Phased arrays or arrayed waveguide gratings (AWG) have proven to be extremely versatile components suitable for multiple applications in multiplexing/demultiplexing, routeing or even as a building block in wavelength selective transmitters. Finally, acousto-optic $LiNbO_3$ filters are particularly versatile, as any combination of wavelength channels to be multiplexed/demultiplexed can be freely chosen, at the expense, however, of a rather high power consumption.

Optical switches are needed and currently used for protection switching, i.e. they assure network survivability in case of various failures. However, they are also of prime importance for routeing of (wavelength) channels in optical networks with a complexity beyond simple optical point-to-point links. It would be desirable to have switches, which exhibit several characteristics simultaneously. These comprise low crosstalk, low insertion loss, low power consumption for switching, and, for certain applications, high switching speed. Switches can be made according to various principles, but none of them exhibits optimum (or close to optimum) performance for a larger number of parameters. Mechanical switches have particularly high crosstalk attenuation and low insertion loss, but moderate switching speed. They are particularly suited for wavelength channel switching; they are widely used in today's optical communication systems and thus they constitute the largest part of Chap. 8. Other switching concepts, which are covered only to a much smaller extent, are still research topics, in particular switches fast enough for optical packet switching, which would raise the (routeing) capacity of the internet tremendously. Finally, another type of device treated in Chap. 8 are $LiNbO_3$-based switches (or modulators) used in high bit rate (10 or 40 Gbit/s) transmitters.

The following chapter describes all-optical time-division multiplexing (OTDM) technology, which is indispensable for the development of very large-capacity optical transmission systems. Focus is placed on the key technologies including ultrashort pulse generation, all-optical multiplexing/demultiplexing, optical timing (clock) extraction and optical pulse-waveform measurement. Their application to all-optical TDM and TDM/WDM transmission systems and the transmission capacity extension into the Tbit/s regime will be given particular attention.

Optical communication owes its rapid progress not only to the vast transmission capacity of optical fibres, but also to the devices enabling the generation, detection and manipulation of optical signals propagating along the fibres, and accordingly the two final book chapters are devoted to the fabrication of optoelectronic (OE) components, modules and optoelectronic integrated circuits (OEICs). Normally they all comprise subcomponents with rather different structure, and as a consequence integration concepts, which have enabled the stupendous progress and success of silicon technology, cannot simply be adapted to optoelectronic devices. Thus the development of hybrid OE modules and that of monolithic OEICs are two approaches complementing each other and which will both be developed further in their own right. This is described in the concluding chapters of the book, which cover the fundamental principles and a survey of the actual status as well.

June 2000   *Herbert Venghaus*
Berlin   *Norbert Grote*

Contents

Abbreviations

AM: amplitude modulation
APD: avalanche photodiode
AR: anti-reflection
APM: additive pulse mode-locking
ASE: amplified spontaneous emission
BER: bit-error rate
BH: buried heterostructure
BR: buried ridge
CET: cooperative energy transfer
CNR: carrier-to-noise ratio
CPM: colliding pulse modelocking
CPW: coplanar waveguide
CSO: composite second-order
CTB: composite triple beat
D: dispersion parameter
DBR: distributed Bragg reflector
DCF: dispersion compensating fibre
DEMUX: demultiplexer
DFB: distributed feedback
DHBT: double heterostructure bipolar transistor
DSC: dispersion-slope compensated/-ion
DSF: dispersion-shifted fibre
DWDM: dense wavelength division multiplexing
EA: electro-absorption
EDF: erbium-doped fibre
EDFA: erbium-doped fibre amplifier
EDFRL: erbium-doped fibre ring laser
FET: field effect transistor
FFP: Fibre Fabry-Perot filter
FHD: flame-hydrolysis deposition
FIT: failures in time
FM: frequency modulation
FOM: figure of merit
FP: Fabry-Perot (filter)

FPR: free propagation region
FSR: free spectral range
FRA: fibre Raman amplifier
FTTF: fibre-to-the-floor
FTTH: fibre-to-the-home
FTTO: fibre-to-the-office
FWM: four-wave mixing
FWHM: full width at half maximum
FXC: fibre crossconnect
GCSOA: gain-clamped semiconductor optical amplifier
GPI: gain-to-pump intensity (ratio)
GPS: global positioning system
GVD: group velocity dispersion
HBT: heterojunction bipolar transistor
HEMT: high electron mobility transistor
HR: high reflection
IBE: ion beam etching
IR: infrared
LAN: local area network
LP: linearly polarized
LSI: large-scale integration
LSHB: longitudinal spatial hole burning
M: multiplication factor
MBE: molecular-beam epitaxy
MCM: multi-chip module (OE-MCM: opto-electronic -)
MCVD: modified chemical vapour deposition
MEMS: microelectromechanical switch
MESFET: metal-semiconductor field effect transistor
MI-WC: Michelson-interferometer wavelength converter
ML: mode-locked
MOCVD: metal-organic chemical vapour deposition
MOMBE: metal-organic MBE
MOVPE: metal-organic vapour-phase epitaxy
(M)QW: (multi-)quantum well
MSL: microstrip line
MSM: metal-semiconductor-metal
MTTF: mean-time-to-failure
NA: numerical aperture
NE: network element
NF: noise figure
NLS: nonlinear Schroedinger equation
NOLM: nonlinear-optical loop mirror
NRZ: non-return to zero
OADM: optical add/drop multiplexer

OADU: optical add/drop unit
OBF/OBPF: optical bandpass-filter
OCH: optical channel protection
OE: optoelectronic
OEIC: optoelectronic integrated circuit
OMS: optical multiplex section
ONU: optical network unit
OT: optical terminal
OTDM: optical TDM
OTM: optical terminal multiplexer
OXC: optical crossconnect
PDL: polarization-dependent loss
PDM: polarization-division multiplexing
PDS: passive double-star
PDSR: passive distribution, switched recombination (switching matrix)
PECVD: plasma-enhanced chemical vapour deposition
PIC: Photonic Integrated Circuit
PIN: p-type – intrinsic – n-type
PLC: planar lightwave circuit
PLL: phase-lock loop
PM: polarization maintaining
PMD: polarization mode dispersion
PRBS: Pseudo-random binary sequence
RE: rare earth
RIE: reactive ion etching
RIN: relative intensity noise
RMS: root-mean-square
RNF: refracted near-field (method)
RX: photoreceiver
RZ: return-to-zero
SAM: separate absorption and multiplication
SAGM: separate absorption, graded and multiplication
SAW: surface acoustic wave
SBS: stimulated Brillouin scattering
SC: supercontinuum
SDH: synchronous digital hierarchy
SDSR: switched distribution, switched recombination (switching matrix)
SFG: sum-frequency generation
SHBT: single heterojunction bipolar transistor
SLALOM: semiconductor laser amplifier in a loop mirror
S/N, SNR: signal-to-noise ratio
SOA: semiconductor optical amplifier
SONET: synchronous optical network
SOP: state of polarization

SPM self-phase modulation
SRS: stimulated Raman scattering
SSC: spot size converter
SSMF: standard single mode fibre
STS: silica-on-terassed-silicon
STF: standard telecommunications fibre
TV: television
TCM: time-compression multiplex
TDM: time-division multiplexing
TDMA: time-division multiple access
TE: transverse electric
TL: transform-limited
TLLM: transmission line laser model
TM: transverse magnetic
TOAD: terahertz optical asymmetric demultiplexer
TSI: time slot interchange
UV: ultra-violet
UTC: uni travelling carrier
VCO: voltage-controlled oscillator
WDM: wavelength-division multiplexing
WGPD: waveguide-integrated photodiode
WINC: wavelength-insensitive coupler
XPM: cross-phase modulation

List of Contributors

Yuji Akahori
NTT Photonics Laboratories
Photonics Integration Project
162 Tokai-Mura
Ibaraki-ken 319-1193
JAPAN

Erland Almström
Plogvägen 6
169 59 Solna
SWEDEN

Heinz-Gunter Bach
Heinrich-Hertz-Institut für
Nachrichtentechnik Berlin GmbH
Einsteinufer 37
10587 Berlin
GERMANY

Herbert Burkhard
Troyesstrasse 46
64297 Darmstadt
GERMANY

Frank Fidorra
Heinrich-Hertz-Institut für
Nachrichtentechnik Berlin GmbH
Einsteinufer 37
10587 Berlin
GERMANY

Philip Garel-Jones
JDS Uniphase Corporation
570 West Hunt Club Road
Nepean (Ottawa)
Ontario K2G 5W8
CANADA

Ken Garrett
4871 Valkyrie Drive
Boulder Colorado 80301
USA

Louis Giraudet
OPTO+
Groupement d'Intérêt Economique
Route de Nozay
91461 Marcoussis Cedex
FRANCE

Norbert Grote
Heinrich-Hertz-Institut für
Nachrichtentechnik Berlin GmbH
Einsteinufer 37
10587 Berlin
GERMANY

Stefan Hansmann
OptoSpeed Deutschland GmbH
Am Kavalleriesand 3
64295 Darmstadt
GERMANY

Helmut Heidrich
Heinrich-Hertz-Institut für
Nachrichtentechnik Berlin GmbH
Einsteinufer 37
10587 Berlin
GERMANY

Harald Herrmann
University of Paderborn
Department of Applied Physics
Warburger Strasse 100
33098 Paderborn
GERMANY

Winfried H.G. Horsthuis
JDS Uniphase Corporation
570 West Hunt Club Road
Nepean (Ottawa)
Ontario K2G 5W8
CANADA

Gerlas van den Hoven
JDS Uniphase Netherlands
Prof. Holstlaan 4 (WAD)
5656 AA Eindhoven
The NETHERLANDS

Sonny Johansson
Ericsson Hungary Ltd.
Laborc út 1
P.O. Box 107
1300 Budapest
HUNGARY

Ronald Kaiser
Heinrich-Hertz-Institut für
Nachrichtentechnik Berlin GmbH
Einsteinufer 37
10587 Berlin
GERMANY

Antoine Kévorkian
Teem Photonics
13, chemin du vieux chêne
38240 Meylan
FRANCE

Ton Koonen
Lucent Technologies
Botterstraat 45
Huizen 1271 XL
The NETHERLANDS

R. Ian MacDonald
JDS Uniphase Corporation
570 West Hunt Club Road
Nepean (Ottawa)
Ontario K2G 5W8
CANADA

Edmond J. Murphy
JDS Uniphase
Electro-optic Products Division
35 Griffin Road South
Bloomfield, CT 06002
USA

Masatoshi Saruwatari
Department of Communications
Engineering
School of Electrical
and Computer Engineering
National Defense Academy
1-10-20 Hashirimizu, Yokosuka-shi
Kanagawa 239-8686
JAPAN

André Scavennec
OPTO+
Groupement d'Intérêt Economique
Route de Nozay
91461 Marcoussis Cedex
FRANCE

Meint K. Smit
Optical Communication Technology
TU Delft, Faculty IST
Department TC
Mekelweg 14
2628 CD Delft
The NETHERLANDS

Wolfgang Sohler
University of Paderborn
Department of Applied Physics
Warburger Strasse 100
33098 Paderborn
GERMANY

Hiroshi Terui
NTT Electronics Corporation
Optical Module Business Promotion
Photonics Business Group
162 Tokai-Mura
Ibaraki-ken 319-1106
JAPAN

Réal Vallée
Laval University
Departement de Physique
Sainte-Foy, Qc, G1K 7P4
CANADA

Herbert Venghaus
Heinrich-Hertz-Institut für
Nachrichtentechnik Berlin GmbH
Einsteinufer 37
10587 Berlin
GERMANY

Carl Michael Weinert
Heinrich-Hertz-Institut für
Nachrichtentechnik Berlin GmbH
Einsteinufer 37
10587 Berlin
GERMANY

Yasufumi Yamada
NTT Electronics Corporation
Product Development Department
Photonics Business Group
162 Tokai-Mura
Ibaraki-ken 319-1106
JAPAN

Mikhail N. Zervas
Optoelectronics Research Centre
Southampton University
Southampton SO17 1 BJ
GREAT BRITAIN

1 Characteristics of Optical Communication Networks

Sonny Johansson and Erland Almström

This chapter will present different network and system solutions built on optical technologies and will discuss associated functionalities. The main focus will be on wavelength division multiplexing (WDM), which has initiated tremendous capacity increase, particularly in the transport area. First, however, some general aspects and features of optical networks will be described. This is then followed by a discussion of WDM networking, where new optical network elements are introduced. WDM networking provides several options in connectivity for a given network topology. The choice of connectivity pattern results in different amounts of wavelengths required and impacts the complexity of the optical network elements. The basics of different connectivity patterns and the requirements on the optical network are discussed in this chapter. Furthermore, protection switching within the optical network is also elaborated, where different protection schemes are presented with particular emphasis on ring protection. Finally, various optical network elements that are useful and important for WDM networking are discussed.

1.1 Optical Network Issues

The most important development in optical communication since the development of low-loss optical fibres is the erbium-doped fibre amplifier (EDFA) [1,2]. The EDFA overcame the earlier power-budget limitation of transmission systems and pushed the transmission limitations forward to become dispersion-limited instead. The EDFA broke through as a commercially mature technology in 1994. At this time most optical system experiments in laboratories exploited EDFAs for various applications. Suddenly several power-consuming techniques became viable, e.g. negative-dispersion fibre for dispersion compensation [3], four-wave mixing [4] or Sagnac-interferometry [5] for optical time-division multiplexing (OTDM), and soliton transmission [6], to name just a few examples, where the EDFA became a crucial element. On the other hand, for coherent transmission [7] the success of the EDFA became a killer. Sensitivity was one of the coherent technology's strongest feature, and by using an EDFA as a preamplifier for a traditional ON/OFF keyed receiver the same sensitivity as that of a coherent receiver could be obtained. As a consequence, a substantial part of the activity in the coherent technology area dropped when the EDFA appeared.

Transparency is a concept that has been a strong selling argument for optical networks. However, the degree of transparency is limited both by transmission rate and by network extension [8]. This leads to difficult compromises when designing a network for different transmission formats, where the transparency decreases with the extension of the network. Different transmission formats have very different characteristics and are differently affected by perturbations. It is a challenge to design an optical network that is fully transparent to transmission format and at the same time obtains a certain coverage of the network. For example, a separate EDFA is transparent to bit rate and code format, but when cascading a number of the same EDFAs in a chain the transparency will be reduced considerably. The transparency is dependent on the EDFA operation conditions, e.g. signal level, gain, and noise accumulation, and these depend on the EDFA span. It is not realistic to change the span of the EDFAs afterwards to adapt the network for new operation conditions such as a higher bit rate for example. Thus an upgrade to higher bit rates will be impossible, if this bit rate was not taken into account in the original design. On the other hand, a network design should be limited to those transmission formats that can be expected within a foreseeable time frame, to avoid over-engineering. Despite the fact that the EDFAs compensate for transmission loss, it is still crucial to keep down this loss between the EDFAs, in order to keep the signal-to-noise ratio (SNR) at an acceptable level. This stresses the importance of minimizing the insertion loss of optical network elements for use in optical networks.

Although the EDFA has considerably extended possible transmission distances, new obstacles have appeared that set new limitations. In addition to chromatic dispersion, polarization-mode dispersion (PMD), polarization-dependent loss, noise accumulation, and gain-spectrum variation are examples of new parameters that limit the bandwidth and transmission distance, and at the same time the scaleability of optical networks. The major limiting factor at higher bit rates and longer transmission distance is chromatic dispersion [9]. However, dispersion compensation methods such as pre-chirping, dispersion-compensated fibre [10], dispersion-compensated grating [11], or mid-span spectral inversion [12] can reduce most of the inter-symbol interference caused by the dispersion. For very long distances and high bit rates PMD is an important limiting factor. Due to its non-deterministic, time-varying characteristic it is difficult to compensate, and to overcome PMD limitations at 40 Gb/s bit rate is a topic of current research. The first choice is to develop and deploy new fibres that are less prone to PMD; the second best solution is the development of compensation techniques [13].

In the case of multi-wavelength networks additional obstacles appear, e.g. stimulated Raman-scattering, cross-phase modulation, and four-wave mixing. For more information about nonlinearities see [14]. These effects set the limit of maximum power and minimum channel spacing allowed in the network. Wavelength stabilization is another issue that is tied to concatenated filtering-effects, i.e. the resulting pass-band of cascaded filters will be narrowed, if the individual filters have non-flat spectral windows. This puts higher requirements on wavelength stabilization of the transmitter terminal laser. Another concatenation effect is the resulting narrow transmission window when cascading EDFAs, which limits the number of wavelengths through the amplifier and has driven the development of gain-flattened EDFAs (see Chap. 5).

1.2 Long-Haul Networks

The first commercial transocean system, which made use of EDFA technology, was installed in 1995 as the Trans-Atlantic Telephone, TAT-12/13 cable network [15]. It had a bit rate capacity of 5 Gb/s on a single wavelength, and was upgraded to 20 Gb/s during 1999 with WDM technology. The TAT 12/13 is configured as a 4-fibre self-healing ring, with 300 ms restoration time. Two nodes are located in the USA, one in the UK and one in France. The longest transparent section over the Atlantic is 5759 km and contains 133 EDFAs, with a span of 45 km. The continuation of the Atlantic cables is TAT-14, which contains four fibres with 16 10-Gb/s channels. Built on the same principles as the TAT 12/13, the Trans-Pacific Cable network, TPC-5, which encircles the Pacific Ocean, was completed the following year [16]. At the same time, the Asia Pacific cable network was put into operation [17]. a year later the Fibre Link Around the Globe (FLAG) cable system came into operation [18], which interconnects the TAT 12/13 with the TPC-5. It runs from the UK to Japan through the Mediterranean, Suez, and the Indian Ocean. The next global submarine project is Oxygen, which spans the world with a length of 16800 km. The total number of access points amounts to 99; there will be four parallel fibres, each carrying 32 channels at 10 Gb/s, which yields an overall capacity of 1.28 Tb/s.

Even higher capacity will be offered by the FLAG Pacific-1 submarine cable system, linking Los Angeles, San Francisco, Seattle, Vancouver and Tokyo and scheduled to go into operation in 2002. The system comprises a northern and a southern cable with eight fibre pairs per cable, dense WDM will ensure a throughput of 64 wavelengths per fibre pair, and with each wavelength carrying 10 Gb/s the total capacity will amount to 5.12 Tb/s.

As indicated, transocean cable networks are generally designed for future capacity upgrade, exploiting the WDM technique. In general, there are two basic means of increasing the capacity of a fibre line-system, either by time-division multiplexing (TDM) or with WDM techniques. Traditionally, TDM was the preferred choice, and as early as 1994 many 100 Gb/s TDM transmission experiments were carried out at laboratories, applying optical means for high-rate TDM, e.g. four-wave mixing or the optical loop mirror approach [19,20]. In the case of a TDM upgrade the bit rate is so high that dispersion becomes the major problem to be overcome. Thus, a dispersion-compensating system must be installed, or the repeater span shortened, in order to upgrade a long-haul line system.

On the other hand, with WDM a lower bit rate is maintained by multiplexing a number of lower rate data on different wavelengths, which limits the effect of dispersion. The commercial breakthrough for the WDM technique came in 1996 due to the development of the flat-gain EDFA [21]. Thus, the capacity of the most advanced WDM systems was attained [22]. Instead of repeaters one single EDFA, which amplifies all wavelengths simultaneously, replaces multiples of repeaters (one for each channel) at each repeater location.

1.3   WDM Networking

Up to now we have only discussed the exploitation of WDM to enhance the transmission capacity on a point-to-point basis. However, a very interesting feature is that WDM usage can be extended to enable optical networking, where the wavelengths are routed to different destinations [23]. Work in this direction was initiated already in the mid 1980s [24]. In order to accomplish this, new types of optical networking elements are needed, which can process separate wavelengths.

As a first step, the wavelength add/drop feature is introduced to provide optical wavelength connections to intermediate nodes along a WDM link. In this way, by adding/dropping a separate wavelength at different locations along this link, while leaving the other wavelengths on the link untouched, a large transport capacity can be obtained without affecting small intermediate nodes. These nodes only have to process the portion of traffic that is carried on the dropped wavelength. In order to realize such a function Optical Add/Drop Multiplexers (OADMs) must be introduced (see Sect. 1.6.2).

A good example of a network that exploits OADMs is the Africa Optical Network (Africa ONE). The Africa ONE is a 40000 km undersea WDM network that encircles the African continent, and provides 54 land drops to 40 different countries [25]. The WDM network is divided into 9 shorter sections, which are terminated by central offices at each end. The link capacity is 2 fibre pairs × 8 wavelengths × 2.5 Gb/s, which gives 40 Gb/s in total. Along each section a number of passive OADMs, built on WDM multiplexers/demultiplexers, are placed on the sea bed, approximately 100 km out from the shore. The placement of multiplexers on the sea bed gives perfect conditions for temperature stabilization, where the temperature is absolutely constant. The function of the OADMs is to add/drop one wavelength from/to a cable station on land. The cable stations, in this case, represent intermediate nodes between two central offices.

In a network with homogeneous traffic demand, i.e. all nodes have a mutual traffic demand with all the other nodes, the traffic pattern becomes meshed. This can be supported within a WDM ring, but the number of wavelengths needed increases rapidly ($\propto N^2$), where N is the number of nodes (see Table 1.1). A mesh WDM ring is therefore not very scaleable, since the WDM technology is still limited by the number of wavelength channels. It counts in tens and not in the thousands that other transport technologies achieve. To meet the requirement of granular optical connectivity for a large network, a mesh-fibre topology is needed. This will reduce the number of wavelengths per link considerably. Unfortunately, the network will be more complex and a new type of optical network element is needed: the optical cross-connect (OXC), which differs from the OADM by cross-connecting separate wavelengths between several fibres (see Sect. 1.6.3).

Figure 1.1 illustrates a metropolitan application of optical networking, exploiting both OADMs and OXCs. In this example the OXCs are used in the mesh backbone and to interconnect local rings. The local rings comprise OADMs, which can also be used, where appropriate, to add/drop wavelengths into/from the backbone. In this way the traffic load is distributed over the network by routeing the wavelengths according to the demand for capacity. However, wavelength channels represent a rather coarse granularity in network provisioning, where each wavelength carries capacity of the order of 0.5–10Gb/s. Thus, finer granularity is provided with a complementary transport technology, represented in Fig. 1.1 by the asynchronous transfer mode (ATM) technology.

Pushing this scenario further, one can imagine a dynamically reconfigurable optical network that optimizes the utilization of the network resources, e.g. ATM switches. In this way, large traffic load changes can be re-balanced

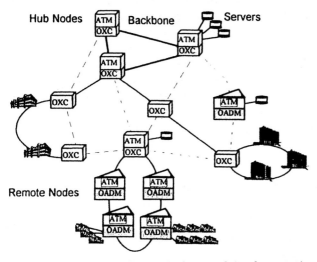

Fig. 1.1. Illustration of an optical network implementation in a metropolitan environment. Optical network elements provide high-capacity routeing, the combination with ATM switches (as an example) assures high granularity

with the optical network layer. For example, one can provide more capacity to industrial areas during working hours, and redirect capacity at night to residential areas to provide video-on-demand, etc.

1.4 Connection Patterns

Given a physical network topology, i.e. node and fibre location, different wavelength connectivity patterns can be established with the help of OADMs. To illustrate some important connectivity patterns we will focus on OADMs in a WDM ring topology. However, the same connectivity patterns can then be adopted for other network topologies as well. The type of connectivity pattern is chosen to meet a certain traffic demand, but will also impact the scaleability of the ring, in terms of number of nodes and available wavelengths. We will assume that the principle of 'wavelength reuse' is exploited, which makes better use of available wavelengths in the network. This means, when a wavelength is dropped and the channel is terminated, a new channel can be sent on the same wavelength. An alternative principle, which is particularly suited for small-scale networks [26], is 'broadcast-and-select'. In this case the wavelengths are broadcast over the entire network and the receiving node selects one. As the same wavelength is also present elsewhere in the network, it cannot be reused.

As mentioned in the previous section, crosstalk can be a severe obstacle that must be carefully suppressed. If the OADM technology used suffers from poor crosstalk performance, the ring could be left open, though. In this case all wavelengths will always be terminated either at the head or at the tail of the open ring. This prevents the light from oscillating, but at the expense of decreased connectivity. Nevertheless, the example of connection patterns that follows below is based on a closed-ring topology.

The most basic is the hub pattern (Fig. 1.2a). In this case, one dedicated wavelength is used to connect each node in the ring to one hub node that is placed somewhere in the same ring. This pattern is simple, because each of

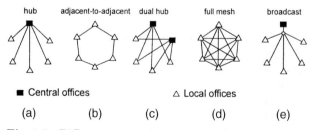

■ Central offices   △ Local offices

(a)   (b)   (c)   (d)   (e)

Fig. 1.2. Different connection patterns that can be established within a WDM ring by proper wavelength routeing. The number of connection lines in (**a**)–(**d**) represents the number of separate wavelengths needed for the pattern. In the case of (**e**) 'broadcast' no back channel is needed

Table 1.1. Number of wavelength connections each OADM must handle (second column) and total number of different wavelengths needed (third column) to interconnect N nodes in different connectivity patterns (first column) in a unidirectional ring (see next section)

Patterns	No. of connections per OADM	Total no. of wavelenghts in the ring
Hub	1	N-1
Dual hub	2	2(N-2)
Adjacent-to-adjacent	2	N
Full mesh	N-1	N(N-1)/2
Broadcast	1/2[a]	1

[a] 1/2 connection: connection in one direction only.

the nodes, except the hub, only handles one wavelength. The hub node, on the other hand, must handle all the wavelengths in the ring and therefore becomes more complex in realization practice. The total number of wavelengths needed in the ring is given in Table 1.1. In the case of a hub pattern, all traffic in the ring must go through the hub node. The hub node has the task of re-grooming traffic to obtain high wavelength utilization, and to serve as a gateway to other sub-networks. The hub pattern is particularly useful when the mutual traffic demand between the nodes in the ring is limited and a large portion is directed out of the ring through the hub.

Another simple pattern is the adjacent-node pattern (Fig. 1.2b). This means that each node has only two connections, one to each of the two neighbours. The traffic is then routed with a multi-hop technique, which means that all nodes within the ring take part in a distributed re-grooming process. The number of wavelengths required in this case is only N, but could be extended with more wavelengths to limit the number of hops. In this case the pattern gradually transforms to a full-mesh pattern, where only single hops occur.

To provide redundancy, dual-homing is often practised (Fig. 1.2c). Then two hub nodes are placed in the same ring to improve the redundancy in case of hub failure. This implies that the other nodes must handle two connections for this case, one to each hub, and the complexity of the ring increases.

The full-mesh pattern is the most complex one (Fig. 1.2d), with all nodes interconnected. In this case the number of wavelengths is consumed rather rapidly (see Table 1.1), and is therefore not as scaleable as the others. However, this pattern is useful when the availability requirement is very high, and the mutual traffic demands between individual nodes are sufficiently high to obtain acceptable wavelength utilization without re-grooming.

A different pattern is the broadcast pattern (Fig. 1.2e). This scheme introduces a new functional requirement at the OADM nodes. The OADM must, in this case, accomplish a drop-and-continue function. This occurs when a wavelength is split in order to be dropped at the same time as it continues around the ring, where it is dropped also at the successive nodes. However, the last node has to prevent the wavelength from making more than one circuit, to avoid interfering with itself.

So far the discussion has assumed that a WDM link/ring is permanently configured. This requires that a network is designed for a certain demand, where the wavelength routeing is set to a pre-defined pattern. However, it is not always possible to predict the future demand and a flexible network solution is desired in order to meet changes in traffic load. This can be achieved with automatic reconfigurable OADM nodes that can select the wavelength to be added and dropped as desired. To give a simple example, consider the case of an adjacent-node pattern WDM ring, which was described above. The traffic is in this case re-groomed at every intermediate node. Let us assume that the traffic flow between two distant (i.e. non-adjacent) nodes increases and becomes the major mutual traffic exchange which these two nodes have with all other nodes. Then it would be better if these two particular nodes were put adjacent to each other, to offload the intermediate nodes with this traffic. In other words, the number of hops would be minimized and traffic load in the client network lowered. This can be achieved by configuring the wavelength settings so that a node receives a direct connection to the most important counterpart, while bypassing intermediate nodes. As a result of this, the node has been logically shifted around the ring to appear next to the node with the highest mutual traffic demand.

However, comparing a fixed adjacent-node pattern of a WDM ring with a similar reconfigurable WDM ring, the latter will consume more wavelengths. This means that a lowered traffic load in the client network is transferred to more wavelengths in the optical network, which is just what one is looking for in some circumstances.

1.5   Optical Network Protection

Network protection against cable break or node failure is very important when planning for robust network solutions. This is a feature that optical networks can provide very efficiently [27]. The idea is to accomplish protection-switching close to the origin of the fault, resulting in a smaller number of actions. This will make the protection-switching fast and effective.

1.5.1   Protection Schemes

There are two basic protection schemes to be considered: dedicated protection and shared protection. 'Dedicated protection' means that a spare route is reserved as a backup for a particular route to be protected. This can be implemented in two ways, denoted $1 + 1$ protection and $1 : 1$ protection. In $1 + 1$ protection the signal is split, at the transmitter end, into two routes directed to the same receiving end. Usually these two routes are different in order to avoid both routes being hit by one cable break. At the receiving end the signal is available from both routes. If the signal from the first route fails, the receiver simply switches over and selects the second route. This scheme is very simple, where the decision of protection switching can be taken locally at the receiver end, without the need of signalling. It only costs a 1×2 splitter at the transmitter end and a 1×2 optical switch at the receiving end, but, unfortunately $1 + 1$ protection allocates two routes (see Fig. 1.3a).

In $1 : 1$ protection, one route is dedicated to backup, but it is utilized for low priority traffic when not activated for protection. Thus the $1 : 1$ protection scheme allocates one route only for the protected traffic. In order to switch in/out low prioritized traffic 2×2 optical switches must be implemented at the transmitter end and at the receiver end as well (Fig. 1.3b). This makes the $1 : 1$ protection scheme a bit more complex than $1 + 1$ protection, because when the receiving end switches over to the protecting route it must signal back to the transmitter end to do the same.

When the probability for failure is low, shared protection is used. This means that several (n) protected units, routes or equipment, share a few (m) protecting units, and this scheme is denoted m:n protection. In the special case, when only one protecting unit is used, it is called $1 : n$ protection. When

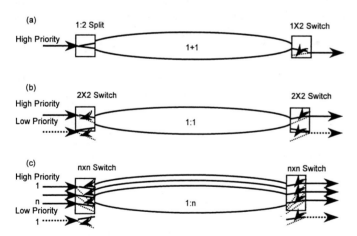

Fig. 1.3. Protection schemes: **(a)** $1 + 1$ dedicated protection: the signal is transmitted on two routes simultaneously; protection-switching is made at the receiving end only. **(b)** $1 : 1$ shared protection: protection switching needed at each end; the protecting route can be utilized for low-priority traffic during normal conditions. **(c)** $1 : n$ shared protection: only one protecting route is used to protect (n-1) other routes

one out of the (n) high-priority units fails, the protecting unit becomes active. However, this means that there are no more spare units available if a second unit also fails (Fig. 1.3c). Therefore, this method is useful only when the probability of more failures is sufficiently low.

So far, protection switching has been discussed for the the space domain only, but it can be also implemented in the wavelength domain. For example, in the case of a 16 WDM transmitter, one transmitter may act as a protecting unit for the other 15 ones. If one of these transmitters breaks, the protecting unit goes active and replaces the failing source by tuning into the right wavelength ('1 : 15 shared protection switching'). This approach gives time to replace the failing unit, while the affected traffic transmits via the protecting unit. Later, when the broken transmitter has been replaced, the traffic is switched back to the original state and the protecting unit is available in case of new failures.

In the case of optical WDM networks there are two more protection principles: optical channel (OCH) protection and optical multiplex section (OMS) protection. In this context an optical channel can be considered to be identical to a wavelength channel, although not conceptually true. OCH protection processes channel-by-channel. This means that different channels, even on the same fibre, can have different protection schemes, protecting routes or no protection at all. OCH protection can be used to protect only certain highly prioritized channels, whereas others are left to their destiny. OMS protection, on the other hand, processes all channels from one and the same fibre as one unit. For example, at cable break all channels are re-routed in a group.

1.5.2   Ring Protection

Optical protection can be accomplished in different fibre topologies, and we will treat the ring structure in more detail here [28]. The ring gives an inherent protection ability and is interesting, because it provides diverse routes for protection and it allows simple protection switching decisions, when "east" or "west" directions are the only options. The ring may look very simple at first glance, but a closer look reveals that it can be designed in different architectural variations. The basic ring architectures are two-fibre unidirectional, two-fibre bidirectional, and four-fibre bidirectional rings. The latter are implemented as two reversed unidirectional rings.

The principal difference between a unidirectional and a bidirectional ring is how a connection is established within the ring. In both cases full duplex connection is supported. In a unidirectional ring all traffic propagates into the same direction, using one fibre. The second fibre is only used as a protecting path in the opposite direction. This means, seen from one node, if the receive path is from "west" the return path will be to "east" (Fig. 1.4a). In the case of a bidirectional ring the return path would be to "west" (in this example), using the second fibre (Fig. 1.4b). Thus, the bidirectional ring

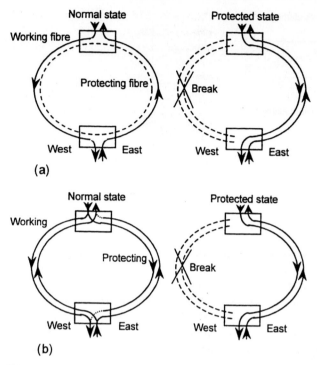

Normal state   Protected state

Working fibre

Protecting fibre   Break

West   East   West   East

(a)

Normal state   Protected state

Working

Protecting   Break

West   East   West   East

(b)

Fig. 1.4. Ring protection: **(a)** unidirectional; **(b)** bidirectional ring protection

uses the shortest sector of the ring. The other sector is used as a protecting connection. This difference between unidirectional and bidirectional rings also gives a different protection scheme.

In the case of unidirectional rings protection switching is accomplished by folding the ring to avoid the faulty section; the two nodes on each side of the fault switch over to the second ring, thus restoring the connection through the second fibre (Fig. 1.4a). The two nodes on each side of the fault will become the head, respectively the tail, of the folded ring. This example illustrates an OMS protection, since the restoration is accomplished on the fibre level independent from individual channels.

When providing a protecting route for a connection in a bidirectional ring, then the ring sector not used for the principal connection is employed as a spare, alternative route. As in the example, this can be implemented as $1 + 1$ OCH protection by splitting the transmitted signal into both fibre rings, to "west" and to "east". At the receiver end, the signal is available both from "west" and "east". Thus, if the signal from "west" fails, the receiver only has to switch over to "east" (Fig. 1.4b). In this way the restoration is accomplished channel by channel. As a consequence this is designated OCH protection in contrast to OMS protection as discussed before, where the focus is on reconnecting a fibre rather than rerouting individual channels.

Wavelength-selective network elements

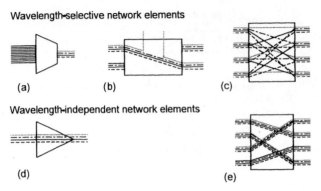

(a)   (b)   (c)

Wavelength-independent network elements

(d)

(e)

Fig. 1.5. (a)-(c) Wavelength-selective, **(d,e)** wavelength-independent optical network elements

1.6   Optical Network Elements

The constitutive parts of any optical network are, besides the interconnecting fibres, a number of functional units, which are generally called optical network elements (NE). These optical NEs can be categorized into two groups: those that operate on individual wavelengths and those that operate on the complete set without any distinction between separate wavelengths. Elements falling into the latter group are the fibre amplifier, which is supposed to amplify all wavelengths simultaneously (and equally), and the fibre cross-connect, which is formed by optical space switches that operate on the fibre base. These devices will not be treated in more detail here, as they are discussed in Chaps. 5 and 9, respectively. Basic network elements, that operate on separate wavelengths, are: the optical terminal multiplexer (OTM), the optical add/drop multiplexer (OADM), and the optical cross-connect OXC (see Fig. 1.5.). However, each of these can be further subdivided to represent different degrees of functionality and thereby complexity in realization.

1.6.1   Optical WDM Terminal Multiplexer

An OTM terminates the optical paths within optical networks [i.e. terminates and (de-)multiplexes wavelengths] and adapts the characteristics of signals interchanged with other networks, e.g. synchronous digital hierarchy (SDH) [29], to the format required for transmission within the optical network. In this case SDH will be the client of the optical network. An OTM may have electrical and optical interfaces, it may be all optical and must have at least one optical interface in order to qualify as an "optical" TM. The OTM includes both transmitter (Chap. 3) and receiver (Chap. 4), plus a duplex interface to the client network. The adaptation function of the OTM can be to set the correct wavelength, but can also imply other signal processing such as: scrambling of signal, change of modulation scheme or adaptation of physical interface. The latter may comprise setting proper output power, linewidth and extinction ratio, for example. The termination function is another functionality of an OTM that extracts a small portion of extra information, overlaid on the signal, which is needed to assess and maintain the transmission through the optical network. This could, for example, be implemented as a pilot tone, which carries an identification code [30].

The main task of the OTM, however, is to generate a good-quality signal that matches the characteristics of the optical network. Three levels of manipulation have to be considered when adapting the characteristics of a signal to the optical network:

- (re)amplification ("1R regeneration")
- amplification + reshaping ("2R regeneration")
- amplification + reshaping + retiming ("3R regeneration").

Amplification (1R regeneration) can be made all-optically, i.e. using a fibre amplifier, or, alternatively, optoelectrically (oe). As 1R regeneration introduces noise to the signal, the number of 1R repeaters that can be cascaded is limited. In order to improve the SNR a reshaping function (2R) can be implemented with a non-linear amplifier. This will open up the signal eye-diagram. Unfortunately, 2R reshaping introduces jitter into the signal instead, which also contributes to a degradation of signal quality. The only means to obtain full recovery of the signal is by 3R regeneration. This requires clock extraction to recover the clock rate from the signal, and a decision circuit to recover the original bit pattern, which makes it more complex in realization. Moreover, the transparency to code format and bit rate is lost using a 3R regenerator, as the code format and bit rate are defined in the design of the regenerator. A repeater which may have different wavelengths at the input and output is normally referred to as a 'transponder'. Transponders based on 1R regenerators (comprising, at least, a detector, a linear amplifier, and a laser as fundamental elements) have limited cascading capability due to noise accumulation, but they are optically transparent. On the other hand, the performance of transponders comprising 3R regeneration is much superior, but at the expense of them being no longer fully optically transparent. Thus attempts are currently being made to improve existing 3R regenerators with respect to bit rate flexibility or in order to get a limited optical transparency. In particular, taking advantage of optical self-pulsations for such purposes is a topic of actual research [31].

Depending on where and how the OTM is operated, there are three basic wavelength allocation schemes for WDM applications:

- single channel with fixed wavelength assignment,
- multi-channel with fixed wavelength assignment,
- single channel with dynamic wavelength assignment.

In the case of a ring/bus network with single hub connectivity, only the single channel OTM with a fixed wavelength setting is needed. This means that the source must be selected to hit the dedicated wavelength, but has no requirement to address also other wavelengths within the network. However, for a mesh connectivity pattern a multi-channel OTM would be useful to handle several connections with different wavelengths. A 100-wavelength channels multiplexing system has been demonstrated [32]. Finally, the OTM with dynamic wavelength assignment, e.g. a tuneable source, will be useful in a network with a dynamic traffic load. This OTM could be reconfigured/retuned to new wavelengths that reach different destinations. New traffic demands can then be met in a flexible manner, so that a better utilization of existing network resources is obtained.

1.6.2 Optical WDM Add/Drop Multiplexers

An OADM comprises one or two optical add/drop units (OADU), at least one transponder or terminal, and optionally a fibre protection unit (see Fig. 1.7, below). The purpose of the OADM is to selectively terminate only one or a few wavelength channels from the network, while forwarding unaltered the remaining wavelength channels that are passed. The channel is extracted from the dropped wavelength and a new channel is inserted on the same wavelength, which is added to the network again. The OADM can either have static wavelength assignment or may be able to dynamically select what wavelength to add/drop.

There are several principles and technology choices [33] for how to add/drop selected wavelengths from an optical network. Technology for wavelength add/drop is discussed in detail in Chap. 7, so we will focus here on the principles only. The OADM enables wavelengths to be reused for different channels in separate parts of the same optical network. However, this requires that the OADU has a high isolation when dropping this wavelength. There is a risk that other wavelengths may leak through, even at a low level, and affect the signal quality negatively. Crosstalk that comes from other wavelengths is often referred to as inter-band crosstalk. A more severe crosstalk component is intra-band crosstalk that appears when a wavelength interferes with itself, or when a dropped channel interferes with the added channel that reuses the same wavelength. This can result in coherent interference between channels, which requires much higher suppression. Therefore, any residual power from a dropped wavelength must be suppressed with an effective stop band. For further references on crosstalk phenomena in optical WDM networks see [34–36].

There are essentially two basic OADU principles. One approach is based on WDM. In this case all wavelengths arriving at the input are demultiplexed, including the wavelengths that are just passing. The drop wavelengths, which could be one or a few, are routed off and terminated. The passing wavelengths are multiplexed again at the output, together with the added wavelength(s) (Fig. 1.6a). In this approach, the OADM must be designed to handle all the wavelengths that may arrive at the node. It will not be possible to introduce

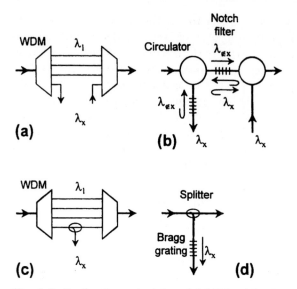

Fig. 1.6. Realization principles of OADUs: **(a)** all wavelengths are demultiplexed and routed separately, **(b)** only one wavelength is selected, the remaining wavelengths are passed through, **(c)** and **(d)** drop-and-continue of one wavelength

a new wavelength to the network later, which the WDMs are not able to demultiplex. This design is flexible, and it is easy to introduce more functions that operate on separate wavelengths, e.g. monitoring. The drawback is that this design is less scaleable towards high numbers of wavelengths.

A second approach, based on notch filters such as fibre Bragg gratings, only separates the wavelength that is dropped. In this way the OADU is transparent to any change in wavelength allocation within the network, as long as it does not affect the particular wavelength that is dropped (Fig. 1.6b). This makes networks based on this type of node easily scaleable with respect to the number of wavelengths. However, in this design the crosstalk component from the dropped channels must be carefully suppressed by the notch filter, since any remaining signal power will lead to interferometric crosstalk. This is the most severe crosstalk component in WDM networks.

Drop-and-continue is a function that is not always required, but is needed for broadcast features or to enable dual homing in the optical network. In those cases, the wavelength is both dropped by the OADU and at the same time passed through towards the next node (Fig. 1.6c,d).

Fibre protection is another essential function for the OADM, which can be realized effectively. However, depending on the protection architecture, as described in Sect. 1.5, the implementation will differ. Figure 1.7 shows three different OADM protection architectures. In all cases optical space switches (see Chap. 8) are needed to accomplish protection switching.

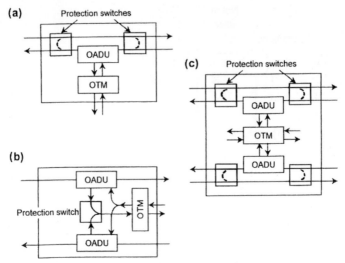

Fig. 1.7. Examples of different OADM protection architectures: **(a)** OADM for OMS protection in a unidirectional ring, **(b)** realization of $1 + 1$ OCH protection in a bidirectional ring, **(c)** duplication of unidirectional ring OMS protection in a four-fibre bidirectional ring implementation

1.6.3   Optical WDM Cross-Connects

An OXC comprises of a cross-connect core unit, which operates on separate wavelengths, and an ingress and an egress part that demultiplexes/multiplexes the wavelengths. Like the OADM, the OXC can do wavelength add/drop as well, simply by connecting OTMs to a few of the OXC ports [37]. In addition, a protection unit can be added to accomplish OMS protection, whereas OCH protection can be handled by the cross-connect core unit. Protection switching was discussed in Sect. 1.5 and will not be further elaborated here. The realization of the core unit can be divided into three basic levels of functionality, which of course will impact the complexity in realization:

- – wavelength routeing, fixed,
- – wavelength switching, blocking,
- – wavelength interchanging/switching, strictly non-blocking.

The simplest form of unit for cross-connecting wavelengths is a fixed configuration wavelength router. It can either be realized by fixed interconnections between a set of WDMs (Fig. 1.8a), or by an arrayed waveguide grating (see Chap. 7 and [38]). The latter is integrated on one chip and realized as one device. Even with a fixed router dynamic routing can be obtained, if OTMs have dynamic wavelength assignment. Then, when shifting the wavelength of an OTM, say symbolically from 'red' to 'green', the transmitted signal will find different outlets of the wavelength router.

WDMs   **Switches**   **WC**

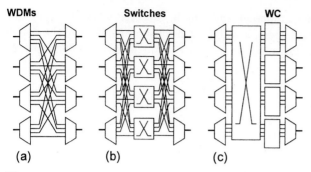

(a)   (b)   (c)

Fig. 1.8. Optical cross-connect architectures illustrated with four input and output ports: **(a)** wavelength routeing, **(b)** wavelength switching and **(c)** wavelength interchanging/switching. *WC*, wavelength converter

The second level of functionality is achieved when replacing the fixed interconnections within the router by space switches (see Chap. 8). Hence, a reconfigurable wavelength router, or in other words an OXC, is obtained. In this case any arriving wavelength from any input port can be cross-connected to any output port. However, two identical wavelengths, e.g. two 'red' ones, from two different input ports cannot be switched to the same output port, which would otherwise lead to severe crosstalk between these 'red' wavelengths. This limitation in cross-connection possibilities makes the OXC wavelength blocking. Under these circumstances the switching architecture of the OXC can also be reduced, having separate switch planes for each wavelength, 'red', 'green', 'blue', etc. (Fig. 1.8b). With N ports at the input and the output, each with M wavelengths, there will be needed M switches of NxN in switch matrix size to realize the OXC.

One way to realize a strictly non-blocking OXC [39] is the introduction of wavelength converters just in front of the output multiplexers (Fig. 1.8c). Now it will be possible to switch any wavelength to any output port, even the two 'red' wavelengths mentioned above. In this case one of the 'red' wavelengths is simply converted to 'green' before being multiplexed at the egress. An OXC with this architecture is non-blocking in the strict sense. However, the enhanced level of functionality puts higher requirements on the switch matrix size, which must be designed as one switch plane with a matrix that is NMxNM strictly non-blocking.

Wavelength converters have already earlier been identified as a key technology to resolve wavelength contention in optical networks. Wavelength conversion can be compared with time slot interchange (TSI) in TDM switching [40], where it is essential to obtain non-blocking circuit switches. However, studies, e.g. [41], have shown that the probability for wavelength contention in core networks is limited and can to a large extent be avoided by appropriate routeing algorithms. An alternative to introducing wavelength converters within the OXC is to drop the blocked wavelength to an OTM that retransmits the signal again on a different wavelength. However, while this approach will save wavelength converters, it will instead consume capacity of the switch core, since the signal is being connected twice through the cross-connect [42].

References

1. T. Li, "The impact of optical amplifiers on long-distance lightwave telecommunications," Proc. IEEE **18**, 1568 (1993)
2. A. Bjarklev, *Optical fiber amplifiers, Design and system applications* (Artech House, Norwood, MA 1993)
3. A.J. Antos and D.K. Smith, "Design and characterization of dispersion compensating fibre based on the LP/sub01/mode," IEEE J. Lightwave Technol. **12**, 1739 (1994)
4. K. Inoue, "Polarization-independent wavelength conversion using fibre four-wave mixing with two orthogonal pump lights of different frequencies," IEEE J. Lightwave Technol. **12**, 1916 (1994)
5. M. Jinno, "All-optical signal regularizing/regeneration using a nonlinear fiber Sagnac interferometer switch with signal-clock walk-off," IEEE J. Lightwave Technol. **12**, 1648 (1994)
6. H. Taga, M. Suzuki, N. Edagawa, H. Tanaka, Y. Yoshida, S. Yamamoto, S. Akiba-S, and H. Wakabayashi, "Multi-thousand kilometer optical soliton data transmission experiments at 5 Gb/s using an electroabsorption modulator pulse generator," IEEE J. Lightwave Technol. **12**, 231 (1994)
7. M.C. Brain, M.J. Creaner, R.C. Steele, N.G. Walker, G.R. Walker, J. Mellis, S. Al-Chalabi, J. Davidson, M. Rutherford, and I.C. Sturgess, "Progress towards the field deployment of coherent optical fiber systems," IEEE J. Lightwave Technol. **8**, 423 (1990)
8. L. Gillner and M. Gustavsson, "Scalability of optical multiwavelength switching networks, power budget analysis," IEEE J. Select. Areas Commun. **14**, 952 (1996)
9. P. Kaiser and D.B. Keck, "Fibre types and their status," in *Optical Fibre Telecommunications II* (S.E. Miller and I.P. Kaminov, eds.), Chap. 2 (Academic Press, New York 1988), p. 29
10. N. Kikuchi, S. Sasaki, and K. Sekine, "10 Gbit/s dispersion-compensated transmission over 2245 km conventional fibres in a recirculating loop," Electron. Lett. **31**, 375 (1995)
11. R. Kashyap, H.-G. Froehlich, A. Swanton, and D.J. Armes, "1.3 m long super-step-chirped fibre Bragg grating with a continuous delay of 13.5 ns and bandwidth 10 nm for broadband dispersion compensation," Electron. Lett. **32**, 1807 (1996)
12. M.C. Tatham, X. Gu, L.D. Westbrook, G. Sherlock, and D.M. Spirit, "Overcoming fibre chromatic dispersion in high bit rate transmission," *Proc. LEOS '94*, 194 (1994)
13. R. Noé, D. Sandel, M. Yoshida-Dierolf, S. Hinz, V. Mirvoda, A. Schöpflin, C. Glingener, E. Gottwald, C. Scheerer, G. Fischer, T. Weyrauch, and W. Haase, "Polarization-mode dispersion compensation at 10, 20 and 40 Gb/s with various optical equalizers," IEEE J. Lightwave Technol. **17**, 1602 (1999)
14. G.P. Agrawal, *Non-linear Fiber Optics*, (Academic, New York 1989)

15. P. Trischitta, M. Colas, M. Green, G. Wuzniak, and J. Arena, "The TAT-12/13 cable network," IEEE Commun. Mag. **34**, 24 (1996)

16. W.C. Barnett, H. Takahira, J.C. Baroni, and Y. Ogi, "The TPC-5 cable network," IEEE Commun. Mag. **34**, 36 (1996)

17. D.R. Gunderson, A. Lecroart, and K. Tatekura, "The Asia Pacific cable network," IEEE Commun. Mag. **34**, 42 (1996)

18. T. Welsh, R. Smith, H. Azami, and R. Chrisner, "The FLAG cable system," IEEE Commun. Mag. **34**, 30 (1996)

19. S. Kawanishi, T. Morioka, O. Kamaatani, H. Takara, and M. Saruwatari, "100 Gb/s, 200 km optical transmission experiment using low jitter PLL timing extraction and all-optical demultiplexing based on polarization-insensitive four-wave mixing," Electron. Lett. **30**, 800 (1994)

20. K. Suzuki, K. Iwatsuki, S. Nishi, and M. Saruwatari, "Error-free demultiplexing of 160 Gb/s pulse signal using optical loop mirror including semiconductor laser amplifier," Electron. Lett. **30**, 1501 (1994)

21. ITU-T Recommendations G.681 "Functional characteristics of inter-office and long-haul line systems using optical amplifiers, including optical multiplexers"

22. H. Taga, "Long-distance transmission experiments using the WDM technology," IEEE J. Lightwave Technol. **14**, 1287 (1996)

23. K. Sato, *Advances in transport network technologies* (Artech House, Norwood, MA 1996)

24. G. Hill, "A wavelength routing approach to optical communication networks," BT Technol. J. **6**, 24 (1988)

25. C. Marra and J. Schesser, "Africa ONE, The Africa Optical Network," IEEE Commun. Mag. **34**, 50 (1996)

26. M.S. Goodman, H. Kobrinski, M.P. Vecchi, R.M. Bulley, and J.L. Gimlett, "The LAMBDANET multiwavelength network, architecture, applications, and demonstrations," IEEE J. Select. Areas Commun. **8**, 995 (1990)

27. T-H. Wu, *Fiber network service survivability* (Artech House, Norwood, MA 1992)

28. A.F. Elrefaie, "Multi-wavelength survivable ring network architectures," *Proc. IEEE ICC '93*, 1245 (1993)

29. ITU-T Recommendation G.957 "Optical interfaces for equipment and systems relating to the synchronous digital hierarchy"

30. G.R. Hill, P.J. Chidgey, F. Kaufhold, T. Lynch, O. Sahlen, M. Gustavsson, M. Janson, B. Lagerstrom, G. Grasso, F. Meli, S. Johansson, J. Ingers, L. Fernandez, S. Rotolo, A. Antonielli, S. Tebaldini, E. Vezzoni, R. Caddedu, N. Caponio, F. Testa, A. Scavennec, M.J. O'Mahony, J. Zhou, A. Yu, W. Sohler, U. Rust, and H. Herrmann, "A transport network layer based on optical network elements," IEEE J. Lightwave Technol. **11**, 667 (1993)

31. C. Bornholdt, B. Sartorius, and M. Möhrle, "All-optical clock recovery at 40 Gbit/s," *Proc. 25th Europ. Conf. Opt. Commun. (ECOC '99)*, Nice, France, PDL-papers, 54 (1999)

32. H. Toba, K. Oda, K. Nakanishi, N. Shibata, K. Nosu, N. Takato, and M. Fukuda, "A 100 channel optical FDM transmission/distribution at 622 Mb/s over 50 km," IEEE J. Lightwave Technol. **8**, 1396 (1990)

33. H. Venghaus, A. Gladisch, B.F. Joergensen, J.-M. Jouanno, M. Kristensen, R.J. Pedersen, F. Testa, D. Trommer, and J.-P. Weber, "Optical add/drop multiplexers for WDM communication systems," *Conf. Opt. Fiber Commun.*

(OFC '97) vol. **6**, OSA Techn. Digest Series (Opt. Soc. America, Washington, DC 1997), p. 280

34. E.L. Goldstein, L. Eskildsen, and A.F. Elrefaie, "Performance implications of component crosstalk in transparent lightwave networks," IEEE Photon. Technol. Lett. **6**, 657 (1994)

35. C.S. Li and F. Tong, "Cross-talk and interference penalty in all-optical networks using static wavelength routers," IEEE Photon. Technol. Lett. **6**, 1120 (1994)

36. J. Zhou, R. Cadeddu, E. Casaccia, C. Cavazzoni, and M.J. O'Mahony, "Crosstalk in multi-wavelength optical cross-connect networks," IEEE Photon. Technol. Lett. **6**, 1423 (1994)

37. S. Johansson, "A transport network involving a reconfigurable WDM network layer - a European demonstration," IEEE J. Lightwave Technol. **14**, 1341 (1996)

38. H. Toba, K. Oda, K. Innoue, K. Nosu, and T. Kitoh, "An optical FDM-based self-healing ring network employing arrayed-waveguide-grating ADM filters and EDFAs level equalizers," IEEE J. Select. Areas Commun./J. Lightwave Technol. **14**, 800 (1996)

39. S. Okamoto, A. Watanabe, and K.-I. Sato, "Optical path cross-connect node architectures for photonic transport network," IEEE J. Lightwave Technol. **14**, 1410 (1996)

40. H. Kobayashi and I. Kaminow, "Duality relationships among 'space', 'time', and 'wavelength' in all-optical networks," IEEE J. Lightwave Technol. **14**, 344 (1996)

41. N. Wauters and P. Demeester, "Wavelength requirements and survivability in WDM cross-connected networks," *Proc. 20th Europ. Conf. Opt. Comm. (ECOC '94)*, Firenze, Italy, 589 (1994)

42. M. Kovacevic and A. Acampora, "Electronic wavelength translation in optical networks," IEEE J. Lightwave Technol. **14**, 1161 (1996)

2 The Optical Fibre

Réal Vallée

2.1 Introduction

The optical fibre now clearly appears to be the best medium for propagating
light signals over long distances. But this now well-established fact has not
always been so obvious and some major technological achievements in glass
fabrication were necessary to finally conceive the idea of optical fibre based
communication systems. As a matter of fact, the idea of guiding light by
total internal reflection inside a cylindrical support is not a new one. In
1870, John Tyndall performed an experiment where a light beam was guided
inside a parabolic stream of water. Following this pioneer experiment, it took
more than a century before some major advances in optical loss reduction
in glass initiated what one could call the modern history of optical fibre.
Actually, the formal proposition of using glass fibres as an optical communication waveguide was made in 1966 by Kao and Hockman. In the decade
following this proposal an extraordinary quest for lower losses took place,
which culminated in 1979 with the value of $0.2\,dB/km$ at $1.55\,\mu m$ reported
by Miya et al. In parallel with this, the laser was invented and it was rapidly
identified as a very convenient tool for injecting light into fibres. Practically,
the advent of the first (ruby) laser meant that light injection into fibres could
be performed very efficiently, whereas the advent of the semiconductor laser
meant that high-speed light modulation was thereafter feasible. Furthermore,
the distributed feedback laser (1975) brought another piece to the puzzle by
adding bandwidth selectivity to the list of attributes of laser sources. In fact,
the affinity between the transverse fundamental mode of the optical fibre
and that of the laser, which makes the energy exchange between the two so
efficient, is somehow underestimated and one could actually wonder what
place the optical fibre would take in today's life without the discovery of the
laser. Accordingly, the progressive shift of the wavelength of the fibre-optic
communication systems from $0.8\,\mu m$ to $1.3\,\mu m$ and then to $1.55\,\mu m$ can be
attributed to significant advances in laser science as well as in fibre-optics
itself. It is therefore not so surprising that the initially parallel development
of these two fields eventually merged in 1987 with the advent of the first fibre
lasers and fibre amplifiers. Nowadays, fibre lasers as well as fibre amplifiers
are basic components of most fibre optic research laboratories as well as most
optical communication systems.

In this chapter we discuss the optical fibre as a transmission medium for optical communications. Our intent is to provide the reader with a practical and consistent treatise on the subject. In Sect. 2.2, the basic equations for step-index fibres are presented. Section 2.3 is devoted to a brief description of the fibre materials. In Sect. 2.4 the basic optical properties of the fibre, including linear losses, dispersion and birefringence are discussed. The problem of the fibre transmission at high intensities will be addressed in Sect. 2.5 with special attention to those phenomena which are detrimental to optical communications. In Sect. 2.6 the problem of pulse propagation and solitons in optical fibres is introduced.

2.2   Waveguiding Properties

2.2.1   Basic Concepts and Parameters

Optical waveguides are dielectric structures which are ideally uniform along the axis of propagation, usually chosen as z. It is therefore their transverse characteristics which basically prescribe their optical properties. In the case of a planar structure for which guidance is taking place along one transverse direction or in the case of an optical fibre which possesses rotational symmetry, a single parameter suffices to describe the so-called refractive index profile of the structure. In fibres, a useful way of describing a wide range of refractive index radial profiles is through the use of the following relation

$$n(r) = n_1 \left[1 - 2\Delta \left(\tfrac{r}{a} \right)^\alpha \right]^{1/2} \quad (0 \leq r \leq a) \tag{2.1}$$
$$= n_2 \qquad\qquad\qquad (a < r)$$

where

$$\Delta = \frac{n_1^2 - n_2^2}{2n_1^2}\ , \tag{2.2}$$

and a is the "core" radius. Various profiles can be obtained by varying the parameter α from one to infinity (see Fig. 2.1).

For $\alpha \to \infty$ one has the special case of a step-index fibre where both the core and its surrounding cladding possess constant refractive indices, i.e. n_1 and n_2 respectively. (Note that the fibre is also surrounded by a protective polymer jacket, but this component does not play a role in the wave guidance). The step-index profile is interesting from the theoretical point of view because it allows for a simpler mathematical treatment. Also, the step-index geometry has become the preferred choice for many applications in telecommunications and sensing. Therefore, we shall mainly concentrate our theoretical study on step-index fibres, reserving Sect. 2.2.3 for a brief discussion of the so-called *graded-index* fibre.

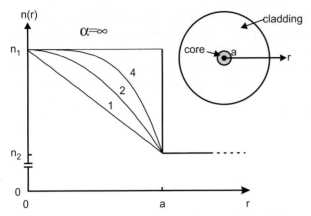

Fig. 2.1. Refractive indices for different values of the parameter α [see (2.1)]. The case $\alpha \to \infty$ corresponds to the step-index fibre

In step-index fibres, the guidance takes place within the core as a consequence of total internal reflection at the core-cladding interface. Total internal reflection occurs whenever a light ray passes from a medium of refraction index n_1 to one with a smaller index n_2 at an angle greater than the critical angle θ_c, which is defined as the angle for which the angle of refraction in the second medium is equal to $90°$. From Snell's law one obtains $\theta_c = \sin^{-1}(n_2/n_1)$, and it appears that for those angles such that $\theta \geq \theta_c$ the modulus of the reflectivity coefficient is equal to one and the field in medium 2 decreases exponentially away from the interface. Applied to the optical fibre, total internal reflection actually ensures that no loss is associated with the guidance itself. Therefore, the only constraint is that n_1 be larger than n_2. Now, although real fibres depart slightly from the nominal step-index profile, for reasons having to do with the fabrication process, it is possible in most practical cases to describe their behaviour with a good accuracy by assuming such an ideal profile.

A good understanding of the physical concepts related to optical guidance can be obtained from elementary geometrical optics. Although such a description can be applied with good accuracy only for waveguides having transverse dimensions significantly larger than the wavelength, it is useful to introduce the basic parameters on such a physical basis. Let us consider first the schematic of Fig. 2.2 showing a light ray impinging at a fibre end:

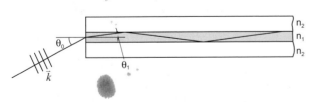

Fig. 2.2. Light injection in a step-index fibre

Using Snell's law and taking into account the constraint imposed by total internal reflection one can easily deduce that light rays incident at the fibre end within the cone of acceptance defined by the maximum angle $\theta_0 = \sin^{-1}(\sqrt{n_1^2 - n_2^2})$ will be trapped within the core. By analogy with a lens systems one can define, on this basis, the numerical aperture of the step-index optical fibre as

$$NA = \sqrt{n_1^2 - n_2^2} \, . \tag{2.3}$$

This parameter, which provides an immediate indication of the characteristics of the light injection into a fibre is also very useful to introduce a very important fibre parameter, namely the V-parameter. Consider first the vacuum wavevector $\boldsymbol{k}$ (of modulus equal to $2\pi/\lambda$, where λ denotes the vacuum wavelength) of an incident beam as depicted in Fig. 2.2. The projection of $\boldsymbol{k}$ on the fibre input face is given by $k \sin \theta_0$. In particular, the projection of the wavenumber corresponding to the maximum accepted ray is simply given by $k\sqrt{n_1^2 - n_2^2}$. Therefore, defining the parameter V as the product of this maximum transverse wavenumber by the fibre core radius a one finally has

$$V = k\sqrt{(n_1^2 - n_2^2)} \, a \, . \tag{2.4}$$

This V-parameter, which is also called the *normalized frequency*, is a key parameter in characterizing optical fibres. It will be shown in Sect. 2.2.2 that the parameter V actually quantifies the fibres' ability to support transverse modes. In fact, the number of transverse modes supported by a fibre is an increasing function of V (*quadratic* in the limit of large V). Now, the concept of modes introduced here obviously needs to be more clearly defined. In fact, although the concept of modes refers to a field description, it is possible to describe in terms of a simple geometrical argument the basic mechanism of wave guidance. To do so, let us first consider the field resulting from the superposition of two plane waves propagating at angles $\pm \sin^{-1}(\kappa/\beta)$, respectively, with respect to z, as shown in Fig. 2.3a. This field is given by

$$E = E_0 \sin(\kappa x) \exp(i\omega t - i\beta z) \, , \tag{2.5}$$

representing a non-uniform (i.e. with a sinusoidal modulation along x) plane wave propagating along z with the propagation constant β. Note that in this case the parameters κ and β can vary continuously and are only restricted by the constraint imposed by the medium of refractive index n, that is

$$k^2 n^2 = \beta^2 + \kappa^2 \, . \tag{2.6}$$

This relation actually illustrates the fact that β and κ correspond to the longitudinal and transverse components of the medium wavenumber, respectively. Now, let us turn to the situation depicted in Fig. 2.3b, where this field is forced to propagate between two conducting planes. The conducting

a)

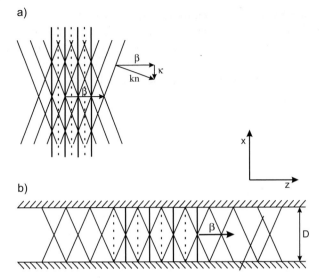

b)

Fig. 2.3. (a) Superposition of two freely propagating plane waves about the z-axis. **(b)** Same superposition but restricted to the space between two conducting planes

walls constrain the field to be zero at their surfaces and therefore impose the following condition on κ:

$$\kappa = \frac{n\pi}{D} \, . \tag{2.7}$$

Consequently, the field has to adjust itself to the walls and is restricted to those discrete values of κ which allow it to respect the limiting conditions imposed by the "guide". Transposing this to the case of a dielectric waveguide, one has a similar constraint, which is now imposed by the requirement that the tangential components of the fields be continuous across the structure. The main difference between the two situations is that in the case of the dielectric waveguide the sinusoidal transverse modulation will be perturbed near its edges to allow for continuity with the evanescent field existing outside the guide.

In an optical fibre, the transverse dependence is actually given by a Bessel function (Sect. 2.2.2) instead of the sinusoid. However, the basic idea that only a discrete set of transverse field configurations is supported by the waveguide is still valid, and this is crucial, for it ensures (from (2.6)) that only a discrete set of propagation constants β will also be allowed. As a matter of fact, some basic results of the wave theory can be interpreted on the basis of this interplay between transverse and longitudinal parameters. Accordingly, it will prove useful to identify a set of interrelated parameters which will play a key role in our wave description. One has first the propagation constant, β, which will be identified further as the eigenvalue of the wave problem. This parameter is often represented in the normalized form

$$b = \frac{\beta^2 - k^2 n_2^2}{k^2 n_1^2 - k^2 n_2^2} \, . \tag{2.8}$$

The inner transverse parameter, κ, can be directly derived from (2.6). It also proves useful to derive a corresponding parameter for the outer region, characterized by the evanescent wave. One has thus the following definitions for the transverse parameters

$$\kappa = \sqrt{k^2 n_1^2 - \beta^2} \quad \text{and} \tag{2.9a}$$

$$\gamma = \sqrt{\beta^2 - k^2 n_2^2} \, , \tag{2.9b}$$

where κ is related to the period of the modulation of the mode inside the core whereas γ corresponds to the rate of decay of the evanescent field outside the core. These parameters often appears under the normalized forms

$$u = \kappa a \quad \text{and} \tag{2.10a}$$

$$w = \gamma a \, . \tag{2.10b}$$

Finally, from the last definitions and from (2.9a,b) and (2.4), one has the important relation

$$V^2 = u^2 + w^2 \, , \tag{2.11}$$

which establishes the conditions under which each guided mode, as defined by its parameters u and w, will adjust itself so as to fit to the fibre constraints, as prescribed by the parameter V.

We have limited the previous discussion to modes that are confined to the core area. There also exist, due to the total internal reflection at the cladding-jacket interface, bounded modes extending over the whole cladding area. Contrary to the core modes these so-called *cladding modes* form a quasi-continuum because of the large size of the cladding compared to the wavelength (the cladding diameter being typically of the order of 125 μm.). Note that these modes are more prone to all kinds of losses, especially scattering at the cladding-jacket interface. Finally, the fibre also supports, in addition to these two categories of bounded modes, a continuum of radiation modes [1]. Figure 2.4 summarizes the situation in terms of the propagation constant ranges corresponding to each of these categories. A complete description of the total field would therefore require that all of these modes be taken into account. However, in most practical cases, one can assume that light was initially injected into a core mode only, so that the other categories need not be considered. In fact, it appears from Fig. 2.4 that for an ideal, straight segment of fibre the coupling between these categories of modes is forbidden since their propagation constants do not overlap. Actually, light coupling between these modes will occur in situations where a perturbation such as a fibre bend, a periodic modulation of the core index, and so on, is present.

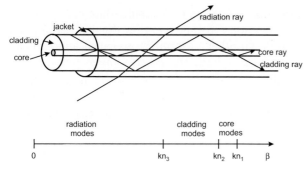

Fig. 2.4. Illustration of the various types of modes along with the corresponding ranges of their propagation constants. The indices n_1, n_2, and n_3 are those of the core, the cladding and the protective jacket, respectively

2.2.2 Basic Equations for the Step-Index Fibre

Weakly Guiding Fibres. For most practical applications the core and cladding have nearly the same refractive indices. Typical values of the parameter Δ (see (2.2)) are of the order of 10^{-3}. Accordingly, one can derive in the limit of small Δ, in which case Δ is equal to about $(n_1 - n_2)/2n_2$, approximate formulas for the basic fibre parameters V and β which will prove to be useful for further theoretical derivations:

$$V \cong kn_2 a\sqrt{2\Delta} \,, \tag{2.12}$$
$$\beta \cong kn_2(1 + b\Delta) \,. \tag{2.13}$$

From the fabrication point of view, a small index difference is advantageous since the core and the cladding are generally made of the same basic material to which is locally added one or a few dopants in order to tailor the refractive index profile. Refractive index differences of this order can be readily obtained without the need for adding high concentrations of dopants, thus preventing significant modification of the optical properties and of the stability of the glass. Another factor for choosing small refractive index differences arises from the fact that single-mode operation requires that the parameter V be as small as possible (actually smaller than 2.405, as we show hereafter). Therefore, the compromise in decreasing V, while maintaining a reasonable size for the core, consists in decreasing the refractive index difference. It also turns out that the resulting *weakly guiding fibres*, to use the term first introduced by Gloge [2], will also be less affected by losses when small index difference and large core diameters are preferred. From the theoretical point of view, the great benefit of such a configuration is that it allows for a significant simplification of the mode description, which is, in general, rather involved.

It is beyond the scope of this chapter to present an exhaustive derivation of the "exact modes" (i.e. EH and HE modes), for such a derivation is rather tedious and can be found in many textbooks [1,3–9]. In fact, the crucial point to recall is that under the weakly guiding conditions, subgroups of modes possess essentially the same propagation constant (i.e. they are degenerate) and are actually superimposed on each other to form the so-called approximate *LP modes*. For most applications, indeed, the description in terms of LP modes is not only sufficient but more practical.

Derivation of the Characteristic Equation for LP Modes. The propagation of electromagnetic waves is governed in general by Maxwell's equations. For a lossless, linear and isotropic dielectric medium like that composing an optical fibre, Maxwell's equations take a simple form from which the wave equation is readily derived. Another simplification for step-index fibres is that one can assume that the permittivity ε_i of each medium (core and cladding) is spatially homogeneous. Further, assuming a temporal dependence of the form $e^{i\omega t}$ for the electric and magnetic fields, the wave equation reduces to the Helmholtz equation. Finally, since the optical fibre is a structure with cylindrical symmetry, it is convenient to express the Helmholtz equation in cylindrical polar coordinates:

$$\frac{\partial^2 \boldsymbol{\Psi}(\boldsymbol{r})}{\partial r^2} + \frac{1}{r}\frac{\partial \boldsymbol{\Psi}(\boldsymbol{r})}{\partial r} + \frac{1}{r^2}\frac{\partial^2 \boldsymbol{\Psi}(\boldsymbol{r})}{\partial \varphi^2} + k^2 n_i^2 \boldsymbol{\Psi}(\boldsymbol{r}) = 0 \;, \tag{2.14}$$

where $\boldsymbol{\Psi}(\boldsymbol{r})$ represents either the electric $\mathcal{E}(\boldsymbol{r})$ or magnetic $\mathcal{H}(\boldsymbol{r})$ field and where the parameters k and n_i, representing the vacuum wavenumber and the refractive index respectively, (with $i = 1$ for the core and 2 for the cladding) are related to the media permittivity ε_i and to the vacuum permeability μ_0 by the usual relation

$$k^2 n_i^2 = \omega^2 \mu_0 \varepsilon_i \;. \tag{2.15}$$

Equation (2.14) actually corresponds to six scalar equations (i.e. one for each of the three components of both the electric and magnetic fields). This system of equations is to be solved in both the core and cladding regions, therefore leading to a boundary-value problem. Now, it follows from Maxwell's equations that out of these six components only two are independent. Furthermore, since our dielectric waveguide is uniform along z and since we are interested in modal solutions (i.e. steady-state transverse fields propagating along z with a propagation constant β) it is convenient to decompose the field $\boldsymbol{\Psi}(\boldsymbol{r})$ as follows:

$$\boldsymbol{\Psi}(\boldsymbol{r}) = \left(\boldsymbol{\Psi}_{\mathrm{t}}(x,y) + \Psi_z(x,y)\boldsymbol{u}_z\right) e^{-i\beta z} \;, \tag{2.16}$$

where $\boldsymbol{u}_z$ represents a unit vector along z and $\boldsymbol{\Psi}_{\mathrm{t}}$ represents a transverse vector field. Substituting (2.16) into (2.14) and referring to the longitudinal components one has

$$\frac{\partial^2 \Psi_z}{\partial r^2} + \frac{1}{r}\frac{\partial \Psi_z}{\partial r} + \frac{1}{r^2}\frac{\partial^2 \Psi_z}{\partial \varphi^2} + K^2 \Psi_z = 0 \;, \tag{2.17}$$

where Ψ_z stands for either E_z or H_z and where the quantity $K^2 = k^2 n_i^2 - \beta^2$ is equal to $-\gamma^2$ in the cladding and to κ^2 in the core (see (2.9)). Thus, from the solution of (2.17) E_z and H_z are obtained, from which the other four components E_r, E_φ, H_r, and H_φ, can be deduced. The choice of the cylindrical system of coordinates, which is adapted to the fibre geometry, allows us to separate the radial and azimuthal dependences, so that the overall solution can be expressed in terms of the product of a function of r by a function of φ. Namely, the azimuthal dependence takes a sinusoidal form whereas the radial one is expressed in terms of Bessel functions. Depending on the region considered (i.e. core vs cladding) and owing to the physical constraint requiring that the fields be finite in the core and have an exponentially decreasing behaviour in the cladding, one is led to the following solutions for the z-components:

$$\Psi_z = A_1 J_\nu(\kappa r) \cos(\nu\varphi + \varphi_\nu) \quad (0 < r < a) , \tag{2.18a}$$

$$\Psi_z = A_2 K_\nu(\gamma r) \cos(\nu\varphi + \varphi_\nu) \quad (a < r < \infty) , \tag{2.18b}$$

where ν is the azimuthal mode number. The function $J_\nu(\kappa r)$ is the Bessel function of the first kind whereas $K_\nu(\gamma r)$ is a modified Bessel function of the second kind. From (2.18a,b) one has four arbitrary amplitudes and two arbitrary phases (i.e. one for E and one for H). Now, recalling that the four components (H_r, H_φ, E_r and E_φ) can be deduced from E_z and H_z, the resulting set of six fields is further subject to the requirement that the tangential components (E_z, E_φ, H_z and H_φ) be continuous at the core-cladding interface. Equating the tangential fields on both sides of the core-cladding interface, one is led to a homogeneous system of four equations and four unknowns. This system can be solved in terms of two characteristic equations corresponding to the two categories of solutions, namely the EH and HE modes. It can be shown that in the weakly-guiding approximation (i.e. assuming that $\varepsilon_1 \approx \varepsilon_2$) these two distinct equations can be recast in terms of a single one:

$$\frac{u J_{l-1}(u)}{J_l(u)} = -\frac{w K_{l-1}(w)}{K_l(w)} , \tag{2.19}$$

which is expressed in terms of a new azimuthal number l. Therefore, in the weakly guiding approximation the whole set of solutions (i.e. for both categories of modes) can be obtained from equation (2.19) and this has an important physical meaning. In fact, from the solution of (2.19), which requires the use of (2.11), the eigenvalue parameter u (or w) is obtained, from which the propagation constant, β, can finally be derived (by use of (2.9) and (2.10)). Thus, in this approximation, the HE and EH modes appear to share the same set of propagation constants, which can be associated with a common azimuthal number l such that

$$l = \begin{cases} \nu - 1 \text{ (for HE modes)} \\ \nu + 1 \text{ (for EH modes)} \end{cases} . \tag{2.20}$$

Accordingly, a new set of modes can be described on the basis of this new azimuthal mode order together with the radial mode order, m, corresponding to the mth root of the characteristic equation (2.19). These modes were called "LP$_{l,m}$ modes" by Gloge [2] and we now show how their corresponding fields can be derived from the superposition of the HE and EH modes.

Field Distributions. The set of modes which are solutions of the fibre boundary-value problem can be classified into two categories, namely the HE and EH modes. These modes are called *hybrid* since they, in general, possess six non-zero components. For EH modes one has $H_z > E_z$ whereas for HE modes one has $H_z < E_z$. Now, although these modes were first derived in terms of cylindrical polar coordinates, they turn out to take a more convenient form when they are converted into Cartesian coordinates:

$$E_x = \mp i\beta a A \left\{ \begin{array}{ll} \dfrac{J_{\nu\mp1}(ur/a)}{u} & (for\ r < a) \\[2mm] \pm\dfrac{J_\nu(u)K_{\nu\mp1}(wr/a)}{wK_\nu(w)} & (for\ r > a) \end{array} \right\} \cos[(\nu\mp1)\varphi]\ , \quad (2.21a)$$

$$E_y = \pm i\beta a A \left\{ \begin{array}{ll} \pm\dfrac{J_{\nu\mp1}(ur/a)}{u} & (for\ r < a) \\[2mm] \dfrac{J_\nu(u)K_{\nu\mp1}(wr/a)}{wK_\nu(w)} & (for\ r > a) \end{array} \right\} \sin[(\nu\mp1)\varphi]\ , \quad (2.21b)$$

$$E_z = A \left\{ \begin{array}{ll} J_\nu(ur/a) & (for\ r < a) \\[2mm] \dfrac{J_\nu(u)K_\nu(wr/a)}{K_\nu(w)} & (for\ r > a) \end{array} \right\} \cos(\nu\varphi)\ , \quad (2.21c)$$

$$\eta H_z = \pm E_z \tan(\nu\varphi)\ , \quad (2.22a)$$

$$\frac{\eta H_y}{E_x} = -\frac{\eta H_x}{E_y} = \left(\frac{kn_i}{\beta}\right)^2\ , \quad (2.22b)$$

where the effective impedance η is defined as

$$\eta = \frac{\omega\mu_0}{\beta} = \frac{k}{\beta}\sqrt{\frac{\mu_0}{\varepsilon_0}} = \frac{k\eta_0}{\beta}\ , \quad (2.23)$$

where the upper sign corresponds to the HE modes, whereas the lower is for the EH modes. Also note that the previous set of equations is expressed on the basis of the cosine dependence of E_z so that a complementary one, where the cosine functions are replaced by sine functions and vice-versa, also exists. It is also important to stress that these solutions are exact and complete. Now, it appears from these equations that the HE$_{l+1,m}$ and EH$_{l-1,m}$ modes take precisely the same form, that is for a given value of l, the HE$_{l+1,m}$ and EH$_{l-1,m}$ modes not only share the same propagation constant (see (2.20)) but also have the same field distribution in Cartesian coordinates. Consequently, coherent superposition between these fields is not only possible but also such that either the x or the y component of the resulting field cancels. Consider, for instance, the superposition $HE_{2,1}$–$EH_{0,1}$ (i.e. $l = 1$). From (2.21), one has

$$E_x = -2i\beta A a \left\{ \begin{array}{c} \dfrac{J_1(ur/a)}{u} \\ \dfrac{J_2(u)K_1(wr/a)}{wK_2(w)} \end{array} \right\} \cos \varphi , \tag{2.24a}$$

$$E_y = 0 \tag{2.24b}$$

which correspond to a field linearly polarized along x. The nomenclature of LP modes arises from this fact. Note that the cancellation of the y component requires the use of the relation

$$\frac{K_{l+1}(w)}{K_{l-1}(w)} \cong -\frac{J_{l+1}(u)}{J_{l-1}(u)} , \tag{2.25}$$

which can be derived with the use of the approximate eigenvalue equation (2.19). Therefore, the LP mode representation is not exact, strictly speaking, but in most practical cases the departure from the perfectly linear polarization can be ignored. However, there is a special case for which the field is actually perfectly polarized. In fact, for $l = 0$, only the HE_{1m} (i.e. $\nu = 1$) modes can be defined and the transverse fields take (from (2.21)) the simple form

$$E_x = \frac{-i\beta a A}{u} J_0(ur/a) , \tag{2.26a}$$

$$E_y = 0 , \tag{2.26b}$$

corresponding to a perfectly linearly polarized field, i.e. without azimuthal dependence. In this case, the linear polarization does not arise from a superposition of two fields but from the cancellation of one of the two components via its azimuthal dependence. Note that, starting from the complementary set of solutions (involving a sine function for E_z), the resulting fields would now be polarized along y. Therefore the HE_{1m} (or LP_{0m}) modes are degenerate with respect to this arbitrariness of the azimuthal dependence only. Of particular interest among these modes is the case $m = 1$ (i.e. the HE_{11} or LP_{01} mode) since, as we show hereafter, it corresponds to the fundamental mode. Other special modes are those corresponding to $\nu = 0$, for which only the EH solutions have a physical meaning in (2.21). Now, depending on the particular set used (i.e. sine or cosine dependence) one obtains solutions having either the H_z or the E_z component equal to zero. This situation corresponds to the TE and TM classes of modes, respectively.

It is also interesting to note that the ratio of the longitudinal to the transverse field (i.e. $E_z/E_{x \text{ or } y}$) is very small in the weakly guiding approximation. From (2.4), (2.9a), (2.10a), (2.12), (2.13) and (2.21) one has indeed

$$\left| \frac{E_z}{E_{x \text{ or } y}} \right| \cong \frac{u}{\beta a} \cong \sqrt{2\Delta} , \tag{2.27}$$

with Δ of the order of 10^{-3}. Therefore, the field propagating in a weakly guiding optical fibre is a nearly *transverse* and *linearly polarized* wave with a propagation constant β. Thus, in most practical cases, the mere expression of the transverse electric field is sufficient. In the LP representation, this transverse field can be simply expressed as

$$E_t = E_0 \left\{ \begin{array}{c} \frac{J_l(ur/a)}{J_l(u)} \\ \frac{K_l(wr/a)}{K_l(w)} \end{array} \right\} \left\{ \begin{array}{c} \cos(l\varphi) \\ \text{or} \\ \sin(l\varphi) \end{array} \right\} , \tag{2.28}$$

where the subscript t stands for either x or y and where the new amplitude factor E_0 includes the factor "$i\beta a/u$". Also note that in order to derive (2.28) we started from the general expression (2.21) and used the approximate eigenvalue equation (2.19). From this equation one sees that each LP mode is characterized by a typical shape which takes (except for the $l = 0$ modes) four different configurations resulting from the product of the two possible polarizations with the two azimuthal dependences (Fig. 2.5). Obviously, such a representation in terms of the LP modes allows for a great simplification of the description of the fibre modes. Actually, the fibre mode description is possible on the basis of (2.19) and (2.28).

Mode Cut-off and Mode Content. For a mode to be actually supported by an optical fibre at a given wavelength its corresponding propagation constant must be a solution of (2.19). Thus, an optical fibre will support

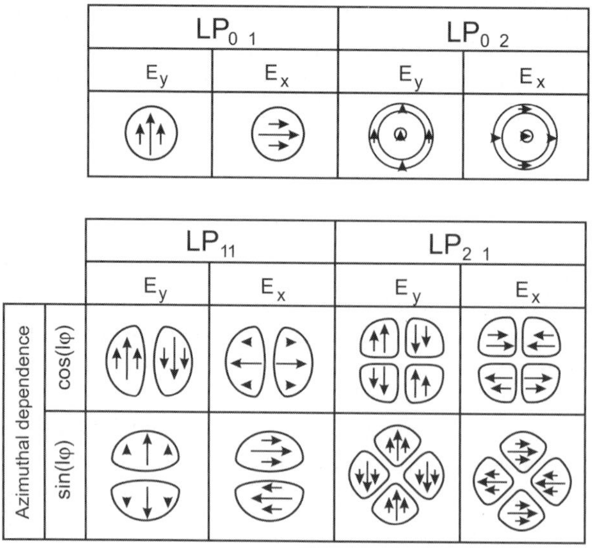

Fig. 2.5. Representation of the possible field distributions of the first four LP modes

only a finite number of modes at a given wavelength. The general trend is for the number of supported modes to increase with the ratio of the fibre transverse dimension to the wavelength (i.e. a/λ), but it is also a function of the numerical aperture. Accordingly, the parameter V previously introduced in Sect. 2.2.1 is actually the best suited for describing the evolution of the fibre mode content. Now, it was shown (see (2.11)) that V is related to the transverse parameters u and w appearing as arguments of the Bessel functions of the modal solutions (2.21). In particular, w is responsible for the radial decrease of the field outside the core so that, as it gets smaller, the field extends more and more into the cladding (see Fig. 2.7, hereafter, and the related discussion on the power fraction carried by the core). The condition $w = 0$ therefore defines the cut-off condition of a mode, that is the point where the mode is not guided anymore, and from (2.11) one also notes that $u_c = V_c$. At the opposite extreme, in the limit where $w \to V$, the mode is said to be well bounded. From this, one can understand better the role of the parameter V in fixing the range over which the transverse parameter w can extend. Transposing these observations to the characteristic equation (2.19), one notes that the condition $w = 0$ yields $J_{l-1}(u) = 0$ whereas for $w \to V \to \infty$, one has $J_l(u) = 0$. Thus, the solutions for u for the LP_{lm} mode will lie between the $\mathrm{m^{th}}$ zero of the $J_{l-1}(u)$ function and the $\mathrm{m^{th}}$ zero of the $J_l(u)$ function, recalling that m represents the radial order. Of particular interest is the mode LP_{01} for which u lies between the first zero of $J_{-1}(u) = -J_1(u) = 0$ and the first zero of $J_0(u) = 2.405$. This mode is the only one having its cut-off at $V = 0$, which means that it has a solution for any value of V i.e. for any wavelength. For this reason, it is called the *fundamental* mode and one sees (Fig. 2.6) that for $V < 2.405$ it is the only one that is supported by the fibre. This range of operation is called the single-mode regime, and we will show its importance in the following sections. On the other hand, for $V > 2.405$ the fibre is said to operate in the multimode regime. It can be shown [1] that in the limit of large V, the number of propagating modes N increases quadratically with V, i.e.

$$N \cong \frac{V^2}{2} .$$
(2.29)

Consequently, for a V value of about 20, corresponding to a core diameter of about 50 μm, there are approximately 200 modes propagating within the fibre. The propagation constants corresponding to such a large number of modes will fill the interval defined by $kn_2 < \beta < kn_1$. Each of these modes will propagate along the fibre with a velocity which is inversely proportional to its β. Therefore, when an optical pulse is incident at one end of such a fibre, each of these modes carries a portion of the pulse energy at its own velocity, which results in "pulse spreading" at the other end. Such a mechanism, named *intermodal dispersion*, is the main factor limiting the fibre bandwidth in step-index multimode fibres. For this reason, multimode step-index fibres

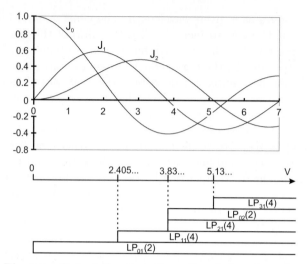

Fig. 2.6. Plot of the first three Bessel functions J_0, J_1, and J_2 along with a schematic showing the order of appearance of the first five LP modes. The degeneracy of each LP mode is shown in parenthesis

have a very limited range of applications in telecommunications and are now mainly used for short-distance light transport applications.

Power Carried by a Mode. Practically, the quantity which is accessible to direct measurement is the optical power. From the Poynting theorem the total power carried by a mode is given by

$$P_{\text{tot}} = P_{\text{core}} + P_{\text{cladding}} = \frac{1}{2} \int_0^a \int_0^{2\pi} \Re(\boldsymbol{E} \times \boldsymbol{H}^*) \cdot \hat{z} r \, dr \, d\varphi$$

$$+ \frac{1}{2} \int_a^\infty \int_0^{2\pi} \Re(\boldsymbol{E} \times \boldsymbol{H}^*) \cdot \hat{z} r \, dr \, d\varphi \, , \tag{2.30}$$

where, as usual, $\Re$ stands for the real part and the asterisk represents the complex conjugate. Now with the help of equations (2.22b), (2.23) and (2.28) and using the approximation $\beta \equiv k n_i$, the expression for the field of a mode linearly polarized along x is

$$\frac{\eta_0}{n_i} H_y = E_x = E_0 \left\{ \begin{array}{c} \frac{J_l(ur/a)}{J_l(u)} \\[2ex] \frac{K_l(wr/a)}{K_l(w)} \end{array} \right\} \cos(l\varphi) \, . \tag{2.31}$$

Introducing this expression of the field into (2.30), one finally obtains after some algebraic simplifications involving (2.19) and recursion formulas for the

Bessel functions

$$P_{\text{tot}} = \frac{V^2}{u^2} \frac{An_i E_0^2}{4\eta_0} \frac{e_l}{\psi_l(w)} ,\qquad (2.32)$$

where $e_l = 2$ for $l = 0$ and $e_l = 1$ for $l \geq 1$, $A = \pi a^2$ and $\psi_l(w)$ is given by

$$\psi_l(w) = \frac{K_l^2(w)}{K_{l+1}(w)K_{l-1}(w)} .\qquad (2.33)$$

By isolating E_0 from (2.32), the modes (as defined by (2.31)) can be expressed in a normalized form involving the power instead of the electric field. The fraction of the total power which is carried by the core can be shown to be given by

$$\frac{P_{\text{core}}}{P_{\text{tot}}} = 1 - \frac{u^2}{V^2}(1 - \psi_l(w)) .\qquad (2.34)$$

This expression has been plotted in Fig. 2.7 for the first four LP modes. From this plot one sees that in order for the core to carry a substantial fraction of the total power, the mode must be as far as possible from its cut-off. This corresponds to a situation where the mode is well confined and is therefore less sensitive to losses, especially those induced by bending (see Sect. 2.4.1). This is why single-mode fibres are usually designed so as to operate in the range $2.0 \leq V \leq 2.405$.

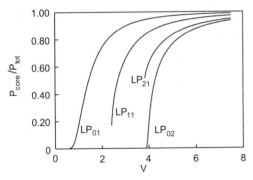

Fig. 2.7. Plot of the ratio $P_{\text{core}}/P_{\text{tot}}$ as a function of the V parameter for the first four LP modes

Approximate Formulas. Before we end this section we introduce some approximate formulas which can be used in place of the formal ones previously derived. First, it can be shown [10] that the transcendent characteristic equation (2.19) can be replaced, with a good level of confidence, by the analytical formula

$$u = u_\infty \frac{V}{V + 1} \left[1 - \frac{u_\infty^2}{6(V + 1)^3} - \frac{u_\infty^4}{20(V + 1)^5} \right] ,\qquad (2.35)$$

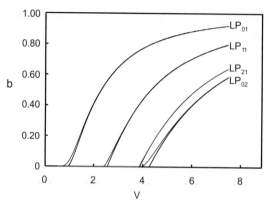

Fig. 2.8. Plot of the normalized propagation constant b as a function of V for the first four LP modes

where u_∞ represents the LP$_{lm}$ mode upper-bound for u, i.e. the mth zero of $J_l(u)$. Note that other approximate formulas with various ranges of validity were derived by various authors but the great benefit of (2.35) is that it applies to all the LP modes and can be refined further with the addition of higher order correction terms. To show the validity of this equation we have plotted, in parallel, the exact (as obtained from (2.19)) and the approached (2.35) values of the normalized propagation constant b (see (2.8)) as a function of V for the first four LP modes (Fig. 2.8). Note that besides a slight departure near the cut-off, the agreement is quite satisfactory for all the modes.

The radial distribution of the fundamental mode appears to be very nearly Gaussian in shape. This suggests that we can approximate the exact field distribution by the following Gaussian function

$$E_t^g = E_0 \exp\left[-\left(\frac{r}{w_0}\right)^2\right] . \tag{2.36}$$

Such a Gaussian function is more practical to use than the exact Bessel function and it also allows us to define a *mode spot size* w_0 which is very useful in problems related to mode matching, either between two fibres, or between a fibre and a Gaussian laser beam. Accordingly, the parameter w_0 can be defined as the value which maximizes the power-launching efficiency as a function of V. On this basis, w_0 can be approximated by the empirical formula [11]

$$w_0 = a\left(0.65 + \frac{1.619}{V^{3/2}} + \frac{2.879}{V^6}\right) , \tag{2.37}$$

which is accurate to better than 1% within the range $2.0 < V < 2.405$. It is interesting to note that, within this range, the ratio w_0/a varies from 1.22 to

1.08, showing that the core radius slightly underestimates the actual mode spot size w_0. Accordingly, it is often convenient to define an effective mode area A_{eff}, which can be simply expressed within the Gaussian approximation as

$$A_{\text{eff}} = \pi w_0^2 \ . \tag{2.38}$$

2.2.3   Graded-Index Fibres

Historically, the first fibres to be developed were single-mode fibres. However, because of the light-coupling and interconnection problems plaguing small-core fibres, researchers concentrated much of their effort in the early 1970s in developing large-core multimode fibres. Therefore, the problem of intermodal dispersion (see Sects. 2.2.2 and 2.4.2) rapidly became a matter of concern. Graded-index fibres were introduced for their ability to significantly reduce intermodal dispersion. However, because of further progress in splicing and light-injection techniques together with the need for increasing further the bandwidth, single-mode fibres were progressively reintroduced in communication systems in the early 1980s. Nowadays, because of their lower loss and their larger bandwidth, single-mode fibres are the medium exclusively used for long-haul communication links, relegating graded-index fibres to short-distance applications like data-link and local area networks, where they still present some advantages in terms of cost and ease of handling.

Graded-index fibres are generally described by the so-called α-profile previously introduced in (2.1). Their basic principle of operation is shown in Fig. 2.9.

Unlike the case of the step-index fibre, the rays are now bent instead of reflected in response to the index gradient which replaces the nominal *interface*. Unlike the step-index fibre, it is not possible to define a single NA for these fibres since the acceptance angle appears to vary from a maximum value of $n_1(2\Delta)^{1/2}$, on the axis, to zero at $r = a$. Moreover, the number of modes, N, that these fibres can support is given by

$$N = \frac{\alpha}{\alpha + 2} \frac{V^2}{2} \ , \tag{2.39}$$

in agreement with (2.29) for the step-index case ($\alpha \to \infty$). From this equation, one notes that a quadratic graded-index fibre will support half the number of modes supported by an equivalent step-index fibre.

From the simple ray description one sees that a ray traveling closer to the axis follows a smaller path but propagates through a medium which has, on the average, a higher index. On this basis, one can imagine that there might exist an index profile such that the optical path differences between the various rays would be minimized, if not reduced to zero. For such a profile, the pulse spreading, which results from the dispersion of the

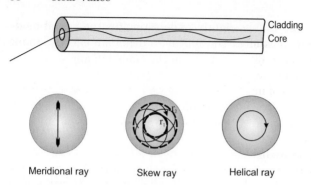

Meridional ray   Skew ray   Helical ray

Fig. 2.9. Ray trajectories in a graded-index fibre

axial velocities among the various rays, would be significantly reduced. To
see if such a profile is conceivable one needs to consider the various types of
rays propagating along the fibre. From the ray-equation analysis, the various
rays can be described, in the special case where $\alpha = 2$, by the parametric
relations [6]

$$r^2(z) = r_1^2 \sin^2(\Omega z) + r_2^2 \cos^2(\Omega z) \,, \tag{2.40a}$$

$$\phi(z) = \arctan \left[\frac{r_2}{r_1} \tan(\Omega z) \right] \,, \tag{2.40b}$$

where $r(z)$ and $\phi(z)$ are the cylindrical coordinates of the ray and where

$$\Omega = \frac{kn_1}{a\beta} \sqrt{2\Delta} \,. \tag{2.41}$$

These equations show that the rays oscillate sinusoidally about the fibre axis
but are confined between two cylindrical surfaces (caustics) of radius r_1 and
r_2 (with $r_1 < r_2$). Two special cases can be derived from these equations,
namely that corresponding to $r_1 \to 0$, which results in *meridional* rays, that
are crossing the fibre axis, and that corresponding to $r_1 = r_2$, for which
the rays appear to remain at a constant distance from the fibre axis, the so-
called *helical* rays. These are two extreme cases corresponding to no azimuthal
variation and no radial variation, respectively, whereas all other rays are *skew*
rays having both azimuthal and radial variations. Depending on the profile
(i.e. depending on α) the propagation equations for the rays will vary but,
basically, the same types of rays will be found. It can be shown that the
optimum profile minimizing dispersion for meridional rays is the hyperbolic
secant, whereas the optimum profile for helical rays is the Lorentzian. Therefore, no profile can minimize dispersion for all the rays at the same time.
However, the Lorentzian and the hyperbolic secant are well approximated by
the parabola so that a parabolic profile is likely to accommodate most of the
rays. Accordingly, it can be demonstrated on the basis of the wave theory
that the optimum profile is in fact *nearly* quadratic if the material dispersion

is taken into account. That is, the optimum profile will vary from fibre to fibre since it also depends on the glass composition. (A detailed description of this theory can be found in [3].)

To conclude here from a practical point of view, graded-index fibres allow, in principle, for three orders of magnitude reduction in intermodal dispersion, compared to equivalent multimode step-index fibres, but tight fabrication conditions must be maintained in order to approach this limit. Typically, they can communicate data at a rate up to 100 Mb/s over distances up to 100 km. Their size have been standardized to 125 μm for the cladding diameter and 50, 62.5, and 85 μm for the core diameter. As for their fabrication cost, they range between the low-cost multimode step-index fibre and the specialized single-mode step-index fibre. Actually, the economy resulting from their use is in good part related to the low cost of the optical sources, connectors and various components used with them. Also, the recent advent of graded-index plastic fibres which are cheaper, more durable and flexible, and easier to install than glass fibres makes them an excellent alternative for local area network (LAN) applications.

2.3   Fibre Materials

A number of requirements must be fulfilled in selecting material for optical fibre fabrication. Material must obviously be a dielectric which is highly transparent over the wavelength range where the fibre is to be used. It must also be capable of being drawn into a fibre form and yet preserve its basic optical and mechanical properties. A good fibre material must also permit the possibility of smoothly varying its refractive index by the addition of dopants. *Glasses* and *plastics* are the two categories of materials that generally satisfy these requirements. Oxide glasses are the most common family of optical glasses. They include the very important silicate glasses that are the most widely used for fibre fabrication. Actually, the availability of ultra-pure and therefore highly-transparent silicate glass fibres has made them the preferred choice for long-haul transmission in the near-infrared. Other glasses with good transmission potential exist, but the constraints imposed by fibre fabrication significantly limit their number. Among them, some halide (especially fluoride) and chalcogenide glasses have attracted much attention recently for their extended transmission window in the infrared [12–14]. Recent advances in plastic fibre fabrication have also made them an excellent choice for short-distance applications, where loss considerations are of lesser concern.

2.3.1   Silicate Glasses

Oxide glasses are amorphous solids made of a combination of network formers such as SiO_2, GeO_2, B_2O_3, Na_2O_3, and P_2O_5. Silicate glasses or silica-based glasses are oxide glasses where SiO_2 is the network former. Silica is made of silicon tetrahedra (SiO_4) which are joined through a *bridging oxygen atom*. Although the tetrahedra are rigid and regular structures, the connection between them allows a degree of freedom, which is basically responsible for the glassy nature of silica. The silica network comprises covalent and directional bonds, which explains its stiffness and viscous nature. Accordingly, silica, like other glasses, does not possess a definite melting point but gradually softens when heated, an important property for fibre fabrication. The refractive index of pure silica can be smoothly varied by the addition of various formers or modifiers; GeO_2, Al_2O_3, P_2O_5 are commonly used to raise the index whereas B_2O_3 is used to lower it. Among the possible dopants, GeO_2, which has the same stoichiometric ratio as SiO_2, has become for this reason the preferred choice for the core. As a matter of fact, each germanium atom can be seen as merely 'replacing' a silicon atom in the glass network, which results in a slight perturbation of the basic structure. The resulting *germano-silicate* glass appears to possess very interesting properties in terms of photosensitivity (see Sect. 2.5.3), of increased rare-earth doping capability [15] and also of increased radiative efficiency [16].

Silicate glasses possess good mechanical properties and are chemically stable. They are also easily drawn into fibre form. Yet, at this moment, their main advantage over all other glasses is their extremely high-transparency in the near-infrared. This property mainly arises from the possibility of producing them from the MCVD (*Modified Chemical Vapour Deposition*) process which is a *soot process* [12] allowing for the production of ultra-pure samples. Their principal drawback is related to their low rare-earth solubility and their relatively high phonon energy, two factors which restrict their use in active fibre devices (fibre lasers and amplifiers). This is why a lot of effort is actually deployed in identifying and producing new families of glasses for these applications.

2.3.2 Plastics

Plastic fibres are the most economical solution to the transmission of visible light over relatively short distances [17]. These fibres transmit through the visible and the near-infrared, making them suitable as optical sensors and illuminators. The first plastic fibres to be commercialized were rather lossy (1000 dB/km at 600 nm wavelength), but nowadays the attenuation can be less than 20 dB/km with typical bandwidth characteristics of 100 MHz·km (for multimode fibres). Plastic fibre cores are made of polystyrene, polymethyl methacrylate (PMMA) or polycarbonate for improved temperature resistance. Plastic fibres are more flexible and less prone to breakage. Weighing only half as much as silica, they can take greater stresses of load, flex and vibrations, and can be fabricated in larger diameters, typically in the range of 100 μm to 1 mm for the core. It is this feature that renders possible easy and therefore cost-effective installation of plastic fibres. Unlike standard glass fibres, the cladding thickness is usually less than one percent. So, they are easier to connect to powerful laser sources. Plastic fibres can withstand continuous service for temperatures ranging from -35 °C to 80 °C but can endure short-term exposures up to 100 °C. Plastic is not as radiation resistant as ultra-pure silica but it can be used far more efficiently than glass in radiation environments (for doses of over 10^6 rad). The use of plastic fibres in communication networks is of great interest for applications where the transmission distances are between tens of centimeters and a few hundred metres. The recent availability of graded-index as well as single-mode plastic fibres is also likely to extend the range of their possible applications.

2.4 Basic Optical Properties

The transmission characteristics of the optical fibre are of utmost importance for optical telecommunication systems. Indeed, as it propagates along the fibre link the optical signal experiences all kinds of losses which demand the use of regenerators or repeaters. Also, when short optical pulses are used, pulse broadening arising from dispersion must be accounted for. Finally, any real optical fibre possesses residual birefringence, which affects the state of polarization of the optical signal.

2.4.1 Losses

Signal attenuation in optical fibres arises from many sources, related either to the material or to the waveguide itself. Depending on the fibre material used, the relative weight of these various contributions will vary as a function of wavelength, leading to qualitatively different spectral characteristics.

Material Absorption

Basic Absorption (Intrinsic Losses). A first source of intrinsic absorption is associated with the tail of the electron transition absorption of the basic constituents, which is located in the UV region of the spectrum. The so-called Urbach tail is actually characterized by an exponential decrease as a function of wavelength and is generally not a predominant source of attenuation in the visible or the near-infrared regions [18] (Fig. 2.10). Another intrinsic absorption arises from the vibrations of the basic molecules, occurring near 9 µm and 13 µm in silica. Through multi-phonon processes these fundamental vibrations are responsible for the sharp long-wavelength cut-off observed near 1.7 µm in silica-based optical fibres. This phenomena is more important in silica than in fluoride or chalcogenide glasses because the phonon energy is higher in silica. Accordingly, the long-wavelength cut-off is translated to around 3 µm in fluoride hosts and beyond 5 µm in some chalcogenides. As pointed out in Sect. 2.3, these two classes of materials are also promising [12] in terms of their projected minimum loss which are of the order of or even

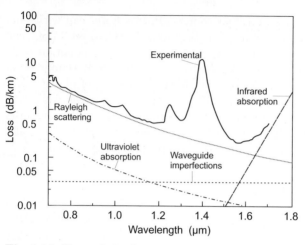

Fig. 2.10. Transmission loss spectra for a silicate glass single-mode fibre (after [17])

smaller than about 10^{-2} dB/km as compared to 10^{-1} dB/km (at 1.5 μm) for fused silica.

Impurities (Extrinsic Losses). The unavoidable presence of impurities in the glass is another source of so-called extrinsic loss. Among the usual impurities are the transition-metal ions like Cr, Mn, Cu, Fe or Ni, which create strong absorption bands in the visible and IR, typically in the range of 1 to 10 dB/km for a concentrations of a few parts per billions. All these impurities must be reduced to concentration much below this value in order to reach minimal attenuation [19]. These impurities can actually be reduced to very small fractions when the glass is fabricated through a vapor deposition process. This is the case for fused-silica but not for fluoride glasses, which are crucible glasses. Another important source of absorption is related to the presence of OH- ions (water). In silica, this leads to three absorption peaks located at 950 nm, 1240 nm, and 1390 nm. Historically, the reduction of the water content in silica fibres has been a crucial step in the development of low-loss telecommunication fibres. Accordingly, the telecommunication window has been established at around 1.3 μm and subsequently 1.55 μm because of the residual OH- absorption peak at around 1.4 μm. This problem appears to have finally been overcome with the advent of the Lucent Technologies *All-Wave* fibre based on a novel fabrication process which practically eliminates the incorporation of OH ions. This fibre, which achieves a 200% reduction of the attenuation at the 1400 nm absorption peak, allows usage over the entire wavelength region extending from 1280 to 1625 nm. In fluoride and chalcogenide glasses, the fundamental vibration mode of the OH$^-$ ion also causes substantial attenuation near 2.9 μm.

Rayleigh Scattering. Rayleigh scattering is the result of microscopic variations in the material density or micro-defects of dimensions well below the operating wavelength. This process, which is characterized by a λ^{-4} dependence, appears to determine the actual theoretical limit of fibre transmission in the visible and in the infrared up to 1.6 μm. The addition of index-modifiers (co-dopants) like germanium or boron into the core leads to an increase in the Rayleigh losses. In the case of germanium the additional loss increases linearly with the concentration. Additional scattering may also result from partial crystallization of the core during its fabrication. This problem is especially important in fluoride-glass optical fibres [12]. Finally, in rare-earth-doped fibre amplifiers the presence of the rare-earth ions in the glass matrix leads to material density fluctuations and therefore to additional scattering. This detrimental effect is enhanced by the formation of clusters occurring in heavily doped fibres [20].

Waveguide Attenuation

Waveguide Imperfections. Scattering can arise from waveguide imperfections or irregularities at the core-cladding interface. Usually, these defects have a typical size which exceeds the wavelength so that Mie scattering is the basic physical process responsible for these additional losses. In most standard telecommunication (silica) fibres, these losses can be kept below 0.05 dB/km (Fig. 2.10) throughout the visible and the near IR regions so that they are not of great concern in general. In single-mode fluoride or chalcogenide fibres produced by a crucible-drawing process, imperfections such as bubbles or microcrystallites often arise at the core-cladding interface, resulting in significant additional losses. Finally, in some special fibres involving *depressed inner cladding*, intrinsic power leakage may also occur because of the spreading of the field in the outer cladding.

Bending Losses. In general, curved waveguides radiate power. Therefore, waveguides can become lossy in reaction to deformations applied to them. Those radiative losses which become significant beyond a critical radius can be qualitatively explained on the basis of the local phase velocity in the outer portion of the core, which exceeds that in the cladding beyond a critical curvature radius. When the curvature radius R is significantly larger than the core diameter we speak of *macro-bending losses*. Various authors have derived curvature loss formula. Referring to Marcuse's approach, one has for the LP$_{01}$ mode in the weakly guiding approximation [21]

$$\gamma = \frac{\sqrt{\pi}(u/a)^2 \exp\left(-\frac{2}{3}(w/a)^3\beta^2 R\right)}{V^2\sqrt{R}(w/a)^{3/2}K_{-1}(w)K_1(w)} \ , \tag{2.42}$$

where γ represents the power loss and the other parameters and functions are as defined in Sect. 2.2. From this expression it can be shown that curvature-induced losses strongly depend on the curvature radius, the index difference, and how far the fibre mode is from cut-off. The critical radius for which losses become appreciable is of the order of a few millimeters, so that for most practical purposes they are not of great concern. Another type of perturbation, referred to as *micro-bending losses*, can also alter the fibre transmission, especially in cable form. These losses arise from the unavoidable random deviation of the fibre core from its ideally straight axis. Although these deviations are relatively small, they are typically associated with sharp local bending, so that they can lead to significant losses. The basic approach to minimizing them in single-mode fibres relies on the improvement of the confinement of the fundamental mode, i.e. operating in the range $2.0 < V < 2.4$. Finally, the various splices and connections between communication links is also a source of loss.

The Loss Coefficient. In this section we have limited our discussion to power-independent or linear losses. (In Sect. 2.5 it will be shown that additional losses also arise as a result of various nonlinear effects). The overall linear losses are generally described through the use of the loss coefficient, α, which allows us to express the exponential decrease of the transmitted power as a function of z

$$P = P_0 \exp(-\alpha z) \,, \tag{2.43}$$

where P_0 is the initial power. From this equation α is formally associated with the inverse of the distance at which the power has decreased by the ratio $1/e$. Alternatively, it is now very common to represent α in units of dB/km. From (2.43) and the definition of the decibel one has $\alpha(\mathrm{dB/km}) = 4.343\alpha$, where z is expressed in kilometres.

2.4.2 Dispersion

Fibre optics communications rely on the use of digital signals. The optical bits or pulses are subject to pulse broadening as they propagate. We have seen in Sect. 2.2 that fibres supporting a large number of modes are affected by *intermodal dispersion*, resulting from the different propagation constants associated with each mode. Accordingly, multimode graded-index fibres were introduced at Sect. 2.2.3 as a very efficient way of reducing this type of dispersion. Strictly speaking, the only way to reduce *intermodal dispersion* to zero is to operate in the single-mode regime. This is not to say that dispersion completely disappears in single-mode fibres, which actually suffer from the so-called *intramodal dispersion*.

The group velocity at which the optical signal propagates is given by the expression ($\bar{n}$ = effective index; β = propagation constant of guided mode)

$$v_g = \frac{c}{\left(\bar{n} + \omega \frac{\mathrm{d}\bar{n}}{\mathrm{d}\omega}\right)} \,, \tag{2.44}$$

which basically shows that this is in general dependent on the frequency, ω (i.e. wavelength). This is a direct consequence of the ω-dependence of $\bar{n}$. Such dependence or *dispersion* is a general property of any dielectric medium. However, in the specific context of a guided wave, dispersion takes on a special character for it also originates from the guidance itself, as we show in the following. Dispersion has detrimental effects on pulse propagation because the different spectral components of a pulse propagate at different velocities, which results in pulse broadening. Accordingly, the so-called *group delay* corresponding to the propagation along a fibre segment of length L is defined as $T_g = L/v_g$.

With the aid of the approximate expression for β (see (2.13)) and neglecting the k-dependence of Δ (i.e. assuming that the dispersion characteristics of the core and the cladding are very similar) one can deduce for the group delay in a weakly guiding fibre

$$T_g = \frac{n_g L}{c} \left\{ 1 + \Delta \frac{\mathrm{d}(Vb)}{\mathrm{d}V} \right\} , \tag{2.45}$$

where a group index n_g has been defined as

$$n_g = \frac{\mathrm{d}(kn)}{\mathrm{d}k} = n - \lambda \frac{\mathrm{d}n}{\mathrm{d}\lambda} , \tag{2.46}$$

with $n \approx n_1 \approx n_2$ standing for either the core or the cladding refractive index. The *dispersion parameter* D is formally obtained by deriving the group-delay T_g with respect to λ:

$$D = \frac{1}{L} \frac{\mathrm{d}T_g}{\mathrm{d}\lambda} = D_\mathrm{M} + D_\mathrm{W} , \tag{2.47}$$

where

$$D_\mathrm{M} = \frac{1}{c} \frac{\mathrm{d}n_g}{\mathrm{d}\lambda} = -\frac{\lambda}{c} \frac{\mathrm{d}^2 n}{\mathrm{d}\lambda^2} \tag{2.48}$$

$$D_\mathrm{W} = \frac{\Delta}{c} \left[\left(\frac{\mathrm{d}n_g}{\mathrm{d}\lambda} \right) \left(\frac{\mathrm{d}(Vb)}{\mathrm{d}V} \right) - \frac{n_g^2}{\lambda n} V \frac{\mathrm{d}^2(Vb)}{\mathrm{d}V^2} \right] . \tag{2.49}$$

It appears from these expressions that D is basically composed of two components: one arising from the material dispersion D_M, the other from the waveguide dispersion D_W. D_W can be seen to be proportional to the small parameter Δ so that it is generally smaller than D_M. However, the material dispersion D_M crosses zero near $1.3\,\mu\mathrm{m}$ in pure silica so that the waveguide dispersion D_W, which can be tailored through an optimization of the waveguide profile, plays a crucial role in defining the exact location of the overall zero dispersion wavelength (Fig. 2.11).

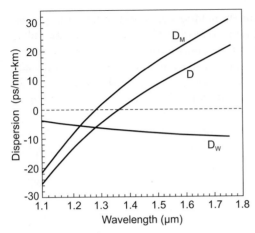

Fig. 2.11. Material (D_M), waveguide (D_W) and overall (D) dispersion curves for a typical single-mode silica fibre

Material Dispersion. Material dispersion arises as a consequence of the wavelength dependence of the refractive index. Such dependence is well described by the Sellmeier equation, which in the case of silica can be expressed as a three-term summation [22]

$$n^2 - 1 = \sum_{j=1}^{3} \frac{A_j \lambda}{\lambda^2 - \lambda_j^2} \, , \qquad (2.50)$$

where the parameters A_j and λ_j, corresponding respectively to the oscillator strength and the resonant wavelength, are compiled in Table 2.1. By deriving this expression with respect to λ, the values of the group index n_g as well as the material dispersion D_M can be obtained. One sees from Fig. 2.11 that for pure silica, D_M crosses zero around 1.27 μm. For germanium-doped silica this value increases almost linearly with the Ge concentration to reach the value of 1.74 μm for pure GeO$_2$ [23].

Table 2.1. Sellmeier coefficients of pure SiO$_2$ and GeO$_2$

	A_1	A_2	A_3	λ_1 (μm)	λ_2 (μm)	λ_3 (μm)
SiO$_2$	0.6961663	0.4079426	0.8974794	0.0684043	0.1162414	9.896161
GeO$_2$	0.8068664	0.7181584	0.8541683	0.0689726	0.1539661	11.841931

Waveguide Dispersion. Waveguide dispersion essentially depends on the parameter V through the first two derivatives of the product "Vb" with respect to V [24]. For the fundamental mode of a step-index fibre, these normalized terms take the universal form illustrated in Fig. 2.12. Thus, unlike material dispersion, which is determined by the glass composition, waveguide dispersion can be modified via the fibre parameters. This can be used to shift the zero of the total dispersion as shown in Fig. 2.11. Indeed, setting $D = 0$ in (2.47) and neglecting the term involving $\mathrm{d}(Vb)/\mathrm{d}V$ in (2.49), which is small compared to D_M, one can obtain the relationship

$$\frac{(n_1 - n_2)}{c\lambda} \left[V \frac{\mathrm{d}^2(Vb)}{\mathrm{d}V^2} \right] \cong D_\mathrm{M} . \tag{2.51}$$

First, we note that the last equation can be satisfied for positive values of D_M only (i.e. for $\lambda > 1.27\,\mu\mathrm{m}$), since the term on the left-hand side is always positive. Thus, waveguide dispersion allows the zero-dispersion to shift towards longer wavelengths. It also appears that both the terms $(n_1 - n_2)$ and $V\,\mathrm{d}^2(Vb)/\mathrm{d}V^2$ should be optimized in order to produce a significant shift. From Fig. 2.12 one sees that $V\,\mathrm{d}^2(Vb)/\mathrm{d}V^2$ has its maximum in the interval $1.0 < V < 1.5$ whereas V should be maintained in the range $2.0 < V < 2.4$ for suitable power confinement of the LP_{01} mode (Sect. 2.2.2). As for the term $(n_1 - n_2)$, a systematic analysis reveals that it should range between 5×10^{-3} and 1×10^{-2} in order to shift the zero in the $1.5\,\mu\mathrm{m}$ region. Now this, along with the condition on V, results in relatively small values of the core diameter as well as a relatively large germanium concentration, both factors contributing to additional losses. Accordingly, it proved rather difficult to fabricate such step-index *dispersion-shifted* fibres and yet to maintain losses at a reasonably low level. This is why special fibres with various refractive index profiles have been developed to solve this problem [25].

Fibre with Modified Dispersion Characteristics. Tailoring the dispersion characteristics in the region $1.3–1.55\,\mu\mathrm{m}$ has been challenging the fibre designers, especially during the 1980s. Now, although the needs have evolved during this period, the efforts have been essentially centreed on two main objectives: flattening the dispersion curve, and shifting its zero toward an optimized wavelength. In both cases, the requirements for specific characteristics in terms of bending losses or mode field diameter resulted in various configurations. The basic idea is to depart from the rather simple step-index profile in order to obtain "b vs V" curves with enhanced dispersion characteristics. In the case of the *dispersion-shifted fibre* (DSF), the index profile generally takes a triangular shape and can be improved with the addition of annular structures around the nominal core or with the use of a depressed inner cladding. In the case of *dispersion-flattened fibres* the general W-profile has been adopted. Multilayer W-profiles have been fabricated such that the dispersion has been reduced to less than $2\,\mathrm{ps/km\text{-}nm}$ over the whole wavelength range extending from 1.3 to $1.6\,\mu\mathrm{m}$ [26]. Generally, fibres with modified dispersion characteristics are more expensive and slightly more

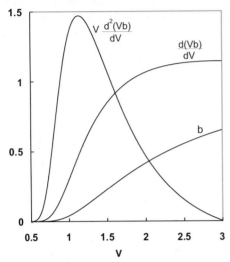

Fig. 2.12. Normalised propagation constant b and its first two derivatives $\mathrm{d}(Vb)/\mathrm{d}V$ and $V\,\mathrm{d}^2(Vb)/\mathrm{d}V^2$ as a function of V

lossy than their step-index counterpart. Consequently, the relevance of their use is analysed on a case-by-case basis as communication systems evolve. Recently, there has been a growing interest in *dispersion management* [27,28] for systems based on soliton transmission (see Sect. 2.6) since it was found that decreasing gradually the dispersion along propagation could help to compensate for the small but non-negligible losses encountered by solitons over long distances. *Dispersion decreasing fibres* [29] were developed for this purpose. It was also proposed to use dispersion-shifted fibre segments in conjunction with standard fibres in a so-called *comblike dispersion profile* [30] configuration, in order to produce trains of optical solitons at high repetition rates.

2.4.3   Polarization

In deriving the fibre modes at Sect. 2.2, it was assumed that the fibre core was perfectly isotropic and circular. Under these conditions it was found that the fundamental HE_{11} mode was composed of two degenerate orthogonal field distributions. If such a fibre should exist, these two components would actually be undistinguishable from each other for they would share exactly the same propagation constant and the same field distribution. Real fibres, however, are subject to small but noticeable fluctuations in the shape of their core along their length. They also experience all kinds of stresses that perturb the nominal circular symmetry and isotropy of their core. For this reason any segment of optical fibre can be regarded as a birefringent element. In terms of the orthogonal components of the HE_{11} mode, this birefringence can be interpreted as raising their degeneracy. Stated differently, each of these two components now possesses its own propagation constant. The resulting modal birefringence B is defined as

$$B = \bar{n}_x - \bar{n}_y = \frac{(\beta_x - \beta_y)}{2\pi/\lambda} , \tag{2.52}$$

where $\bar{n}_x$ and $\bar{n}_y$ are the effective indexes related to the propagation constants corresponding to the fibre principal axes. A typical value of B for a standard single-mode fibre is 10^{-6}. In practice, after propagation along a few metres of such a fibre, the light initially polarized at an angle with respect to the principal axis will pass through various states of polarization and will exit in general with an elliptic polarization. Thus, a standard single-mode fibre segment behaves much like a multiple-order birefringent plate.

However, the evolution of the polarization state along the fibre is completely irregular because both the modal birefringence and the local principal axis vary randomly as a function of z. Therefore, it is not possible to predict what the polarization state will be after a given length of fibre. Moreover, because the stress-induced birefringence is in general temperature dependent, any change in the fibre ambient conditions is likely to modify its polarization characteristics. In most applications where a photodiode is used to detect the intensity of the signal transmitted through the fibre, such polarization effects are irrelevant. However, whenever a coherent detection scheme is involved at the fibre output, these effects need to be taken into account. Another problem that also arises from the fibre birefringence is the so-called *birefringence dispersion* affecting short pulses propagating over relatively long distances. Such dispersion arises when an input pulse excites both polarizations, which propagate at different velocities because of birefringence.

Polarization detrimental effects can be overcome, based on the fact that the stochastic fluctuations leading to polarization disturbance have a spectral power density with low-pass filter characteristics. This means that the period associated with these perturbations is larger than a critical period Λ_c, which is typically of the order of a few millimeters. Historically, a lot of effort has been devoted in order to increase this critical length via better control of the fabrication process, but the task appeared to be excessively demanding and therefore not practical. Alternatively, it was found that inducing a large and regular birefringence such that its associated beat length would be smaller than Λ_c would overcome the stochastic effects. Various types of *highly birefringent* fibres based on this principle are now commercially available. One method of producing such fibres consists in making the core elliptical in shape [31]. However, the birefringence resulting from core ellipticity is proportional to $(n_1 - n_2)^2$, so that significant birefringence can only be obtained for relatively small cores if the conditions for single-mode operations are to be maintained. This gives rise to various problems in terms of ease of handling and fabrication. This is why a more practical approach consists of introducing materials in the preform with different thermal expansion coefficients on both sides of the core region in order to induce a large lateral stress on it. Various geometries (e.g. [32,33]) have been successfully demonstrated and typical values of B are of the order of 4×10^{-3}. Note that such fibres behave exactly like a standard wave plate, i.e. they maintain a linear state of polarization, provided that light is initially injected along one of their two principal axes. However, special highly birefringent fibres have also been designed such that one of the two propagating modes is preferentially attenuated. Large extinction ratios can be achieved after only a few metres of propagation in such *polarizing fibres* [34]. Alternatively, it is possible to produce low-birefringence fibres, i.e. with almost no internal birefringence, by rotating the preform at several thousand revolutions per minute during drawing [35]. The basic idea behind this approach is to rotate the principal axis of the intrinsic birefringence so rapidly that the light cannot follow it and actually sees a circularly symmetric core. All of these fibres have found interesting applications, especially in optical sensing where devices usually involve relatively short fibre lengths (i.e. 100 m or less) and are polarization sensitive. However, for many applications it is not possible to use these special fibres and one has to turn to other ways to control the polarization within the fibre actually at hand. Accordingly, it should be stressed that, as long as the degree of polarization is preserved – which is the case in telecommunications where narrow linewidth optical sources are used – there is essentially no difference between propagating over 100 m or over 100 km in that, in both cases, an arbitrary and unknown elliptic polarization will result. Consequently, all that is required to control the state of polarization (SOP) is a device capable of transforming an arbitrary input polarization state into a given polarization state at the fibre output. Basically, this can be done by inducing a lateral stress over a fibre segment. Such a stress can be directly induced by the use of electromagnetic squeezers but the problem of glass fatigue plagues this approach. Practically, the most preferred approach to SOP control relies on bending-induced birefringence and essentially consists in forming fibre loops of radius R such that a given wave retardance (typically $\lambda/4$) is produced. It can be shown that for a single-loop-based quarter-wave device one has [36]

$$R = kCr^2 \, , \tag{2.53}$$

where, as before, $k = 2\pi/\lambda$, C is a constant equal to 0.133 for silica, and r is the fibre (cladding) radius. On this basis, a rotatable fibre coil can be constructed allowing complete control of the SOP.

2.5   Nonlinear Optical Properties

Nonlinear properties of glass fibres mainly arise from the third-order susceptibility term, χ_3, which is the first non-zero nonlinear term in an isotropic medium. Self-phase modulation as well as stimulated four-photon processes are related to the non-resonant real part of χ_3 whereas stimulated Raman and Brillouin scattering are linked to its imaginary part. Now the strength of a nonlinear process is basically determined by the product of three parameters, namely the nonlinear coefficient (related to χ_3), the interaction length L, and the pump intensity I. Silica can be classified among the weakly nonlinear materials since its nonlinear coefficients are several orders of magnitude smaller than those of the properly called nonlinear materials. However, because of the tremendous enhancement of the two other parameters resulting from its waveguiding properties, silica fibres show very unique nonlinear properties. In fact, compared to the bulk geometry at the same power level, the optical fibre allows an enhancement of the product "LI" which is of the order of the ratio L/a. This ratio is only limited by losses, so that in a standard single-mode fibre it can be as large as 10^9, which compensates largely for the weak nonlinear coefficients of silica. Consequently, optical fibres permit a significant reduction of the thresholds of various stimulated nonlinear processes which are practically impossible to observe in bulk media. Such effects may be detrimental or beneficial depending on the application as we show in Sects. 2.5.1 and 2.5.2.

Permanent structural change of glasses can occur upon exposure to radiation of sufficiently high energy. Such changes were found to occur in germano-silicate fibres exposed to UV radiation in the wavelength range of $240-260$ nm. This photosensitivity induces modifications of the optical properties of germano-silicate glasses that can be used for fabricating intra-core fibre gratings. Although such an effect is not, strictly speaking, a nonlinear effect, it will be discussed in Sect. 2.5.3.

2.5.1   Stimulated Scattering Processes

Light scattering is a consequence of the inherent lack of perfect homogeneity of any material medium. Unlike Rayleigh scattering, which results from fixed or non-propagating density fluctuations, Raman and Brillouin scattering arise from propagating fluctuations and are consequently associated with frequency shifts. For this reason, Raman and Brillouin scattering are called inelastic scattering processes since they convert part of the incident optical beam to a frequency-downshifted (*Stokes*) beam. (Note that a so-called *anti-Stokes* (i.e. frequency-upshifted) beam is also produced, but its amplitude at room temperature is much smaller than that of the Stokes beam and can be neglected for most applications). More specifically, Brillouin scattering is the scattering of light from sound waves (acoustic phonons), whereas Raman scattering is the scattering of light by the vibrational modes of the molecules constituting the medium (optical phonons). Now, although these two processes have essentially the same origin, in practice they lead to rather different phenomena.

Stimulated Raman Scattering. The *spontaneous Raman scattering* is a rather weak process characterized by a Stokes wave growing linearly with the propagation distance. However, in optical fibres the beam intensities that are readily attained are responsible for an enhanced version of this process where the Stokes wave now appears to grow (at least initially) exponentially as it propagates. Physically, this situation corresponds to one where the pump beam not only interacts with the medium but also modifies its optical properties. This parametric process, called *stimulated Raman scattering*, can be mathematically described by the differential equation

$$\frac{\mathrm{d}I_s}{\mathrm{d}z} = g_\mathrm{R} I_\mathrm{p} I_\mathrm{s} \, , \tag{2.54}$$

where I_p and I_s represent the pump and signal (Stokes) intensities respectively, and g_R is the Raman gain coefficient, which is related to the nonlinear susceptibility χ_3. Note that (2.54) is valid under cw or quasi-cw conditions, that is for optical pulses whose spectral bandwidth is smaller than that of the Raman gain. The coefficient g_R has been measured experimentally as a function of the Stokes shift for germano-silicate fibres [37] (Fig. 2.13). It is characterized by a wide peak extending over 40 THz and showing a maximum at 13.2 THz ($440\,\mathrm{cm}^{-1}$). Such a wide spectrum is typical of an amorphous material, such as fused silica, for which the usual wavevector selection rules do not hold. The results shown in Fig. 2.13 were obtained for a pump at 795 nm wavelength, but it can be shown that g_R obeys an inverse-law dependence as a function of the wavelength

$$g_\mathrm{R}(\mathrm{max}) = \frac{9.75 \times 10^{-14}}{\lambda\,(\mu\mathrm{m})} \quad (\mathrm{m/W}) \, , \tag{2.55}$$

where $g_\mathrm{R}(\mathrm{max})$ stands for the peak value of the Raman gain. It should also be stressed that g_R is very sensitive to the presence of dopants. In a standard Ge-doped silica fibre, g_R is typically five times larger than in pure silica.

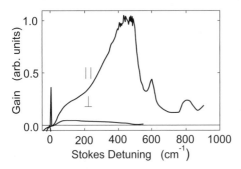

Fig. 2.13. Raman gain as a function of the Stokes detuning (after [43]). The upper curve corresponds to the pump and the signal with parallel polarizations. The lower curve is for perpendicular polarization (extracted from [37])

The physical meaning of (2.54) is that any frequency component lying within the Raman bandwidth of a given pump signal will experience gain as it propagates along with this pump signal. There is usually only one beam (i.e. one frequency component) at the fibre input which acts as a pump, so that the stimulated Stokes signal actually grows from the background spontaneous Raman scattering. Practically, it is the frequency component shifted by 13.2 THz with respect to the pump which will actually build up, because it experiences more gain than the other components. According to (2.54), this signal will grow exponentially as long as the pump is not depleted. Now, one is generally interested to determine the value of the *critical power* P_R^{crit} such that, at the output of the fibre segment of length L, the generated Stokes signal becomes appreciable. Defining this critical power as the input pump power such that the Stokes and pump power are equal at the fibre output, it can be shown that [38]

$$P_R^{crit} \cong 16 \frac{A_{eff}}{g_R(\max)L_{eff}} \; , \tag{2.56}$$

where A_{eff} is the mode effective area (see (2.38)) and L_{eff} is a fibre effective length defined as

$$L_{eff} = \frac{1}{\alpha}(1 - \exp(-\alpha L)) \; , \tag{2.57}$$

where α is the absorption coefficient previously defined in (2.43). Note that for a short segment such that $\alpha L \to 0$, L_{eff} reduces to L whereas for a long segment it is approximately given by $1/\alpha$. Equation (2.56) provides an excellent guideline to estimate the onset of *stimulated Raman scattering* (SRS). For a standard communication fibre with $\alpha \approx 0.2$ dB/km (at 1.55 μm), one has $L_{eff} \approx 20$ km and the critical power is about 500 mW, which is quite high for a cw beam. Actually, it will be shown in the next section that the threshold for *stimulated Brillouin scattering* (SBS) is significantly smaller than that for SRS, so that the onset of the latter is actually inhibited by the former for a cw input. It should also be pointed out that (2.56) was derived for the case where the two beams propagate in the same direction i.e. *forward SRS*. The corresponding threshold for *backward SRS* leads to a similar result with the exception of the numerical factor 16, which is replaced by 20. Although this may appear to be a slight difference, it turns out that backward SRS is, for this reason, hindered by forward SRS for pulse inputs in the nanosecond regime (1−100 ns). Backward SRS is also inhibited by SBS for cw or quasi-cw inputs (i.e. pulse widths longer than 100 ns). Forward SRS can be readily observed with pulse trains provided that the pulses are long enough (i.e. typically longer than 1 ns) to prevent the phenomenon of *pulse walk-off* between the Stokes and pump beams which results from dispersion. Theory predicts a complete power transfer from the pump to the co-propagating Stokes beam, provided that the input power is slightly larger than P_R^{crit}.

Moreover, when the input power is significantly larger than P_R^{crit}, one assists to a cascade process where the newly generated Stokes beam acts as a pump for a subsequent Stokes beam and so on until the power is reduced below the critical power for a further shift. Such a cascade has been observed using high-power pulses from a YAG laser at 1.06 μm wavelength [39].

Because of its large gain bandwidth, SRS was identified early on as a good candidate for the development of optical amplifiers [40,41]. The implementation of fibre-Raman amplifiers is rather straightforward with comparable results for both the forward and backward configurations. The main difference with the SRS process previously described is that a seed signal lying close to the maximum Raman-gain of the pump beam is now injected at one fibre end in addition to this pump beam. Thus, instead of being initiated from the spontaneous Raman scattering, the process rapidly starts up from this seed signal. Typically, in order to reach a 30 dB amplification factor, fibre-Raman amplifiers require a 1 W pump power or more. Therefore, although these amplifiers possess many attractive features such as low noise and polarization insensitivity of their gain, they could not until recently be pumped by compact laser-diode sources and they had to rely on less convenient sources like the Nd:YAG laser. This is why the Raman-gain amplifiers were largely supplanted in fibre-based communication systems by the more versatile rare-earth-doped amplifiers discovered in the late 1980s (see Chap. 5).

Another interesting application of SRS is the fibre-Raman laser. In this case the fibre segment is terminated at both ends by highly reflecting elements in order to form a Fabry–Perot cavity. The main consequence of this is a substantial reduction of the SRS threshold compared to the single-pass configuration. Various configurations have been successfully demonstrated, based on either linear or loop cavities. In the hybrid version, involving bulk mirrors, a prism is generally used as the intra-cavity tuning element. Tunability over a 10 nm bandwidth is typically obtained with such a configuration when lasing is restricted to one Stokes shift. However, one can also take advantage of the cascade SRS previously discussed to produce several wavelengths simultaneously. Now the recent availability of high power Nd^{3+} or Yb^{3+} cladding-pumped fibre lasers together with the development of intra-core fibre Bragg gratings has opened up new horizons for the fibre-Raman lasers. A good example of this is the five-shifts-cascaded Raman-laser shown in Fig. 2.14. This laser produces a single-mode fibre output power of 1.7 W based on the principle of five embedded laser cavities which efficiently convert the 1117 nm pump power from a Yb^{3+} fibre laser into a 1480 nm output [42].

Stimulated Brillouin Scattering. The process of *stimulated Brillouin scattering (SBS)* is qualitatively similar to the SRS process just discussed. However, the quantitative differences between these two processes have important practical implications in optical fibres. Based on the classical picture, SBS can be viewed as a self-induced scattering process where a pump

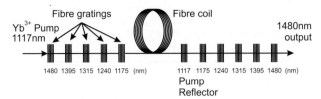

Fig. 2.14. Cascaded fibre Raman laser. The output at 1480 nm is obtained after five consecutive Raman shifts of the pump at 1117 nm wavelength

wave generates, through the process of electrostriction, an acoustic wave which in turns generates an index grating which scatters the pump wave. According to the quantum-mechanical picture, this process corresponds to a pump photon being annihilated at the profit of a down-shifted photon and a phonon. Because of the usual constraints imposed by energy and momentum conservation, the Doppler shift associated with this process is found to be a maximum in the backward direction and zero in the forward. Thus, one major difference with the nearly isotropic SRS process is that SBS occurs in the backward direction only. Another important difference which also arises from the different properties of optical versus acoustic phonons is that the frequency shift associated with SBS is approximately three orders of magnitude smaller than that of SRS. As for the bandwidth $\Delta\nu_B$, it is related to the phonon lifetime T_B through the relation $\Delta\nu_B = 1/\pi T_B$. In silica, one has $T_B \approx 16\,\text{ns}$ (at $1.5\,\mu\text{m}$ wavelength), which implies $\Delta\nu_B \approx 20\,\text{MHz}$. However, the actual Brillouin bandwidth appears to vary significantly from one fibre to another depending on its inhomogeneity or on its germanium content. Therefore, the approximate value of 50 MHz can be reasonably put forward, which implies a five orders of magnitude reduction compared to the 5 THz bandwidth associated with the SRS process. On the other hand, the Brillouin gain g_B appears to be significantly larger than the Raman gain and also independent of the wavelength. A comparative summary of the previous parameters is shown in Table 2.2.

Following an analysis similar to that presented for SRS, it can be shown that the critical pump power for the onset of SBS is given by

$$P_B^{\text{crit}} = 21\frac{A_{\text{eff}}}{g_B(\text{max})L_{\text{eff}}}\,, \tag{2.58}$$

which is similar to the expression derived for SRS. However, because g_B is approximately three orders of magnitude larger than g_R, the critical power associated with SBS will be typically three orders of magnitude smaller than that of SRS for a cw input. Thus, in many practical situations involving cw beams, the onset of SBS hinders that of SRS. The situation is drastically changed when pulses with duration significantly shorter than the phonon

Table 2.2. Parameters associated with stimulated Raman scattering (SRS) and stimulated Brillouin scattering (SBS) for fused silica at 1.55 μm wavelength

	SRS	SBS
Stokes shift	13 THz	11 GHz
$\Delta\nu$	≈ 5 THz	≈ 50 MHz
g^{MAX}	6.3×10^{-14} m/W	$\approx 5 \times 10^{-11}$ m/W
P_{crit}	≈ 500 mW	≈ 1 mW

lifetime T_{B} are used. For a pulse width smaller than 1 ns, the Raman gain actually exceeds the effective Brillouin gain. It should also be pointed out that (2.58) assumes that the polarization vectors of the pump and Stokes waves remain parallel during propagation. In general, the value of $P_{\mathrm{B}}^{\mathrm{crit}}$ must be increased by a factor lying between 1 and 2 to take the polarization scrambling into account.

Similarly to SRS, SBS has been used to make fibre-lasers and fibre-amplifiers. However, because of the significantly reduced Brillouin-gain bandwidth as well as the low achievable output power, the practical interest of these devices is rather limited. Actually, in the context of optical telecommunications, SBS is more readily associated with harmful effects. In fact, standard telecommunications fibres operating at 1.55 μm typically have background losses, α , of the order of 0.2 dB/km, i.e. $L_{\mathrm{eff}} \approx 20$ km. From (2.58) with $A_{\mathrm{eff}} = 50\,\mu\mathrm{m}^2$, the critical power corresponding to such a length is of the order of 1 mW. Thus, the onset of SBS is very likely to occur in long-haul communication links based on cw or quasi-cw signals if special attention is not paid to avoid it. Practically, SBS is harmful, not only because of its associated loss but also because any backward-propagating wave generated in a fibre device is directly fed back into the laser source, which results in instability problems even at very low intensities of the retroreflected signals. Therefore, various ways have been developed to prevent the onset of SBS. The basic idea is naturally to increase the bandwidth of the signal. This can be performed by using short pulses, as previously pointed out, but also by modulating the phase or the amplitude of the input signal. Now, although the general trend is for the Brillouin threshold to increase with the bit rate, it appears that the exact dependence also depends on the data transmission format used. Therefore, a careful analysis must be performed in order to establish the practical power limitations imposed by SBS in a given system [43].

2.5.2   Third-Order Nonlinear Parametric Processes

The modal solution of Sect. 2.2.2 was derived under the assumption that the medium responded linearly to the applied electric field. At the intensities which are readily attained in optical fibres, this assumption must be reconsidered. In general, the polarization of a medium can be expressed as a power series in the field strength. However, since optical fibres are generally made of an amorphous material, the first non-zero nonlinear term (in the dipole approximation) and the only one that generally needs to be considered, is the cubic one. Furthermore, since nonlinearity in glass arises from the nonlinear response of bound electrons it can be assumed to be instantaneous so that $\boldsymbol{P}^{\mathrm{NL}}$ takes the form

$$\boldsymbol{P}^{\mathrm{NL}}(\boldsymbol{r},t) = \varepsilon_0 \chi^{(3)} \vdots \boldsymbol{E}(\boldsymbol{r},t)\boldsymbol{E}(\boldsymbol{r},t)\boldsymbol{E}(\boldsymbol{r},t) , \tag{2.59}$$

where the dots indicate the tensorial operation on the field. The electric field is generally expressed for such third-order process as the sum of three distinct frequency components

$$\boldsymbol{E}(\boldsymbol{r},t) = \frac{\hat{x}}{2} \left[E_1 \exp(-\mathrm{i}\omega_1 t) + E_2 \exp(-\mathrm{i}\omega_2 t) + E_3 \exp(-\mathrm{i}\omega_3 t) + c.c. \right] , \tag{2.60}$$

where it is assumed for simplicity that the three components are linearly polarized along the x axis. Introducing the expression of this field into (2.59) results in 216 (6^3) product terms which can be regrouped in terms of 44 different frequency components ω_{NL}. Symbolically, such frequency mixing can be written as:

$$\omega_{\mathrm{NL}} = \pm\omega_i \pm \omega_j \pm \omega_k \tag{2.61}$$

where $i, j, k = 1, 2, 3$. Therefore, when three optical waves co-propagate in an optical fibre, 44 different frequency components are produced through the third-order nonlinear process involving $\chi^{(3)}$. However, these secondary frequency components are intrinsically weak and can only reach significant amplitudes if they are allowed to build up during propagation along the fibre, and this requires that they be in phase with one of the source terms. Such a phase-matching condition is satisfied only for a few special situations in optical fibres.

Nonlinear Refraction. Phase-matching will automatically occur for the polarization terms such that $\omega_{\mathrm{NL}} = \omega_1$ where ω_1 is chosen arbitrarily to represent one of the input waves. Physically, the corresponding nonlinear term $P^{\mathrm{NL}}(\omega_1)$ can be interpreted as a nonlinear contribution to the refractive index experienced by the optical wave at ω_1. Now, from (2.61), it appears that there exists only three possibilities for this situation to occur, i.e. $\omega_1 + \omega_1 - \omega_1$,

$\omega_1 + \omega_2 - \omega_2$ and $\omega_1 + \omega_3 - \omega_3$, which correspond (from (2.59) and (2.60)) to the field products $|E_1|^2 E_1$, $|E_2|^2 E_1$, and $|E_3|^2 E_1$, respectively. The first of these terms, which involves the field E_1 only, is referred to as the *self-phase modulation* (SPM) term, whereas the two others, which are functionally identical to each other, are referred to as the *cross-phase modulation* terms. Note that, in general, for n input frequencies the total cross-phase modulation contribution to ω_1 would simply be expressed as a sum over $(n-1)$ terms, i.e. $\sum_{j=2}^{n} |E_j|^2 E_1$. Thus, for the sake of simplicity, we will restrict our discussion to the two-wave case in the following.

The refractive index change at the frequencies ω_1 and ω_2 resulting from self- and cross-phase modulation can be expressed as [44]

$$\Delta n_j = n_2(I_j + 2I_{3-j}) , \tag{2.62}$$

where $j = 1, 2$ and I_j is the (average) mode intensity which is equal to P_j/A_{eff} where P_j is the power and A_{eff} is the effective mode area. The nonlinear-index coefficient, n_2, is related to $\chi^{(3)}$ by

$$n_2 = \frac{3}{4n^2 c\varepsilon_0}\chi^{(3)}_{xxxx} . \tag{2.63}$$

The first consequence of this refractive index change is to affect the wave-guiding conditions, that is to perturb the modal solution previously derived in Sect. 2.2. However, the nonlinear-index coefficient n_2 is usually quite small ($3.0 \times 10^{-20}\,\text{m}^2/\text{W}$ in fused silica), so that the nonlinearly induced index change can be considered as a small perturbation to the linear solution. From a standard perturbation analysis, it is shown that, to first order, the field distribution is not perturbed by the index change whereas the correction term to the propagation constant $\Delta\beta$ is simply given by the product $k\Delta n_j$. From this and using (2.62), the expression for the nonlinear phase shift can be derived [44]:

$$\Delta\phi_j = \int_0^L \Delta\beta\,\mathrm{d}z = kn_2 L_{\text{eff}}(I_j + 2I_{3-j}) , \tag{2.64}$$

where L_{eff} represents, as before, the effective length which accounts for the decrease of the power due to loss and I_j is the intensity at the fibre input.

Under cw or quasi-cw conditions, such a phase shift is an important power-limiting factor for systems based on phase-sensitive detection. In multi-channel (i.e. Wavelength Division Multiplexing (WDM) systems) the problem becomes even worse, for the nonlinear phase shifts arising from each channel add up to the total phase shift through cross-phase modulation.

For short optical pulses, the rapid temporal variation of the intensity induces an equivalent variation of the nonlinear phase-shift. This temporal variation of the phase within the pulse envelope actually implies that the instantaneous frequency of the pulse varies across the pulse, that is, a *frequency chirp* is produced. From (2.64) and discarding the cross-phase term for simplicity, the frequency chirp is given by

$$\Delta\omega(t) = -\frac{\partial\Delta\phi}{\partial t} = -kn_2 L_{\text{eff}} \frac{\partial I_j}{\partial t} . \tag{2.65}$$

For a Gaussian pulse envelope of width t_0 (i.e. $I = I_0 \exp(-t^2/t_0^2)$) the chirp takes the form

$$\Delta\omega(t) = \frac{2kn_2 L_{\text{eff}} I_0}{t_0} \left(\frac{t}{t_0}\right) \exp(-t^2/t_0^2) . \tag{2.66}$$

Thus, at the center of the pulse ($t = 0$) one has $\Delta\omega(t) = 0$, which means that the instantaneous frequency is equal to the carrier frequency ω_0 (see Sect. 2.6). However, for $t > 0$, corresponding to the trailing half of the pulse, one has a positive (blue) shift, whereas for $t < 0$ a negative (red) shift is imposed on the pulse. One also notes that the shift is proportional to the pulse intensity and inversely proportional to its width. Thus, the frequency chirp induced by self-phase modulation rapidly becomes significant for short and intense pulses. Accordingly, the frequency spectrum of these pulses rapidly broadens to a point where dispersion starts playing a significant role. In fact, a major consequence of SPM in the context of short and intense pulses is their enhanced temporal broadening. Furthermore, very interesting new features arise from the interplay between self-phase modulation and dispersion, as we show in Sect. 2.6.

Nonlinear Birefringence. Similarly to what occurs for two optical waves at different frequencies but identical polarization, one has, for two orthogonally polarized waves at the same frequency, the nonlinearly-induced index change

$$\Delta n_x = n_2 \left(I_x + \frac{2}{3} I_y\right) , \tag{2.67}$$

where I_x and I_y are the intensities along the axis x and y respectively. The index change along y, i.e. Δn_y is simply obtained by interchanging x and y in (2.67). Thus, a net birefringence $\Delta n_x - \Delta n_y$ is nonlinearly induced whenever I_x is different from I_y. The similarity between (2.67) and (2.62) basically arises from the tensor nature of $\chi^{(3)}$, which discriminates between the two polarizations and mixes them as well. Practically, this self-induced birefringence changes the state of polarization (SOP) of a given input wave if it is elliptically polarized at the fibre input, that is, if it has two orthogonal polarization components. Moreover, for fibres with intrinsic (linear) birefringence the nonlinear birefringence can lead to instability problems where a small change in the input SOP results in large variations of the output SOP [45]. On the other hand, nonlinear birefringence has been used for some practical applications including an optical shutter [46] and a pulse shaper [47].

Four-Wave Mixing. Besides the frequency mixing terms leading to self-phase modulation and cross-phase modulation, one can identify two categories of mixing terms from (2.61). The first one corresponds to the case where the three signs are identical, i.e. $\omega_i + \omega_j + \omega_k$ or $-\omega_i - \omega_j - \omega_k$. This kind of mixing corresponds, for instance, to the third-harmonic generation (e.g. $3\omega_1$) or frequency up-conversion (e.g. $2\omega_1 + \omega_2$) and it is generally not possible to satisfy the phase-matching requirement for these processes in optical fibres, where no angular adjustment is allowed. The other category corresponds to the situation where one sign in (2.61) is different from the two others (e.g. $\omega_i + \omega_j - \omega_k$). In this case one has: $\omega_{\mathrm{NL}} + \omega_k = \omega_i + \omega_j$ and the nonlinear mixing can be interpreted as a process where two photons are annihilated at the profit of two others. A particular case of interest in this category is that for which $\omega_i = \omega_j$, where phase-matching requirements are more easily satisfied. Now setting for definiteness $i = j = \mathrm{p}$ (for pump), $k = \mathrm{s}$ (for Stokes) and $NL = \mathrm{a}$ (anti-Stokes) one has

$$\omega_{\mathrm{p}} - \omega_{\mathrm{s}} = \omega_a - \omega_{\mathrm{p}} = \Omega_{\mathrm{s}} , \tag{2.68}$$

where, by analogy with the Brillouin or Raman-stimulated scattering processes, one has a situation where a pump wave at ω_{p} creates two sidebands shifted by Ω_{s} symmetrically with respect to ω_p. Such partially degenerate four-wave mixing therefore provides gain at the Stokes and anti-Stokes frequencies which can grow either from noise or from a weak seed signal. The phase-matching requirement for such a process can be expressed as

$$\Delta\beta = (\bar{n}_{\mathrm{a}}\omega_{\mathrm{a}} + \bar{n}_{\mathrm{s}}\omega_{\mathrm{s}} - 2\bar{n}_{\mathrm{p}}\omega_{\mathrm{p}})/c = 0 , \tag{2.69}$$

where $\bar{n}_{\mathrm{a}}$, $\bar{n}_{\mathrm{s}}$ and $\bar{n}_{\mathrm{p}}$ represent the mode effective indexes at the anti-Stokes, Stokes and pump frequencies, respectively. Because of dispersion, this phase-matching condition cannot be satisfied in general. Now, recalling that dispersion originates from three sources, material, waveguide and also nonlinearity (through the self- and cross-phase modulation terms), it is rather easy in multimode fibres to have intermodal (waveguide) dispersion to compensate for the two others so that the first experimental demonstration of four-wave mixing in optical fibres was based on this idea [48]. In single-mode fibres, different approaches can be put forward, based on the fact that the waveguide contribution is in general negligible except near the zero-dispersion wavelength. The self-phase matching requirement can also be satisfied by operating in the anomalous dispersion regime [49] or with the help of polarization-preserving fibres [50].

2.5.3   Photosensitivity

When a germano-silicate glass fibre is exposed to UV irradiation near to 242 nm wavelength, many of its optical properties are permanently modified, such as the refractive index, the absorption, the birefringence and the nonlinear susceptibility. This *photosensitivity* has been the subject of a tremendous research effort since its discovery [51] in 1978, and although it has already found a large number of applications, the exact mechanism of its origin is still a matter of debate [52].

Basic Mechanism. It is generally accepted that defects resulting from the incorporation of dopants in the silica–glass matrix are the origin of its photosensitivity. In the case where germanium is the main dopant, a type of defect characterized by an oxygen-deficient bond (O-D-B) has been identified as being responsible for the observed strong absorption band at 242 nm. A common form of O-D-B is shown in Fig. 2.15a. It is characterized by the absence of a bridging oxygen atom between a germanium atom and its neighboring germanium (or silicon) atom. Upon exposure at 242 nm wavelength this defect is the site of a photochemical reaction (Fig. 2.15b) resulting in the creation of a so-called GeE' colour centre together with the ejection of an electron in the conduction band. Once the electron has reached the conduction band it can either be recaptured by its original sites or migrate and be trapped elsewhere, in which case a permanent change of the local glass properties occurs. This photochemical reaction has been shown to be the origin of various phenomenological changes, namely a permanent increase of the refractive index, a permanent partial bleaching of the 242 nm absorption band accompanied by an increase of the 195 nm absorption band, a permanent increase of the absorption in the visible, and so on. A model based on a three-level system with a fundamental singlet level located at the mid-band-gap of SiO_2 can be used to interpret and correlate these changes quite satisfactorily [53]. Furthermore, the assumption that the photoionisation of the defects involves a thermally driven step has led to a model which provides a good explanation of the stretched-exponential dependence of the parameters associated with photosensitivity.

Although it is now well established that the photoinduced index change is related to the ionization of defects in the glass structure, the exact mechanism linking the two is not yet fully understood. On the one hand, it was suggested

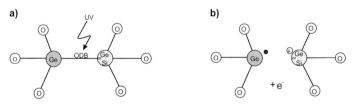

Fig. 2.15. Possible scenario for the creation of a GeE' centre: (**a**) An oxygen-deficient bond (ODB) is ionized by a UV photon; (**b**) a GeE' centre is produced along with a positively charged Si (or Ge) centre and a free electron

that the refractive-index change mainly arose from the relocalisation of electrons at different absorbing sites in the glass. According to this *colour centre model* [54], the index change relates to the absorption change $\Delta\alpha(\lambda)$ through the well-known Kramers-Kronig relation. From this relation, a net positive index change can be deduced on the basis of the measured $\Delta\alpha$, which is in qualitative agreement with the experimental results. On the other hand, it is now generally accepted that structural changes of the glass resulting from the ionisation of defects plays a role in the refractive index change [55]. According to this model the region of the glass exposed to UV irradiation would densify and therefore exhibit an increase of its refractive index, whereas in the unexposed regions the refractive index would, in reaction, tend to diminish. The relative contributions of these two phenomena has yet to be determined precisely but it is already clear that it depends on the nature and the concentration of the dopants as well as the conditions under which the fibre was fabricated and subsequently treated. For instance, it is now rather well established that the main contribution to the index change in hydrogen-loaded germano-silicate fibres can be explained on the basis of the colour-centre model. Now, it appears that in most standard fibres, the overall index change produced by either induced absorption or structural change is of the order of $1-5 \times 10^{-4}$, which is not sufficient for many Bragg grating applications. This is why enhancement techniques were developed.

Enhancement of Photosensitivity. Generally speaking, the number of defects and therefore the photosensitivity of germano-silicate fibres increases with the germanium concentration. However, the photosensitivity appears to depend also on the fabrication process, so that different fibre samples with the same germanium content can exhibit different photosensitivities. In fact, applying reducing conditions during the fabrication process leads to a substantial increase of the photosensitivity. Alternatively, the reduction can be performed after the fibre has been pulled by the *flame-brushing* technique, which consists in exposing a segment of fibre to a hydrogen flame. Another approach consists of adding boron as a co-dopant, which allows the Ge content to be increased without increasing the refractive index. Ideally, though, one is interested in producing index changes in a standard telecommunication fibre. For this purpose the technique of *hydrogen loading* has been developed [56] where a fibre segment is placed in a chamber filled with H_2 at a temperature below 125 °C and at high pressure (typically 200 bar) for a few days (typically 10 days at room temperature). The benefit of this approach as compared to the flame-brushing technique is that hydrogen now diffuses into the glass mainly in the molecular form so that the creation of OH^- bonds, resulting in additional losses at around 1.4 µm wavelength, is substantially reduced. Therefore, at the price of a reasonable increase in absorption, the photosensitivity can be increased by a factor of up to one hundred in most fibres by this technique. Consequently, refractive-index modulations of the order of 10^{-3} can be produced, which allows for the manufacture of very efficient intra-core Bragg gratings.

2.6 Pulse Propagation in Optical Fibres

2.6.1 Derivation of the Wave Equation for the Pulse Envelope

In Sect. 2.2.2, a (z,t)-dependence of the form $\exp(\mathrm{i}\beta z - \mathrm{i}\omega t)$ was assumed in order to derive the modal solution. Thus, our solution was derived for a monochromatic wave, i.e. under cw conditions. In order to allow for the existence of a pulsed signal, an envelope function $A(z,t)$ (which is assumed to vary slowly with respect to electric field oscillating at frequency ω_0) must be introduced so that the overall dependence on z and t can be expressed as

$$E(z,t) = A(z,t) \exp\left[\mathrm{i}(\beta z - \omega_0 t)\right] . \tag{2.70}$$

In the frequency domain this implies that the optical wave now extends over a finite bandwidth centreed about a carrier frequency ω_0. Now, since the wave equation (2.17) was derived for a monochromatic wave, it cannot be applied to a general (non-monochromatic) wave because the refractive index appearing in it is a function of ω. Formally, the problem must be approached in the frequency domain and the rigorous derivation of the wave equation for the envelope function is rather tedious [57]. It is possible, however, to derive this equation on an heuristic basis [58], which we show here.

Dispersive Effects. Because of dispersion, the propagation constant, β, is a function of ω. It is thus convenient to expand $\beta(\omega)$ in a Taylor series about the carrier frequency

$$\beta - \beta_0 = (\omega - \omega_0)\beta_1 + \frac{1}{2}(\omega - \omega_0)^2\beta_2 + \frac{1}{6}(\omega - \omega_0)^3\beta_3 + \cdots , \tag{2.71}$$

where

$$\beta_i = \left.\frac{\mathrm{d}^i\beta}{\mathrm{d}\omega^i}\right|_{\omega=\omega_0} . \tag{2.72}$$

Interestingly, the first terms of this series have the following physical meanings:

$\beta_0 = \beta(\omega_0)$: Propagation constant at ω_0.

$\beta_1 = \left.\dfrac{\mathrm{d}\beta}{\mathrm{d}\omega}\right|_{\omega_0} = \dfrac{1}{v_g}$: Inverse of the group velocity (at ω_0)

$\beta_2 = \left.\dfrac{\mathrm{d}^2\beta}{\mathrm{d}\omega^2}\right|_{\omega_0} = \dfrac{1}{v_g^2}\dfrac{\partial v_g}{\partial \omega}$: Group velocity dispersion (GVD)

$\beta_3, \beta_4,$ etc : higher order dispersive terms.

In particular, the GVD term, β_2, is related to the dispersion parameter D derived previously (2.47) by

$$\beta_2(\text{ps}^2/\text{km}) = \frac{-\lambda^2}{2\pi c} D(\text{ps}/\text{nm–km}) , \tag{2.73}$$

where the usual units for these parameters appear between parenthesis. Now, since the frequency bandwidth of the optical pulses is usually small compared with its carrier frequency, the GVD term in (2.71) is the last one that needs to be kept in the expansion. Also, since the envelope is a slowly varying function of z and t we can Fourier transform it in the two-dimensional frequency space defined by the variables $\Delta\omega = (\omega - \omega_0)$ and $\Delta\beta = (\beta - \beta_0)$:

$$\tilde{A}(\Delta\omega, \Delta\beta) = \int_{-\infty}^{\infty} \int_{-\infty}^{\infty} A(z,t) \exp\left[i(\Delta\omega t - \Delta\beta z)\right] dz\, dt \tag{2.74}$$

$$A(z,t) = \frac{1}{(2\pi)^2} \int_{-\infty}^{\infty} \int_{-\infty}^{\infty} \tilde{A}(\Delta\omega, \Delta\beta)$$
$$\times \exp\left[-i(\Delta\omega t - \Delta\beta z)\right] d(\Delta\beta)\, d(\Delta\omega) . \tag{2.75}$$

From these equations it appears that the partial derivatives of $A(z,t)$ with respect to z and t can be associated with the variables $\Delta\beta$ and $\Delta\omega$, i.e. $\Delta\beta A \Leftrightarrow -i\partial A/\partial z$ and $\Delta\omega A \Leftrightarrow i\partial A/\partial t$. Transposing this to the equation (2.71) one has:

$$\left(\frac{\partial A}{\partial z} + \beta_1 \frac{\partial A}{\partial t}\right) + \frac{i\beta_2}{2} \frac{\partial^2 A}{\partial t^2} = 0 , \tag{2.76}$$

which is the wave equation for the envelope in the linear and dispersive regime. One sees that for $\beta_2 = 0$ (i.e. no pulse spreading) the familiar one-dimensional wave equation is obtained, having d'Alembert's solutions of the form $A \equiv A(z \pm v_g t)$ with $v_g = 1/\beta_1$. Accordingly, it is customary to use a new coordinate system, the so-called *retarded time-frame*, defined as: $z \Leftrightarrow z$, $T \Leftrightarrow (t - \beta_1 z)$, so that equation (2.76) becomes

$$\frac{\partial A}{\partial z} + \frac{i\beta_2}{2} \frac{\partial^2 A}{\partial T^2} = 0 . \tag{2.77}$$

Nonlinear Effects. The propagation constant needs a correction term $\delta\beta$ at high intensity because of the self-phase modulation phenomenon (Sect. 2.5.2). Recalling that this correction term is simply given by the product $k\Delta n$ with Δn given by (2.62), one has for $\delta\beta$ (ignoring the cross-phase modulation term)

$$\delta\beta = \frac{2\pi}{\lambda} n_2 I . \tag{2.78}$$

We shall assume that this small variation of the propagation constant, arising from nonlinear effects, can be simply added to the dispersion-variation terms appearing on the right-hand side of (2.71). Following the same procedure as before, a modified version of (2.77), which now includes the nonlinearity, can be derived:

$$\frac{\partial A}{\partial z} + \frac{\mathrm{i}\beta_2}{2}\frac{\partial^2 A}{\partial T^2} - \mathrm{i}\frac{2\pi n_2 I}{\lambda}A = 0 \ . \tag{2.79}$$

Now, since $I = P/A_{\mathrm{eff}}$ (where P is the optical power and A_{eff} is the effective area (Sect. 2.5.2)), the amplitude A can be normalized in units corresponding to the square root of the power (i.e. $W^{1/2}$), so that (2.79) finally becomes

$$\mathrm{i}\frac{\partial A}{\partial z} - \frac{1}{2}\beta_2\frac{\partial^2 A}{\partial T^2} + \gamma|A|^2 A = 0 \ , \tag{2.80}$$

with

$$\gamma = \frac{kn_2}{A_{\mathrm{eff}}} \ , \tag{2.81}$$

which describes the propagation of an optical pulse in a lossless optical fibre. Because of its similarity with the famous Schroedinger equation, this equation is usually called the nonlinear Schroedinger equation, or NLS in short. For completeness, a loss factor of the form $\mathrm{i}\alpha A/2$ should be added to it which would make it depart slightly from the standard form. However, in optical fibres, the losses are so small that dispersive as well as nonlinear effects usually arise before linear losses become appreciable, so that one can ignore them in our discussion of the basic concepts. In fact, the great merit of equation (2.80) is to possess a very special type of solution called a soliton. These solitons are pulses which propagate with virtually no deformation. From a practical point of view this property makes them very attractive candidates for long-haul communication links, according to the original suggestion by Hasegawa and Tappert [59].

2.6.2   Solution of the Envelope Wave Equation: The Soliton

Equation (2.80) was derived under the assumption that $\Delta\omega \ll \omega_0$, which means practically that it is valid for pulse widths down to 100 fs. For shorter pulses, additional terms must be included which perturb the ideal soliton solution. For a complete description of these terms, which lead to phenomena like *self-frequency shift*, *self-steepening*, *cubic dispersion* and so on, the reader is referred to the book by Agrawal [44]. For definiteness, let us assume an incident pulse of peak power P_0 and pulse width T_0. This allows us to normalize further (2.80) by setting $\tau = T/T_0$ and $A(z,\tau) = P_0^{1/2}U(z,\tau)$ where $U(z,\tau)$ is a normalized pulse envelope. Substituting into (2.80), we obtain

$$\mathrm{i}\frac{\partial U}{\partial z} = \frac{\mathrm{sgn}(\beta_2)}{2L_D}\frac{\partial^2 U}{\partial \tau^2} - \frac{1}{L_{NL}}|U|^2 U \ , \tag{2.82}$$

where two important parameters having the dimension of a distance have been defined:

$$L_D = \frac{T_0^2}{|\beta_2|} \, , \; L_{NL} = \frac{1}{\gamma P_0} \, . \tag{2.83}$$

These two parameters, which essentially scale the length over which SPM and GVD become appreciable, provide an effective means of analyzing the pulse propagation qualitatively [44]. In fact, depending on the fibre length L (as compared to L_D and L_{NL}) different behaviours will be found. For a fibre length, L, much smaller than both L_D and L_{NL}, no appreciable effect, dispersive or nonlinear, will occur and the pulse shape will remain the same. For instance, for a standard telecommunication silica fibre operating around 1.5 μm wavelength, one has $|\beta_2| \cong 20\,\mathrm{ps}^2/\mathrm{km}$ and $\gamma \cong 3\,\mathrm{km}^{-1}\mathrm{W}^{-1}$, so that pulse distortion will be negligible over 50 km or less if $T_0 > 100\,\mathrm{ps}$ and $P_0 < 1\,\mathrm{mW}$. On the other hand, when the fibre length is longer or comparable to either L_D or L_{NL}, dispersive or nonlinear effects start to arise. Dispersive effects will be dominant whenever $L_D \ll L_{NL}$ since the second term on the right-hand-side of (2.82) is then negligible compared to the first one. The resulting equation is then easily integrated and the resulting solution is characterized by a temporal broadening of the pulse. When $L_D \gg L_{NL}$, the first term of (2.82) can be neglected, so that the resulting equation can again be readily solved. In this case it is found that the envelope of the pulse is not affected, whereas its phase suffers a nonlinear shift resulting in the creation of a frequency chirp. Interestingly, though, in the purely dispersive regime it is the temporal pulse shape which suffers broadening, whereas in the purely nonlinear regime it is the frequency spectrum which is broadened. In the most general case, where $L_D \cong L_{NL}$, both terms must be taken into account and a numerical integration of (2.82) is required. This regime is characterized by a very special behavior resulting from the dynamical interplay between the dispersive and nonlinear effects. Actually, depending on the sign of the parameter β_2, the SPM and GVD can be regarded as acting conjointly or in opposition. In the normal dispersion regime (i.e. $\beta_2 > 0$, which occurs for $\lambda < 1.3\,\mu\mathrm{m}$ in standard silica fibres (see Sect. 2.4.2)) the nonlinearly induced chirp is such that dispersion acts on it so as to enhance pulse broadening. Stated differently, nonlinearity induces a red frequency-shift at the pulse leading edge and a blue one at its trailing edge so that dispersion, by postponing the blue component, enhances their spreading. For negative values of β_2, in the so-called *anomalous dispersion regime*, one has a different situation where the SPM-induced frequency chirp can be counterbalanced by the GVD chirp under special conditions. Specifically, for $L_D = L_{NL}$ there exists a pulse shape such that the SPM-induced frequency chirp, as given by (2.65), can be perfectly countered by the chirp produced by the negative group dispersion. This special pulse shape known as the fundamental soliton takes the form

$$u(\xi, \tau) = \mathrm{sec}\,\mathrm{h}(\tau)\exp(\mathrm{i}\xi/2) \, , \tag{2.84}$$

where the normalized distance, $\xi = z/L_D$, has been defined. Note that the square modulus of this expression, i.e. the pulse intensity, is simply given by the function $\text{sech}^2(\tau)$, which is independent of ξ. Hence, this pulse possesses the interesting property of propagating *without* distortion. By direct substitution, one sees that (2.84) is indeed a solution of (2.82) (with the distance normalized in units of L_D). More rigorously, it can be shown by the use of the inverse scattering method [60] that the NLS equation supports a family of solutions to which the solution (2.84) belongs. For communication purposes, though, it is the fundamental or first-order soliton which is of most practical interest.

A very interesting property of the fundamental soliton is its stability to perturbation, also referred to as its robustness. Stated differently, the soliton is an attractive solution in that a pulse with a slightly different initial shape at the fibre input progressively adjusts its shape so as to evolve into it. Furthermore, solitons can also be resistant to perturbation while they propagate. This basic adaptation mechanism enables the soliton to adjust its width so as to maintain the basic condition $L_D = L_{NL}$. For instance, in the presence of losses the soliton pulse width appears to increase more or less linearly as a function of the distance at a rate which is proportional to the absorption coefficient. Obviously, there are practical limits to what the solitons can endure, but fundamentally, the soliton's relative insensitivity to initial conditions and to small perturbations makes it a *practical* solution for long-haul communication systems [61].

References

1. D. Marcuse, "Theory of dielectric waveguides," in *Quantum Electronics - Principles and Applications* (Y-H Pao and P. Kelley, eds.) (Academic Press, New York 1974)
2. D. Gloge, Appl. Opt. **10**, 2252 (1971)
3. T. Okoshi, *Optical Fibres*, (Academic Press, New York 1982)
4. L.B. Jeunhomme, "Single-mode fibre optics: principles and applications," in *Optical Engineering, Vol. 4* (B.J. Thompson, ed.) (Marcel Dekker, Inc., New York 1983)
5. P.A. Bélanger, "Optical fibre theory", in *Series in Optics and Photonics, Vol. 5* (S.L. Chin, ed.) (World Scientific, Singapore 1993)
6. D. Marcuse, *Light transmission optics, 2nd ed.* (Van Nostrand Reinhold, New York 1982)
7. N.S. Kapany and J.J. Burke, *Optical waveguides* (Academic Press, New York 1972)
8. J.E. Midwinter, *Optical fibres for transmission* (John Wiley & Sons, New York 1979)
9. E.G. Neumann, "Single-mode fibres," in *Springer Series in Optical science, Vol. 57* (T. Tamir, ed.) (Springer-Verlag, Heidelberg 1988)

10. M. Miyagi and S. Nishida, "An approximate formula for describing dispersion properties of optical dielectric slab and fibre waveguides," J. Opt. Soc. Am. **69**, 291 (1979)
11. D. Marcuse, Bell System Tech. J. **56**, 703 (1977)
12. *Optical fibre amplifiers: Materials, devices and applications* (S. Sudo, ed.) (Artech House, Boston 1997)
13. *Optical fibre lasers and amplifiers* (P.W. France, ed.) (Blackie, Glasgow and London 1991)
14. I.D. Aggarwal and G. Lu, *Fluoride glass fibre-optics* (Academic Press, Boston 1991)
15. S. Takahashi, S. Shibata, and M. Yasu, "Low-loss and low-OH content soda-lime-silica glass fibre," Electron. Lett. **14**, 151 (1978)
16. H. Takahashi, I. Sugimoto, and T. Sato, "Germanium-oxide glass optical fibre prepared by the VAD method," Electron. Lett. **18**, 398 (1982)
17. A.M. Loireau-Lozac'h, M. Guittard, and J. Flahaut, Mater. Res. Bull. **11**, 1489 (1976)
18. T. Miya, Y. Terunuma, T. Osaka, and T. Miyashita, "Ultimate low-loss single-mode fibre at 1.55 µm," Electron. Lett. **15**, 106 (1979)
19. J.A. Buck, *Fundamentals of optical fibres* (John Wiley and Sons, New York 1995)
20. S.B. Poole, D.N. Payne, and M.E. Fermann, "Fabrication of low-loss optical fibres containing rare-earth ions," Electron. Lett. **21**, 737 (1985)
21. D. Marcuse, "Curvature loss formula for optical fibres," J. Opt. Soc. Am. **66**, 216 (1976)
22. I.H. Malitson, J. Opt. Soc. Am. **55**, 1205 (1965)
23. J.W. Fleming, "Dispersion in GeO_2–SiO_2 glasses," Appl. Opt. **23**, 4487 (1984)
24. D. Gloge, Appl. Opt. **10**, 2442 (1971)
25. B.J. Ainslie and C.R. Day, "A review of single-mode fibres with modified dispersion characteristics," J. Lightwave Technol. **LT-4**, 967 (1986)
26. V.A. Bhagavatula, M.S. Spotz, W.F. Love, and D.B. Keck, "Segmented-core single-mode fibres with low loss and low dispersion," Electron. Lett. **19**, 317 (1983)
27. N.J. Smith, N.J. Doran, W. Foryasiak, and F.M. Knox, "Soliton transmission using periodic dispersion compensation," J. Lightwave Technol. **15**, 1808 (1997)
28. A. Hasegawa, Y. Kodama, and A. Maruta, "Recent progress in dispersion-managed soliton transmission technologies," Opt. Fibre Technol. **3**, 197 (1997)
29. K. Tajima, "Compensation of soliton broadening in nonlinear optical fibres with loss," Opt. Lett. **12**, 54 (1987)
30. S.V. Chernikov, J.R. Taylor, and R. Kashyap, "Comblike dispersion-profiled fibre for soliton pulse train generation," Opt. Lett. **19**, 539 (1994)
31. R.B. Dyott, J.R. Cozens, and D.G. Morris, "Preservation of polarization in optical fibre waveguide with elliptical cores," Electron. Lett. **15**, 380 (1979)
32. Y. Sasaki, K. Okamoto, T. Hosaka, and N. Shibata, "Polarization-maintaining and absorption-reducing fibres," *Proc. Optical Fibre Communications Conf.*, Phoenix, USA, paper ThCC6 (Washington 1982) p. 54
33. R.D. Byrch, D.N. Payne, and N.P. Varnham, "Fabrication of polarization-maintaining fibres using gas-phase etching," Electron. Lett. **18**, 1036 (1982)
34. M.P. Varnham, D.N. Payne, R.D. Byrch, and E.J. Tarbox, "Single polarization operation of highly birefringent low-tie optical fibres," Electron. Lett. **19**, 247 (1983)

35. A.J. Barlow, D.N. Payne, M.R. Hadley, and R.J. Mansfield, "Production of single-mode fibres with negligible intrinsic birefringence and polarization-mode dispersion," Electron. Lett. **17**, 725 (1981)
36. H.C. Lefèvre, "Single-mode fibre fractional wave devices and polarization controllers," Electron. Lett. **16**, 778 (1980)
37. D.J. Dougherty, F.X. Kartner, H.A. Hauss, and E.P. Ippen, "Measurement of the Raman gain spectrum of optical fibres," Opt. Lett. **20**, 31 (1995)
38. R.G. Smith, Appl. Opt. **11**, 2489 (1972)
39. L.G. Cohen and C. Lin, "A universal fibre-optic (UFO) measurement system based on a near-IR fibre Raman laser," IEEE J. Quantum Electron. **QE-14**, 855 (1978)
40. C. Lin and R.H. Stolen, "Backward Raman amplification and pulse steepening in silica fibres," Appl. Phys. Lett. **29**, 428 (1976)
41. Y. Aoki, "Properties of fibre Raman amplifiers and their applicability to digital optical communication systems," J. Lightwave Technol. **6**, 1225 (1988)
42. S.G. Grubb et al., "High power 1.48 µm cascaded Raman laser in germanosilicate fibres," *Proc. Optical Amplifiers and their Applications*, Davos, Switzerland 1995, paper SaA4
43. G.P. Agrawal, "Fibre-optic communication systems", in *Wiley Series in Microwave and Optical Engineering*, (K. Chang, ed.) (John Wiley & Sons, New York 1992)
44. G.P. Agrawal, *Nonlinear fibre optics, 2nd ed.* (Academic Press, New York 1995)
45. H.G. Winful, "Polarization instabilities in birefringent nonlinear media: application to fibre-optic devices," Opt. Lett. **11**, 33 (1986)
46. N. Finlayson, B.K. Nayar, and N.J. Doran, "Ultrafast multibeatlength alloptical fibre switch," Electron. Lett. **27**, 1209 (1991)
47. B. Nikolaus, D. Grischkowsky, and A.C. Balant, "Optical pulse reshaping based on the nonlinear birefringence of single-mode optical fibres," Opt. Lett. **8**, 189 (1983)
48. R.H. Stolen, J.E. Bjorkholm, and A. Ashkin, Appl. Phys. Lett. **24**, 308 (1974)
49. Z. Su, X. Zhu, and W. Sibbett, "Conversion of femtosecond pulses from the 1.5- to the 1.3-µm region by self-phase-modulation-mediated four-wave mixing," J. Opt. Soc. Am. B **10**, 1050 (1993)
50. K. Stenerson and R.K. Jain, "Small-Stokes-shift frequency conversion in singlemode birefringent fibres," Opt. Commun. **51**, 121 (1984)
51. K.O. Hill, Y. Fujii, D.C. Johnson, and B.S. Kawasaki, "Photosensitivity in optical fibre waveguides: Application to reflection filter fabrication," Appl. Phys. Lett. **32**, 647 (1978)
52. M. Douay, W.X. Xie, T. Taunay, P. Bernage, P. Niay, P. Cordier, B. Poumellec, L. Dong, J.F. Bayon, H. Poignant, and E. Delvaque, "Densification involved in the UV-based photosensitivity of silica glasses and optical fibres," J. Lightwave Technol. **15**, 1329 (1997)
53. M. Gallagher and U. Österberg, "Spectroscopy of defects in germanium-doped silica glasses," J. Appl. Phys. **74**, 2771 (1993)
54. D.P. Hand and P.St.J. Russell, "Photoinduced refractive-index changes in germano-silicate fibres," Opt. Lett. **15**, 102 (1990)
55. P.Y. Fonjallaz, H.G. Limberger, R.P. Salathé, F. Cochet, and B. Leuenberg, "Correlation of index change with stress changes in fibre containing UV-written Bragg gratings," *ECOC '94*, (Sept. 1994), Genova, Italy, Vol. **2**, 1005 (1994)

56. P.J. Lemaire, R.M. Atkins, V. Mizrahi, and W.A. Reed, "High pressure H_2 loading as a technique for achieving ultrahigh UV photosensitivity and thermal sensitivity in GeO_2 doped optical fibres," Electron. Lett. **29**, 1191 (1993)
57. Y. Kodama and A. Hasegawa, "Nonlinear pulse propagation in a monomode dielectric guide," IEEE J. Quantum Electron. **QE-23**, 510 (1987)
58. A. Hasegawa, *Optical solitons in fibres, 2nd enlarged ed.* (Springer-Verlag, Berlin 1990)
59. A. Hasegawa and F. Tappert, "Transmission of stationary nonlinear optical pulses in dispersive dielectric fibres," Appl. Phys. Lett. **23**, 142 (1973)
60. V.E. Zakharov and A.B. Shabat, "Interaction between solitons in a stable medium," Sov. Phys. JETP **34**, 62 (1972)
61. R.J. Essiambre and G.P. Agrawal, "Soliton communication systems," Prog. Opt. **37**, 185 (1997)

3 Transmitters

Herbert Burkhard and Stefan Hansmann

3.1 Introduction

The performance of a fibre-optic communication system is essentially determined by the specific features of the transmitters applied to the conversion of the electrical data stream into optical signals. Semiconductor lasers are attractive coherent light sources for this purpose which combine excellent modulation properties, high efficiency and reliability with compact size enabling good fibre coupling and integration.

Basically, a laser consists of an active medium amplifying the optical field by stimulated emission and a resonant cavity selecting an optical mode with respect to wavelength and direction. In a semiconductor laser (Fig. 3.1) the gain is provided by electrons and holes injected into an active layer via a forward-biased pn-transition. The carriers can recombine there radiatively or nonradiatively, where the radiative transitions have a sufficient probability in semiconductors with a direct bandgap due to ease of momentum conservation.

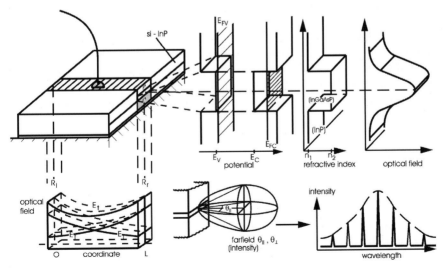

Fig. 3.1. Principle of Fabry–Perot double heterostructure laser

If the external voltage is increased and the carrier density exceeds a critical value, the rate of photon emission exceeds that of absorption and an incident wave is amplified coherently due to the process of stimulated emission. It turns out that a net gain occurs if the separation between the quasi-Fermi levels describing the filling of the conduction and valence bands is larger than the photon energy. The carrier density required to achieve this population inversion as well as the resulting gain values (in the range of $10^3 \, cm^{-1}$) is extremely high in direct semiconductor materials due to the high density of energy levels involved in the lasing process. Thus, the carriers must be confined to a small volume in order to achieve a low threshold current and continuous operation at room temperature. In double-heterostructure lasers the low-bandgap active layer is sandwiched between n- and p-doped cladding layers with higher bandgap epitaxially grown on a substrate. The injected carriers are captured in this potential well due to the difference in bandgap energies.

In addition, the low-bandgap active layer has a higher refractive index than the surrounding cladding layers, thus, acting as an optical waveguide confining the generated optical field to the gain region. Efficient waveguiding action parallel to the pn-junction (lateral) can also be obtained by embedding the active layer laterally between material with a lower index, thus yielding an index-guided laser structure. The optical feedback, necessary for laser operation is provided by the cleaved facets forming a Fabry–Perot (FP) cavity. This resonator selects the photons generated by stimulated emission with respect to direction and wavelength. The lightwave travelling perpendicularly to the facets is amplified if the wavelength matches a longitudinal mode of the resonator. The lasing process starts if the gain experienced during one round trip in the cavity equals the losses caused by absorption, scattering and the light output through the facets. Since the transitions occur between a pair of bands rather than well-defined states, the gain curve of a semiconductor laser turns out to be spectrally broad (roughly 50 nm at 1550 nm wavelength) compared to the typical wavelength differences (0.5 nm) between the longitudinal modes of a cavity with a typical length of 200 μm. Thus, single-mode operation, which is desirable for long-distance fibre transmission, can only be obtained using more complicated laser structures with an integration of distributed feedback (DFB) gratings.

3.2   Theory

3.2.1   Rate Equations for Single-Mode Operation

The fundamental static and dynamic properties of semiconductor laser diodes can be modelled using a set of rate equations describing the interaction of electron-hole pairs and photons in the active layer. We consider a strongly index-guiding double-heterostructure supporting a single optical mode travelling as a plane wave axially (z, see Fig. 3.1) in the cavity. The current I is assumed to be uniformly injected into the active layer with a volume V and recombines there completely, neglecting leakage currents. The carrier density within the active layer is treated as homogeneous in the transverse and lateral directions, since the respective inhomogeneities of photon density are small and the resulting gradients in carrier density are smoothed out by diffusion. In FP lasers with a sufficiently high mirror reflectivity, the axial variations of the photon density s and the carrier density n can be neglected and the rate equations [31] can be written as

$$\frac{dn}{dt} = \frac{I}{eV} - \frac{n}{\tau_{nr}} - Bn^2 - Cn^3$$
$$- v_g \frac{N_{act}}{N_{eff}} g_{act}(n,s)\, s + \frac{F_n(t)}{V}, \tag{3.1}$$

$$\frac{d}{dt}\left(\frac{s}{\sqrt{K_z}}\right) = v_g \left\{ \Gamma_{act} \frac{N_{act}}{N_{eff}} g_{act}(n,s) - g_{th} \right\} \frac{s}{\sqrt{K_z}}$$
$$+ \frac{\Gamma_{act}\,(R_{sp} + F_s(t))}{V}, \tag{3.2}$$

$$\frac{d\Phi}{dt} = \frac{1}{2}\, \alpha\, v_g\, \Gamma_{act} \frac{N_{act}}{N_{eff}} g_{act}(n,s) + F_\Phi(t), \tag{3.3}$$

where v_g is the group velocity of the waveguide, e the electron charge, g_{act} is the material gain of the active layer, Γ_{act} is the optical confinement factor of the active layer and $1/\tau_{nr}$, B, C are the parameters describing nonradiative, bimolecular and Auger recombination, respectively. Φ is the phase of the complex electric field E which is connected with the photon number $S = sV/\Gamma_{act}$ via

$$E(t) = \sqrt{S(t)} \exp(i\Phi(t)). \tag{3.4}$$

The waveguiding effect on the modal gain [6] is taken into consideration using the correction factor $\frac{N_{act}}{N_{eff}}$, where N_{eff}, N_{act} denote the effective indexes of the waveguide and the refractive index of the active layer, respectively. The threshold modal gain g_{th} of the lasing mode consists of the total internal optical losses α_s of the waveguide and the mirror losses α_m, which are assumed to be homogeneously distributed in the resonator of length L. The internal optical losses are caused by optical scattering from imperfections in the bulk media or at interfaces and free carrier absorption in the active and cladding layers.

R_{sp} is the time-averaged rate of spontaneous emission into the lasing mode. $F_s(t)$, $F_n(t)$ and $F_\Phi(t)$ represent Langevin noise sources taking into account the statistical nature of the spontaneous emission and the shot-noise character of the carrier recombination and generation. The Langevin forces leading to fluctuations of carrier density and photon density are correlated and have zero mean value.

The longitudinal excess factor K_z accounts for the enhancement of the spontaneous emission noise due to the axial dependence of the complex electric field $E(z,t)$:

$$K_z(t) = \frac{\left|\int_0^L |E(z,t)|^2 \, \mathrm{d}z\right|^2}{\left|\int_0^L E^2(z,t) \, \mathrm{d}z\right|^2}. \tag{3.5}$$

In the case of a transverse single-mode index-guided FP laser with the power reflectivities of the end facets R_l and R_r, the factor K_z is given by

$$K_z^{\mathrm{FP}} = \left[\frac{(\sqrt{R_l} + \sqrt{R_r})(1 - \sqrt{R_l R_r})}{\sqrt{R_l R_r} \, ln(1/R_l R_r)}\right]^2. \tag{3.6}$$

The correction factor $1/\sqrt{K_z(t)}$ in the rate equation for the photon density must be taken into account if the longitudinal field distribution changes with time [4].

The first rate equation (3.1) can be formally derived from the quantum-mechanical density-matrix formalism. It can be interpreted as a balance of carriers which are injected as a current I and contribute to stimulated emission or are lost for the lasing process via the different recombination processes. The second and third equations (3.2) and (3.3) can be derived from Maxwell's equations with the rotating wave and slowly varying amplitude approximations.

The asymmetry of equations (3.1) and (3.2) with respect to the confinement factor Γ_{act} is removed if the equations are expressed in terms of the photon number sV/Γ_{act} and carrier number nV. The rate equations represent a classical model neglecting the quantization of the carriers and the electromagnetic field. Thus, the number of photons and carriers are floating-point values.

The timescale for the calculation of the evolution of the photon and carrier populations in the use of these rate equations is limited to a lower bound. The physical basis of the rate equation for the optical field requires time steps that are not shorter than the round-trip time of the photons in the cavity, which is in the range of picoseconds. This means that a single passage of a plane wave through an active medium cannot be treated with this model.

To include the axial variations, the photon density is divided into counterpropagating waves $s^\pm$ with $s = s^+ + s^-$ travelling in the $\pm z$-direction. In addition, the total derivative $\frac{\mathrm{d}}{\mathrm{d}t}$ in equation (3.2) is replaced by $\frac{\partial}{\partial t} \pm v_g \frac{\partial}{\partial z}$ and the travelling wave rate equations for the photon and carrier densities are obtained with the abbreviation $\tilde{g}_{\mathrm{act}} = \frac{N_{\mathrm{act}}}{N_{\mathrm{eff}}} g_{\mathrm{act}}$ neglecting the time dependence of K_z:

$$\frac{dn}{dt} = \frac{I}{eV} - \frac{n}{\tau_{nr}} - Bn^2 - Cn^3 - v_g \tilde{g}_{act}(n,s)s + \frac{F_n(t)}{V}, \quad (3.7)$$

$$\frac{\partial s^+}{\partial t} + v_g \frac{\partial s^+}{\partial z} = v_g \{ \Gamma_{act}\, \tilde{g}_{act}(n,s) - \alpha_s \} s^+$$

$$+ \frac{\Gamma_{act}}{V} \left(\frac{R_{sp}}{2} + F_s^+(t) \right), \quad (3.8)$$

$$\frac{\partial s^-}{\partial t} - v_g \frac{\partial s^-}{\partial z} = v_g \{ \Gamma_{act}\, \tilde{g}_{act}(n,s) - \alpha_s \} s^-$$

$$+ \frac{\Gamma_{act}}{V} \left(\frac{R_{sp}}{2} + F_s^-(t) \right). \quad (3.9)$$

The boundary conditions define the mirror losses located at the end facets with the power reflectivities R_l, R_r

$$s^+(0,t) = R_l\, s^-(0,t) \quad\quad (3.10)$$
$$s^-(L,t) = R_r\, s^+(L,t). \quad\quad (3.11)$$

The mirror losses α_m assumed to be homogeneously distributed inside the cavity in the rate equations (3.1) and (3.2) are related to the counter-propagating photon densities via a round-trip condition

$$\alpha_m = \frac{T_l s^-(z=0) + T_r s^+(z=L)}{\int_0^L [s^-(z) + s^+(z)]\, dz}, \quad\quad (3.12)$$

where T_l, T_r denote the power transmission coefficients of the end facets.

The travelling wave equations can be used for the description of modelocking phenomena or in the case of devices with a small mirror reflectivity revealing a strong axial inhomogeneity of the photon distribution. The equations for the counterpropagating complex field amplitudes $E(z,t)^\pm$ corresponding to equations (3.8) and (3.9) look similar but also contain additional terms, since the phase information is included.

The strongly inhomogeneous axial photon distributions $s(z)$ occurring in some devices lead to an inhomogeneous current injection $I(z)$, if a constant electrostatic potential at the contacts is assumed. In such cases, the uniform current density appearing in the carrier rate equation must be replaced by a term containing the applied voltage, the serial resistance and the Fermi voltage [3].

An additional diffusion term $D\, \partial^2 n / \partial z^2$ with the diffusion constant D is neglected in equation (3.7) assuming that the typical diffusion length in III/V semiconductors is small compared to the spatial extension of the axial uniformities.

There are two other processes that can be treated by rate equations:

1. Optical feedback from a distant reflector with the amplitude reflection coefficient r_{ext}. The right-hand side of (3.2) and (3.3) has to be completed by a feedback term [10]

$$\frac{ds}{dt} : \quad \frac{2\kappa_{\text{ext}}}{\tau_L} \sqrt{s(t)\, s(t - \tau_{\text{ext}})}$$
$$\times \cos[\omega_{\text{th}}\tau_{\text{ext}} + \Phi(t) - \Phi(t - \tau_{\text{ext}})]\,, \tag{3.13}$$

$$\frac{d\Phi}{dt} : \quad -\frac{\kappa_{\text{ext}}}{\tau_L} \sqrt{\frac{s(t - \tau_{\text{ext}})}{s(t)}}$$
$$\times \sin[\omega_{\text{th}}\tau_{\text{ext}} + \Phi(t) - \Phi(t - \tau_{\text{ext}})]\,, \tag{3.14}$$

where τ_{ext} denotes the round-trip delay through the external cavity of length L_{ext}, Φ is the phase of the light reflected externally, τ_L is the laser cavity round-trip delay and the coupling coefficient κ_{ext} to the external cavity is given by

$$\kappa_{\text{ext}} = r_{\text{ext}} \frac{1 - R_r}{\sqrt{R_r}}\,. \tag{3.15}$$

This result is obtained under the assumption that the feedback light is subject to only one round trip in the external cavity formed by one laser facet (R_r) and the external amplitude reflection r_{ext} with the round-trip time τ_{ext}. If the feedback is very strong ($r_{\text{ext}} \geq 0.1$), more round trips have to be considered [11].

2. Optical injection with frequency detuning $d = f_M - f_S$ between a master- and slave-laser. In this case, we have to add the following expressions to the right-hand side of equations (3.2) and (3.3):

$$\frac{ds}{dt} : \quad \frac{2(1 - R_r)}{\tau_L \sqrt{R_r}} \sqrt{s_{\text{inj}}(t)\, s(t)} \, \cos[\Phi(t) - \Phi_{\text{inj}}(t)]\,, \tag{3.16}$$

$$\frac{d\Phi}{dt} : \quad -2\pi d - \frac{1 - R_r}{\tau_L \sqrt{R_r}} \sqrt{\frac{s_{\text{inj}}(t)}{s(t)}} \, \sin[\Phi(t) - \Phi_{\text{inj}}(t)]\,. \tag{3.17}$$

The rate equation for the carrier density remains unchanged in both cases.

The properties of the semiconductor material and the laser structure are described by phenomenological parameters which can be measured or calculated. The dependence of these material parameters, e.g. on carrier density, photon density, will be treated in the following.

3.2.2   Material Properties

Optical Gain. Laser action requires optical gain for overcoming the losses of the cavity. The gain is created by excess electrons and holes in the semiconductor which can be generated by processes like absorption of photons with sufficient energy to excite electrons from the valence to the conduction band or by injection at pn junctions. The excess carriers can then recombine through different channels either without or with phonon participation. If the radiative transition across the bandgap E_g does not involve a change in the carrier momentum, this process is called direct. If, however, a transition is considered between states of differing momentum belonging to the lowest energy separation, the transition can only take place if a phonon is simultaneously emitted or absorbed, since a photon alone cannot support enough momentum k to enable the process by momentum conservation. Indirect semiconductors are the group-IV materials silicon and germanium and also the III-V semiconductor GaP with the minimum of the conduction band situated outside the center of the Brillouin zone ($k = 0$). Since the transition probability for electrons at the lowest energy gap of these materials is considerably lower due to the involvement of phonons in this band-to-band-transition (pair formation process), gain production is relatively poor compared to direct semiconductors such as GaAs, InP and quaternary mixtures thereof (Fig. 3.2). Until now no indirect semiconductor has been reported to have achieved lasing action, thus we will restrict ourselves in the following to direct semiconductors with transitions at $k = 0$.

By sufficiently high carrier injection into the bands of a direct semiconductor, a population inversion can be created. Within the conduction- and valence-bands independent quasi-equilibrium ensembles of the respective injected carriers will build up within the timescale of the intraband relaxation time of roughly 100 fs. Under these conditions only two quasi-Fermi levels E_{Fc} and E_{Fv} need to be taken into consideration instead of considering 10^{23} cm^{-3} states of the semiconductor. If the difference of the two energies equals or is larger than the bandgap energy, the material gain produced becomes positive

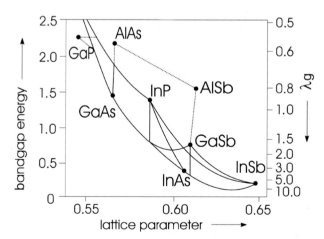

Fig. 3.2. Bandgap and lattice constant for various III/V compound semiconductors [5]

and depends on the injection via the occupation of the density of states [8,9]:

$$g_{act}(E) = \frac{\pi e^2 \hbar}{\varepsilon_0 N_{act} c m_0^2 E}$$
$$\times \int_0^\infty \sum_{cv} \rho_{cv}(\acute{E}) M_{cv}^2 \cdot [f_c(\acute{E}) - f_v(\acute{E} - E)] \cdot L(E, \acute{E}) \, d\acute{E} \quad (3.18)$$

where $\rho_{cv}(\acute{E})$ denotes the joint density of states, ε_0 the vacuum dielectric constant, M_{cv} the transition matrix element, c the speed of light, m_0 the electron mass, E the photon energy and f_c, f_v are the Fermi distributions for the conduction and valence bands, respectively. The energy $\acute{E}$ is measured from the edge of the conduction band. The Lorentzian broadening function

$$L(E, \acute{E}) = \frac{\hbar}{\pi \tau_{in}} \frac{1}{(\acute{E} - E)^2 + \left(\frac{\hbar}{\tau_{in}}\right)^2} \quad (3.19)$$

accounts for the intraband relaxation time τ_{in}. The Lorentzian lineshape function overestimates, however, the homogeneous broadening since the decay described by the function is too slow. The consequence is a finite absorption at photon energies lower than the bandgap energy. A more realistic lineshape function is given by the expression

$$L(E, \acute{E}) = \frac{\hbar}{\pi \tau_{in}} sech \left(\frac{\acute{E} - E}{\hbar/\tau_{in}}\right) \quad (3.20)$$

As the carrier density is increased (typically in the range of 10^{18} cm^{-3}) inversion is obtained if the condition $E_{Fc} - E_{Fv} > E_g$ is fulfilled and the gain becomes positive within an increasing wavelength range. For a fixed energy the dependence of the material gain on the carrier density is approximately linear in bulk materials:

$$g_{act}(n) = \frac{dg}{dn} (n - n_{tr}), \quad (3.21)$$

with the differential gain $\frac{dg}{dn}$ and the transparency carrier density n_{tr}. Since the peak of the gain shifts to higher energy with increasing injection due to band filling effects, the differential gain is higher on the high-energy tail of the gain curve [35].

The variation of the gain with carrier density depends strongly on the densities of states of the different types of semiconductor material, namely bulk (3D), quantum well (2D), quantum wire (1D) or even quantum dots (0D). The energy dependence of the density of states for the different materials with parabolic bands are

$$\rho_{3D} = \frac{1}{4\pi^2}\left(\frac{2m^*}{\hbar^2}\right)^{3/2} \tag{3.22}$$

$$\rho_{2D} = \frac{m^*}{2\pi\hbar^2}\frac{1}{L_z}\sum_l H(E - E_l); \text{ with } H(E - E_l)$$

$$= \begin{cases} 0 & for \quad E < E_l \\ 1 & for \quad E \ge E_l \end{cases} \tag{3.23}$$

$$\rho_{1D} = \frac{1}{2\pi L_x L_y}\left(\frac{2m^*}{\hbar^2}\right)^{1/2}\sum_{l,m}(E - E_{l,m})^{-1/2}, \tag{3.24}$$

$$\rho_{0D} = \frac{1}{L_x L_y L_z}\sum_{l,m,n}\delta(E - E_{l,m,n}). \tag{3.25}$$

The dependences are shown in Fig. 3.3.

In quantum wells (QW) the thickness of the active layer is comparable to the de Broglie wavelength of the confined carriers whose energy, corresponding to a motion normal to the well, becomes quantized. The resulting two-dimensional Fermi gas with discrete energy levels exhibits a staircase density of states in each band which leads to a higher differential gain at low injection and a saturation of the gain at high injection [25]. To avoid this gain saturation, multiquantum well structures (MQW) can be used. For a number of N_q quantum wells the gain vs. carrier density can be approximated by a logarithmic formula for positive gain values :

$$g(n) = N_q g_0 \ln\frac{n}{n_{tr}}, \qquad (g \ge 0). \tag{3.26}$$

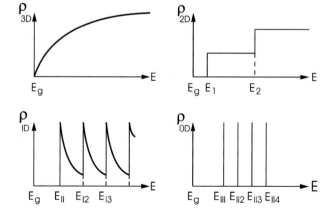

Fig. 3.3. Density of states for bulk, quantum well, quantum wire and quantum dot material

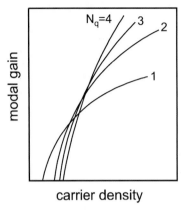

Fig. 3.4. Modal gain as a function of carrier density for a varying number of quantum wells N_q

Positive and negative gain values can be described using the following empirical relationship:

$$g(n) = N_q \mathring{g}_0 \ln \frac{n + n_s}{n_{tr} + n_s} , \qquad (g \geq 0) \tag{3.27}$$

with an additional parameter n_s.

The nonlinear dependence of the modal gain of a MQW structure on the carrier density is shown in Fig. 3.4 for a varying number of quantum wells. The transparency density as well as the differential gain dg/dn increases with growing number of quantum wells.

The gain as a function of the energy is given in Fig. 3.5 for bulk (broken line) and quantum-well (full line) structures.

The gain also depends on the photon density via spectral hole-burning, which is a reduction in gain at the energy corresponding to the lasing wavelength. The number of occupied conduction-band states and empty valence-band states is reduced by stimulated emission with increasing photon density. The recombined carriers are replenished from the large pool of electrons and holes in the bands via intraband scattering mechanisms, e.g. electron–electron and electron–phonon collisions, leading to a relaxation of the non-equilibrium distribution to a Fermi distribution. If the time constant of the stimulated emission becomes comparable to the intraband relaxation time, a gain reduction with an approximate Lorentzian shape occurs near the oscillation wavelength. This gain reduction can be written approximately as

$$g_{act}(n, s) = \frac{g(n)}{1 + \varepsilon s} , \tag{3.28}$$

where the gain compression constant ε is in the range $10^{-17}\,\mathrm{cm}^3$. Spectral hole-burning effects are essential for the interpretation of the dynamic behaviour of semiconductor lasers [26].

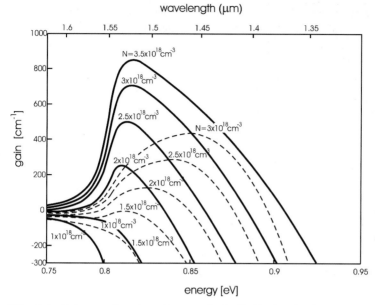

Fig. 3.5. Gain as a function of energy for bulk (*broken line*) and quantum well (*full line*) material

Spontaneous Emission Rate. The noise characteristics of a semiconductor laser are mainly determined by the spontaneous emission, which is generated isotropically and statistically. The rate of spontaneous emission coupled into one mode can be obtained using the Einstein relation, which equates the gain spectrum to the spontaneous emission spectrum:

$$R_{sp} = v_g g_{th}\, n_{sp}\, K_z\,, \qquad (3.29)$$

with the population inversion factor

$$n_{sp} = \left[1 - \exp\left(\frac{\hbar\omega - \Delta E_F}{k_B T}\right)\right]^{-1} \qquad (3.30)$$

defining the ratio of spontaneous and stimulated recombination rates. ΔE_F is the energy separation between the Fermi levels and k_B is the Boltzmann factor. For an axially homogeneous system with a threshold carrier density n_{th} the population inversion factor is approximately given by $n_{sp} \approx n_{th}/(n_{th} - n_{tr})$.

In direct semiconductors the total spontaneous emission rate R_{sp} into all modes has the form $R_{sp} = Bnp$, where B is the bimolecular radiative recombination coefficient, which typically is in the range 10^{-10} cm^{3}s^{-1} and n, p are the densities of electrons in the conduction band and of holes in the valence band, respectively. If the active material is doped, the total spontaneous emission rate can be composed of the spontaneous emission rate at thermal equilibrium $R_{\mathrm{sp}}^{(0)}$ and a rate $R_{\mathrm{sp}}^{(\mathrm{ex})}$ due to injected excess carriers: $R_{\mathrm{sp}} = R_{\mathrm{sp}}^{(0)} + R_{\mathrm{sp}}^{(\mathrm{ex})}$. The excess carrier density Δn defines a radiative carrier lifetime τ_{r} through $R_{\mathrm{sp}}^{(\mathrm{ex})} = \Delta n / \tau_{\mathrm{r}}$. For $\Delta n = \Delta p$ the spontaneous recombination rate of the excess carriers becomes

$$R_{\mathrm{sp}}^{(\mathrm{ex})} = B\,\Delta n\,(p_0 + n_0 + \Delta n)\,, \tag{3.31}$$

where p_0 and n_0 are the doping concentrations. The corresponding lifetime can be written as $\tau_{\mathrm{r}}^{-1} = B(p_0 + n_0 + \Delta n)$. For low injected carrier density Δn ($\Delta n \ll p_0, n_0$) τ_{r} is constant ($[B(p_0 + n_0)]^{-1}$) and for high injection it is given by $(B\Delta n)^{-1}$. This value is typically in the range of a few nanoseconds for carrier densities above $10^{18}\ \mathrm{cm}^{-3}$.

Nonradiative Recombination. Electrons and holes can also recombine non-radiatively. Defects, surfaces and hetero-interfaces can act as centres of non-radiative recombination usually described by a recombination rate $R_{\mathrm{sp}} = A\,n$. In long-wavelength semiconductors, however, the Auger recombination dominates. In this process the energy of an electron–hole recombination is transferred via Coulomb interaction to another carrier. This carrier is excited to a higher energy in the band and then relaxes by emitting phonons. The recombination rate for the different types of Auger processes (band-to-band, phonon-assisted, trap-assisted) can be expressed for low doping as

$$R_{\mathrm{sp}} = C\,n^3\,. \tag{3.32}$$

To obtain a linear analysis, it is useful to define a differential carrier lifetime

$$\tau_{\mathrm{d}}^{-1} = \frac{\partial R_{\mathrm{sp}}}{\partial n} = \frac{1}{\tau_{\mathrm{nr}}} + 2Bn + 3Cn^2\,. \tag{3.33}$$

Refractive Index Change. In semiconductors the real part of the refractive index depends on the carrier density on account of various physical mechanisms. With increasing injection the band-to-band absorption is reduced due to bandfilling effects. In addition, the absorption increases due to the reduction of the bandgap (bandgap renormalization resulting from many-body effects) and the absorption also increases due to the increasing absorption of free carriers. The resulting total change of the real part of the refractive index related to the gain-spectrum via the Kramers-Kronig relation depends on the wavelength relative to the gain maximum. In the case of $1.5\,\mu\mathrm{m}$ wavelength InGaAsP the refractive index decreases with injection. The dependence of the refractive index on the carrier density is theoretically described by the effective linewidth enhancement or Henry factor

$$\alpha = \frac{\partial N_{\mathrm{eff}}}{\partial \gamma_{\mathrm{eff}}}\,, \tag{3.34}$$

where the complex effective refractive index is defined by $N_{\text{eff}} - i\gamma_{\text{eff}}$. The change of the effective index with carrier density can be written as

$$\delta N_{\text{eff}} \cong \Gamma_{\text{act}} \frac{N_{\text{act}}}{N_{\text{eff}}} \delta N_{\text{act}} = -\Gamma_{\text{act}} \frac{N_{\text{act}}}{N_{\text{eff}}} \frac{\alpha\lambda}{4\pi} \frac{\partial g_{\text{act}}(n)}{\partial n} \delta n. \tag{3.35}$$

The linewidth enhancement factor is important for the treatment of linewidth and frequency chirp under modulation [30]. Typically it ranges from 3 to 5, decreasing from the long-wavelength to the short-wavelength side of the gain curve [35].

3.2.3   Steady-State Characteristics

The single-mode rate equations can be used to analyse the steady-state behaviour of a semiconductor laser. Setting the time derivative to zero in (3.2), we obtain an implicit expression for the photon number S in the case of continuous-wave (cw) operation:

$$S = \frac{R_{\text{sp}}}{v_{\text{g}}[g_{\text{th}} - \Gamma_{\text{act}}\tilde{g}_{\text{act}}(n, s)]}. \tag{3.36}$$

The number of photons increases as the gain value asymptotically approaches the losses g_{th}. The small gain difference is compensated for by spontaneous emission, which provides the noise input amplified by stimulated emission. Below threshold the photon density is small and (3.1) gives a linear increase of the carrier density according to $n \cong I\tau_{\text{d}}/(eV)$. Above threshold the gain is approximately clamped at $g(n_{\text{th}}) = g_{\text{th}}$ and the corresponding threshold current which is defined in the limiting case of vanishing spontaneous emission ($R_{\text{sp}} = 0$) becomes

$$I_{\text{th}} = eV R_{\text{sp}}(n_{\text{th}}). \tag{3.37}$$

Using equation (3.1) the photon number above threshold can then be written as

$$S = \frac{I - I_{\text{th}}}{ev_{\text{g}}g_{\text{th}}}. \tag{3.38}$$

Since the carrier density is clamped at threshold, all injected carriers in excess of the threshold current contribute to stimulated emission and the number of photons increases proportionally to $(I - I_{\text{th}})$. The resulting output power $P = v_{\text{g}}\hbar\omega\alpha_{\text{m}}S$ emitted from both facets becomes

$$P = \frac{\hbar\omega}{e} \eta_{\text{i}} \frac{\alpha_{\text{m}}}{g_{\text{th}}} (I - I_{\text{th}}), \tag{3.39}$$

where we assume that only a fraction η_{i} of the external drive current reaches the active region and the remaining fraction $(1-\eta_{\text{i}})$ is lost via leakage current or non-radiative recombination. Thus, the P–I curve of a semiconductor laser diode is a straight line above threshold with a slope defined by the external quantum efficiency

$$\eta_{\text{ext}} = \frac{\mathrm{d}P}{\mathrm{d}I}\frac{e}{\hbar\omega} = \eta_i\frac{\alpha_{\text{m}}}{g_{\text{th}}}\,, \tag{3.40}$$

which can be be interpreted as the ratio of the number of emitted photons to the number of injected electrons per time. The sharpness of the transition from spontaneous emission below threshold to stimulated emission above threshold depends on the amount of spontaneous emission into the lasing mode. Spectral hole-burning neglected so far leads to a weak increase of the carrier density with increasing injection resulting in a slight bending of the P–I curve.

The threshold current of a semiconductor laser depends on the temperature T, which can be described phenomenologically by

$$I_{th}(T) = I_0 \exp\frac{T}{T_0}\,, \tag{3.41}$$

where T_0 is the characteristic temperature, which typically ranges between 40 and 90 K for semiconductor lasers emitting at around 1550 nm wavelength.

The lasing spectrum of an FP semiconductor laser can be derived from a round-trip condition for the counter-propagating electric field amplitudes $E^{\pm}(z) = E^{\pm}\exp[\mp ik_0(N_{\text{eff}} - i\gamma_{\text{eff}})z]$, where the imaginary part of the effective index γ_{eff} is related to the intensity gain $g = -2k_0\gamma_{\text{eff}}$. In the case of a stationary laser oscillation, the complex electric field remains unchanged after one complete round trip in the cavity:

$$\sqrt{R_1R_r}\,\exp\left[2ik_0(N_{\text{eff}} - i\gamma_{\text{eff}})L\right] = 1\,. \tag{3.42}$$

The real part of this equation gives the mirror loss α_{m} of the FP cavity

$$\alpha_{\text{m}} = \frac{1}{2L}\ln\left(R_1R_r\right) \tag{3.43}$$

and the solution for the imaginary part yields the emission frequencies ν_{m} of the optical modes :

$$\nu_{\text{m}} = \frac{mc}{2LN_{\text{eff}}}\,, \qquad m = integer\,. \tag{3.44}$$

For the calculation of the frequency spacing $\delta\nu$ between the modes, the dispersion of the effective index described by the effective group index N_{g} must be taken into account:

$$N_{\text{g}} = N_{\text{eff}} + \nu\frac{\mathrm{d}N_{\text{eff}}}{\mathrm{d}\nu}\,, \tag{3.45}$$

yielding

$$\delta\nu = \frac{c}{2LN_{\mathrm{g}}}.$$
(3.46)

Because the mode spacing of a typical FP semiconductor laser ($\approx 0.5\,\mathrm{nm}$) is small compared to the width of the gain curve ($\approx 50\,\mathrm{nm}$) and the mirror loss of the various modes is independent of the wavelength, multimode operation is expected. The number of modes which start to lase above threshold depends on details of the gain curve and the gain-broadening mechanism.

3.2.4   Small-Signal Modulation Characteristics

Semiconductor lasers are interesting devices for optical communication purposes since the optical output signal can be directly modulated by the drive current. Unfortunately, it is not possible to obtain an analytical solution of the nonlinear rate equations (3.1) and (3.2) in the case of an arbitrary modulation of the injection current. A small-signal analysis of the linearized rate equations, however, yields analytical expressions describing how the laser behaves dynamically in response to small perturbations. For this purpose, the injection current is assumed to be sinusoidally modulated with the circular frequency ω_{m} around the mean value $\bar{I}$ according to $I(t) = \bar{I} + \Delta I \exp(\mathrm{i}\omega_{\mathrm{m}}t)$. For a small modulation amplitude $\Delta I << \bar{I}$ the rate equations can be linearized and the photon density as well as the carrier density n can be assumed to vary sinusoidally around their static values $\bar{s}$ and $\bar{n}$, respectively:

$$s(t) = \bar{s} + \Delta s \exp\left(\mathrm{i}\,\omega_{\mathrm{m}}t\right),$$
(3.47)

$$n(t) = \bar{n} + \Delta n \exp\left(\mathrm{i}\,\omega_{\mathrm{m}}t\right).$$
(3.48)

These expressions are inserted into the rate equations and all products of small quantities, e.g. $\Delta n \Delta s$, are neglected. We assume a gain model according to $g(n,s) = \frac{\mathrm{d}g}{\mathrm{d}n}(n - n_{\mathrm{tr}})/(1 + \varepsilon s)$, a constant K_z and use the following approximation for the nonlinear term describing the gain compression:

$$\frac{1}{1 + \varepsilon\left(\bar{s} + \Delta s\, e^{\mathrm{i}\omega_{\mathrm{m}}t}\right)} \approx \frac{1}{1 + \varepsilon\bar{s}} - \Delta s\, e^{\mathrm{i}\,\omega_{\mathrm{m}}t}\left[\frac{\varepsilon}{1 + \varepsilon\bar{s}} - \frac{\varepsilon^2\bar{s}}{(1 + \varepsilon\bar{s})^2}\right]$$

$$= \mathcal{A} - \Delta s\, e^{\mathrm{i}\,\omega_{\mathrm{m}}t}\,\mathcal{B}.$$
(3.49)

After subtraction of the stationary solution, we obtain the following linearized equations

$$\mathrm{i}\omega_{\mathrm{m}}\Delta n = \frac{\Delta I}{eV} - \frac{\Delta n}{\tau_{\mathrm{d}}} - v_{\mathrm{g}}\frac{\tilde{\mathrm{d}g}}{\mathrm{d}n}\left[\Delta n \mathcal{A}\bar{s} + \Delta s\left(\bar{n} - n_{\mathrm{tr}}\right)(\mathcal{A} - \mathcal{B}\bar{s})\right],$$
(3.50)

$$\mathrm{i}\omega_{\mathrm{m}}\Delta s = v_{\mathrm{g}}\,\Gamma_{\mathrm{akt}}\frac{\tilde{\mathrm{d}g}}{\mathrm{d}n}\bar{s}\left[\mathcal{A}\Delta n - \Delta s \mathcal{B}\left(\bar{n} - n_{tr}\right)\right] - \frac{\Gamma_{\mathrm{akt}}R_{\mathrm{sp}}\Delta s}{V\bar{s}},$$
(3.51)

with the abbreviation $\frac{\tilde{dg}}{dn} = \frac{dg}{dn}\frac{N_{act}}{N_{eff}}$. The solution of these equations is the modulation of the carrier density and photon density

$$\frac{\Delta n}{\Delta I} = \frac{1}{eV\omega_r^2}\left| i\,\omega_m + \frac{\Gamma_{akt}R_{sp}}{\bar{s}V} + v_g\Gamma_{akt}\bar{s}\frac{\tilde{dg}}{dn}\mathcal{B}\left(\bar{n}-n_t\right)\right|\mathcal{H}(\omega_m)\,, \quad (3.52)$$

$$\frac{\Delta s}{\Delta I} = \frac{v_g\Gamma_{akt}\frac{\tilde{dg}}{dn}\bar{s}\mathcal{A}}{eV\omega_r^2}\mathcal{H}(\omega_m)\,. \quad (3.53)$$

The transfer function $\mathcal{H}(\omega_m)$ is given by

$$\mathcal{H}(\omega_m) = \left|1 + \frac{i\,\omega_m\gamma_d}{\omega_r^2} - \frac{\omega_m^2}{\omega_r^2}\right|^{-1}\,, \quad (3.54)$$

where the relaxation resonance frequency ω_r and the damping constant γ_d are defined as

$$\omega_r^2 = \left[\tau_d^{-1} + v_g\frac{\tilde{dg}}{dn}\mathcal{A}\bar{s}\right]\left[\frac{\Gamma_{akt}R_{sp}}{V\bar{s}} + v_g\Gamma_{akt}\frac{\tilde{dg}}{dn}\bar{s}\mathcal{B}\left(\bar{n}-n_t\right)\right]$$

$$+ v_g^2\frac{\tilde{dg}}{dn}^2\Gamma_{akt}\bar{s}\mathcal{A}\left(\bar{n}-n_t\right)(\mathcal{A}-\mathcal{B}\bar{s})\,, \quad (3.55)$$

$$\gamma_d = \tau_d^{-1} + \frac{\Gamma_{akt}R_{sp}}{V\bar{s}} + v_g\frac{\tilde{dg}}{dn}\bar{s}\left[\Gamma_{akt}\mathcal{B}\left(\bar{n}-n_t\right)+\mathcal{A}\right]\,, \quad (3.56)$$

The damping parameter γ_d divided by the square of the resonance frequency $(\omega_r/2\pi)^2$ is termed the K-factor and turns out to be structure dependent.

Amplitude Modulation. The modulation of the photon density can be written in terms of a power modulation

$$\frac{\Delta P}{\Delta I} = \frac{P}{\bar{s}}\frac{\Delta s}{\Delta I}\,. \quad (3.57)$$

The DC response of the laser is equivalent to the slope of the static P–I curve above threshold and can be calculated from (3.57) and (3.53) using $\mathcal{H}(0) = 1$. Neglecting spontaneous emission ($R_{sp} = 0$) as well as gain compression ($\varepsilon = 0$) the resulting efficiency dP/dI corresponds to equation (3.40). The frequency dependence of the amplitude modulation response is shown in Fig. 3.6 for various bias currents. For low frequencies the amplitude response is relatively flat and its value corresponds to the slope of the static P–I curve. At the resonance frequency ω_r a pronounced peak occurs. With increasing injection this resonance peak shifts to higher frequencies and decreases in height.

This resonant behaviour is caused by an oscillation of the excitation energy between the carrier and photon populations. According to equation (3.55) the frequency of this relaxation oscillation is mainly governed by

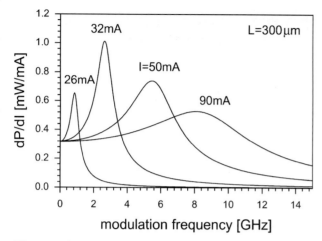

Fig. 3.6. Amplitude modulation response for various bias currents. The parameters of the FP laser are: $L = 300\,\mu\text{m}$, $\frac{dg}{dn} = 3 \cdot 10^{-16}\,\text{cm}^2$, $\varepsilon = 2 \cdot 10^{-17}\,\text{cm}^3$, $\alpha_\text{s} = 10\,\text{cm}^{-1}$, $b_\text{act} = 2\,\mu\text{m}$, $d_\text{act} = 80\,\text{nm}$, $I_\text{th} = 24\,\text{mA}$

the differential gain $\frac{dg}{dn}$, the static photon density $\bar{s}$ and the threshold gain $g_\text{th} = \Gamma_\text{akt} \frac{dg}{dn} (\bar{n} - n_\text{tr})/(1 + \varepsilon \bar{s})$ affecting the response of the stimulated emission after carrier injection. The resonance peak is damped because of carrier recombination, spontaneous emission and spectral hole-burning (equation 3.56) since these processes lead to a gain reduction. Experimental results show that the resonance damping in semiconductor lasers is dominated by the gain compression parameter ε. The damping limits the modulation speed of a semiconductor laser since it governs the high-frequency decay of the response and leads to the disappearance of the resonance peak at high photon densities.

In addition to the intrinsic optical response modelled by the small-signal analysis, the high-frequency behaviour of the laser is also affected by the electrical parasitics of the laser structure. The simplest equivalent-circuit model consisting of a series resistance and a capacitance parallel to the pn junction yields an electrical transfer function $\mathcal{H}_\text{el}(\omega_\text{m})$ with a 3dB-cutoff circular frequency $\omega_\text{3dB} = (R_\text{s}C_\text{p})^{-1}$

$$\mathcal{H}_\text{el}(\omega_\text{m}) = \frac{1}{1 + \omega_\text{m}^2/\omega_\text{3dB}^2} . \tag{3.58}$$

The total amplitude modulation response is given by the product of the intrinsic response $dP/dI(\omega_\text{m})$ and the electrical transfer function $\mathcal{H}_\text{el}(\omega_\text{m})$.

Frequency Modulation. In a semiconductor laser the modulation of the carrier density gives rise to a modulation of the optical frequency due to the amplitude-phase coupling expressed by the linewidth enhancement factor α.

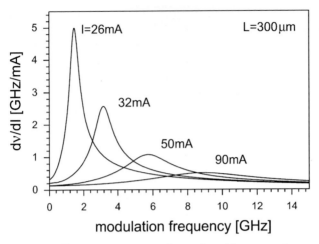

Fig. 3.7. Frequency response for various bias currents

The variation of the emission frequency $\delta\nu$ resulting from a change of the carrier density can be estimated from equation (3.52) assuming a linear dependence of the gain on the carrier density and taking into account the dispersion according to $\frac{dN_{eff}}{dn} = \frac{N_g^{eff}}{\lambda}\frac{d\lambda}{dn}$:

$$\frac{d\nu}{dI} = -\frac{c_0}{\lambda^2}\frac{d\lambda}{dI} \simeq -\frac{c_0}{\lambda}\frac{1}{N_g^{eff}}\frac{dN_{eff}}{dn}\frac{dn}{dI}$$

$$\simeq -\frac{v_g}{\lambda}\frac{N_{act}}{N_{eff}}\Gamma_{akt}\frac{dN_{akt}}{dn}\frac{dn}{dI} \simeq \frac{v_g}{4\pi}\frac{\tilde{d}g}{dn}\alpha\,\Gamma_{akt}\frac{dn}{dI}. \tag{3.59}$$

Figure 3.7 shows the FM response calculated from equation (3.59) as a function of the modulation frequency. The frequency modulation characteristic is relatively flat for $\omega \ll \omega_r$ and peaks at the relaxation oscillation frequency. The low-frequency limit of the FM response FM(0) corresponds to the static tunability of the laser. At frequencies higher than the resonance frequency, the roll-off of the FM response $(d\nu/dI \propto \omega_m^{-1})$ is weaker than in the case of the AM response $(dP/dI \propto \omega_m^{-2})$. In the low-frequency regime (kHz–MHz) thermal effects govern the FM response (thermal dip) since the current modulation leads to a temperature variation which modulates the refractive index and thereby the emission frequency. At higher frequencies, however, the electronic carrier effects dominate. The frequency modulation can be utilized, for example, in coherent optical fibre transmission systems. For intensity modulation applications, however, the frequency chirping is disadvantageous since it broadens the output spectrum and limits the transmission capacity for dispersive fibres.

The expressions (3.52) and (3.53) are derived assuming uniform distributions for the carriers and photons. In real devices the axial distributions are inhomogeneous and modified by carrier injection, which gives rise to additional contributions to the FM- and AM-response.

Being useful for predicting the parameter dependence, the small-signal analysis is only valid for very small modulation amplitudes. Considering the current amplitudes used in realistic system applications, a numerical analysis of the rate equations becomes necessary. Due to the highly nonlinear properties of the devices, the response to the rectangularly shaped current pulses turns out to be very complex with respect to frequency and amplitude.

3.2.5 Noise Properties

The light emitted from a semiconductor laser fluctuates with respect to intensity and phase due to the quantization of the spontaneous emission and the discrete nature of the generation and recombination of carriers.

Relative Intensity Noise. In a semiclassical approach these statistical processes are described in the time domain by the Langevin noise sources $F_s(t)$ and $F_n(t)$, respectively. Since the noise input is usually small, the statistical change of the intensity and phase can be calculated using a small signal analysis with constant injection current and the Langevin noise terms as driving forces. For this purpose, the term $\Delta I/eV$ in equation (3.50) is replaced by $\tilde{F}_n(\omega_m)/V$ and in equation (3.51) the term $\gamma_{act}\tilde{F}_s(\omega_m)$ is added, where $\tilde{F}_n(\omega_m)$ and $\tilde{F}_s(\omega_m)$ denote the Fourier-transformed Langevin noise sources. The fluctuations of the photon density and carrier density resulting from the noise input can be calculated in a way similar to the calculation of the AM- and FM-response, yielding the relative intensity noise (RIN), which is defined as the time average of the square of the photon density fluctuation $<|\Delta s(\omega_m)|^2>$ divided by the square of the photon density $\bar{s}^2$. The time average of the square of the Langevin noise sources is given by $<|\tilde{F}_s(\omega_m)|^2> = 2R_{sp}\,\bar{s}\,V/\Gamma_{akt}$ and $<|\tilde{F}_n(\omega_m)|^2> = 2(R_{sp}\,\bar{s}\,V/\Gamma_{akt}+\bar{n}V/\tau_d)$. Then the RIN becomes

$$RIN = \frac{\langle|\Delta s(\omega_m)|^2\rangle}{\bar{s}^2} \tag{3.60}$$

$$= \frac{2\Gamma_{akt}R_{sp}}{\bar{s}V}\frac{|\mathcal{H}(\omega_m)|^2}{\omega_r^4}$$

$$\times \left[\omega_m^2 + \left(\frac{1}{\tau_d}+v_g\frac{\tilde{dg}}{dn}\mathcal{A}\bar{s}\right)^2 + v_g^2\frac{\tilde{dg}}{dn}^2\mathcal{A}^2\bar{s}^2\left(1+\frac{\bar{n}\Gamma_{act}}{\bar{s}\tau_d R_{sp}}\right)\right]. \tag{3.61}$$

The spectrum of the relative intensity noise (Fig. 3.8) shows a maximum at the resonance frequency, demonstrating that the laser responds to a noise input in a way similar to an external modulation. Since the resonance frequency is proportional to the square root of the photon density, the RIN amplitude decreases with the laser power P as P^{-3}. Usually the influence of the carrier

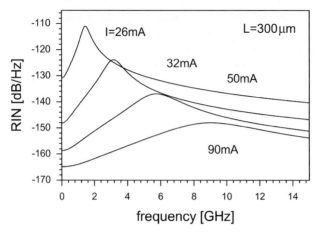

Fig. 3.8. Frequency dependence of the relative intensity noise for various bias currents

shot noise $F_n(t)$ is small in comparison with the spontaneous emission noise $F_s(t)$.

At low frequencies the RIN of a single-mode laser is sensitively affected by side modes even if their intensity is small compared to the main mode. In the case of multimode operation, the so-called 'mode partition noise' also contributes to the RIN of the different lasing modes. By measuring the frequency dependence of the RIN, it is possible to extract important material parameters describing the intrinsic response of the laser, e.g. $\frac{dg}{dn}$ and ε, without suffering from parasitic electrical effects and the problems of high-frequency current modulation.

Phase Noise. Two mechanisms contribute to the fluctuations of the phase of the optical field, giving rise to a frequency noise [34]. First, the phase is directly changed since the spontaneous emission is incoherent. This can be described in the frequency domain by a Langevin noise source $\tilde{F}_\Phi(\omega_m)$. Second, the phase changes, since the spontaneous emission of a photon alters the carrier density, which in turn changes the refractive index of the active layer via amplitude-phase coupling. This indirect effect is determined by the Langevin noise sources $\tilde{F}_n(\omega_m)$ and $\tilde{F}_s(\omega_m)$ leading to fluctuations of the carrier density. Both effects yield a total phase change $\Delta\dot{\varphi} = 2\pi\,\Delta\nu$ which can be simulated using a small-signal analysis analogous to the RIN calculation. Using $<|\tilde{F}_\varphi|^2> = R_{sp}/(2S)$ and neglecting correlations between the Langevin noise terms, the spectral density of the frequency noise $S_{\dot\varphi} = <|\Delta\dot\varphi|^2>$ can be written as

$$S_{\dot\varphi} = \frac{R_{sp}}{2\,S}\left[1 + \alpha^2\mathcal{H}(\omega_m)\right] + S_{nn}\;. \tag{3.62}$$

The first frequency-independent term corresponds to the noise resulting directly from spontaneous emission and leading to an instantaneous phase change. The second term expresses the phase noise created by spontaneous emission and amplitude phase coupling which is retarded due to the interaction between carriers and photons. The last term S_{nn} turns out to be small and describes the noise resulting from carrier shot noise which also gives rise to a phase change due to amplitude-phase coupling. The frequency-noise spectrum is flat at low frequencies and exhibits a maximum at the resonance frequency.

The phase noise at zero frequency corresponds to the width $\Delta\nu$ of the Lorentzian shaped laser line, which is an important parameter in coherent optical fibre communication systems:

$$\Delta\nu = \frac{S_{\dot{\varphi}}(\omega_m = 0)}{2\,\pi} = \frac{R_{sp}}{4\pi\,S}\left(1 + \alpha^2\right). \tag{3.63}$$

This modified Schawlow–Townes formula is valid for cw operation above threshold if the laser emits in a single longitudinal mode and contributions from $1/f$ noise, carrier shot-noise and phase noise of the side modes can be neglected. The linewidth is determined by the linewidth enhancement factor α as well as the spontaneous emission rate $R_{sp} = v_g g_{th} n_{sp} K_z$ and is inversely proportional to the output power. Experimental results, however, often reveal a rebroadening of the linewidth at high ouput powers.

3.3   Basic Design of Semiconductor Laser Structures

3.3.1   Concepts of Lateral Confinement

In all practical types of semiconductor lasers, lateral single-mode waveguides are utilized to enable the device to emit light from an aperture in the micron range with a very stable output power into a narrow aperture. In most cases, however, the emerging optical beam has too large an angular width to couple the light efficiently into a fibre which has a narrow beam width, if no additional optical components are inserted. The best coupling efficiency between laser and fibre without coupling optics is achieved if the spot sizes of the field distributions in the respective waveguides are identical. Then, simultaneously, the alignment tolerance is largest. In matching a roughly $10\,\mu m$ diameter spot size of a single mode fibre to that of a $2\,\mu m$ spot of a laser, design problems for the laser are unavoidable. Thus, the laser is optimized to a particular task and additional optical elements are utilized leading to severe coupling problems and costs. A different solution is the integration of a spot-size transformer onto the laser chip.

Lateral patterning of the active region of the laser is required for photon and carrier confinement. In addition, one has to facilitate lateral confinement of the current to avoid leakage currents bypassing the active region. The confinement of photons, carriers and current has been implemented in many ways, reflecting the specific purpose of the laser. Here we focus only on a few of the most frequently used ones.

For the technological implementation of the laser shown in Fig. 3.9, first the total structure is grown by MOCVD or MBE (or equivalent). Then the approximately 2 μm wide ridge for the definition of the active zone and waveguide is etched. This etching is preferentially done by a dry process (RIE, IBE etc.) to guarantee sidewalls as smooth as possible for low optical scattering losses. The possible damage or deposition of residues can be removed by a subsequent chemical treatment. Thereafter a regrowth with semi-insulating (si) InP planarizes the structure and buries the active layer completely in a higher bandgap material to facilitate photon, carrier and current confinement. The highly p-doped top InGaAs layer serves as a means to get low ohmic contact resistance for a minimum heat production limiting high output power operation. Structure 2 is constructed in a similar way except that the si–InP is replaced by a current-blocking pnp–InP sequence. In structure 3 the growth of the layers is stopped after the active layer, which

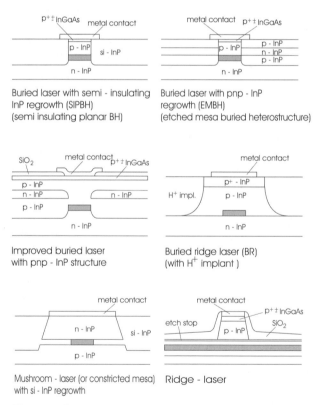

Fig. 3.9. Schematic cross-section of different types of semiconductor laser structures

is then etched as a stripe; the subsequent growth step planarizes the structure again, first with a p-InP and then an n-InP layer. This last layer is opened in the form of a stripe parallel to the active waveguide and regrown again to the final structure; obviously this is quite a complicated procedure. The buried ridge (BR) laser 4 has no epitaxial blocking layers and, thus, it requires a well-controlled p-doping of the confining layer $(7 \cdot 10^{17}$–$1.5 \cdot 10^{18} \, \mathrm{cm}^{-3})$ to facilitate current confinement and limitation of the losses due to free carrier absorption. A proton implantation near the contact limits additional leakage currents.

Laser 5 is called a mushroom laser or constricted mesa due to its cross-sectional shape [28]. It combines easy production due to the narrow stripe definition by wet-chemical undercutting with excellent high-speed performance. There is only one regrowth process necessary and the ohmic contacts can be made large to yield a very low series resistance. Types 1 and 5 are the only structures without extended pn junctions connected to the active layer, which cause a large parasitic capacitance and, thus, limit the high-frequency modulation performance.

Laser 6 is called a ridge laser. Waveguiding is accomplished by a ridge on top of the active layer which is separated in a proper way to achieve stable lateral single-mode (TE) operation and simultaneously keep the lateral leakage currents limited. Typical geometrical parameters as well as refractive index data for a laser with an emission wavelength of 1550 nm in the InGaAsP/InP-system are: ridge width 3 µm, total active waveguide thickness 200 nm (either bulk or, e.g., 10 QWs plus barriers and confinement layers) and thickness of the separating layer laterally adjacent to the ridge 250–400 nm. This structure is, to a large extent, index guiding, but not completely like all the other structures described here. In particular, the polarization properties of this type of lasers also have to be taken into consideration. In MQW structures – if no strain is present – TE material gain prevails and, thus, also prevails in the laser cavity. The disadvantage of this structure is that the threshold current is higher than in BH lasers and the far-field is elliptic, thus complicating fibre coupling. Advantages are, however, the easy processing, the excellent high-power performance and the fact that during laser processing the active layer never needs to be exposed to the ambient atmosphere, which leads to excellent ageing properties.

In all these lasers the active layer can be a bulk layer or may be composed of a stack of quantum wells or even quantum wires embedded in confining material. The utilization of QWs as a gain medium in lasers offers a number of additional design parameters that can be tailored to achieve better performance for specific applications. Some of these parameters are

- thickness of QWs and composition, number of QWs, barrier thickness and composition
- strain in QWs and barriers, composition, strain compensation [27].

Due to the nonlinear dependence of gain on carrier density in QWs, however, the design procedure for the simultaneous optimization of one or more special laser properties turns out to be more complicated [29].

3.4 Single-Mode Laser Structures

Transmitters used in optical fibre communication systems should emit light predominantly in a single longitudinal mode since the presence of side-modes limits the transmission capacity due to pulse broadening caused by the chromatic dispersion of the fibre. Semiconductor lasers with an FP resonator usually exhibit multimode operation since the gain spectrum is wider than the longitudinal mode spacing and the broadening of the gain profile, which due to spectral hole burning is not perfectly homogeneous, offers several modes with enough gain to oscillate. The techniques to achieve a reliable longitudinal mode control even under high bit-rate modulation can be categorized into two main groups:

- Short lasers
  The discrimination against side-modes in FP resonators can be enhanced by reducing the cavity length L. If the mode spacing $\delta\nu \propto L^{-1}$ becomes comparable to the width of the gain curve, only one mode will oscillate near the gain peak. To obtain stable single-mode operation, however, the lasers must be extremely short. This requires a very good reflectivity of the end facets to overcome the high mirror losses $\alpha_m \propto L^{-1}$ leading to high threshold current densities. The problems of fabricating very short semiconductor devices can be solved using a vertical cavity surface emitting laser (VCSEL) structure.

- Frequency-selective feedback
  The second method to obtain single-mode operation is to incorporate a frequency-selective element in the resonator structure. This can be realized by using coupled cavities, an external grating or a Bragg grating:
  - Coupled cavities
    If one or more additional mirrors are introduced in the FP resonator, the boundary conditions added due to the reflections at each interface severely limit the number of longitudinal modes. To achieve single-mode operation, however, it is often necessary to tune the resonator by changing the drive current or the temperature. Usually the single-mode regime is small so that such structures can only be modulated over a limited current range without mode jumps. In addition, the reproducible fabrication of nearly identical devices turns out to be difficult since the spectral properties strongly depend on the exact lengths of the sections.

– External grating

The frequency selection can also be realized by an external grating outside the resonator. The mechanical stability of such lasers, however, is a critical point since the grating is not integrated on the wafer. Consequently, lasers with external gratings are expensive devices which are less suitable for fibre-optic communication purposes.

– Bragg grating

The method most frequently used to achieve single-mode emission is to incorporate a Bragg grating which creates a periodic variation of the complex refractive index and distributes the feedback throughout the cavity. Dynamic single-mode operation is achieved if the threshold gain for the oscillating mode is significantly smaller than the threshold gain for the other modes. The devices employing Bragg gratings can be classified roughly into three categories: distributed Bragg reflector (DBR), distributed feedback (DFB, Fig. 3.10) and gain-coupled (GC) lasers.

In DBR lasers the Bragg grating is etched in passive regions near the cavity ends. The index grating (variation of the real part of the refractive index) acts as an effective mirror with wavelength-dependent reflectivity and surrounds the central part of the cavity which is active and remains uncorrugated. The longitudinal mode with a wavelength located near the reflectivity maximum of the grating is selected. Since

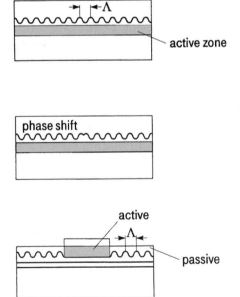

Fig. 3.10. Grating structure of standard DFB (*top*), phase shifted DFB (*middle*) and DBR laser (*bottom*)

a DBR laser is formed by replacing the mirrors by passive gratings, the properties can be described by an effective mirror model. The transition between the active section and the passive gratings usually complicates the technological realization of in-plane DBR lasers. An important advantage of DBR lasers is that the wavelength can be changed if the grating regions are equipped with separate electrodes which can tune the Bragg frequency via the carrier-induced refractive index change.

In DFB lasers the index grating covers the entire resonator length. At the wavelength corresponding to the corrugation period of the grating, the forward- and backward-travelling waves created by the Bragg scattering are confined in the central part of the cavity so that the mirror losses become a function of the wavelength. The longitudinal mode with the lowest mirror losses corresponding to the most effective concentration of photons in the resonator is selected.

In gain-coupled devices a periodic variation of gain or loss is used to favour a longitudinal mode of the Fabry–Perot resonator. In the ideal case, there is no Bragg scattering at the gain grating, and the longitudinal photon distribution as well as the mirror losses are unchanged compared to the Fabry–Perot cavity. The overlap with the loss- or gain grating, however, varies between the different longitudinal modes of the FP resonator. The mode experiencing the largest overlap with the gain grating (or minimum overlap with a loss grating) is selected.

3.4.1 Coupled-Mode Theory

The spectral properties of DFB lasers are essentially determined by the integrated Bragg grating. The waveguiding in such periodic structures can be analysed by the coupled-mode theory [1] giving approximate analytical solutions which are important for understanding Bragg gratings. In this one-dimensional approach, the lateral and transverse structure of the waveguide is taken into consideration by using the effective index N_{eff}. The propagation of plane TE waves travelling in the direction of propagation z can be described by the time-independent wave equation

$$\frac{\mathrm{d}^2 E_y(z)}{\mathrm{d}z^2} + k_0^2\, \tilde{N}^2(z)\, E_y(z) = 0\,, \tag{3.64}$$

where the complex refractive index $\tilde{N} = N - \mathrm{i}\gamma$ of the waveguide is assumed to be periodically modulated: $\tilde{N}(z) = \tilde{N}_{\mathrm{eff}} + \Delta\tilde{N}\, cos(2\beta_0\, z)$. Restricting the analysis to small variations of the refractive index yields

$$k_0^2\, \tilde{N}^2(z) \approx \beta^2 + \mathrm{i}g_{\mathrm{eff}}\beta + 4\beta\kappa\cos(2\,\beta_0\, z)\,, \tag{3.65}$$

where $g_{\mathrm{eff}} = -2k_0\gamma_{\mathrm{eff}}$ denotes the intensity mode gain and $\beta = 2\pi N_{\mathrm{eff}}/\lambda_0$ is the propagation constant of the wave. The Bragg wave number is $\beta_0 = m\pi/\Lambda$

for a grating with a corrugation period Λ and the order $m = 1, 2....$ The complex coupling coefficient $\kappa = \pi \Delta N / \lambda_0 + i \Delta g / 4$ is proportional to the variation of the refractive index ΔN, the gain variation Δg and the number of corrugations per length. As a general solution for the contra-directional coupling considered here we assume right- and left-propagating waves with slowly varying amplitudes $\mathcal{E}_+(z)$ and $\mathcal{E}_-(z)$, respectively, which are coupled by the scattering at the index modulation:

$$E_y(z) = \mathcal{E}_+ \, e^{i\beta_0 z} + \mathcal{E}_- \, e^{-i\beta_0 z} \, . \tag{3.66}$$

The phase mismatch $\delta = \beta - \beta_0$ of the interfering waves is small if the analysis is restricted to the wavelength region near the Bragg wavelength defined by $\beta = \beta_0$.

For a weak z-dependence of the field amplitudes $\mathcal{E}_+(z)$ and $\mathcal{E}_-(z)$ their second derivatives can be neglected and we obtain

$$\frac{d\mathcal{E}_-}{dz} - \left(\frac{g_{\text{eff}}}{2} - i\delta\right) \mathcal{E}_- = -i\kappa \, \mathcal{E}_+ \, , \tag{3.67}$$

$$\frac{d\mathcal{E}_+}{dz} + \left(\frac{g_{\text{eff}}}{2} - i\delta\right) \mathcal{E}_+ = i\kappa \, \mathcal{E}_- \, . \tag{3.68}$$

These coupled-mode equations describe the light propagation in waveguides with a periodic variation of the complex refractive index by counter-propagating modes exchanging energy by scattering. The strength of the interaction and the amount of feedback in the grating structure are determined by the coupling coefficient κ. Using an exponential function for the amplitudes $\mathcal{E}_+$ and $\mathcal{E}_-$ and considering the boundary conditions at the end facets, we obtain a dispersion relation which defines the wavelength and the threshold gain of the various modes. In the following, some analytical solutions will be given for an axially homogeneous first-order grating ($m = 1$) of length L and a perfect antireflection coating on the facets ($R_l = R_r = 0$).

In the case of pure index coupling ($\Delta N_{\text{eff}} \neq 0$, $\Delta g = 0$) the transmission spectrum turns out to be symmetrical with respect to the Bragg wavelength $\lambda_B = 2N_{\text{eff}}\Lambda$ where oscillation is forbidden. For a small coupling coefficient, the mode spacing approximately takes the value of an FP resonator $\Delta\lambda = \lambda^2/2\tilde{N}_{\text{eff}}L$ but, in contrast to an FP cavity, the threshold gain of the modes is wavelength-dependent and increases with growing distance from the Bragg wavelength. A strong coupling produces a transmission stopband with the width $\Delta\lambda \cong \kappa\lambda_B^2/(\pi N_{\text{eff}})$ centered at the Bragg wavelength in which transmission is strongly damped. The two modes with the lowest threshold gain $g \cong 2\pi^2/(\kappa^2 L^3)$ are located at the edges of the stopband symmetrical to the Bragg wavelength.

In the case of pure gain coupling ($\Delta g \neq 0$, $\Delta N_{\text{eff}} = 0$) the mode degeneracy is removed, which means that the mode with the lowest threshold gain oscillates at the Bragg wavelength symmetrically surrounded by the other modes with higher threshold gain values. The mode spacing is

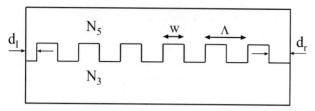

Fig. 3.11. Schematic sketch of a DFB grating with the duty cycle $\frac{w}{\Lambda}$

$\Delta\lambda = \lambda^2/2N_{\text{eff}}L$ and no stopband occurs since there is no backscattering at index steps in the grating. The mode selection is due to the different overlap of the standing waves in the FP resonator with the gain grating.

In second-order gratings ($m = 2$) additional scattering occurs in the transversal direction leading to higher losses. In addition, the coupling coefficient of second-order gratings depends more sensitively on the exact shape of the grating so that it becomes difficult to control. That is why first-order gratings predominate although the corrugation periods are smaller ($\Lambda \cong 240\,\text{nm}$ for $\lambda = 1.55\,\mu\text{m}$).

The definition of the coupling factor κ can be modified for gratings with the order m and an arbitrary shape by multiplication with the Fourier coefficient $\mathcal{F}_m$. In the case of a pure index grating with a rectagular shape and the duty cycle w/Λ, one obtains

$$\kappa = \frac{4\Delta N_{\text{eff}}}{\lambda_{\text{B}}} \sin\left(m\frac{w}{\Lambda}\pi\right) . \tag{3.69}$$

The effective refractive index step ΔN_{eff} in the longitudinal direction can be calculated from the transversal and lateral solution of the wave equation. With the approximation $\Delta\tilde{N}_{\text{eff}} \cong \Delta\epsilon_{\text{eff}}/(2\tilde{N}_{\text{eff}})$ and using the notation given in Fig. 3.11 the coupling factor becomes

$$\kappa = \frac{\Gamma_{\text{g}}\,(N_3^2 - N_5^2)}{\lambda_{\text{B}}\,N_{\text{eff}}} \sin\left(m\frac{w}{\Lambda}\pi\right) . \tag{3.70}$$

The optical confinement factor of the grating Γ_{g} is calculated from the lateral and transverse structure of the waveguide where the refractive index N_4 in the grating area is obtained by averaging the respective dielectric susceptibilities:

$$N_4 = \sqrt{N_5^2 + \frac{w}{\Lambda}\,(N_3^2 - N_5^2)}. \tag{3.71}$$

3.4.2 Basic Properties of Index-Coupled DFB Lasers

The spectral properties of two typical DFB laser structures with reflecting facets ($L = 250\,\mu\text{m}$, $\Lambda = 240\,\text{nm}$, $N_{\text{eff}} = 3.24$, $R_{\text{r,l}} = 0.28$) are summarized in Fig. 3.12. The left hand side shows the transmission spectra calculated below

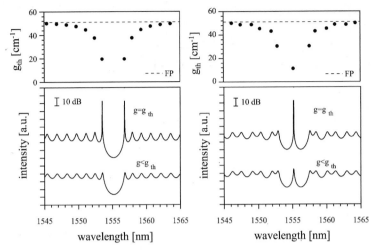

Fig. 3.12. Left-hand side: Transmission spectra (*bottom*) of a DFB laser ($\kappa = 100\,\mathrm{cm}^{-1}$, $L = 250\,\mu\mathrm{m}$, $\Lambda = 240\,\mathrm{nm}$, $\Phi_{\mathrm{l,r}} = 180°$) calculated below ($g < g_{\mathrm{th}}$) and at threshold ($g = g_{\mathrm{th}}$) *Top*: Corresponding mirror losses of the longitudinal modes. The value for an FP resonator is given in broken lines. Right-hand side: Transmision spectra (*bottom*) and corresponding mirror losses (*top*) for a $\lambda/4$ phase shifted DFB laser

and at the threshold revealing the stopband at the Bragg wavelength resulting from the backscattering in the grating. At threshold the transmission for the two degenerate lasing modes at the edges of the stopband becomes infinite in theory. The mirror losses of the modes given in the upper part of Fig. 3.12 increase with growing distance from the Bragg wavelength, approaching the value for an FP resonator (broken line) This mode degeneracy in index-coupled DFB laser structures is usually removed by incorporating a $\lambda/4$ phase shift in the grating. This can technologically be implemented by inserting an additional section of length $\lambda_0/(4N_{\mathrm{eff}}) = \Lambda/2$ in the middle of the grating. The transmission spectra and the corresponding mirror losses are plotted on the right-hand side of Fig. 3.12. The introduction of the $\lambda/4$ phase shift selects the Bragg mode in the middle of the stopband revealing the lowest mirror losses so that single-mode operation is obtained.

The axial distribution of the light intensity in the cavity is connected with the mirror losses α_{m} via the round-trip condition (3.12). Thus, a decrease of the mirror losses for the longitudinal modes in a DFB grating is equivalent to an increasing longitudinal optical confinement, which means that the photons are concentrated inside the cavity and only a small fraction of the light intensity leaves the resonator through the end facets. The longitudinal photon distribution for the Bragg mode of a DFB grating shows this confinement of the photons in the cavity (Fig. 3.13). The axial profiles for the intensity of the right- and left-travelling part of the light are plotted in dotted and

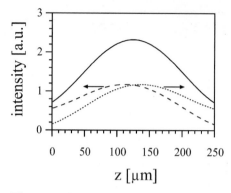

Fig. 3.13. Longitudinal photon distribution for the Bragg mode of a DFB grating. The intensity distributions of the left- and right-travelling waves are given in broken and dotted lines, respectively

broken lines, respectively. The intensity distribution for the Bragg mode of a $\lambda/4$ phase-shifted DFB grating is given in Fig. 3.14 for various coupling coefficients. Even for moderate coupling coefficients, the curves show a strong maximum in the position of the phase shift. The case of $\kappa = 0$ corresponds to an FP cavity without a grating.

The strong inhomogeneity of the photon density distribution in DFB lasers leads to an inhomogeneous carrier density distribution above threshold due to recombination by stimulated emission. With increasing injection the carrier density is depleted in places with a high photon density, which is shown in Fig. 3.15 for a $\lambda/4$ phase-shifted DFB laser. This phenomenon is called longitudinal spatial hole burning (LSHB) and has several important consequences for the static and dynamic behaviour of DFB lasers above threshold [24]. First, the mode discrimination is influenced since a variation of the

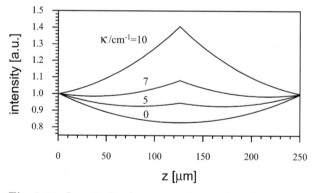

Fig. 3.14. Longitudinal intensity distribution for the Bragg mode of a $\lambda/4$-phase shifted DFB grating for various coupling coefficients κ

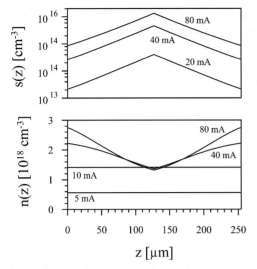

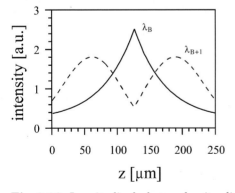

Fig. 3.15. Longitudinal distributions of the photon density (*top*) and the carrier density (*bottom*) for a $\lambda/4$ phase-shifted DFB laser for various bias currents

carrier density distribution changes the round-trip gain of the various modes having different photon density distributions. For a $\lambda/4$ phase-shifted DFB laser, for instance, the photon density distribution of the long-wavelength side-mode exhibits a minimum in the phase shift position in contrast to the Bragg mode (Fig. 3.16). If the Bragg mode lases above threshold, the inhomogeneous carrier density created by LSHB favours the side-mode with increasing injection, since it profits by the high carrier density at the facets. Thus, the side-mode suppression degrades with increasing output power due to LSHB. Second, the mode-wavelengths change even above threshold, since

Fig. 3.16. Longitudinal photon density distribution of the Bragg mode (λ_B) and the long-wavelength side-mode (λ_{B+1}) of a $\lambda/4$ phase-shifted DFB laser

the inhomogeneity of the carrier density caused by LSHB leads to an inhomogeneous axial distribution of the effective refractive index. This effect is utilized in several types of tunable DFB lasers.

The phase relation between the grating and the end facets is difficult to control during the cleaving process of DFB lasers, since the corrugation period for a first-order grating ($\lambda = 1.55\,\mu m$) is typically 240 nm. The end facet phases at the left and right facets are defined as

$$\Phi_{l,r} = d_{l,r} \cdot 360°/\Lambda, \tag{3.72}$$

where the thickness $d_{l,r}$ is illustrated in Fig. 3.11. Experimental and theoretical investigations show that all static and dynamic optical properties of as-cleaved DFB lasers are strongly influenced by Φ_l and Φ_r. The mirror losses of the various longitudinal modes, the mode discrimination, the intensity distribution, the optical spectra and the dynamic and noise characteristics vary considerably as a function of the end facet phases [32]. Since these end facet phases are distributed randomly after the cleaving process, the yield of good DFB devices is limited. The problem of the uncertain end facet phases can be eliminated by appropriate antireflection coatings.

3.4.3   Advanced DFB Laser Structures

Various DFB lasers with a more complex grating structure have been developed in order to obtain a high yield of single-mode devices and a flat axial photon distribution with reduced LSHB. Some examples are

- $2 \times \lambda/8$-phase shifts. The distance between the phase shifts, however, must be optimized to achieve a high yield [18].
- Corrugation pitch modulation (CPM). The DFB grating is divided into three sections. The corrugation period in the central section is slightly higher than in the outer sections so that the phase shift is quasi-continuously distributed along the cavity [19].
- Axial variation of duty cycle by using, e.g., a holographic double exposure technique [20].
- Axial variation of coupling coefficient by variation of the etch depth [21] or sampled gratings [22].
- Bent waveguides superimposed on homogeneous grating fields can be used to obtain quasi-continuously and arbitrarily chirped gratings with high spatial resolution [7].
- Axially inhomogeneous injection using a three-electrode structure. Spatial hole burning can be compensated for if the injected current density is higher in the central section near the peak of the photon density, thus reducing the gain of the side-modes suffering from the lower current in the outer section [23].

3.4.4 Gain-Coupled Lasers

In gain-coupled lasers the longitudinal single-mode operation is obtained by an axial variation of gain or loss. The gain grating can be implemented, for example, by etching the active layer [14] or corrugating a pnp-layer structure above the active layer so that the current injected into the active layer and the gain is periodically varied [15]. The grating can also be formed in an additional layer absorbing at the emission wavelength of the active layer. In this case, a loss grating is created, acting as a saturable periodic absorber [13]. Due to the Kramers-Kronig relations, the gain grating is inevitably accompanied by an index grating. Depending on the phase relation between the real and imaginary parts of the resulting complex coupling coefficient, in-phase and anti-phase gratings can be distinguished.

In contrast to index-coupled DFB lasers, there is almost no reflection of the lightwave in a gain grating. Therefore, the longitudinal intensity distribution and the spectral positions of the various modes are equal to an FP resonator. The mode selection is caused by the different overlap of the longitudinal modes with the gain grating. In addition, the influence of the end facet phases is strongly reduced in comparison with index-coupled DFB lasers since the interference of the reflection in the grating and at the facets is avoided. Therefore, the single-mode yield of gain-coupled lasers with uncoated facets is usually higher than in the case of index-coupled DFB lasers. The degrading effects of longitudinal spatial hole-burning can also be reduced in gain-coupled lasers if the residual index coupling is kept small.

3.4.5 Modelling of DFB Lasers

Various methods have been developed to model the static and dynamic characteristics of DFB laser structures. Besides the semi-analytic coupled-mode theory already presented above, many simulation programs are based on the transfer matrix (TM) or the transmission-line (TLM) technique, which will be outlined in the following.

Transfer Matrix Method. The DFB laser is divided axially into a series of homogeneous stratified layers whose complex refractive index $\tilde{N}(z)$ takes into account the lateral and transversal structure of the waveguide [2,16,17]. In the case of a grating with a rectangular shape, each corrugation is composed of two layers. A TE-light wave polarized in the y-direction and propagating in the z-direction perpendicular to the layer stack (Fig. 3.17) is described by the wave equation and an expression for the magnetic field

$$\frac{d^2 E_y}{dz^2} = k_0^2\, \tilde{N}^2(z)\, E_y \,, \tag{3.73}$$

$$i\,\omega\mu_0\, H_x = \frac{dE_y}{dz} \,. \tag{3.74}$$

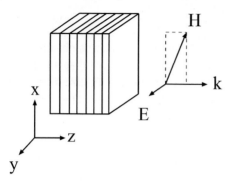

Fig. 3.17. Propagation of a TE lightwave in a one-dimensional layer stack

The remaining field components vanish. Within a homogeneous layer j with complex refractive index $\tilde{N}_j = N - i\gamma$ and thickness d_j the solution is given by two counter-propagating monochromatic plane waves

$$E^{\pm}(z) = E_{y,j}^{\pm} \exp\left[\mp i\, k_0\, \tilde{N}_j\, (z - z_j)\right], \tag{3.75}$$

$$\begin{pmatrix} E_y(z) \\ H_x(z) \end{pmatrix} = \begin{pmatrix} 1 & 1 \\ \dfrac{-k_0\, \tilde{N}_j}{\omega\mu_0} & \dfrac{k_0\, \tilde{N}_j}{\omega\mu_0} \end{pmatrix} \begin{pmatrix} E^+(z) \\ E^-(z) \end{pmatrix}. \tag{3.76}$$

The continuity of the tangential field components required at the interface $z = z_j$ between two adjacent layers can then be expressed in a matrix notation

$$\begin{pmatrix} E_{y,j} \\ H_{x,j} \end{pmatrix} = \begin{pmatrix} \cos(k_0\tilde{N}_j d_j) & \dfrac{\omega\mu_0}{ik_0\tilde{N}_j} \sin(k_0\tilde{N}_j d_j) \\ \dfrac{k_0\tilde{N}_j}{i\omega\mu_0} \sin(k_0\tilde{N}_j d_j) & \cos(k_0\tilde{N}_j d_j) \end{pmatrix} \begin{pmatrix} E_{y,j+1} \\ H_{x,j+1} \end{pmatrix}, \tag{3.77}$$

where $\tilde{k}_j = 2\pi\,\tilde{N}_j/\lambda$ denotes the complex wave number with the vacuum wavelength λ. $E_{y,j}$, $H_{x,j}$ are the electric and magnetic fields at the interface $z = z_j$ of the layer j, respectively. The matrix in equation (3.77), called the transfer matrix $\mathcal{M}_j$, completely describes the optical properties of a planar homogeneous layer in this one-dimensional model. Due to energy conservation, the transfer matrix is unimodular ($\det \mathcal{M}_j = 1$). For a multilayer structure the total transfer matrix $\mathcal{M}$ is obtained as the product of the matrices $\mathcal{M}_j$ of all $j = 1...p$ layers the system is composed of

$$\begin{pmatrix} E_{y,1} \\ H_{z,1} \end{pmatrix} = \mathcal{M} \begin{pmatrix} E_{y,p+1} \\ H_{z,p+1} \end{pmatrix},$$

with $\mathcal{M}$ defined by $$\mathcal{M} = \prod_{j=1}^{q} \mathcal{M}_j, \tag{3.78}$$

where $(E_{y,1}, H_{z,1}), (E_{y,p+1}, H_{z,p+1})$ denote the fields at the beginning and at the end of the multilayer stack located between two semi-infinite media with real refractive indices N_0, N_{p+1}. If we assume an incident wave $E_{y,0}^+$ coming from the left-hand side and take into account the boundary condition $E_{y,0}^\pm = E_{y,1}^\pm$ at the interface $z = 0$, there is no backward-travelling wave in the right outer space: $E_{y,p+1}^- = 0$. The fields on the left- and right-hand sides of the multilayer stack are connected via the transfer matrix $\mathcal{M}$

$$\begin{pmatrix} E_{y,1} \\ H_{x,1} \end{pmatrix} = \begin{pmatrix} m_{11} & m_{12} \\ m_{21} & m_{22} \end{pmatrix} \begin{pmatrix} 1 \\ -\frac{k_0 \tilde{N}_{p+1}}{\omega \mu_0} \end{pmatrix}, \tag{3.79}$$

where the normalization $E_{y,p+1} = 1$ was used. The power of the lightwave is given by the z-component of the Poynting vector $S_z^\pm = E_y^\pm \cdot H_x^{\pm *} = k_0 N/(\omega \mu_0) |E_y^\pm|^2$, where S_z^- corresponds to a wave travelling in the z-direction. From this, the reflection and transmission coefficients for the intensity of the lightwave can be derived

$$\mathcal{R} = \frac{|E_{y,0}^-|^2}{|E_{y,0}^+|^2} = \frac{|E_{y,1} + \omega \mu_0 H_{x,1}/(k_0 N_0)|^2}{|E_{y,1} - \omega \mu_0 H_{x,1}/(k_0 N_0)|^2} \tag{3.80}$$

$$= \frac{|m_{11} - m_{12} k_0 N_{p+1}/(\omega \mu_0) + \omega \mu_0 (m_{21} - m_{22} k_0 N_{p+1}/(\omega \mu_0))/(k_0 N_0)|^2}{|m_{11} - m_{12} k_0 N_{p+1}/(\omega \mu_0) - \omega \mu_0 (m_{21} - m_{22} k_0 N_{p+1}/(\omega \mu_0))/(k_0 N_0)|^2} \tag{3.81}$$

$$\mathcal{T} = \frac{N_{p+1} |E_{y,p+1}^+|^2}{N_0 |E_{y,0}^+|^2} = \frac{4 N_{p+1}}{|E_{y,1} - \omega \mu_0 H_{x,1}/(k_0 N_0)|^2} \tag{3.82}$$

$$= \frac{4 N_{p+1}}{|m_{11} - m_{12} k_0 N_{p+1}/(\omega \mu_0) - \omega \mu_0 (m_{21} - m_{22} k_0 N_{p+1}/(\omega \mu_0))/(k_0 N_0)|^2} . \tag{3.83}$$

Using these equations, it is possible to calculate the transmission and reflection spectra of a DFB laser described as a multilayer stack. The transmission of a lasing mode tends to infinity if the net gain equals the threshold gain g_{th} and the wavelength corresponds to a mode wavelength λ_{th} of the cavity, indicating self-oscillation of the device

$$\mathcal{T}(\lambda_{th}, \gamma_{th}) \to \infty . \tag{3.84}$$

Thus, the threshold can be found using an algorithm which maximizes the transmission by an iterative variation of gain and wavelength. At threshold condition the longitudinal intensity distribution of the lasing mode can also be obtained by the transfer matrix algorithm. The emission spectrum emitted through the left and right mirrors of the cavity can be calculated as $|\mathcal{R}_{l,r} + \mathcal{T}_{l,r} - 1|$, where $\mathcal{R}_l$ denotes the reflection of the multilayer stack for a wave coming from the left-hand side. Spatial hole-burning effects are taken into account by subdividing the cavity into a sufficient number m of sections, each characterized by a constant carrier density, photon density, effective refractive index, corrugation period, phase shift and coupling coefficient of the DFB grating. Within a section i the local interaction between the carrier density n_i and the photon densities $s_{i,j}$ of the longitudinal mode j are usually described by the rate equations.

The transfer matrix method does not rely on the approximations of the coupled-mode theory and can, therefore, be applied to arbitrary one-dimensional DFB laser structures with axially varying parameters. In addition, the possibility of an iterative multiplication of the transfer matrix with itself in long periodic structures enables an efficient and numerically accurate computation which is easy to implement due to the matrix formulation.

Transmission Line Laser Model. An alternative to the method mentioned above is the scattering matrix or more commonly called the Transmission Line Laser Model (TLLM) [36], which does not suffer from the time step limitation of the rate equation (3.2). In the TLLM the incident waves are scattered to produce outgoing reflected waves. This scattering is described by scattering matrices. After a propagation of the field amplitudes through a section of the cavity (separated into, for example, N sections of equal length) the reflected waves impinge on adjacent sections, where they are scattered again, becoming incident waves. Since the propagation time delay represented by a transmission line between the scattering events is regarded as equal to the time step between the iterations, all incident waves will arrive simultaneously. Thus, the solution is simply the repetition of the scattering process in every section (scattering matrix) at every iteration step. The time steps are calculated by $\delta l/v$, where δl is the section length and v the phase velocity of the propagating fields. The utilization of the phase velocity once more reflects the propagation of single plane waves rather than wave groups which would require the group velocity v_g. The group velocity comes into play due to the presence of dispersion in the waveguide. Dispersion is, however, not directly introduced into the model. Dispersion acts indirectly via the time-domain filtered gain and propagation constant. The filter-function can be chosen such that the gain has the desired width in the spectral domain. Similar arguments are valid for the propagation constant, i.e. the refractive index. In a semiconductor laser both values – phase- and group-index – differ significantly: the effective phase velocity is typically $v = c_o/3.23$ whereas the effective group velocity roughly amounts to $v_g = c_o/3.7$. The use of the incorrect value directly reflects a similar relative error, e.g. in static and dynamic behaviour. In a Fabry–Perot laser the mode spacing depends on the group index, thus, the model can directly be checked.

The TLLM implicitly includes power transfer between modes by intermodulation, which may occur by gain saturation and the inhomogeneously distributed carrier population along the cavity, as all modes share the same carrier density within a section. The change of the carrier density at an iteration due to the stimulated emission is calculated from the sum of the squares of forward- and backward-propagating fields carrying the energy of all modes and their beating. The propagating waves become amplitude- and phase-modulated by the carrier density variations, resulting in a power transfer across the lasing spectrum. This effect makes the model suitable for the simulation of the high-speed dynamics of lasers. The main difference of the TLLM from some travelling-wave models is that the optical field rather than the photon density is transmitted between the sections.

The model for the fields is solved in the time domain. Hence, the spectral properties can be found by a Fourier transformation of the outgoing field over a number of iterations. In this way all features – including nonlinear effects such as optical frequency mixing which broadens the spectra – can be described. The bandwidth of the model is extremely large since the time steps can be in the 100 fs regime or lower (corresponding to $\geq 10\,\text{THz}$). In many practical cases, the bandwidth has to be restricted to a desired bandwidth. Although the model is solved in the time domain, the bandwidth can easily be tailored by digital filtering as described below.

There is a big advantage of this model: the propagation delays between the scattering events allow a very complicated laser cavity to be split up into smaller models that are solved independently and coupled together. In setting up this TLLM, we start by defining the time-dependent optical field amplitudes $A_k(t)$ in each section k of the cavity under consideration (Fig. 3.18). Then we have a temporal evolution of the field in the N sections according to

$$
\begin{aligned}
A_k(t+\mathrm{d}t) = &\left[t_{k-1}A_{k-1}(t)(f_{k-1}/f_k)^{1/2} + r_{k-1}B_{k-1}(t)\right] \\
&\times \exp(\gamma_k \delta l) + \rho_F \delta l\,.
\end{aligned} \tag{3.85}
$$

A_k and B_k are the slowly varying complex amplitudes of a field propagating forward and backward in the cavity according to

$$
E(z,t) = A_k(z,t)\exp(-\mathrm{i}\beta_0 z + \mathrm{i}\omega_o t) + B_k(z,t)\exp(\mathrm{i}\beta_0 z + \mathrm{i}\omega_o t)\,, \tag{3.86}
$$

where $\beta_0 = \pi/\Lambda$ and ω_0 are the reference wavenumber of the Bragg reflection and its angular frequency, respectively. The boundary conditions of the cavity are: $A_1 = r_1 B_2$ and $B_{N+1} = r_N A_N$

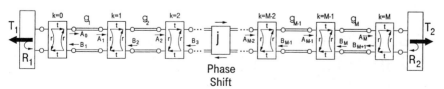

Phase
Shift

Fig. 3.18. Definition of a laser in the TLLM model

The reflection and transmission coefficients in section k are r_k and t_k, respectively. The propagation constant is γ_k in section k with a length δ_l (with $N \cdot \delta_l = L$), and ρ_F is the Langevin noise source in section k. For the case of the field changing its spot size along the cavity as, for example, in a taper, the parameter $(f_k)^{1/2}$ is employed to describe the change of the field in one dimension such that f_{k-1}/f_k describes the relative change of the area of the photon flux, which depends on the field amplitudes as

$$S_k(t) = A_k(t) \cdot A_k^*(t) + B_k(t) \cdot B_k^*(t) . \tag{3.87}$$

The propagation constant for the waves in the cavity in section k is

$$\gamma_k(t) = \frac{1}{2}\Gamma_k \left[g_k(t) + i\alpha \frac{dg}{dn}(n_k(t) - n_{\mathrm{tr}}) \right] - \alpha_s/2 . \tag{3.88}$$

Note that the imaginary part in the rectangular bracket does not contain the gain compression factor $(1 + \varepsilon s)$. The imaginary part results solely from the refractive index change ΔN due to the injected carriers. This index change $\Delta N = \frac{dN}{dn}[n_k(t) - n_{\mathrm{tr}}]$ is then transformed into $\alpha \frac{dg}{dn}[n_k(t) - n_{\mathrm{tr}}]$. The introduction of $(1 + \varepsilon s)$ in the expression would mean that there is self-phase modulation through the presence of photons in the cavity (Kerr effect). But this effect is not considered here.

The gain in section k is commonly described by

$$g_k(t) = \frac{dg}{dn}(n_k(t) - n_{\mathrm{tr}})/(1 + \varepsilon s) , \tag{3.89}$$

where Γ_k is the confinement factor in section k. Since the carrier density producing the gain and phase changes varies only slowly as compared to the fields, a rate equation may be employed for this quantity:

$$\frac{dn_k(t)}{dt} = \eta_{\mathrm{int}} I_k(t)/eV_k - v_g g_k(t) S_k(t) - R_{\mathrm{sp}}^k(t) , \tag{3.90}$$

where $R_{\mathrm{sp}}^k(t) = n_k(t)/\tau_{\mathrm{nr}} + Bn_k^2(t) + Cn_k^3(t)$.

The reflection and transmission coefficients can be calculated on the basis of the coupled-mode analysis for any active or passive waveguide grating. The result for the stationary state is

$$r_k = \frac{i\kappa\sinh(\gamma_k\delta_l)}{\gamma_k\cosh(\gamma_k\delta_l) + i\Delta\beta\sinh(\gamma_k\delta_l)} \quad \text{with} \quad \gamma_k^2 = k^2 - (\Delta\beta)^2 , \tag{3.91}$$

$$t_k = \frac{\gamma_k}{\gamma_k\cosh(\gamma_k\delta_l) + i\Delta\beta\sinh(\gamma_k\delta_l)} , \tag{3.92}$$

where $\Delta\beta = \beta - \beta_0$ is the deviation of the propagation constant from the Bragg condition.

The utilization of these values for r_k and t_k is a very good solution to the problem. However, in each section k, r_k and t_k are different from the previous ones due to the axial variations of many parameters, and thus excess computation time may result since r_k and t_k have to be recalculated again and again. Results obtained in this way, however, do not differ in a significant way from those where fixed r_k and t_k values have been taken. A very good approximation to this problem is to take the r_k and t_k values at the Bragg frequency $(\Delta\beta = 0)$

$$r_k = \mathrm{i}\tanh(k\delta l)\,, \tag{3.93}$$
$$t_k = 1/\cosh(k\delta l)\,. \tag{3.94}$$

These values are constant and have only to be recalculated for those sections where the coupling constant κ varies.

The Langevin noise source $\rho_{F,R}(z,t)$ for the optical field is calculated from the bimolecular recombination contribution to the spontaneous emission rate R_{sp}. $\rho_{F,R}$ is the noise contribution to the complex optical field in each section with length δl for forward and backward directions independently

$$\rho_{F,R} = \left(\eta R_{sp}^{B}\right)^{1/2}\cdot(\mathrm{rand}\,x + \mathrm{i}\,\mathrm{rand}\,y)/\sqrt{2}\,, \tag{3.95}$$

with $\eta = \Gamma_k\frac{\mathrm{dg}}{\mathrm{dn}}\tau_{\mathrm{d}}/V_{\mathrm{act}}$ and $R_{sp}^{B} = Bn^2/L$.

The random numbers x and y (rand x and rand y) obey Gaussian statistics:

$$\langle\mathrm{rand}\,x\rangle = \langle\mathrm{rand}\,y\rangle = 0 \tag{3.96}$$
$$\langle(\mathrm{rand}\,x)^2\rangle = \langle(\mathrm{rand}\,y)^2\rangle = 1\,. \tag{3.97}$$

So far, the bandwidth of the model has not been restricted and only depends on the time step $\delta l/v$, which is in the range of several THz. A realistic laser, however, has a limited bandwidth. Thus, excluding spurious modes, we can introduce a digital filter to confine the bandwidth according to

$$A_k(t) = A_k(t - \mathrm{dt})\cdot f + A_k^v(t)(1 - f)\,, \tag{3.98}$$
$$B_k(t) = B_k(t - \mathrm{dt})\cdot f + B_k^v(t)(1 - f)\,. \tag{3.99}$$

The filtered amplitudes are calculated from the normal unfiltered ones at time t weighted by $(1 - f)$ and the filtered amplitudes at a timestep dt earlier weighted by f, where f denotes a complex number with $|f| \ll 1$. For $f = 0$ the spectrum is flat, an increasing real part of f narrows the spectrum, and a positive imaginary part of f shifts the maximum power of the spectrum to longer wavelengths relative to the Bragg wavelength, and vice versa. The digital filter additionally introduces the effect of dispersion into the laser cavity, since it limits the bandwidth. The utilization of a realistic filter bandwidth, corresponding to the frequency dependence of the gain curve, leads to a calculated mode spacing in a Fabry-Perot laser corresponding to an effective group index which is approximately 10% higher than the effective phase index, which is taken as an input parameter. Without dispersion the resulting mode spacing would correspond to the effective phase index. These simulation results compare well with experimental findings.

There may also be a contribution to the stationary deviation $\Delta\beta$ of the propagation constant from the Bragg condition due to the instantaneous frequency deviation $\Delta\omega$ from ω_0 according to $\Delta\beta = \Delta\omega/v_g$. This frequency deviation $\Delta\omega$ can be calculated by extrapolating the phase to the point in time of interest; the derivative then represents the instantanteous angular frequency ω, and $\Delta\beta$ is determined accordingly. This effect is, however, not considered in most instances, so that computation time may be saved by dropping this term in $\Delta\beta$.

A comparison of this model with experimental observations in the static as well as dynamic regimes of many kinds of lasers shows extremely good agreement.

3.5   Tunable Lasers

A controlled variation of the emission wavelength can be achieved by changing the effective refractive index of the cavity or part of it. In practice, at least two independent control currents are necessary: one in the active part and one in the tuning section for variation of the effective refractive index. The main physical mechanisms for tuning are the plasma effect due to the carrier injection, the Franz-Keldysh effect in depleted regions and thermal effects caused by ohmic heating.

In the next section some of the most frequently utilized types of tunable lasers are briefly characterized and their potential is outlined.

3.5.1   External Cavity Laser (ECL)

Tuning is achieved in these lasers by an external frequency-selective element like a rotatable grating. The tuning range is in this case limited either by the width of the gain curve or the spectral filter width of the frequency-selective element, whichever is smaller. The tuning range may be up to 10% of the emission frequency, the linewidth is in the range of 10 kHz. Since tuning is based on mechanical movements of a part of the resonator, the tuning speed is in the millisecond range. Faster tuning (μs scale) can be achieved by the use of electro-optic or acousto-optic elements in the resonator.

3.5.2   Thermal Tuning

Tuning by heating is based on the temperature dependence of nearly all laser parameters. It is achieved either by changing the temperature of the whole laser including the submount or by a stripe heater placed on the chip near the active waveguide. Thermal tuning depends on heat dissipation and, thus, is a slow process with a time constant in the millisecond range.

3.5.3   Multisection DFB Laser

In these lasers the cavity is subdivided into two or more sections for independent current injection (Fig. 3.19). By proper current control it is possible to
vary the lasing wavelength without changing the output power. The structure
can also be employed for low-chirp direct intensity modulation. Switching is
fast since all laser parts are active. The tuning range is 1–2 nm for 2-section
lasers and 2–3 nm for 3-section lasers.

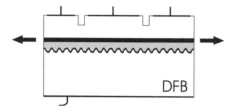

Fig. 3.19. Schematic cross-section of a multi-section DFB laser

3.5.4   DBR Laser

This structure mostly consists of 3 or 4 sections (Fig. 3.20). One section is for
gain production, one for the Bragg reflection and one for phase matching. The
tuning mechanism is more similar to an ECL than in the case of a multisection
DFB laser. The continuous tuning range is up to 5 nm, with mode jumps up
to 9 nm.

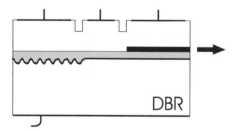

Fig. 3.20. Schematic-cross section of a DBR laser

3.5.5   Tunable Twin-Guide Laser (TTG)

In this case, the three functions in the device of gain production, Bragg
reflection and phase matching are vertically integrated [33]. A conductive
intermediate layer separates gain production and Bragg reflection on the one

hand from phase tuning on the other hand and allows separate control of the respective currents (Fig. 3.21). With only one control current, a continuous tuning of up to 7 nm is possible; the total tuning amounts to 15 nm. Unfortunately, the shot noise in the common conductive control electrode is rather large and, thus, the linewidth of the emitted light is affected in a negative way. Thus, it is not possible to maintain a low linewidth if the wavelength is tuned.

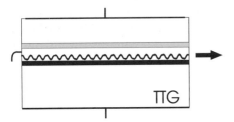

Fig. 3.21. Schematic cross-section of a TTG laser

3.5.6   Codirectionally Coupled Lasers (CCL)

In CCLs two waveguides with different propagation constants are vertically integrated (Fig. 3.22). One contains the gain section and the Bragg grating. The latter has the function of coupling the propagating field of the selected wavelength to the neighbouring passive waveguide for coupling out. The active waveguide is blocked so that it has no output. The grating period is larger than in a DFB laser by 1–2 orders of magnitude. The tuning range is also larger by this factor. More than 50 nm of discontinuous tuning have been achieved. The side-mode suppression ratio of such lasers, however, has not so far been very satisfactory (−20 dB). Improvements may be achieved

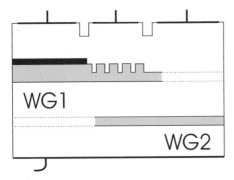

Fig. 3.22. Schematic cross-section of a CCL laser

by properly shaping the grating groves since direct reflections may occur at the interfaces and, thus, disturb the Bragg effect.

3.5.7   Y-Laser

This laser has a y-shaped waveguide with arms of different length (Fig. 3.23). The fact that 3 or more control currents are needed makes this device difficult to operate. However, 40 nm of discontinuous tuning can be achieved. Despite the complicated operating conditions for single-mode emission, this type of laser may also be used for other WDM-network functions like wavelength conversion.

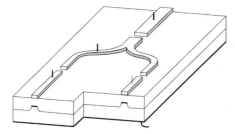

Fig. 3.23. Schematic cross-section of a Y-laser

3.5.8   Superstructure Grating DBR Laser
or Sampled Grating Laser (SSG-Laser)

This laser is a strongly modified version of a DBR laser consisting of a front-SSG, a rear-SSG, an active, and a phase-control section (Fig. 3.24). The two SSG mirrors have a periodic reflection spectrum. The wavelength spacing between two reflection peaks is inversely proportional to the length of one sampled grating interval. The two mirror spectra need a slightly different

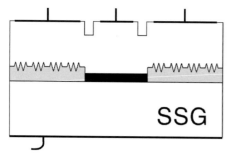

Fig. 3.24. Schematic cross-section of a SSG-DBR-laser

period. Lasing occurs at the wavelength where both mirrors show a reflection maximum at the same time and only one pair of reflection maxima is aligned. For tuning, two different reflection maxima have to be aligned. The lasing wavelength then changes discontinuously. Tuning ranges of up to 100 nm are possible [12].

3.5.9   Bent-Waveguide DFB Laser (BWL)

The advantages of multi-section DFB lasers with respect to high output power, tuning speed, linewidth, and modulation bandwidth are maintained in the BWL. This type contains a bent waveguide over a homogeneous Bragg grating [7]. Thus, the grating period varies along the bent waveguide according to the local angle of the waveguide relative to the grating. This technique enables a continuously varying grating pitch for the design of a wide tuning range. The switching speed is in the ns range.

References

1. H. Kogelnik and C.V. Shank, "Coupled-Wave Theory of Distributed Feedback Lasers," J. Appl. Phys. **43**, 2327 (1972)
2. G. Björk and O. Nilsson, "A new exact and efficient numerical matrix theory of complicated laser structures: properties of asymmetric phase-shifted DFB lasers," J. Lightwave Techn. **5**, 140 (1987)
3. H.E. Lassen, H. Wenzel, and B. Tromborg, "Influence of series resistance on modulation responses of DFB lasers," Electron. Lett. **29**, 1124 (1993)
4. U. Bandelow, R. Schatz, and H.-J. Wünsche, "A correct single-mode photon rate equation for multi-section lasers," IEEE Photon. Technol. Lett. **8**, 614 (1996)
5. J.E. Bowers and M.A. Pollack, "Semiconductor lasers for telecommunications," in *Optical Fibre Telecommunications II* (S.E. Miller and I.P. Kaminov, eds.), (Academic Press, San Diego 1988) p. 512
6. S. Asada, "Waveguiding effect on modal gain in optical waveguide devices," IEEE J. Quantum Electron. **27**, 884 (1991)
7. H. Hillmer, K. Magari, and Y. Suzuki, "Chirped gratings for DFB laser diodes using bent waveguides," IEEE J. Photonics Technol. Lett. **5**, 10 (1993)
8. B. Zee, "Broadening mechanism in semiconductor (GaAs) lasers: limitations to single mode power emission," IEEE J. Quantum Electron. **QE14**, 727 (1978)
9. R.H. Yan, S.W. Corzine, L.A. Coldren, and I. Suemune, "Corrections to the expression for gain in GaAs," IEEE J. Quantum Electron. **26**, 213 (1990)
10. N. Schunk and K. Petermann, "Numerical analysis of the feedback regimes for a single-mode semiconductor laser with external feedback," IEEE J. Quantum Electron. **24**, 1242 (1988)
11. D.-S. Seo, J.-D. Park, J. McInerney, and M. Osinski, "Multiple feedback effects in asymmetric external cavity semiconductor lasers," IEEE J. Quantum Electron. **25**, 2229 (1989)
12. Y. Tohmori, Y. Yoshikuni, H. Ishii, F. Kano, T. Tamamura, and Y. Kondo, "Over 100 nm wavelength tuning in superstructure grating (SSG) DBR lasers," Electron. Lett. **29**, 352 (1993)

13. B. Borchert, B. Stegmüller, and R. Gessner, "Fabrication and characteristics of improved strained quantum-well GaInAlAs gain-coupled DFB lasers," Electron. Lett. **29**, 210 (1993)

14. Y. Nakano, Y. Luo, and K. Tada, "Facet reflection independent, single longitudinal mode oscillation in a GaAlAs/GaAs distributed feedback laser equipped with a gain-coupling mechanism," Appl. Phys. Lett. **55**, 1606 (1989)

15. C. Kazmierski, D. Robein, D. Mathoorasing, A. Ougazzaden, and M. Filoche, "1.5-μm DFB laser with new current-induced gain gratings," IEEE J. Select. Topics Quantum Electron. **1**, 371 (1995)

16. M. Born and E. Wolf, *Principles of Optics*, (Pergamon Press, Oxford) p. 50

17. S. Hansmann, "Transfer matrix analysis of the spectral properties of complex distributed feedback laser structures," IEEE J. Quantum Electron. **28**, 2589 (1992)

18. J.E.A. Whiteaway, B. Garrett, and G.H.B. Thompson, "The static and dynamic characteristics of single and multiple phase-shifted DFB laser structures," IEEE J. Quantum Electron. **28**, 1277 (1992)

19. M. Okai, M. Suzuki, and T. Taniwatari, "Strained multiquantum-well corrugation-pitch-modulated distributed feedback laser with ultranarrow (3.6 kHz) spectral linewidth," Electron. Lett. **29**, 1696 (1993)

20. A. Talneau, J. Charil, A. Ougazzaden, and J.C. Bouley, "High-power operation of phase-shifted DFB lasers with amplitude-modulated coupling coefficient," Electron. Lett. **28**, 1395 (1992)

21. Y. Kotaki, M. Matsuda, T. Fujii, and H. Ishikawa, "MQW-DFB lasers with nonuniform-depth $\lambda/4$-shifted grating," *17th European Conference on Optical Communications (ECOC '91)*, Paris, p. 137

22. S. Hansmann, H. Hillmer, H. Walter, H. Burkhard, B. Hübner, and E. Kuphal, "Variation of coupling coefficients by sampled gratings in complex coupled distributed feedback lasers," IEEE J. Select. Topics Quantum Electron. **1**, 341 (1995)

23. M. Usami and S. Akiba, "Suppression of longitudinal spatial hole-burning effect in $\lambda/4$-shifted DFB lasers by nonuniform current distribution," IEEE J. Quantum Electron. **25**, 1245 (1989)

24. H. Soda, Y. Kotaki, H. Sudo, H. Ishikawa, S. Yamakoshi, and H. Imai, "Stability in single longitudinal mode operation in GaInAsP/InP phase-adjusted DFB lasers", IEEE. J. Quantum Electron. **23**, 804 (1987)

25. Y. Arakawa and A. Yariv, "Quantum well lasers - gain, spectra, dynamics," IEEE J. Quantum Electron. **22**, 1887 (1986)

26. J. Manning, R. Olshansky, D.M. Fye, and W. Powazinik, "Strong influence of nonlinear gain on spectral and dynamic characteristics of InGaAsP lasers," Electron. Lett. **21**, 496 (1985)

27. L.F. Tiemeijer, P.J.A. Thijs, P.J. de Waard, J.J.M. Binsma, and T. v. Dongen, "Dependence of polarization, gain, linewidth enhancement factor, and K factor on the sign of the strain of InGaAs/InP strained-layer multiquantum well lasers," Appl. Phys. Lett. **58**, 2738 (1991)

28. H. Burkhard and E. Kuphal, "InGaAsP/InP mushroom stripe lasers with low cw threshold and high output power," Jpn. J. Appl. Phys. **22**, L721 (1983)

29. P.W.A. McIlroy, A. Kurobe, Y. Uematsu, "Analysis and application of theoretical gain curves to the design of multi-quantum-well lasers," IEEE J. Quantum Electron. **21**, 1958 (1985)

30. C.H. Henry, "Theory of spontaneous emission noise in open resonators and its application to lasers and optical amplifiers," J. Lightwave Technol. **4**, 288 (1986)

31. K. Petermann, *Laser Diode Modulation and Noise*, (Kluwer Academic Publishers, Dordrecht 1988)

32. T. Matsuoka, H. Nagai, Y. Noguchi, Y. Suzuki, and Y. Kawaguchi, "Effect of grating phase at the cleaved facet on DFB laser properties," Jpn. J. Appl. Phys. **23**, 138 (1984)

33. M.-C. Amann, S. Illek, C. Schanen, and W. Thulke, "Tunable twin-guide laser: a novel laser diode with improved tuning performance," Appl. Phys. Lett. **54**, 2532 (1989)

34. B. Tromborg, H. Olesen, and X. Pan, "Theory of linewidth for multielectrode laser diodes with spatially distributed noise sources," IEEE J. Quantum Electron. **27**, 178 (1991)

35. L.D. Westbrook and B. Eng, "Measurements of $\mathrm{d}g/\mathrm{d}N$ and $\mathrm{d}n/\mathrm{d}N$ and their dependence on photon energy in $\lambda = 1.5\,\mu\mathrm{m}$ InGaAsP laser diodes," IEEE Proc. J. **133**, 135 (1986)

36. M. Akbari, H.J. Schöll, G. Faby, and H. Burkhard, "Very fast computation of dynamic characteristics of injection-locked DFB-laser diodes for optical system applications," *Digest International Conference on Semiconductor and Integrated Optoelectronics, SIOE*, Cardiff, UK, p. 24 (1997)

4 Optical Photodetectors

André Scavennec and Louis Giraudet

4.1 Introduction

Photodetectors are used to convert the modulated optical signal transmitted by the optical fibre into an electronic signal. Although the conversion efficiency actually depends on the complete photoreceiver rather than on the photodetector only, several parameters are used to characterize the ability of the photodetector to meet the requirements of optical fibre transmission: responsivity, at the wavelength of interest; bandwidth; and noise characteristics; see e.g. [1,2]. For specific applications, the photodetector has to withstand high optical power, for instance in analogue multichannel transmission; linearity is then a must.

A family of photodetectors is now well accepted for those applications: semiconductor photodiodes, with or without internal gain, offer attractive detection properties, together with characteristics such as low bias voltage, small dimensions, good reliability and most often low cost.

In this chapter, the photoelectric effect appearing in semiconductor photodiodes will first be described, from the photon absorption generating charge carriers to the flow of those carriers through the load circuit. The main characteristics of photodiodes will then be presented, for simple devices as well as for devices operating in the avalanche-gain mode. Finally, the association of photodiodes and low-noise preamplifiers will be discussed, including their monolithic integration into a receiver forming an optoelectronic integrated circuit (OEIC).

4.2 The PIN Photodiode

A photodiode is basically a p-n (or p-i-n, with i standing for intrinsic or undoped) semiconductor junction which is sensitive to incident light through the absorption of photons and generates a photocurrent imaging this photon flux.

4.2.1 PIN Photodiode Operation

When illuminating a semiconductor material with photons characterised by a photon energy ($h\nu$) larger than the material bandgap (E_G), photon absorption occurs, with an associated generation of electron–hole pairs. The light intensity exhibits an exponential decay behaviour, characterised by the material absorption coefficient $\alpha(\lambda)$ at the considered wavelength. The optical power at depth x is given by

$$P(x) = P(0) \exp(-\alpha x) , \qquad (4.1)$$

where $P(0) = rP_{\mathrm{opt}}$ is the optical power at the surface, with r being the reflection coefficient and P_{opt} the incident optical power.

The associated spatial electron–hole pair generation rate (each absorbed photon of energy $h\nu$ is expected to create a pair) is

$$G(x) = \frac{P(0)}{h\nu}\alpha \exp(-\alpha x) . \qquad (4.2)$$

For wavelengths below the absorption threshold where the material starts to be absorbing, α is usually in the 10^4 cm^{-1} range (Fig. 4.1), so that most of the photons are absorbed within a few µm from the surface. This is especially valid for direct gap materials like GaAs and InGaAs, but does not hold for indirect gap materials, and absorption length can increase to a few tens of µm when approaching the absorption edge (0.85 µm for Si and 1.6 µm for Ge).

When electron–hole pairs are created in the vicinity of a junction, as a result of the absorption of photons, minority carriers can recombine, or, through

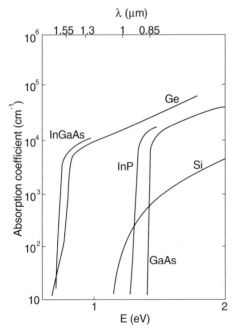

Fig. 4.1. Absorption coefficients of several photodetector materials of interest for optical fibre transmission

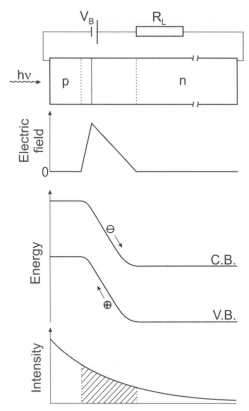

Fig. 4.2. Schematic representation of a pin photodetector. From *top* to *bottom*: pin photodiode with bias source V_B and load resistor R_L; shown as *dotted lines* is the extension of the space-charge depletion region; extension of the electric field in the structure; conduction and valence band diagrams, showing photogenerated electron and hole drift in the depletion region; decrease of the optical intensity in the photodiode; the shaded area indicates the ratio of photons absorbed in the depletion region

a drift or diffusion/drift process across the space-charge region, minority electrons and holes are collected by the n- and p-type regions respectively (Fig. 4.2). Electrical neutrality in the semiconductor is then recovered by a current flowing through the bias/load circuit. Only photocarriers drifting across the space-charge region contribute to the external current. Since the collected photocurrent i_S is proportional to the electron–hole pair generation rate, that is to the optical power, a photodiode is a quadratic detector:

$$i_S \propto P_{\text{opt}} . \tag{4.3}$$

This current is superimposed on the 'dark' current of the junction (see (4.4)), which is usually reverse biased, although a photodiode can be operated even without bias.

$$i(V_B) = i_0[\exp(qV_B/kT) - 1] - i_S ,\tag{4.4}$$

with i_0 the dark current of the photodiode and V_B the bias voltage.

4.2.2 PIN Photodiode Characteristics

Photodiodes are optoelectronic devices in which the generation of electron–hole pairs is governed by the optical absorption properties of the semi-conductor, while their photocurrent-handling characteristics are governed by the usual current and carrier continuity equations of electronic devices. Most photodiodes are p-n diodes with their junction located close ($\sim$ 0.5 to 1 μm) to the surface (photosensitive area) on which the light impinges, the photocurrent being collected by an annular ohmic contact surrounding the photosensitive area and a second one located on the back surface.

Responsivity. In a photodiode, as sketched in Fig. 4.2, with photons impinging from the left onto the photodiode perpendicular to the junction plane, one can identify three different regions (the same would apply to a photodiode with an n-type front region and p-type base region) as follows.

(i)   p-type front region without electric field.
      Electrons generated through the absorption process can recombine in the same region with majority holes, or recombine at the surface after diffusion (surfaces are often characterized by a high recombination velocity (Table 4.1)); they can also diffuse towards the depletion region, and drift through this region before reaching the n-type base. The thinner the front region, with respect to the electron diffusion length, and the lower the surface recombination velocity, the more likely is this latter process to occur. Actually, the front region is usually about 0.5 μm thick, and the surface recombination can be made very small by special passivation measures or by the use of heterostructures.

(ii) Depletion region.
      Carriers generated in this region are subjected to a large electric field,

Table 4.1. Recombination velocity (S) for different air–semiconductor and semiconductor–semiconductor interfaces

	Si	Ge	GaAs	InP	InGaAs	AlGaAs/ GaAs	InP/ InGaAs
S (cm s^{-1})	1	10	10^6	10^3	10^4	10^3	10^3

with electrons and holes drifting towards the n- and p-type neutral region, respectively. Since their transit through the depletion region is very fast, and their density usually low enough, carriers experience almost no recombination.

(iii) n-type base region.

Minority holes generated there can either recombine with majority electrons or, following a diffusion process, drift through the depletion region back to the p-type region.

Neglecting surface recombination, one can express the conversion efficiency from photon to electron–hole pairs, known as the quantum efficiency, by the parameter

$$\eta_i = \frac{n_c}{n_{ph}} , \tag{4.5}$$

where n_c is the sum of the rate (s^{-1}) of charges diffusing from the homogeneous n- and p-type regions to the depletion region and the rate of pairs photo-generated in this depletion region; n_{ph} is the rate of absorbed photons.

The quantum efficiency of the photodiode, representing the ratio between the number of charges flowing through the external circuit to restore the device electrical neutrality to the number of absorbed photons, can be expressed as the combination of the efficiencies of the three regions (see Fig. 4.2):

$$\eta_i = \int G(\alpha, x)\Re(x)\,\mathrm{d}x , \tag{4.6}$$

with $\Re(x) = 1$ for carriers generated in the depletion region.

The quantum efficiency is usually quite high (about 90% for photodiodes in the visible and infrared region) for a photodiode with a structure optimized for a given wavelength. This internal efficiency can be significantly lower than the external efficiency (η_e), referring to the number of incident photons, depending on the surface reflection (for semiconductors with a refractive index in the 3.5–4 range the surface reflection coefficient r is about 30%). Antireflection coatings made from $\lambda/4$ dielectric layers can eliminate this surface reflection, so that η_e and η_i are almost identical; however this can be obtained only over a limited wavelength range.

Rather than using the quantum efficiency, one usually characterizes a photodiode through its responsivity R_0, which relates the photocurrent to the incident optical power. Responsivity and quantum efficiency are obviously related:

$$R_0 = \left(\frac{q}{h\nu}\right)\eta_e . \tag{4.7}$$

Since the efficiency can be close to 100% for well-designed photodiodes, the above relation shows that one expects responsivity to increase with wavelength: responsivity is about 0.5 A/W for Si photodiodes operating at 0.85 μm, and can be over 1 A/W for InGaAs photodiodes at 1.55 μm.

So far, only conventional photodiodes have been considered, with a pin structure and front illumination. In order to improve responsivity, wavelength range or bandwidth, other structures can be used, for instance with a Schottky junction or illumination in the plane of the junction. Such structures will be discussed when appropriate.

Response Time. The response time of a photodiode is set by the following different contributions: some are related to the internal process of carrier collection, while others reflect the influence of the bias/load circuit.

(i) Collection time, which includes two contributions.
  - The transit time of carriers through the depletion region (depth: w_d); when injected in this region, electrons and holes are subjected to a large electric field and drift towards the n- and p-type region at their respective drift velocity ($v_{sl} \approx 5 \times 10^6 - 10^7$ cm/s, depending on material and carrier type). The transit time is then given by

$$\tau_T = \frac{w_d}{v_{sl}} . \tag{4.8}$$

    τ_T can range from a few ps to hundreds of ps, depending on the depletion layer depth (0.5 μm for some high-speed InGaAs photodiodes to 40 μm for Si photodiodes designed for 0.85 μm wavelength applications). This transit time usually sets the bandwidth limitation for photodiodes used in optical fibre communications.
  - The diffusion time of carriers generated in the neutral front and base regions. For a well-designed photodiode with negligible generation in the base region (this may occur however with devices operated close to their absorption threshold, e.g. Ge photodiodes at 1.55 μm), and assuming no surface recombination and no dopant gradient in the front layer, the associated time constant is given by

$$\tau_{diff} = \frac{w_p^2}{2D_n} , \tag{4.9}$$

    with D_n, the diffusion constant of electrons in the front layer and w_p the front layer depth.
    τ_{diff} is usually in the range of hundreds of ps. To be contrasted with the situation for Si and Ge ($D_n \approx D_p \approx 2$ cm^2/s), there is a large difference between the electron and hole diffusion constants in III-V materials ($D_n \approx 25$ cm^2/s $\gg D_p$). For that reason, III-V photodiodes are usually designed with a p-type front region (actually both responsivity and response time are simultaneously improved through this choice).

(ii) Circuit contribution.
    The dynamic equivalent circuit of a photodiode (Fig. 4.3) is mainly comprised of a photocurrent generator and a parallel combination of the

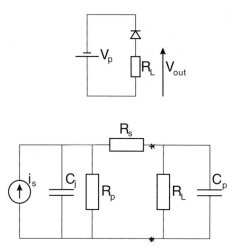

Fig. 4.3. Dynamic equivalent circuit of a photodiode

junction capacitance (C_j) and high value resistor (R_p) characteristic of
a reverse-biased diode. A series resistor (R_S) takes into account the resistance of the semiconductor, in particular the base (substrate) resistance
and the (thin) front-region resistance, as well as the contact resistances.
To complete this equivalent circuit, one should add the load resistor
across which the photocurrent is monitored. An important parameter
controlling the response time of the photodiode is then the junction
capacitance, related to the geometrical characteristics of the depletion
region: in general the photodiode's front region is highly doped, while
the underlying intrinsic region has a very low doping level. Assuming
the latter to be constant, the capacitance can then be written as

$$C_j = \frac{\varepsilon_r \varepsilon_0 A}{w_d} = \left[\frac{q \varepsilon_r \varepsilon_0 N_B}{2(V_{bi} - V_B)} \right]^{1/2} A \,, \qquad (4.10a)$$

where A is the junction area, V_{bi} and V_B the junction built-in voltage
and bias voltage, respectively, and N_B the intrinsic region doping level.
When V_B is large enough to deplete the intrinsic layer (which is generally
the case in well-designed pin photodiodes) the junction capacitance is set
by the intrinsic region depth, w_i:

$$C_j = \frac{\varepsilon_r \varepsilon_0 A}{w_i} \,. \qquad (4.10b)$$

Figure 4.4 gives typical junction capacitance characteristics of InGaAs
photodiodes. Worth noting is the influence of parasitic capacitances (C_p)
associated with the bonding pads or with the package which can severely
degrade the response time. Considering the transimpedance of the circuit

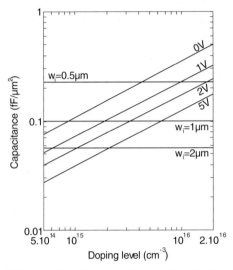

Fig. 4.4. Specific capacitance of an InGaAs photodiode as a function of the extension of the depletion region set by the doping level and bias voltage (0, 1, 2 and 5 V) or by the depth (0.5, 1 and 2 μm) of the 'intrinsic' region

in Fig. 4.3, one can write

$$\frac{V_L}{i_S} = \frac{R_L}{1 + j\omega(R_S C_j + R_L C_p + R_L C_j) - R_S R_L C_p C_j \omega^2} \ . \tag{4.11}$$

The time constant controlling the circuit response is thus given by

$$\tau_c = R_S C_j + R_L C_p + R_L C_j \ . \tag{4.12}$$

(iii) Response time.

The global response time of a photodiode is then the result (convolution) of the above two contributions: depending on the wavelength, the material, the sensitive area, or the load resistor (often a 50 Ω load is considered), the response speed can vary widely. Moreover, since the photodiode is not a first-order system, its response time cannot be quoted simply, requiring much care when characterizing a photodiode.

Conflicting requirements often appear for optimizing simultaneously responsivity and bandwidth, since the latter parameter calls for a thin depleted region, while the former implies this region to be wide enough for absorbing most of the light.

Linearity and Response Homogeneity. The linearity of photodiodes can be considered as very good at low optical power, even at high frequency. However under high-power illumination conditions (as found in multichannel analogue transmission for TV broadcasting or phased-array antenna feeders), large densities of carriers are created which modify the voltage applied to the photodetector through the load-resistance feedback, and disturb the electric-field distribution over the depletion region. The electric-field screening degrades the photodiode bandwidth, and sometimes the D.C. responsivity. The threshold for this saturation phenomenon has been investigated, pointing to carrier drift velocity, depletion layer width and bias field as the main parameters [3,4].

The dynamic response of a photodiode is also dependent on the location in the photosensitive area where optical absorption occurs. In particular, this dependence can be observed when the optical beam, characterized by a width smaller than the photosensitive area, is scanned across this area. With light absorbed close to the ohmic contact, the bandwidth is the largest, while illumination in the centre requires that excess carriers collected by the front layer flow along that resistive layer (sheet resistance typically $100\,\Omega/\mathrm{sq}$) before reaching the ohmic contact. This additional series resistance can degrade somehow the bandwidth. The small-diameter photodiodes used in optical fibre transmission are usually not prone to this degradation owing to their small sensitive area, matched to the fibre mode.

Sensitivity – Signal-to-Noise Ratio. Sensitivity is a most important parameter used to characterize the ability of a photodetector to detect small-power signals. The sensitivity is the minimum signal power a photodiode can detect for a given signal bandwidth; it is related to the signal-to-noise ratio (S/N), which is limited by the responsivity and the circuit noise. In a simple way, one can express the S/N considering a simplified photoreceiver circuit comprising a photodiode and the associated resistive load (Fig. 4.5). Two noise sources have to be considered:

(i)   The first is the photodiode junction shot noise; the (photo)current flowing through the junction generates a white noise (constant spectral density), known as Schottky noise. Its spectral density is given by

$$S_{\mathrm{iS}}(f) = 2qi = 2q(i_{\mathrm{S}} + i_0)\,. \tag{4.13}$$

(ii) The second is associated with the thermal noise of the load resistor:

$$S_{\mathrm{iR}}(f) = \frac{4kT}{R_{\mathrm{L}}}\,. \tag{4.14}$$

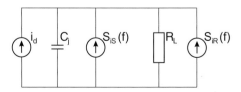

Fig. 4.5. Representation of the photoreceiver noise sources for the simple circuit of Fig. 4.3

The photoreceiver S/N can then simply be expressed as

$$S/N = \frac{i_S^2}{2q(i_S + i_0) + 4kT/R_L}\Delta f \tag{4.15a}$$

$$= \frac{(R_0 P_{opt})^2}{2q(R_0 P_{opt} + i_0) + 4kT/R_L}\Delta f , \tag{4.15b}$$

with Δf being the bandwidth over which the signal is detected.

The quantum limit, which characterizes the noise contribution from the signal itself with neglecting the other noise sources, is then defined by

$$S/N = \frac{R_0 P_{opt}}{2q\Delta f} . \tag{4.16}$$

In real systems, additional noise sources have to be considered, including the series resistance and $1/f$ noise; but, to first order, expression (4.15) shows the influence of the main parameters, including the dark current and the load resistor value, which are discussed below.

(i) Dark current.

For actual photodiodes, the dark current is mainly characteristic of the semiconductor material and junction technology. The reverse current of the photodiode is generally comprised of several contributions: bulk diffusion and generation current, related to the material quality and doping level, but also surface generation current, dependent on the process technology and quality of the surface passivation, and tunnelling current in small bandgap materials. A general feature is an increase of the dark current as the semiconductor material bandgap decreases (Fig. 4.6). For

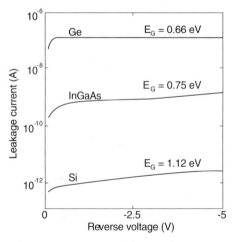

Fig. 4.6. Typical dark (leakage) current characteristics of Si, Ge and InGaAs photodiodes with junction area in the $50-100\,\mu\text{m}$ diameter range

optical-fibre digital transmission, the optical power at the receiver is usually in the 10^{-2}–$10^2\,\mu\mathrm{W}$ range, meaning that the dark-current of a Si photodiode does not impair the receiver sensitivity, while this can be the case for InGaAs and Ge photodiodes, in particular at low bit rates. Also important to notice is the large dark current activation with temperature ($i_0 \propto n_i$ or n_i^2 or, expressed differently, $i_0 \propto \exp(E_G/kT)$ or $\exp(E_G/2kT)$). For instance, the dark current of a Ge photodiode increases by two orders of magnitude when the temperature increases from 20 to 60 °C.

(ii) Load resistor.

To reduce the thermal noise, the load resistance has to be as large as possible; from the simplified circuit of Figs. 4.3 and 4.5, one can derive a maximum value for the load resistance set by the photodiode capacitance C_d (including the parasitic capacitance of the circuit) and the bandwidth Δf required by the signal spectrum: $R_L C_d = 1/2\pi\Delta f$. Expression (4.15) can then be written in the form

$$S/N = \frac{i_S^2}{2q(i_S + i_0)\Delta f + 8\pi kTC_d\Delta f^2} \, . \tag{4.15c}$$

Although the dark current may limit the S/N in small-bandwidth applications, it is the thermal noise of the load which usually governs the sensitivity of the receiver of Fig. 4.5 associating a photodiode and a load resistor. With such a receiver, and assuming a responsivity of $1\,\mathrm{A/W}$ and a photodiode capacitance of $0.3\,\mathrm{pF}$ (no parasitics), one would expect a S/N ratio of unity for a 622 Mbit/s signal with $0.1\,\mu\mathrm{W}$ power.

To improve the sensitivity of such receivers, various directions have been explored since the advent of optical fibre communications:

- Optical preamplification of the signal before photodetection: this technique is now widely used, following the development of low-noise EDFAs (see Sect. 5.1) in the 1980s, and provides the best sensitivity.
- Amplification of the photocurrent within the photodiode through the multiplication phenomenon (avalanche multiplication) associated with impact ionization.
- Reduction of thermal noise of the photoreceiver preamplifier. The two latter ways to improve the sensitivity are presented in Sects. 4.3 and 4.5, respectively.

4.2.3 Edge-Illuminated PIN Photodiodes

While photodiodes are usually fabricated from a shallow junction, with light impinging perpendicularly to the junction, there are advantages in using photodiodes with illumination parallel to the junction. Actually this scheme was demonstrated to extend the response of silicon photodiodes beyond $1\,\mu\mathrm{m}$ by

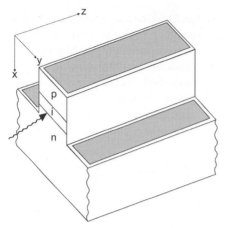

Fig. 4.7. Schematic view of an edge-illuminated pin photodiode with a ridge structure

lengthening the absorption region. More recently, this type of structure was studied anew to bring an answer to a number of problems encountered with conventional (InGaAs) photodiodes: limited bandwidth-responsivity product, lack of compatibility with semiconductor laser geometry preventing a common optical coupling approach with waveguides or fibres, limited saturation power, etc. For those reasons, a large interest in exploiting the specific features of edge-illuminated pin photodiodes has developed since the late 1980s [5,6], in parallel with evanescently coupled structures discussed in Chap. 11.

Let us consider the structure of Fig. 4.7, comprising an intrinsic absorbing layer surrounded by two wider bandgap (transparent) p and n regions. Providing the refractive indices are such that $n_{core} > n_{p,n}$, the stucture is an absorbing waveguide confining light propagating along the z axis: in particular, this situation opens a new way to handle, almost independently, the response time associated with the carrier transit time through the depletion region and the responsivity.

Responsivity. Responsivity is limited by the way the optical power, injected for instance from the fibre, is coupled to the absorbing core. Depending on the structure and materials characteristics, the waveguide is monomode or multimode; its modal geometric structure will set the overlap with the mode coupled from the fibre $\psi(x, y)$ and basically will define the optical power coupling efficiency η_c. Assuming the waveguide along the y-axis to be larger than the coupled optical mode, η_c is given by

$$\eta_c = \sum_i \left| \int \psi_x(x)\varphi_{xi}(x)\,\mathrm{d}x \right|^2 , \qquad (4.17)$$

with i representing the waveguide mode order along the x-axis.

A second factor limiting the responsivity is related to the device length and to the way the waveguide confines the optical power within the absorbing core: the confinement factor Γ_i, related to mode i, defines the absorption length of the optical power in the mode, so that the effective absorption coefficient is now

$$\alpha_i = \Gamma_i \alpha \; , \; \text{where} \; \Gamma_i = \int\limits_{core} |\varphi_{xi}|^2 \, dx \; . \tag{4.18}$$

The overall responsivity is then dependent on the pin photodiode length (since $\alpha_i \ll \alpha$), overlap of the mode coupled from the fibre and modal structure of the photodiode, and input facet reflectivity.

Response Time. As in conventional photodiodes, response time is limited by the transit of carriers through the depletion region (assuming no absorption in the confining layers), and by the junction capacitance and series resistance: when the absorbing core becomes very thin, the capacitance becomes a real limitation, due to the load resistor (usually $50\,\Omega$ adding to the series resistance of the photodiode). Depending on applications, response speed can be traded for saturation power or alignment tolerances.

4.2.4   Metal-Semiconductor-Metal Photodiodes

Less conventional structures than p-n junctions can be used as photodetectors; in particular photoconductors (semiconductor resistors), Schottky diodes and Metal-Semiconductor-Metal (MSM) structures have been considered for optical fibre applications. Actually the latter ones exhibit attractive characteristics which motivate a continuing interest [7]. Figure 4.8 shows a sketch of a MSM photodiode. A bias voltage is applied to two interleaved sets of electrodes, forming Schottky contacts on the semiconductor surface. Provided that both the distance between the electrodes and the semiconductor doping level are small enough, an electric field extends into the semiconductor from one set of electrodes to the other (corresponding to the reach-through voltage) to collect the photocarriers generated by illuminating the MSM device. The main characteristics of MSM photodiodes have been investigated and compared to those of pin photodiodes.

The dark current is mainly given by the leakage current of the Schottky electrodes, which is mainly related to the barrier height of the metal/semiconductor interface; generation current can also contribute to this dark current in the case of a low bandgap material (InGaAs on InP on which devices with $1-10\,\text{nA}$ dark current have been reported).

One attractive feature of the MSM photodiode is its low junction capacitance (by a factor of 5 lower with respect to a pin photodiode of similar sensitive area): for instance, an InGaAs MSM of $70 \times 70\,\mu\text{m}^2$ active area exhibits a capacitance of only $150\,\text{fF}$. As for a pin photodiode, the responsivity

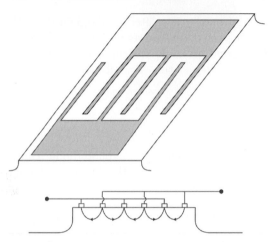

Fig. 4.8. Schematic representation of a Metal-Semiconductor-Metal (MSM) photo-diode with an interdigitated finger structure. The lower figure shows a cross-section of the device, with the electric field penetrating through the absorbing top region

is limited by the reflection coefficient and represents the efficiency of the depletion region. Obviously, the electrodes should be as narrow as possible to minimize the amount of reflection for a surface-illuminated MSM photodiode (back illumination has been experimentallly demonstrated in InGaAs/InP MSM photodiodes to partly avoid this problem [8]). An unexpected current gain mechanism is sometimes observed on MSM photodiodes, in spite of the Schottky contacts. This has been identified as resulting from surface effects, in a way similar to photoconductors, and possibly to a modification of the Schottky barrier characteristics [9].

Finally, the intrinsic response time of a MSM photodiode depends on the transit time of carriers through the depletion region. Due to the intrinsic two-dimensional geometry of the MSM photodiode, a bias voltage usually higher than for a pin photodiode is required to ensure a large enough electric field extends throughout the active part of the structure (Fig. 4.9). Moreover, the frequency response proves to be more complex than in conventional pin photodiodes. Very large bandwidth can be obtained (70 GHz in [11]) provided that the absorption region is thin enough (0.5 μm, for instance) and the electrodes close to each other (< 1 μm), such that the transit length remains short. This is not always compatible with a very good responsivity [12], except for waveguide-illuminated MSM photodiodes.

Worth noting is the very low series resistance of MSM photodetectors compared to pin photodiodes. This behaviour is associated with the fact that there is no semiconductor series resistance present. As a result, a very low RC time constant is obtained for MSM photodiodes.

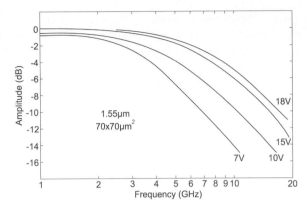

Fig. 4.9. Frequency response of an MSM InGaAs photodiode for different bias voltages; the finger width and spacing are 1 and 2 μm, respectively (taken from [10])

4.3   The Avalanche Photodiode (APD)

While pin photodiodes perform adequately for low-bandwidth signals, an improvement of their S/N, when the bandwidth extends beyond 1 MHz, can be obtained by taking advantage of the effect of avalanche multiplication: when a junction is reverse-biased with a high enough electric field, impact ionization can be used to multiply the current flowing through the junction. The impact ionization process takes place when carriers gain enough energy from the electric field, typically an energy larger than the bandgap energy, to ionize atoms, thereby generating secondary electron–hole pairs. This process usually occurs only over a limited depth of the depletion layer in the vicinity of the p-n interface, where the electric field is the highest (Fig. 4.10). The

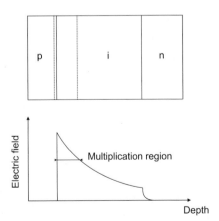

Fig. 4.10. Extension of the multiplication region in a pin avalanche photodiode: only a limited part of the depletion region provides multiplication

avalanche multiplication process has been extensively investigated in bulk materials. In particular, under constant electric field conditions, it has been shown that the process characteristics are mainly governed by the ratio between the ionization rates of electrons, α_e, and holes, β_h (characterizing their respective ability to induce ionization) and by the type of carrier which is predominently injected in the multiplication region.

4.3.1 Characteristics of APDs

Analysis of the multiplication phenomenon by McIntyre, in particular with respect to the influence of the ratio of the ionization rates on noise [13], led to the identification of specific characteristics which can be summarized as follows, considering that only one type of carrier is injected in the avalanche region, and assuming the electric field to be constant.

(i)   Injected carriers have the higher ionization rate (preferred situation): Multiplication proceeds as the carriers drift across the avalanche region, with the secondary carriers of the other type drifting back without creating secondary pairs. The multiplication process is characterized by a multiplication factor M (mean value of the ratio of collected to injected carriers). At large M values, the transit-time-limited response time is increased by a factor of two with respect to the $M = 1$ regime (assuming equal drift velocity for electrons and holes) due to the secondary carriers drifting back (Fig. 4.11).

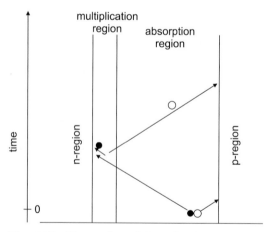

Fig. 4.11. Illustration of the collection response time in an avalanche photodiode. The secondary pair transit time leads to an increase of the total transit time, by a factor of about two, if the photogenerated carriers injected in the multiplication region have an ionisation rate much larger than the other carrier type

(ii) Injected carriers have an ionization rate similar to that of the other carriers:
Multiplication first associated only with the injected carriers also develops through ionization induced by the carriers of the other type; this leads to secondary carriers drifting in one direction or the other, depending on their type. A first consequence is a more abrupt I–V characteristic than in the former case; a second one is an increase of the response time (the collection time is basically multiplied by M, which may not be a problem provided that the avalanche region is very small compared to the total depletion region width).

(iii) Injected carriers have a lower ionization rate than that of the other carriers:
Multiplication occurs mainly for a very limited number of injected carriers with which a large number of secondary pairs are associated; the response time, as in case (i), is only degraded by a factor of two.

4.3.2 APD Noise

An important feature of the avalanching process is the increased shot-noise density. Because of the random occurrence of the impact-ionization process, i.e. primary carriers experience different multiplication factors, the shot noise density is now given by

$$S_{iS}(f) = 2qi \langle M^2 \rangle = 2qi \langle M \rangle^2 F(M) , \qquad (4.19)$$

where $\langle M \rangle$ is the mean value of M, and $F(M)$ is the excess-noise factor due to the multiplication effect.

An analytic formula has been derived for $F(M)$; actually $F(M)$ depends on M, but also on the α_e/β_h ratio and the type of injected carrier (Fig. 4.12). In order to have a low excess noise, an APD should exhibit a large difference in ionization rates for holes and electrons (see Table 4.2), and have mainly carriers with the higher ionization rate injected in the avalanche region. For that reason, specific designs are mandatory for APDs, taking into account the material properties (ionization rate ratio, absorption coefficient, etc.). One should also note that the dark current may experience a different noise factor to that for the photocurrent.

Table 4.2. Ionization-rate ratio for different semiconductor materials of interest for the fabrication of APDs (the electric field at which this ratio is given is such that the higher of α_e or β_h is close to $10^4 \, \text{cm}^{-1}$)

Material	Si	Ge	GaAs	InP	GaSb	InGaAs	AlInAs
α_e/β_h	20	0.7	2	0.5	1.4	3	2.5

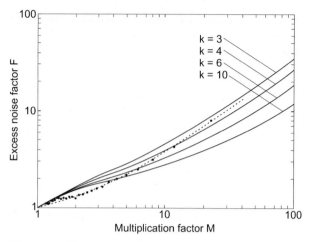

Fig. 4.12. Excess-noise factor vs multiplication factor in a multiple quantum well (MQW) AlInAs/AlInGaAs APD. For reference, theoretical curves by McIntyre [13] corresponding to different ionization rates ratios k ($= \alpha/\beta$ or β/α, depending on whether electrons or holes are injected in the multiplication region) are given together with a fit (*dotted line*) using Teisch's noise model, which is more appropriate for multiplication in quantum wells [20]; data taken from [21]

From Table 4.2, one expects Si to exhibit low excess-noise characteristics, as well as a large multiplication gain-bandwidth product, in contrast with most other materials that are characterized by similar ionization rates. This means that low noise APDs can be envisioned at a wavelength of $0.85\,\mu$m using Si, but that only moderate performance can be expected at longer wavelengths, where InP and GaSb and their related compounds are used. Actually, lower excess-noise figures than expected from the bulk ionization-rate ratio have been reported for devices with a very short avalanche-region length in which the avalanche effect may not be considered as a purely random phenomenon [14,15]. This obviously offers room for some future improvement of APD characteristics.

4.3.3   Structures for Improved Noise Characteristics

In order to obtain avalanche materials more suitable than the ones of Table 4.2 for low noise operation at $1.3\,\mu$m and $1.5\,\mu$m wavelength, the following approaches have been explored, aiming at enhancing the difference in ionization rates.

Spin-Orbit Resonance. For specific ternary or quaternary material compositions a resonance appears between the bandgap energy and the valence-band split-off energy. This resonance induces an increase in the ionization rate of holes, which results in $\alpha_e/\beta_h \ll 1$. This phenomenon, first reported for

$Al_xGa_{1-x}Sb$ with $x_{Al} = 0.07$ [16], has also been shown to lead to improved α_e/β_h ratio in HgCdTe APDs [17]. However, in both cases the insufficient material quality or properties prevented the fabrication of high-performance, high-reliability APDs.

Multiple Quantum Well (MQW) APDs. First proposed by Chin [18] and further investigated by Capasso [19], the avalanche effect is expected to be enhanced in MQW or in staircase structures for one of the carrier types, depending on the asymmetry in the conduction and valence band discontinuities. Let us consider, for instance, the AlInAs/InGaAs system, with a conduction-band discontinuity of 0.5 eV, considerably larger than the valence-band discontinuity of 0.2 eV (Fig. 4.13). An electron drifting from an AlInAs barrier to a InGaAs well will benefit from an extra energy of 0.5 eV which, added to its kinetic energy at the output of the barrier, helps this electron to reach the ionization threshold energy (between E_G and $3E_G/2$, with E_G the bandgap of the quantum well material). At the same time, a hole would receive only an added energy of 0.2 eV upon entering the InGaAs well. Actually an improvement from 2 to 20 in α_e/β_h has been reported for such MQWs compared to bulk InGaAs. However, the exact physics behind this improvement is still under investigation: transitions between conduction bands, influence of the barrier and well lengths, electron trapping in the well, tunneling through the barrier, amongst others, need to be further studied.

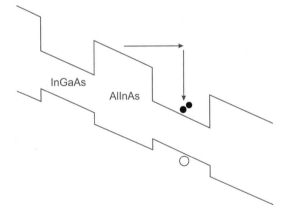

Fig. 4.13. Impact ionization in an AlInAs/InGaAs Multi-Quantum-Well structure characterized by a large asymmetry in conduction and valence band discontinuities. In such a structure, ionization by electrons is favoured owing to the excess energy gained by an electron crossing the AlInAs/InGaAs interface

4.4   Photodiodes

Following the evolution of optical transmission, from multimode fibres operating at 0.85 µm wavelength to monomode fibres at 1.3 and 1.55 µm, various types of photodiodes have been developed; presently, photodiodes made of Si and InGaAs/InP are predominantly employed in the first and second/third windows, respectively.

4.4.1   Silicon Photodiodes

Benefiting from the well-established microelectronics technology, pin photodiodes able to operate at 0.85 µm have been available for a long time, with a planar technology on n-type substrate, with or without a guard ring, as sketched in Fig. 4.14. A specific characteristic of the wafers used for the fabrication of 0.85 µm wavelength photodiodes is the thick non-intentionally doped epitaxial layer (20−50 µm) needed for a good responsivity.

The Si APD illustrates the specific design necessary to comply with requirements for low noise and large gain-bandwidth product. The structure is quite complex, to cope with the low absorption coefficient and with the requirement of pure electron injection in the multiplication region, which is necessarily situated close to the surface at the p-n junction. In the so-called pπpπn structure of Fig. 4.15, a p-type substrate is needed so as to allow for an n-type front layer. A p-type sheet charge is implanted in the epitaxial layer (which has a low-level p-type doping, hence the π designation) below the n-π junction. At low reverse voltage, the depletion layer is limited to the π region between the front n-region and the p-type sheet charge. The operating range of the APD is for bias voltages beyond punch-through, that is when the p-type sheet charge becomes fully depleted and a drift field develops over the

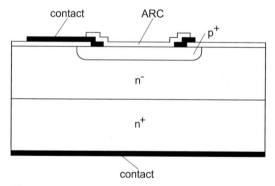

Fig. 4.14. Schematic cross-section of a Si pin photodiode. Actual structures may differ significantly depending on the application: for optical fibre applications in the 0.85 µm window, a thick intrinsic (n⁻) region is needed (in the 10−40 µm thickness range); *ARC*, antireflection coating

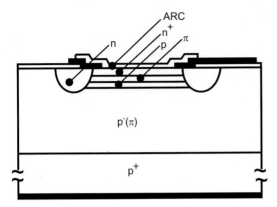

Fig. 4.15. Cross-section of a Si APD with the so-called pπpπn structure (π standing for low p⁻doping level). This structure provides a rather uniform electric field in the multiplication region, and is designed such that the photocarriers crossing the multiplication region are mainly electrons

whole absorption region. Multiplication factors of several 10s, even up to 100, can be used at bias voltages between 100 and 200 V. To withstand the high avalanche electric field, a very good material quality of both the substrate and the epitaxial layer is mandatory, as the junction diameter is usually rather large to accommodate the beam size of multimode fibre.

4.4.2   InGaAs Photodiodes

While Ge photodiodes were the components of choice for the earlier fibre transmissions at 1.3 μm, their degraded performance at 1.55 μm and the development of an InP-based optoelectronic technology led to the present situation with InGaAs-based photodetectors presently covering most demands of long-wavelength transmission.

Conventional InGaAs PIN Photodiodes. Lattice-matched to InP, $In_{0.53}Ga_{0.47}As$ has an absorption threshold wavelength of 1.67 μm at room temperature and absorption coefficients of $10^4\,cm^{-1}$ and $0.7 \times 10^4\,cm^{-1}$ at 1.3 μm and 1.55 μm, respectively. While the first devices just mimicked Si photodiodes, InGaAs pin photodiodes have later been refined by taking advantage of specific features of the InP heterostructure technology, as summarized below:

- Transparent wide-bandgap 'window' front layers are widely used to improve both responsivity and response time, since there is no absorption in the front p-type region, while decreasing the surface leakage currents.
- Semi-insulating substrates allow for the fabrication of low-capacitance bonding pads: this helps to keep the benefit of the small junction capacitance associated with a small photosensitive area [22].

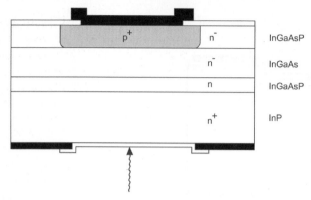

Fig. 4.16. Example of an InGaAs photodiode with back-illumination. The InGaAsP layers, as well as the InP substrate, are transparent layers

- Transparent substrates offer the possibility of illumination through the substrate in flip-chip devices; this is a way to improve the photodiode bandwidth without compromising the responsivity, by reducing the thickness of the absorbing depletion layer: Actually light crosses this depleted layer twice, owing to the reflection provided by the front metallization (Fig. 4.16).

A drawback of this heterostructure approach is related to the trapping of carriers due to energy-band discontinuities. In particular, due to the large valence-band offset in the InP/InGaAs system, holes can be trapped at such interfaces before being thermionically re-emitted, which eventually degrades the device response speed. Specific designs are then needed to avoid such effects, incorporating for example quaternary transition layers for bandgap grading.

Many different devices are now available, exploiting one or several of the features described above to satisfy application-dependent requirements: For high-sensitivity photoreceivers, photodiodes with a photosensitive area of $30\,\mu m$ in diameter and a total capacitance of about $100\,fF$ are available, with a responsivity close to $1\,A/W$ at $1.55\,\mu m$, thus being suitable for digital transmission rates of 2.5 or 10 Gbit/s. Figure 4.17 shows the responsivity and frequency response of an InGaAs photodiode as a function of the thickness of the absorption layer. It is evident that both parameters cannot be optimized independently of each other.

Edge-Illuminated InGaAs PIN Photodiodes. To overcome the responsivity-bandwidth limitation of front- or back-illuminated photodiodes (Fig. 4.17), new structures have been investigated, including resonant-cavity photodetectors [23] and edge-illuminated devices of the refracting-facet [24] or waveguide type. Double-heterostructure multimode photodiodes were demon-

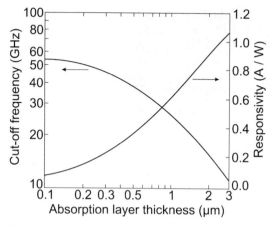

Fig. 4.17. Responsivity and frequency bandwidth of an InGaAs/InP photodiode at 1.55 μm as set by the the InGaAs absorption layer thickness (limitation due to the junction capacitance is not taken into account)

strated, providing a responsivity of 0.6 A/W together with a bandwidth of 110 GHz [25]. In such structures, quaternary layers with a small refractive index difference to the absorbing core are used to obtain a good coupling efficiency between the optical fibre mode and the 20−100 μm-long photodiode.

Recent developments have been conducted to further improve the performance of edge-illuminated photodiodes in terms of bandwidth or power-handling capability:

(i) Travelling-wave photodiodes: edge-illuminated photodiodes can be designed such that the electrode (coplanar) geometry is velocity-matched to the propagating optical signal [26]. By choosing a low absorption coefficient, it is then possible to distribute absorption over an extended photodiode length. The distributed carrier generation is collected in-phase, leading to a large bandwidth (the limitation is mainly set by the velocity mismatch and bandwidth of the electrode design) and high saturation power device [27]. A limitation on the high-frequency responsivity is given by the fact that at each point two signals are generated which propagate in opposite directions along the electrode structure.

(ii) Uni Travelling Carrier photodiode: this device includes, next to the transparent drift region, a thin p-type absorption layer where electron–hole pairs are photogenerated. Only electrons are injected in the wide-bandgap drift region, with no holes present to induce space-charge effects in this region. Actually saturation and the associated nonlinearities usually result in conventional photodiodes from the poor hole velocity-field characteristics, which are much less favourable than those of electrons. Since the electron diffusion time in the p-type layer is negligible, this structure allows for a very large bandwidth without saturation up to tens of mW. An output voltage signal as high as 1.9 V has been reported with a 110 GHz bandwidth and a responsivity of about 0.15 A/W for a back-illuminated UTC photodiode [28], with UTC waveguide structures providing better responsivity.

Edge-illuminated photodiodes are also attractive for their compatibility with lasers for in-plane fibre pigtailing and for flip-chip assembling on SiO_2/Si mother-boards. Devices with a large alignment tolerance, up to 5 μm in the x- and y-directions, have been demonstrated.

InGaAs Avalanche Photodiodes. Due to its low bandgap and direct-type transition, InGaAs is not suited as avalanche multiplication material. In fact, the tunnel current is of the order of 1 A/cm^2 when avalanche occurs. For that reason the following new structures were designed, called Separate Absorption and Multiplication (SAM)-APDs, in which absorption takes place in an InGaAs layer, while multiplication occurs in a wider bandgap material.

(i) InGaAs/InP-SAGM: Since $\beta_h > \alpha_e$ in InP, the best structure calls for an injection of holes in the InP avalanche region. This determines the main characteristics of the structure, including the choice of an n-type substrate. The basic design was improved by the insertion of a grading layer between the avalanche and absorption layers, hence the acronym SAGM-APD, where G stands for grading. The purpose of this layer is to smooth out the valence-band discontinuity, which was found to trap holes at room temperature [29] and even to prevent operation at low temperature. Two important features need to be optimized to achieve high gain-bandwidth product APDs with low noise.

- The layer design should enable depletion of the InGaAs absorber with an electric field below the onset of tunnel currents (typically at 70 kV/cm, although 150 kV/cm is accepted at the very interface), and a large enough electric field at the InP avalanching junction (about 400 kV/cm), together with a thin n-type InP layer to keep the transit time acceptable. The product of the doping concentration and the depletion width in this layer has to be of the order of $2{-}3 \times 10^{12}$ cm^{-2} to produce APDs with a gain-bandwidth product of 60 GHz [14,30,31].

- Implementation of an efficient guard-ring structure to prevent edge-breakdown and inhomogeneous response, using for instance Cd or Be implants of low doping levels.

(ii) SAGM with MQW avalanche region: various SAM or SAGM structures have been investigated, employing lattice-matched AlInAs/InGaAs MQWs or MQWs with AlInGaAs or InGaAsP quaternary materials in the wells in order to reduce the dark current [32,33]. Very promising results have been demonstrated with these structures in terms of the gain-bandwidth product reaching about 150 GHz. This resulted in record receiver sensitivity of $-29\,\mathrm{dBm}$ at $10\,\mathrm{Gbit/s}$ [34]. Current developments are aimed at demonstrating passivated reliable devices (with guard-ring or equivalent structure and established MQW stability under $400\,\mathrm{kV/cm}$ field).

(iii) Other structures: AlInAs has also been studied as a multiplication layer with respect to high gain-bandwidth product APDs. Some advantages might result from a somewhat more favourable ratio of the ionization rates in this material compared to InP, but p-n junctions in InAlAs have not yet reached the quality of InP ones.

APDs are mostly interesting for applications in the $622\,\mathrm{Mbit/s}$–$10\,\mathrm{Gbit/s}$ range, where they offer a sensitivity margin of about 4–$7\,\mathrm{dB}$ over conventional pin photoreceivers. This is still less than the gain an optical preamplifier can provide (about 12–$15\,\mathrm{dB}$ in the same range), but at a lower cost. In order to extend the MQW-SAGM concept to bit rates higher than $10\,\mathrm{Gbit/s}$, edge-illuminated photodiodes have been demonstrated incorporating a thin absorption-layer structure. These devices exhibit high responsivity and large gain-bandwidth product ($160\,\mathrm{GHz}$ [35]), and thus open the way to APDs able to operate at $40\,\mathrm{Gbit/s}$.

4.5 Photoreceivers

The photoreceiver is made of two equally important parts: the photodiode and the preamplifier. The specific design of the preamplifier sets the main operating characteristics of the photoreceiver: bandwidth, sensitivity, dynamic range, etc. [36,37]. While one or another feature may be emphasized depending on the application, the sensitivity at a given bit rate is usually the most important one. For a given photodiode, the sensitivity depends on both the noise properties of the electronic amplifier technology and the parasitics associated with the packaging technology (Fig. 4.18).

4.5.1 Conventional Photoreceivers

The design of a conventional high-sensitivity photoreceiver is usually a two-step process. In the first step, the sensitivity is optimized by choosing those amplifier operating parameters giving the best S/N ratio at low optical power. This usually leads to a high input impedance of the preamplifier. Adjusting of the photoreceiver characteristics for the proper bandwidth then comes as the second step. This two-step procedure helps reach an input noise density much lower than the value set by the thermal noise given by (4.15c).

Signal-to-Noise Ratio. Provided that the input stage of the amplifier has a large enough gain, the S/N figure of the receiver is mainly determined by the characteristics of the photodiode and the input amplifying stage. The

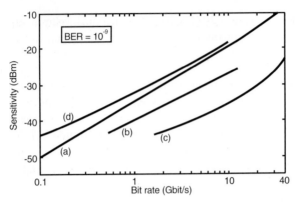

Fig. 4.18a–d. Typical present sensitivity trends for photoreceivers (RX) operating at 1.55 μm: (**a**) pin RX; (**b**) APD RX; (**c**) optical preamplifier + pin RX; (**d**) OEIC RX

amplifier can be represented as a noiseless ideal voltage-gain preamplifier with an equivalent noise source (with a spectral density $S_{iA}(f)$) and $R_{in}C_{in}$ dipole at the input. Then the photoreceiver can be sketched by the equivalent circuit of Fig. 4.19, and (4.15) can be written as

$$S/N = \frac{(MR_0P_{opt})^2}{\int_{\Delta f} [2qi_{0S} + 2q(i_{0B} + R_0P_{opt})M^2F(M) + 4kT/R_L + S_{iA}(f)]\, df} \quad . \quad (4.20)$$

A number of assumptions are made, including the fact that the same multiplication factor applies to the bulk dark current (i_{0B}) and to the photocurrent, while the surface current (i_{0S}) goes unmultiplied.

The preamplifier input noise density $S_{iA}(f)$ can be calculated for both a bipolar and a FET input stage. Basically this input noise source is comprised of two components, assumed to be uncorrelated for simplification, one being related to the input junction, the other to the main collector or drain current.

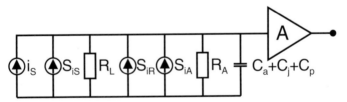

Fig. 4.19. Noise modelling of a photoreceiver: the noise equivalent source of the preamplifier appears as a current noise generator referred to the input for direct comparison with the signal current i_s and other noise sources. The total input capacitance includes the photodiode, parasitic and amplifier contributions

The equivalent input noise density for a bipolar input stage is then given by

$$S_{iA}(f) = 2qI_B + \frac{2kT(C_T\omega)^2}{g_m} .$$

(4.21)

In this expression C_T stands for the total input capacitance of the photoreceiver, including the photodiode and parasitic capacitances. The two terms on the right-hand side of the expression show inverse dependence on I_B. Following the required bandwidth of the photoreceiver, I_C ($= \beta I_B$, with β the current gain) can be tuned to minimize the total noise power of the receiver.

The bipolar receiver noise power can then be written in the form

$$\int_{\Delta f} S_{iA}(f)\,df = 8\pi \frac{C_T}{\sqrt{3\beta}} \Delta f^2 .$$

(4.22)

Worth noting is the fact that the noise increases with the square of the bandwidth, meaning that for a given signal optical power the S/N decreases quite rapidly with the required bandwidth. A figure of merit of the bipolar photoreceiver appears in the above relation as $C_T/\sqrt{\beta}$, which should be kept minimal for optimum sensitivity. This figure of merit, which points to the total input capacitance (pin + bipolar + interconnection) as a major parameter controlling the sensitivity, applies as long as the base resistance can be neglected.

For an FET the input noise density is given by the relationship

$$S_{iA}(f) = 2qI_G + \frac{2kT\gamma(C_T\omega)^2}{g_m} .$$

(4.23)

The adjustable parameter γ depends on the material and the structure, but remains close to unity. For small bandwidth applications ($< 1\,\mathrm{GHz}$), and provided the gate shot-noise can be neglected, the input noise of the FET amplifier is usually lower than that of a bipolar one; at higher frequency, since this input noise increases as Δf^3, FET amplifiers tend to become as noisy as their bipolar counterparts. In this frequency range, the figure of merit of the FET preamplifier is $C_T/\sqrt{g_m}$.

Photoreceiver Architecture. When choosing the best bias conditions for low noise, the preamplifier exhibits a high input impedance which, combined with the high impedance of the photodiode, usually leads to a low-bandwidth receiver. The proper bandwidth, matched to the signal spectrum, can then be retrieved by adding an equalizer stage behind the amplifier. Alternatively, a transimpedance amplifier can be used (Fig. 4.20), at the expense of some noise degradation resulting from the combined effect of the feedback resistor and its associated parasitic capacitance. The transimpedance amplifier usually offers the advantage of an enhanced dynamic range.

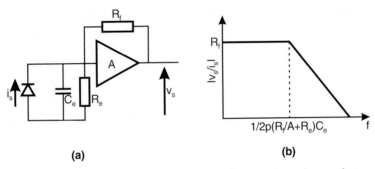

(a)   **(b)**

Fig. 4.20. (a) Schematic representation of a transimpedance photoreceiver with a bandwidth set by the input capacitance and the feedback resistor (R_f); (b) amplifier gain contribution

4.5.2   Specific Photoreceivers

The characteristics required from a photoreceiver depend on the application. For instance, sensitivity may be traded for dynamic range or increased linearity. In access passive optical networks or in future networks with packet switching, operation of receivers in the burst mode is required adding new demands. Of specific interest are the very broadband photoreceivers required to handle the high bit rates of 10 and 40 Gbit/s per channel. For such photoreceivers, the approach of Sect. 4.5.1 has to be modified.

In order to reduce the noise contribution from the drain or collector referred to the input (last term of (4.21) and (4.23)), a series inductance can be inserted between the photodiode and the input transistor to compensate their capacitances. The resulting resonance amplifies the signal at the transistor input, increases the gain and reduces the noise in the upper part of the bandwidth.

For the higher bit rates (20−40 Gbit/s), when the impedances in the amplifier are getting close to 50 Ω, the use of a distributed architecture is well suited [38], in particular for a low-capacitance FET amplifier. The distributed gate capacitance is inserted in a wide bandwidth transmission line loaded by its characteristic impedance, close to 50 or 100 Ω. Although such a low impedance makes the preamplifier quite noisy in the lower part of the spectrum, the average input noise density remains quite low (around 10 pA/$\sqrt{\text{Hz}}$) [39].

4.5.3   OEIC Photoreceivers

Since the early 1980s, monolithically integrated photoreceivers (OptoElectronic Integrated Circuit Photoreceivers) have been identified as important components for optical-fibre communications [40,41]. There are several reasons for this interest: compactness, lower cost, increased sensitivity through the suppression of interconnection parasitics at the photodiode/bias resistor/

transistor input node, and flatter response in the case of very large bandwidth photoreceivers. In order to integrate such OEIC photoreceivers, suitable devices and process technology have been identified and tested.

Devices for Integration. Si pin photodiodes and bipolar transistors have been integrated in phototransistors for several decades. When considering optical-fibre communications and III-V devices, the choice of the devices to be integrated is not straightforward (in particular similar sensitivities are expected from the use of FET and bipolar preamplifiers, see Fig. 4.21) and depends largely on the substrate, on the maturity of the technology and the integration technology (Figs. 4.22 and 4.23). Integrating devices with different thicknesses and different requirements in terms of lithography without compromising the fabrication yield is a real challenge.

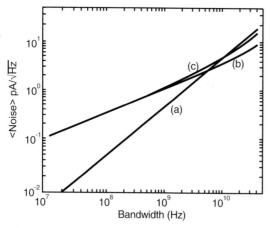

Fig. 4.21a–c. Noise density referred to the preamplifier input for a pin photodiode with $C_d = 0.1\,\mathrm{pF}$ and $R_s = 10\,\Omega$; (**a**) pin-FET with $f_t = 100\,\mathrm{GHz}$ and $C_{gs} = 0.1\,\mathrm{pF}$; channel noise factor $\gamma = 1.5$; (**b**) pin-HBT with $\beta = 80$, $C_{be} = 80\,\mathrm{fF}$ and $R_b = 0$; (**c**) same with $R_b = 50\,\Omega$. Low-frequency noise contribution is neglected; taken from [42]

OEIC Receivers on GaAs. While the main interest is presently focused on the $1.3-1.6\,\mu\mathrm{m}$ wavelength range, it is worth considering how the GaAs OEIC receiver technology evolved. First, OEIC photoreceivers incorporated pin photodiodes and MESFETs, pioneering the recessed substrate approach to planarize the wafer and ease the processing. It later appeared that using MSM photodiodes was much simpler, since such devices can be fabricated in the GaAs buffer layer (or semi-insulating undoped substrate). Moreover, this is at no additional cost since the MSM electrode can be deposited at the

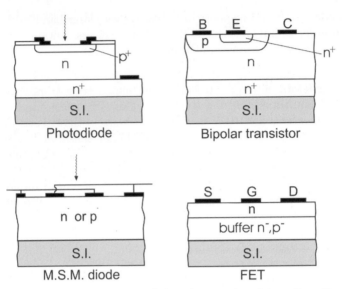

Fig. 4.22. Devices for monolithic photoreceiver integration. One should notice the similarity of MSM and FET structures (the n-channel layer of the latter one can be selectively fabricated, by ion-implantation for instance), while the base–collector junction of a bipolar transistor looks similar to a pin photodiode

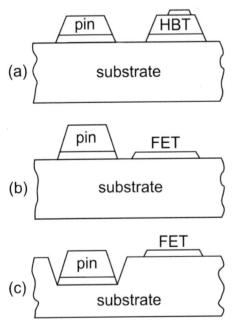

Fig. 4.23. Integration schemes for integrated photoreceivers: (**a**) pin-HBT; (**b**) pin-FET on a planar substrate; (**c**) pin-FET on a recessed substrate

same time as the gate electrodes. Good performance have been reported for such OEIC receivers [43].

OEIC Receivers on InP. The same approach is not applicable to InP-based OEIC photoreceivers intended for detection at 1.3−1.6 µm, as the InP substrate is transparent to these wavelengths. Moreover, the technology of pin photodiodes is better controlled than that of MSMs in the InP/InGaAs system. The simplest integration scheme in the InP system is based on the Heterojunction Bipolar Transistor (HBT) technology in that an InGaAs PIN photodiode can be implemented in the base–collector junction of the HBT preamplifier. Excellent performance (sensitivity of −17 dBm at 20 Gbit/s [44], operation at 40 Gbit/s [45]) has been reported using this scheme. The only drawback of this approach is that the responsivity of the photodiode has to be compromised with the HBT cut-off frequency (an f_t of 100 GHz implies a collector thickness lower than 400 nm; a 2 µm-thick collector would limit f_t to 30 GHz). Other integration schemes require the growth of different layers for the photodetector and the preamplifer, using selective epitaxy or a recessed substrate approach [46,47]. High-bandwidth photoreceivers exploiting this scheme have been demonstrated [48,49]. Indeed, for quite a long time the sensitivity of such receivers has been limited at moderate bit-rates by the large $1/f$ noise of InP-based FETs [50]. Present progress with both HBTs and HEMTs enables high performance OEIC photoreceivers to be fabricated. Their development is now pending the availability of a large-enough market and the demonstration of their reliability.

4.6 Conclusion

In the past decades, photodetectors for optical fibres have evolved from simple p-n junctions to sophisticated structures with improved performance in terms of responsivity, speed, and linearity, able to fulfill the more and more stringent requirements of transmission systems. In the past decade manufacturability and reliability of InGaAsP/InP pin and avalanche photodiodes was a main issue. More recently, characteristics such as speed and power handling capability have received much attention. On the other side, avalanche multiplication, actively studied in the early 1980s, lost part of its importance with the development of optical amplifiers. This situation may change again if the pressure on links' power budget increases. Indeed, more development efforts will be needed to accommodate the needs of systems the complexity of which keeps steadily increasing. Wavelength selective and agile photodiodes, for instance, are expected to become important devices for future networks. Improved knowledge of material physics, bandgap engineering, and integration technology will help coping with these new requirements.

148   A. Scavennec, L. Giraudet

References

1. J. Müller, "Photodiodes for optical communication," Adv. Electron. Electron, Phys. **55**, 189 (1981)
2. T. Pearsall, "Photodetectors for communication by optical fibres," in *Optical Fibre Communications* (M.J. Howes and D.V. Morgan, eds.), Chap. 6 (Wiley, New York 1980)
3. M. Dentan and B. De Crémoux, "Numerical simulation of the nonlinear response of a p-i-n photodiode under high illumination," J. Lightwave Technol. **8**, 1137 (1990)
4. K.J. Williams, R.D. Esman, and M. Dagenais, "Nonlinearities in p-i-n microwave photodetectors," J. Lightwave Technol. **14**, 84 (1996)
5. J.E. Bowers and C.A. Burrus, "Ultrawide-band long-wavelength p-i-n photodetectors," J. Lightwave Technol. **5**, 1339 (1987)
6. K. Kato, S. Hata, K. Kawano, J. Yoshida, and A. Kozen, "A high-efficiency 50 GHz InGaAs multimode waveguide photodetector," IEEE J. Quantum Electron. **28**, 2728 (1992)
7. J.B.D. Soole and H. Schumacher, "InGaAs metal-semiconductor-metal photodetectors for long wavelength optical communications," IEEE J. Quantum Electron. **27**, 737 (1991)
8. M.C. Hargis, S.E. Ralph, J. Woodall, D. McInturff, A.J. Negri, and P.O. Haugsjaa, "Temporal and spectral characteristics of back-illuminated InGaAs metal-semiconductor-metal photodetectors," IEEE J. Photon. Technol. **8**, 110 (1996)
9. J. Burm and L.F. Eastman, "Low-frequency gain in MSM photodiodes due to charge accumulation and image force lowering," IEEE Photon. Technol. Lett. **8**, 113 (1996)
10. A. Temmar, J.P. Praseuth, J.F. Palmier, and A. Scavennec, "Photodiode de type métal-semiconductor-métal (MSM) sur substrat d'InP," J. Phys. III **6**, 1059 (1996)
11. E. Dröge, E.H. Böttcher, D. Bimberg, O. Reimann, and R. Steingrüber, "70 GHz InGaAs metal-semiconductor-metal photodetectors for polarization-insensitive operation," Electron. Lett. **34**, 1421 (1998)
12. I.S. Ashour, H. El Kadi, K. Sherif, J.P. Vilcot, and D. Decoster, "Cut-off frequency and responsivity limitation of AlInAs/GaInAs MSM PD using a two dimensional bipolar physical model," IEEE Trans. Electron Devices **42**, 231 (1995)
13. R.J. McIntyre, "Multiplication noise in in uniform avalanche diodes," IEEE Trans. Electron Devices **13**, 164 (1966)
14. J.C. Campbell, S. Chandrasekhar, W.T. Tsang, G.J. Qua, and B.C. Johnson, "Multiplication noise of wide-bandwidth InP/InGaAs/InGaAs avalanche photodiodes," IEEE J. Lightwave Technol. **7**, 473 (1989)
15. C. Hu, K.A. Anselm, B.G. Streetman, and J.C. Campbell, "Noise characteristics of thin multiplication region GaAs avalanche photodiodes," Appl. Phys. Lett. **69**, 3734 (1996)
16. O. Hildebrand, W. Kuebart, and M.H. Pilkuhn, "Resonant enhancement of impact in $Ga_{1-x}Al_xSb$," Appl. Phys. Lett. **37**, 801 (1980)
17. B. Orsal, R. Alabedra, A. Maatougui, and J.C. Flachet, "$Hg_{0.56}Cd_{0.44}Te$ 1.6–2.5 μm avalanche photodiode and noise study far from resonant inpact ionization," IEEE Trans. Electron Devices **38**, 1748 (1991)

18. R. Chin, N. Holoniak Jr, G.E. Stillman, J.Y. Tang, and K. Hess, "Impact ionization in multilayered heterojunction structures," Electron. Lett. **16**, 467 (1980)

19. F. Capasso, W.T. Tsang, A.L. Hutchinson, and G.F. Williams, "Enhancement of electron impact ionization in a superlattice: a new avalanche photodiode with a large ionization rate ratio," Appl. Phys. Lett. **40**, 38 (1980)

20. M.C. Teich, K. Matsuo, and B.E. Saleh, "Excess noise factors for conventional and superlattice avalanche photodiodes and photomultiplier tubes," IEEE J. Quantum Electon. **22**, 1184 (1986)

21. T. Barrou, "Photodiode à avalanche à multi-puits quantiques AlInAs/GaInAs," Thesis (Orsay, France 1995)

22. D. Wake, R.H. Walling, and I.D. Henning, "Planar-junction, top-illuminated GaInAs/ InP pin photodiode with bandwidth of 25 GHz," Electron. Lett. **25**, 967 (1989)

23. J.C. Campbell, "Resonant-cavity photodetectors," *Proc. International Electron Device Meeting*, 575 (1995)

24. H. Fukano, K. Kato, O. Nakajima, and Y. Matuoka, "Low-cost, high-speed and high-responsivity photodiode module employing edge-illuminated refracting-facet photodiode," Electron. Lett. **35**, 842 (1999)

25. K. Kato, A. Kozen, Y. Muramoto, Y. Itaya, T. Nagatsuma, and M. Yaita, "110 GHz, 50%-efficiency mushroom-mesa waveguide p-i-n photodiode for a 1.55 µm wavelength," IEEE Photon. Technol. Lett. **6**, 719 (1994)

26. K.S. Giboney, R.L. Nagarajan, T.E. Reynolds, S.T. Allen, R.P. Mirin, M.J.W. Rodwell, and J.E. Bowers, "Travelling-wave photodetectors with 172 GHz bandwidth and 76 GHz bandwidth-efficiency product," IEEE Photon. Technol. Lett. **7**, 412 (1995)

27. V.M. Hietala, G.A. Vawter, T.M. Brennan, and B.E. Hammons, "Traveling-wave photodetectors for high-power, large-bandwidth applications," IEEE Trans. MTT **43**, 2291 (1995)

28. N. Shimizu, N. Watanabe, T. Furuta, and T. Ishibashi, "InP-InGaAs uni-traveling-carrier photodiode with improved 3-dB bandwidth of over 150 GHz," IEEE Photon. Technol. Lett. **10**, 412 (1998)

29. S.R. Forrest, O.K. Kim, and R.G. Smith, "Optical response time of $In_{0.53}Ga_{0.47}As/InP$ avalanche photodiodes," Appl. Phys. Lett. **41**, 95 (1982)

30. K. Taguchi, T. Torikai, Y. Sugimoto, K. Makita, and H. Ishihara, "Planar-structure InP/InGaAsP/InGaAs avalanche photodiodes with preferential lateral extended guard ring for $1.0-1.6\,\mu m$ wavelength optical communication use," J. Lightwave Technol. **6**, 1643 (1988)

31. J. Yu, L.E. Tarof, R. Bruce, D.G. Knight, K. Visvanatha, and T. Baird, "Noise performance of separate absorption, grading, charge and multiplication InP/InGaAs avalanche photodiodes," IEEE Photon. Technol. Lett. **6**, 632 (1994)

32. T. Kagawa, Y. Kawamura, and H. Iwamura, "A wide-bandwidth low-noise InGaAsP/InAlAs superlattice avalanche photodiode with a flip-chip structure for wavelengths of 1.3 and 1.55 µm," IEEE J. Quantum Electron. **29**, 1387 (1993)

33. I. Watanabe, M. Tsuji, M. Hayashi, K. Makita, and K. Taguchi, "Design and performance of InAlGaAs/InAlAs superlattice avalanche photodiodes," J. Lightwave Technol. **15**, 1012 (1997)

34. T.Y. Yun, M.S. Park, J.H. Han, I. Watanabe, and K. Makita, "10 Gigabit-per-second high sensitivity and wide dynamic range APD-HEMT optical receiver," IEEE Photon. Technol. Lett. **8**, 1232 (1996)
35. C. Cohen-Jonathan, L. Giraudet, A. Bonzo, and J.P. Praseuth, "Waveguide AlInAs/ GaAlInAs avalanche photodiode with a gain-bandwidth product over 160 GHz," Electron. Lett. **33**, 1492 (1997)
36. R.G. Smith and S.D. Personnick, *Receiver design for optical fibre communication systems*, Chap. 4 (Springer, New York 1980)
37. T. van Muoi, "Receiver design for high-speed optical-fibre systems," J. Lightwave Technol. **2**, 243 (1984)
38. Y. Miyamoto, M. Yoneyama, Y. Imai, K. Kato, and H. Tsunetsugu, "40 Gbit/s optical module using a flip-chip bonding technique for device interconnection," Electron. Lett. **34**, 493 (1998)
39. E. Legros, S. Vuye, L. Giraudet, and C. Joly, "High-sensitivity 40 Gbit/s photoreceiver using GaAs PHEMT distributed amplifiers," Electron. Lett. **34**, 1351 (1998)
40. O. Wada, T. Sakurai, and T. Nakagami, "Recent progress in optoelectronic integrated circuits (OEIC's)," IEEE J. Quantum Electron. **22**, 805 (1986)
41. S.W. Bland, "InP in integrated opto-electronics," in *Properties of Indium Phosphide* (Inspec IEE, London 1991) p. 467
42. A. Scavennec, L. Giraudet, and E. Legros, "InP electronic preamplifiers in photoreceiver applications", *Proc. InP and Related Materials*, Schwäbisch Gmünd, Germany, 427 (1996)
43. D.L. Rogers, "Integrated optical receivers using MSM detectors," J. Lightwave Technol. **9**, 1635 (1991)
44. L.M. Lunardi, S. Chandrasekhar, A.H. Gnauck, C.A. Burrus, and R.A. Hamm, "20-Gb/s monolithic p-i-n/HBT photoreceiver module for 1.55-µm applications," IEEE Photon. Technol. Lett. **7**, 1201 (1995)
45. M. Bitter, R. Bauknecht, W. Hunziker, and H. Melchior, "Monolithically integrated 40 Gb/s InP/InGaAs PIN/HBT optical receiver module," *Proc. 11th International Conference on InP and Related Materials* Davos, Switzerland, 381 (1998)
46. H. Yano, G. Sasaki, M. Murata, and H. Hayashi, "An ultra-high-speed optoelectronic integrated receiver for fibre-optic communications," IEEE Trans. Electron Devices **39**, 2254 (1992)
47. W. Kuebart, J.-H. Reemtsma, D. Kaiser, H. Grosskopf, F. Besca, G. Luz, W. Körber, and I. Gyuro, "High sensitivity InP-based pin-HEMT receiver-OEIC's for 10 Gb/s," IEEE Trans. MTT **43**, 2334 (1995)
48. A. Umbach, S. Van Waasen, U. Auer, H.G. Bach, R.M. Berttenburg, V. Breuer, W. Ebert, G. Janssen, G.G. Mekonnen, W. Passenberg, W. Schlaak, C. Schramm, A. Seeger, F.-J. Tegude, and G. Unterbörsch, "Monolithic pin-HEMT 1.55 µm photoreceiver on InP with 27 GHz bandwidth," Electron. Lett. **32**, 2142 (1996)
49. K. Takahata, Y. Muramoto, H. Fukano, K. Kato, A. Kozen, O. Nakajima, and Y. Matsuoka, "46.5-GHz-bandwidth monolithic receiver OEIC consisting of a waveguide p-i-n photodiode and a HEMT distributed amplifier," IEEE Photon. Technol. Lett **8**, 1150 (1998)
50. A. Scavennec, M. Billard, P. Blanconnier, E. Caquot, P. Carer, L. Giraudet, L. Nguyen, F. Lugiez, and J.P. Praseuth, "InGaAs/InP monolithic photoreceivers for 1.3−1.5 µm optical fibre transmission," Proc. SPIE **39**, 168 (1989)

5   Optical Amplifiers

Mikhail N. Zervas and Gerlas van den Hoven

Introduction – Need for Amplification

The development of low-loss optical silica fibres and reliable semiconductor
laser sources in the early 1970s have enabled the practical realisation and
massive deployment of optical-fibre communications worldwide. However,
high bit-rate transmission is severely limited by the fibre residual loss and
dispersion [1]. Fibre non-linearities, on the other hand, can further limit
high bit rates over long distances [2]. Figure 5.1 shows a typical attenuation
and dispersion spectrum of single-mode, standard-telecommunication fibres
(STF) and dispersion-shifted fibres (DSF). In the second telecommunications
window (centered around 1.31 μm), STFs exhibit very low dispersion and
propagation losses of about 0.35 dB/km. Use of DSFs can shift the zero-
dispersion point in the third telecommunication window (centered around
1.55 μm), where the propagation loss is about 0.2 dB/km.

The transmission limits due to propagation loss and chromatic dispersion
in STFs and DSFs are shown in Fig. 5.2, where the non-regenerated (i.e.
non-amplified, non-dispersion-compensated) distance is plotted as a function

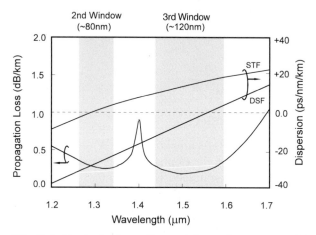

Fig. 5.1. Typical attenuation and dispersion spectra of single-mode, standard telecom (STF) and dispersion shifted (DSF) fibres

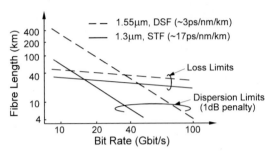

Fig. 5.2. Transmission-distance limits, imposed by transmission loss and chromatic dispersion in STFs and DSFs, as a function of data bit-rate

of the data bit rate. It is shown that for bit rates lower than 10 Gbit/s, the non-regenerated transmission distance in both windows is limited primarily by the fibre loss. At higher bit-rates, fibre chromatic dispersion is proven to be the dominant limiting factor. From Fig. 5.2, it is obvious that it is impossible to transmit high bit-rate data over long distances without proper amplification and dispersion compensation schemes.

This chapter will concentrate on the various fibre amplification schemes. Rare-earth-doped fibre amplifiers can be used to overcome propagation losses at both the 2nd and 3rd telecommunication windows. Erbium-doped fibre amplifiers have been successfully used for amplification in the 3rd telecommunication window, while praseodymium-and neodymium-doped fibre amplifiers can be used in the 2nd window. In addition, nonlinear optical fibre amplifiers, such as Raman amplifiers, can in principle be used throughout the optical spectrum. These fibre-based optical amplifiers will be treated in Sect. 5.1.

Another class of optical amplifiers that can potentially be used for amplification at both telecom windows relies on semiconductors. Such devices, commonly referred to as semiconductor optical amplifiers (SOAs) will be covered extensively in Sect. 5.2.

5.1   Optical Fibre Amplifiers

5.1.1   Erbium-Doped Fibre Amplifiers

The invention of the erbium-doped fibre amplifier (EDFA) in 1987 [3,4] was one of the most important recent developments that has revolutionized optical telecommunications. Diode-pumped EDFAs [5,6], in particular, have accelerated the deployment of high-capacity optical fibre terrestial and undersea-cable systems worldwide. The basic EDFA configuration is shown in Fig. 5.3. The diode pump output and the incoming signal are combined through a dichroic coupler and are launched into a short length of optical fibre containing a low concentration of erbium ions. Pump photons excite

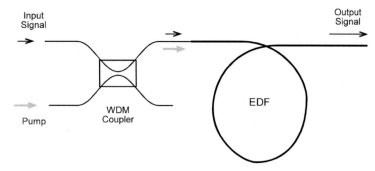

Fig. 5.3. Basic EDFA configuration

the rare-earth ions which, in turn, give efficient signal amplification in the 1.55 μm wavelength region, provided that adequate population inversion is achieved.

EDFAs are shown to provide extremely high gain, low (quantum-limited) noise figure and high saturation output powers over a wide bandwidth. The polarization sensitivity is extremely small and the signal cross-talk insignificant. These general characteristics render EDFAs ideal for use as boosters, inline amplifiers and preamplifiers. On the other hand, EDFAs are transparent to the data format and can simultaneously accommodate a large number of wavelength-division-multiplexed (WDM) channels. The use of EDFAs in terrestial networks and undersea systems can significantly increase the transmission capacity, operational flexibility and additional network functionality and reliability.

Energy Levels of Er^{3+} in Fused Silica. The observed infrared and visible optical spectra of Er^{3+} ions result primarily from electron transitions between 4f states [7–9]. The energy level diagram of 'free' Er^{3+} ions is shown in Fig. 5.4a. Each level is labeled with the corresponding Russell-Saunders coupling term $(^{2S+1}L_J)$. The metastable level $(^{4}I_{13/2})$ is responsible for the optical gain in the 3rd telecommunication window centered around 1550 nm. The higher energy levels correspond to different optical pumping wavelengths.

In a multi-electron atom, the triplet (S, L, J) corresponds to the total spin, total orbital angular momentum and total angular momentum, respectively, of a particular energy level. The quantum number L is denoted by the letter S, P, D, F, G, H, I, ... for $L = 0, 1, 2, 3, 4, 5, 6, ...$, respectively. Each energy level is comprised of $2J+1$ degenerate quantum states (level multiplicity). The number of spin configurations, on the other hand, is $2S+1$ (spin multiplicity). All the energy levels of the 'free' ion, shown in Fig. 5.4a, have the same parity, which implies that electric-dipole transitions between these levels are totally forbidden. Weak magnetic-dipole transitions, however, can still take place.

When an Er^{3+} ion is embedded in a crystalline host material, it is subjected to electric fields due to the neighbouring atoms of the crystalline lattice.

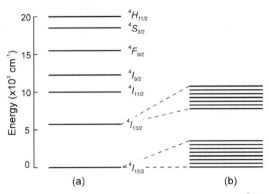

Fig. 5.4. (a) Energy level diagram of 'free' Er^{3+} ions, (b) Stark-split sublevels for ground-state and metastable levels

These electric fields are known as *crystal* or *ligand fields*. There are two main effects associated with the presence of the crystal fields. Firstly, due to the Stark effect, the $2J + 1$ degenerate quantum states of the SLJ energy level are split into a number of sub-levels, called Stark levels. The exact number of Stark levels depends on the symmetry of the crystal field, the maximum number being $g = J + 1/2$. In the presence of a magnetic field, each Stark level can be further split into two sub-levels (Zeeman effect), and, in that way, the energy-level degeneracy is fully lifted. Secondly, the crystal field breaks the inversion symmetry of the ion's environment and enables electric-dipole transitions (typically of the order of 10^{-6}) to now take place between Stark levels of different SLJ energy levels. These two effects are quite important from the optical-amplification point of view. The latter enables radiative transitions between various energy levels and, therefore, amplification to take place. The first effect, on the other hand, guarantees that amplification occurs over certain (wide) bandwidth.

For erbium ions in a crystalline material, the ground-state $(^4I_{15/2})$ and metastable $(^4I_{13/2})$ levels are split into a maximum of 8 and 7 Stark sub-levels, respectively (see Fig. 5.4b). The separation between the various sub-levels is quite small ($\sim 20-80\,cm^{-1}$) and lie well within the phonon spectrum of most host materials. At room temperature, all the Stark levels within a given multiplet are phonon-coupled, enabling fast transitions (in a fs timescale) between Stark sub-levels to take place (thermalisation effect). The fast thermalisation effect results in a population distribution within the ground-state and metastable manifolds that follow Boltzmann temperature-dependent statistics [10]. The strong phonon coupling between the various Stark-split sub-levels results in *homogeneous broadening* of the particular energy level.

When rare-earth ions, such as Er^{3+}, are incorporated in an amorphous material, such as fused silica, the various energy-level multiplets are subjected to another important broadening mechanism, known as *inhomogeneous broadening*. This type of broadening reflects spectroscopic differences between individual rare-earth ions that result from the fact that each ion occupies a unique site in the glass-host matrix and is, therefore, subjected to a different crystal (ligand) field.

Both types of line broadening have important implications on the amplifier spectral-gain shape and saturation performance, as well as, the laser quantum efficiency. Homogeneous broadening results in a uniform saturation thoughout the entire amplifier-gain spectrum and lower ASE build-up (higher efficiency). Homogeneous broadening also results in power amplifiers and lasers with near-quantum-limited performance [8]. Inhomogeneous broadening, on the other hand, may result in increased pump-power requirements, pump- and/or signal-wavelength dependent gain spectrum, as well as more gradual large-signal saturation.

Both types of broadening will be affected by the composition of the host material. For Al^- and Al/Ge-codoped silica glasses, for example, the lower Stark sub-levels of the ground-state and metastable levels are split by $\sim 20-50\,cm^{-1}$. The inhomogeneous broadening linewidth is $\sim 40-50\,cm^{-1}$, while, at room temperature, the homogeneous broadening linewidth is $\sim 50\,cm^{-1}$. This implies that a behaviour with mixed homogeneous and inhomogeneous characteristics is expected. However, in most applications, it has been proven experimentally that the $^4I_{15/2} - {}^4I_{13/2}$ erbium transition is predominantly homogeneously broadened.

There are two important parameters characterising the spontaneous emission from a certain atomic or ionic level to lower ones, in the absence of any parasitic nonradiative effect. These are the radiative lifetime of this level and the corresponding branching ratio. The radiative lifetime is related to spontaneous emission processes. The probability of a spontaneous transition between a high manifold a to a low one b is given by the Einstein coefficient $A_{a,b}$ [7,8,10]. Transitions are quite often expressed in terms of the oscillator strength $f_{a,b}$, a dimensionless parameter which is proportional to $A_{a,b}$. The oscillator strength of a fully allowed electric-dipole transition is about one. In the case of erbium ions, however, where, as already mentioned, emission and absorption effects are due to weak electric-dipole (comparable to magnetic-dipole) transitions, the oscillator strength is about 10^{-6} [7]. In the case where there are a number of lower LSJ manifolds into which an excited ion can decay radiatively, the radiative (fluoresence) lifetime of the upper manifold a is given by

$$\tau_a = \frac{1}{\Gamma_a}\,, \tag{5.1}$$

where

$$\Gamma_a = \sum_f A_{af} \tag{5.2}$$

is the total spontaneous emission rate out of manifold a. The branching ratio β_{ab} to a lower level b is given by

$$\beta_{ab} = \frac{A_{ab}}{\Gamma_a} = A_{ab}\tau_a \tag{5.3}$$

The branching ratio gives the probability for a radiative decay from the upper manifold a is to a lower manifold b. In erbium ions, all the radiative transitions take place between the metastable and ground manifolds and, therefore, the branching ratio is one. Each radiative transition to a lower state is accompanied by the spontaneous emission of a photon at a frequency corresponding to the energy gap between the two levels. The branching ratio affects significantly the efficiency of an amplifier and the threshold of a laser.

Absorption and Emission Cross-Sections. In addition to spontaneous emission, active ions can provide stimulated emission and stimulated absorption of photons. The interaction of light of certain frequency and the active rare-earth ions is fully described in terms of absorption and emission cross-sections. The probability per unit time that an ion in a lower manifold b absorbs a photon of frequency ν and is excited to an upper manifold a is given by

$$P_{ba} = \sigma_{ba}(\nu)\Phi(\nu) , \tag{5.4}$$

where $\sigma_{ba}(\nu)$ and $\Phi(\nu)[= I_{\mathrm{p}}(\nu)/h\nu]$ is the absorption cross-section and photon flux at frequency ν, respectively. The cross-section appearing in (5.4) is obtained by summing up the strengths of all the individual Stark transitions, weighted by the occupancy probability of each Stark-split sublevel [8]. If N_b is the ion density (population) of level b, the absorption rate, i.e. the total number of absorptions per unit time, between manifolds a and b is given by

$$R_{ba}(\nu) = N_b P_{ba}(\nu) = N_b \sigma_{ba}(\nu)\frac{I_{\mathrm{p}}(\nu)}{h\nu} , \tag{5.5}$$

where $I_{\mathrm{p}}(\nu)$ is the pump light intensity at frequency ν. Incoming signal photons, on the other hand, can de-excite ions by forcing an electron transition from level a back to level b. Such an electron transition is accompanied by a stimulated photon emission at the signal frequency. The probability per unit time that an ion in an excited level a will emit a photon and be de-excited to lower level b, through a stimulated process, is given by

$$P_{ab}(\nu) = \sigma_{ab}(\nu)\Phi_{\mathrm{s}}(\nu) , \tag{5.6}$$

where $\sigma_{ab}(\nu)$ and $\Phi_{\mathrm{s}}(\nu)[= I_{\mathrm{s}}(\nu)/h\nu]$ is the emission cross-section and signal photon flux at frequency ν, respectively. The emission cross-section appearing in (5.6) is again a weighted average over the strengths of all the individual Stark transitions [8]. If N_a is the ion density (population) of level a, the stimulated emission rate, i.e. the total number of stimulated emissions per unit time, between manifolds a and b is given by

$$W_{ab}(\nu) = N_a P_{ab}(\nu) = N_a \sigma_{ab}(\nu)\frac{I_{\mathrm{s}}(\nu)}{h\nu} , \tag{5.7}$$

where $I_s(\nu)$ is the signal light intensity at frequency ν. Stimulated emission cross-sections are related to the fluoresence lifetime and branching ratio of the excited level a through the relation

$$\frac{\beta_{ab}}{\tau_a} = \frac{8\pi n^2}{c^2}\nu^2\sigma_{ab}(\nu)\,d\nu = 8\pi n^2 c\frac{\sigma_{ab}(\lambda)}{\lambda^4}\,d\lambda\ . \tag{5.8}$$

Exact knowledge of the absorption and emission cross-section spectra is essential for the accurate description and detailed evaluation of the performance of optical amplifiers and lasers [10–13]. Emission and absorption cross-sections can be accurately determined by direct gain-loss and saturation power measurements [14]. Absolute cross-sections and accurate spectral information can also be obtained by using the McCumber theory [15]. Use of the Ladenburg-Fachtbauer relation [16], on the other hand, has been shown to give quite inaccurate cross-section values [14,15]. Figure 5.5 shows typical Er^{3+} normalized emission and absorption cross-section spectra pertinent to a germano-silicate host. The normalisation is with respect to the maximum absorption cross-section, which occurs at around 1532 nm and has a typical value of 7.5×10^{-25} m^2. Host material is shown to have a significant effect on the absolute values and shapes of both spectra [14].

Two-Level Model – Propagation Equations. In Fig. 5.6 a simplified energy level diagram for erbium ions in silica-based hosts is shown, which involves only the lowest and most important pump and signal energy levels. When a 1480 nm pump is used, due to the fast intraband thermalisation process, the amplification process can be fully described with a two-level model. Strictly speaking, pumping into the 980 nm absorption band will require a three-level-model description of the amplification process. However, for average pump powers less than 1 W, the fast non-radiative decay from the $^4I_{11/2}$ manifold to the $^4I_{13/2}$ manifold results in a negligible population

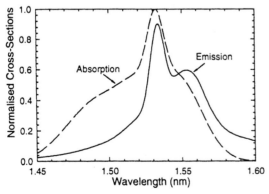

Fig. 5.5. Typical normalized emission and absorption erbium cross-sections

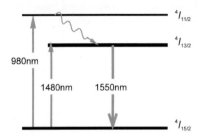

Fig. 5.6. Simplified energy levels for erbium ions

for the $^4I_{11/2}$ manifold ($N_3 \approx 0$). Therefore, in all practical situations, the amplification process can be fully described with a two-level model for both the 980 nm and 1480 nm pumping wavelengths.

Assuming homogeneous broadening and erbium ions uniformly distributed across the fibre core, the wavelength-dependent amplifier performance is greatly simplified by introducing two principal fibre parameters, namely, the fibre absorption (α_k) and gain (g_k) spectra [11], which are given by

$$\alpha_k = \sigma_{ak}\Gamma_k N_t \, , \tag{5.9a}$$

$$g_k = \sigma_{ek}\Gamma_k N_t \, , \tag{5.9b}$$

where σ_{ak} and σ_{ek} are the absorption and emission cross-sections, respectively, at frequency ν_k. N_t is the total erbium-ion concentration and Γ_k is the overlap integral between the dopant and optical mode distributions, namely

$$\Gamma_k = \int_0^{2\pi} \int_0^b I_k(r, \phi) r \, dr \, d\phi \, . \tag{5.10}$$

$I_k(r, \phi)$ is the normalized electric field intensity and b is the core radius. Under steady-state conditions of operation, the wavelength-dependent EDFA response is given by the following spectrally resolved propagation equations:

$$\frac{dP_k^\pm}{dz} = \pm(\alpha_k + g_k)\frac{N_2}{N_t}P_k(z) \pm g_k\frac{N_2}{N_t}mh\nu_k\Delta\nu_k \mp (\alpha_k + \alpha_k^*)P_k(z) \, , \tag{5.11}$$

where

$$\frac{N_2}{N_t} = \frac{\sum_k \frac{\alpha_k}{\alpha_k+g_k}\frac{P_k(z)}{P_k^{\text{sat}}}}{1 + \sum_k \frac{P_k(z)}{P_k^{\text{sat}}}} \tag{5.12}$$

gives the population of the metastable level as a function of the local signal, pump and ASE powers. The population of the ground state is $N_1 = N_t - N_2$. $P_k(z)$ is the power at frequency ν_k at certain position z along the amplifier length. The second term in (5.12) represents the spontaneous emission generated by the excited ions over a bandwidth $\Delta\nu_k$. α_k^* is the background propagation loss. The parameter m gives the total number of spatial and polarization guided modes supported by the fibre core. For single-mode fibres, therefore, $m = 2$. The $\pm$ sign in (5.11) corresponds to forward and backward propagation, respectively. The power at frequency ν_k is normalised to the corresponding intrinsic saturation power P_k^{sat}, given by

$$P_k^{\text{sat}} = \frac{h\nu_k A_{\text{eff}} N_t}{(\alpha_k + g_k)\tau} ,\tag{5.13}$$

where $A_{\text{eff}} = \pi b^2$ and τ is the metastable level lifetime. Using the effective overlap approximation, the EDFA performance can be fully described by knowing α_k, g_k, α_k^* and P_k^{sat}. The intrinsic saturation power can be determined experimentally by single-frequency transmission measurements [14,17,18]. Equations (5.12) and (5.13) apply to pump and signal wavelengths and fully describe the evolution of forward and backward ASE [19]. They are valid under any pumping and saturation conditions and they are numerically integrated by standard numerical methods. The full EDFA bandwidth (BW) is sliced into M equal segments $\Delta\nu_k = BW/M$. Typically, $\Delta\nu_k = 125\,\text{GHz}$ ($\Leftrightarrow 1\,\text{nm}$). In case of only one pump and one signal wavelength, the numerical model involves the integration of $2(M+1)$ equations. When the exact ASE spectral shape is not of primary interest, the number of propagation equations can be reduced considerably by introducing an effective ASE bandwidth and considering the effect of the ASE separately from that of the signal. In this case, the number of differential equations reduces to four [11,20–22].

In cases of moderate gain (usually $< 20\,\text{dB}$) or relatively strong input signal power (usually $> -20\,\text{dBm}$) the ASE is negligible and the gain performance of the EDFA can be accurately described by a much simpler analytical model [17,18]. Under these gain conditions and negligible background loss, the propagation equations (5.11) can be easily re-arranged and integrated by parts to give the output power at frequency ν_k, namely

$$P_k^{\text{out}} = P_k^{\text{in}} \exp\{-\alpha_k L\} \exp\left\{\frac{P_{\text{in}} - P_{\text{out}}}{P_k^{\text{sat}}}\right\} ,\tag{5.14}$$

where P_k^{in} is the input power at the same frequency (see Fig. 5.7) and L is the erbium-doped fibre length. P_{in}, P_{out} are the total input and output powers,

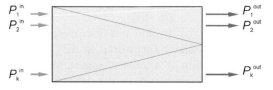

Fig. 5.7. Black-box representation for the analytic-model description of an EDFA

respectively, expressed as

$$P_{\text{in}} = \sum_{k=1}^{N} P_k^{\text{in}}, \quad P_{\text{out}} = \sum_{k=1}^{N} P_k^{\text{out}}. \tag{5.15}$$

By summing up equations (5.14) over k, a single transcendental equation is obtained, namely

$$P_{\text{out}} = \sum_{k=1}^{N} P_k^{\text{in}} \exp\{-\alpha_k L\} \exp\left\{\frac{P_{\text{in}} - P_{\text{out}}}{P_k^{\text{sat}}}\right\}, \tag{5.16}$$

which can be easily solved to obtain the total output power P_{out}. By substituting back into (5.14), the output power at each different frequency is finally obtained. While the propagation-equation model (5.11, 5.12) gives the detailed evolution of the various EDFA parameters, i.e. pump, signal(s), ASE, population inversion, local gain, etc., along the amplifier length, the analytic model (5.14)–(5.16) provides pump and signal(s) powers at the output terminals only (black-box representation). However, the accuracy of both models relies on the exact knowledge of the characteristic parameters α_k and P_k^{sat} at each wavelength λ_k. In the case that only one (either pump or signal) wavelength is transmitted through the EDFA, equations (5.14)–(5.16) reduce to a single equation giving the transmissivity at this wavelength, namely

$$T_k = \exp\{-\alpha_k L\} \exp\left\{\frac{(1 - T_k)P_k^{\text{in}}}{P_k^{\text{sat}}}\right\}, \tag{5.17}$$

where $T_k = P_k^{\text{out}}/P_k^{\text{in}}$. Both α_k and P_k^{sat} can be obtained simultaneously by fitting (5.17) into single-wavelength transmission experimental data [17]. For small input signals ($P_k^{\text{in}} \ll P_k^{\text{sat}}$), (5.17) reduces to $T_k^{\text{min}} \approx \exp\{-\alpha_k L\}$, which implies that α_k can be determined directly from the small-signal transmissivity value. For input power equal to the intrinsic saturation power ($P_k^{\text{in}} = P_k^{\text{sat}}$), (5.17) reduces to $T_k = \exp(-\alpha_k(L))\exp(1 - T_k)$. For high signal absorption (i.e. $\alpha_k L \gg 1$), which implies that $T_k \ll 1$, the transmissivity is finally approximated by $T_k \approx T_k^{\text{min}}e$. Under these conditions, the intrinsic saturation power corresponds to the input power that results in a transmissivity increase by a factor of e (or 4.34 dB), compared with the small-signal value.

The meaning and implications of the intrinsic saturation power differ in the case of small absorption (i.e., $\alpha_k L \ll 1$). In the latter case, the transmissivity is quite high ($T_k \approx 1$) and, for $P_k^{\text{in}} = P_k^{\text{sat}}$, (5.17) reduces to $T_k \approx \exp(-1/2\alpha_k L)$ or, equivalently, T_k (in dB) $= 1/2T_k^{\text{min}}$ (in dB). In the small absorption case, the intrinsic saturation power corresponds to the input power that results in a transmissivity (expressed in dB) increase by a factor of 2. In this case the inversion is almost constant along the entire amplifier length. Using equations (5.11) and (5.12), it can be easily shown that, in this case, the constant populations of the metastable

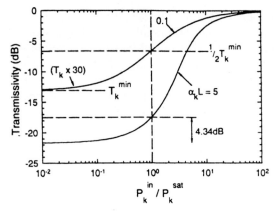

Fig. 5.8. Transmissivity versus normalised input power (P_k^{in}/P_k^{sat}) for $\alpha_k L = 0.1$ and 5

and ground levels are given, respectively, by $N_2/N_t = \alpha_k/2(\alpha_k + g_k)$ and $N_1/N_t = (\alpha_k + g_k)/2(\alpha_k + g_k)$. At $\lambda_k = 980\,\text{nm}$, where $g_k = 0$, the populations become equal $(N_2 = N_1 = N_t/2)$ [14]. Equation (5.17) is plotted in logarithmic scale in Fig. 5.8 for two extreme cases corresponding to low $(\alpha_k L = 0.1)$ and high $(\alpha_k L = 5)$ small-signal absorption. In the former case, the transmissivity is multiplied by a factor of 30 to facilitate visualization of the P_k^{sat} effect.

Noise Figure. Doped fibre amplifiers are linear quantum devices whose performance is inevitably limited by noise generation. In linear quantum amplifiers, the dominant noise component is quantum mechanical in nature and appears as a direct consequence of the uncertainty principle [23,24]. Starting with the photon statistics master equation [25], and assuming a coherent optical signal with mean photon number $\langle n(0) \rangle$, at the input of the amplifier, the mean and variance of the photon number, at the output of a lossless amplifier, are given by [26,27]

$$\langle n(z) \rangle = G(z) \langle n(0) \rangle + mN(z) , \tag{5.18}$$

$$\sigma(z)^2 = \langle n(z)^2 \rangle - \langle n(z) \rangle^2 = G(z)\langle n(0) \rangle + mN(z)$$
$$+ 2G(z)\langle n(0) \rangle N(z) + mN(z)^2 , \tag{5.19}$$

where

$$G(z) = \exp\left\{ \int_0^z [\sigma_e N_2(\zeta) - \sigma_a N_1(\zeta)] \, d\zeta \right\} \tag{5.20}$$

is the amplifier gain, and

$$N(z) = G(z) \int_0^z \frac{\sigma_e N_2(\zeta)}{G(\zeta)} \, d\zeta = [G(z) - 1] + G(z) \int_0^z \frac{\sigma_a N_1(\zeta)}{G(\zeta)} \, d\zeta \tag{5.21}$$

is the ASE output photon number per mode. The signal is supposed to be single-moded while the ASE populates the total number (m) of propagating spatial and polarization eigenmodes (note that $m = 2$ for single mode fibres). The first and second terms in the photon number variance (5.19) represent the signal and ASE shot noise, respectively. The third and fourth terms, on the other hand, give the signal-spontaneous (s-sp) and spontaneous-spontaneous (sp-sp) beat noise components, respectively. Beat noise is a term borrowed from the semi-classical description of noise generation and results from the signal-ASE interference at the end of the amplification process, at the (ideal) square-law detector [28]. In the quantum description, however, these terms are of a more fundamental character, reflecting the statistical nature of the stimulated and spontaneous emission processes and, in that respect, they occur throughout the amplification process [27]. It is important to stress that the $2G\langle n(0)\rangle N$ 'beat' term appears even in the ideal case of amplification without spontaneous emission ('noiseless' amplifier) and gives the output photon number variance increase due to the stimulated emission process [27].

The amplifier noise performance is usually characterized in terms of the spontaneous emission factor n_{sp}, which is defined as

$$
n_{sp}(z) = \frac{N(z)}{G(z) - 1} = \frac{G(z)}{G(z) - 1} \int_0^z \frac{\sigma_e N_2(\zeta)}{G(\zeta)} \, d\zeta
$$

$$
= 1 + \frac{G(z)}{G(z) - 1} \int_0^z \frac{\sigma_a N_1(\zeta)}{G(\zeta)} \, d\zeta \; . \tag{5.22}
$$

In the special case of constant population inversion (N_1 and N_2 const.) along the entire amplifier length, it is easily shown that the spontaneous emission factor takes the form

$$
n_{sp} = \frac{\sigma_e N_2}{\sigma_e N_2 - \sigma_a N_1} = 1 + \frac{\sigma_a N_1}{\sigma_e N_2 - \sigma_a N_1} \; . \tag{5.23}
$$

The noise power per mode, over a bandwidth B_0, introduced during the amplification process, is given by

$$
P_N = h\nu N(z) B_0 = n_{sp} h\nu \left[G(z) - 1\right] B_0 \; . \tag{5.24}
$$

For a fully inverted medium ($N_1 = 0$), however, $n_{sp} = 1$ and the noise power reduces to a minimum value $P_N^{min} = h\nu(G - 1)B_0$. This is known to result from zero-point energy amplification and it is the absolute minimum noise power generated by a quantum amplifier. It is further shown that such minimum noise power corresponds to both the input and output signals being characterized by the minimum uncertainty relation ($\Delta n \Delta \varphi = 1/2$) [23]. In a fully inverted medium, amplification results from stimulated emission only and that results in minimum added noise.

For a partially-inverted amplifying medium ($N_1 \neq 0$ and ($\sigma_e N_2 > \sigma_a N_1$), on the other hand, the spontaneous emission factor is $n_{sp} > 1$, which results in an increased noise performance. In a partially inverted medium, the amplification process involves both stimulated emission ($\sigma_e N_2$) and stimulated absorption ($\sigma_a N_1$) of signal photons. From (5.22) and (5.23) it is obvious that the increase in n_{sp} results from the stimulated absorption contribution. For a certain input-signal photon number $\langle n(0) \rangle$ and amplifier gain G, a partially inverted medium will require an increased number of stimulated absorptions and emissions which will both contribute to an increased noise term $N(z)$ and photon-number variance (through the term $2G\langle n(0) \rangle N$), even in the absence of spontaneous emission. The unavoidable presence of spontaneous emission affects the amplifier noise performance indirectly in that it reduces significantly the amplifier mean population inversion. In case of high gain ($G \gg 1$), the noise power can be considered as resulting from the amplification of n_{sp} input photons (equivalent input noise).

The noise performance of an optical amplifier is characterized by a measurable quantity referred to as the optical noise figure ($\mathrm{NF_{opt}}$). $\mathrm{NF_{opt}}$ is referenced to an ideal photodetector and gives the signal-to-noise ratio (SNR) change due to the amplification process alone. It is defined as the ratio

$$\mathrm{NF_{opt}} = \frac{\mathrm{SNR_{in}}}{\mathrm{SNR_{out}}}, \tag{5.25}$$

where $\mathrm{SNR_{out}}$ ($= G^2 \langle n(0) \rangle^2 / \sigma(z)^2$) and $\mathrm{SNR_{in}}$ ($= \langle n(0) \rangle^2 / \sigma(0)^2$) are the signal-to-noise ratios at the output and input of the amplifier, respectively. After substituting (5.18) and (5.19) into (5.25), the following expression is obtained for the noise figure

$$\mathrm{NF_{opt}} = 2n_{sp} \frac{[G(z) - 1]}{G(z)} + \frac{1}{G(z)}, \tag{5.26}$$

where $n_{sp} = P_N / [(G - 1)h\nu B_0]$ and $P_N = P_{\mathrm{ASE}}^{\mathrm{total}} / m$. $P_{\mathrm{ASE}}^{\mathrm{total}}$ is the total ASE power distributed over all propagating modes (for single mode fibres $m = 2$) [29]. In deriving (5.26), we have assumed that the input signal is coherent (governed by Poisson statistics) and quite strong so that the spontaneous shot noise and spontaneous-spontaneous beat noise components can be neglected. In reality, the amplifier output is optically filtered before being detected, which further reduces the significance of these terms. The remaining first and second terms in (5.26) correspond to output signal-spontaneous beat and output signal shot noise, respectively. For high-gain amplifiers ($G \gg 1$), the optical noise figure is approximated by

$$\mathrm{NF_{opt}} = 2n_{sp} \tag{5.27}$$

which implies that $\mathrm{NF_{opt}} \geq 2$. The minimum value $\mathrm{NF_{opt}^{min}} = 2$ ($\rightarrow 3\,\mathrm{dB}$) is obtained with high-gain, fully inverted ($n_{sp} = 1$) amplifiers and is usually referred to as the 3-dB quantum limit. It should be stressed that, as (5.26) implies, noise figures below the 3-dB quantum limit can be achieved only at the expense of reduced gain.

In the case of unsaturated constant population inversion, taking into account the pump- and signal-power relative contributions into populations N_1 and N_2, (5.23) and (5.27) result in the following simplified expression for the high-gain optical noise figure, namely

$$\mathrm{NF}_{\mathrm{opt}} = \frac{2}{1 - \frac{\sigma_{ep}\sigma_{as}}{\sigma_{ap}\sigma_{es}}} \tag{5.28}$$

where σ_{ei} and σ_{ai} are the emission and absorption cross-sections for the pump $(i = p)$ and signal $(i = s)$, respectively. From (5.28) it is evident that a high gain amplifier can reach the 3-dB quantum noise limit only when pumped at 980 nm, where $s_{ep} = 0$. Pump wavelengths corresponding to the $^4I_{13/2}$ band have non-zero emission cross-sections and, therefore, give $\mathrm{NF}_{\mathrm{opt}} > 3\,\mathrm{dB}$. The $\mathrm{NF}_{\mathrm{opt}}$ quantum limit at the 1480 nm pump wavelength is about 4−5 dB depending on the signal wavelength (longer signal wavelengths show, in general, smaller noise figure). In addition, for the same gain level, forward-pumping configurations result in smaller $\mathrm{NF}_{\mathrm{opt}}$ compared with backward-pumping schemes [30].

From the discussion so far, it has become clear that the unavoidable generation of quantum noise during the amplification process degrades the SNR at the output of the amplifier, resulting always in $\mathrm{NF}_{\mathrm{opt}} > 1$. From this point of view, therefore, the use of an optical amplifier proves to be disadvantageous. This paradox arises from the fact that the optical noise figure is referenced to an ideal (noiseless and unity quantum efficiency) photodetector. In real systems, however, optical receivers are characterized by reduced (< 1) quantum efficiency and the presence of electrical shot noise. In this case, the use of high-gain optical amplifiers is always beneficial, resulting in large enhancement of the receiver minimum-detectable power or sensitivity [28,31–33].

Gain and Saturation Characteristics. From the propagation equations (5.11) and (5.12) it is evident that the local population inversion is determined by the combined effect of the pump, signal and ASE± powers, which makes their power evolution interdependent. Such dependency is shown schematically in Fig. 5.9. The pump and signal co-propagate and their respective launched powers are 20 mW and −30 dBm. The pump and signal wavelengths are 980 nm and 1532 nm, respectively. The erbium concentration is $1.46 \times 10^{24}\,\mathrm{m}^{-3}$, which corresponds to a signal absorption of 2 dB/m. It is shown that the pump power is absorbed at different rates along the fibre length, depending on the relative local powers of the signal and ASE($\pm$). The increased pump absorption at the EDFA input is entirely due to the strong backward ASE($-$). At the output end, the performance is dominated by the presence of the strong signal and the forward ASE($+$).

Figure 5.10 shows the variation of the EDFA gain and noise figure as a function of the fibre length, for pump powers of 10 mW and 50 mW. The other parameters are similar to those of Fig. 5.9. It is shown that, for each

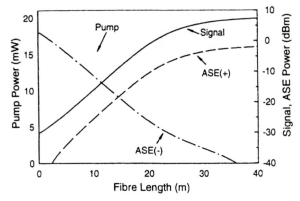

Fig. 5.9. Evolution of pump, signal and ASE± powers along EDFA length

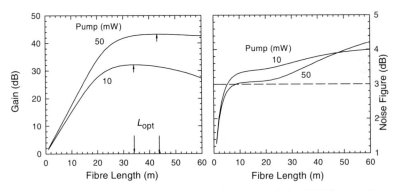

Fig. 5.10. Gain and noise figure as a function of the EDFA length at different pump powers. The arrows indicate the optimum length

pump power, the optical gain increases nonlinearly with the fibre length. It is also shown that the gain attains its maximum value at an *optimum length* (L_{opt}) which varies with the input pump power. At the optimum length, the population inversion is such that the local gain coefficient $\gamma(L_{\mathrm{opt}}) = \sigma_{\mathrm{es}} N_2 - \sigma_{\mathrm{as}} N_1 = 0$. Beyond the optimum length, $\gamma < 0$ and the optical signal is re-absorbed and attenuated. For short fibre lengths the noise figure is below the 3-dB quantum limit due to reduced gain. It should be noted however that, at optimum length, the noise figure is about 1 dB above the quantum limit. This is primarily due to incomplete population inversion at the EDFA front end, which is caused by the large backward ASE($-$) [34].

The amplifier gain and gain efficiency vary nonlinearly with the input pump power. For a given pump power, the gain efficiency (in dB/mW) is defined as the ratio of the attained optical gain (in dB) over the input pump power (in mW). Figure 5.11 shows the gain and gain efficiency of an EDFA as a function of the input pump power, for different fibre lengths. It is shown that for every length there is a pump power that results in transparency, i.e.

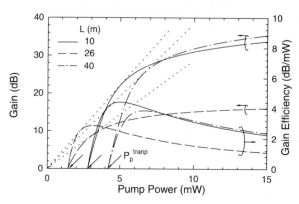

Fig. 5.11. Gain (*left axis*) and gain efficiency (*right axis*) as a function of input pump power, for erbium-doped fibre lengths 10 m, 26 m and 40 m. The input signal is -30 dBm at 1532 nm wavelength

$G = 1$. Such pump power is called *transparency pump power* ($P_{\mathrm{p}}^{\mathrm{transp}}$) and increases with the fibre length. The gain efficiency varies non-monotonically with the input pump power. For $P_{\mathrm{p}} > P_{\mathrm{p}}^{\mathrm{transp}}$, an initial sharp increase in gain efficiency is followed by a gradual decrease, which signifies the onset of saturation effects. For every length, the maximum gain efficiency is given by the tangent from the origin of the axes to the corresponding gain curve (dotted lines).

The optimum length (see Fig. 5.10) and transparency pump power (see Fig. 5.11) are characteristics of a two- or three-level system, where the laser transition ends at the ground level, and reflect the fact that the signal and pump are in direct competition. Obviously, these effects are not observed in four-level systems, like Nd^{+3} and Pr^{+3}, where the final laser-transition and ground levels are different.

It can be easily envisaged that for each pump power the maximum gain and gain efficiency are attained with the corresponding optimum EDF length. Figure 5.12a shows the gain (left axis) and the maximum gain efficiency (right axis) as a function of the input pump power, for input signal powers -50 dBm, -30 dBm and -10 dBm at 1532 nm wavelength. The pump wavelength is 980 nm and the fibre length is always optimized. Figure 5.12b shows the corresponding optical noise figure. It is shown that the gain efficiency varies in the same qualitative way as in the case of fixed length (see Fig. 5.11). However, increasing the input signal power results in reduced gain and gain efficiency due to increased saturation. For all input signal powers, the maximum gain efficiency is achieved at low pump powers (~ 5 mW) and corresponds to relatively low optical gains. It should be also stressed that, under the condition of maximum gain efficiency, the corresponding noise figure is considerably higher than the 3-dB quantum limit. It is then evident that operating the EDFA at the maximum gain-efficiency point is not beneficial. From Fig. 5.12b

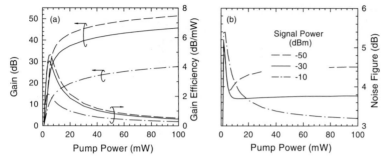

Fig. 5.12. (a) Gain (*left axis*), maximum gain efficiency (*right axis*), and (b) corresponding noise figure versus pump power (at optimum fibre length), for input signal powers −50 dBm, −30 dBm and −10 dBm at 1536 nm wavelength

it is obvious that optimizing the fibre length always results in noise figure higher than the quantum limit. For input signal powers around −10 dBm and high pump powers, however, the noise figure approaches the 3-dB limit due to self-saturation effects (see discussion below). The maximum gain efficiency increases with fibre NA and dopant confinement [35].

However, the maximum gain efficiency is still a useful figure of merit and is frequently used to compare the various pumping wavelengths, erbium hosts and fibre geometries. The best reported maximum gain efficiencies, to date, at the two main pump bands of 980 nm and 1480 nm, are 11 dB/mW [36] and 6.3 dB/mW [37], respectively.

The gain and noise performance of an EDFA are greatly affected by the input signal power. Figure 5.13a,b show the output signal power and noise figure, respectively, as a function of the input signal power, for pump powers of 50 mW and 100 mW. The signal wavelength is 1532 nm. For each input signal power the EDF length is optimised and is shown in Fig. 5.13c. Figure 5.13d shows the corresponding forward- and backward-propagating ASE. From Figs. 5.13a,b it is obvious that there are three different regimes of EDFA operation. First, *in the small-input-signal* or *linear regime* ($P_s^{in} <\sim$ −35 dBm), the output signal power increases linearly with the input signal power. The optical gain in this regime is constant (see also Fig. 5.14), and the noise figure departs considerably from the 3-dB quantum limit. In this regime, the EDFA operation is dominated by the presence of strong backward- and forward-propagating ASE, as is shown in Fig. 5.13d, that degrades the population inversion at the amplifier front end, compromising the gain and noise performance. For input signal powers in the range of about $\sim$ −35 dBm $< P_s^{in} <\sim$ 0 dBm, the signal is strong enough to saturate the gain medium and reduce the optical gain (smaller output–input curve slope). This range of input signal power is usually referred to as the *self-saturation regime.* In this regime, self-saturation results in appreciably smaller backward and forward ASE and, therefore, in improved population inversion throughout the

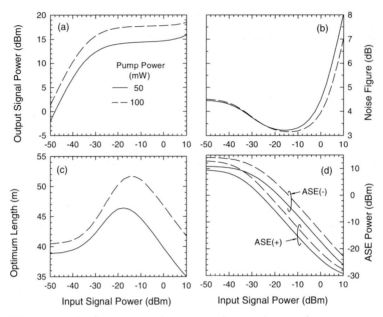

Fig. 5.13a–d. Output signal power and optical noise figure versus input signal (linear, self-saturation and deep saturation regimes)

EDF length. As a result, the optimum length increases and, more importantly, the noise figure reduces closer to the 3-dB quantum limit. Finally, for input signal powers $P_s^{in} >\sim 0\,\mathrm{dBm}$, the amplifier gain reduces asymptotically to unity. The amplifier is then said to be in the *deep saturation regime*. In this regime, due to extremely high signal power and despite the negligible ASE power, the optical noise figure increases dramatically.

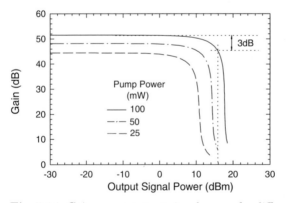

Fig. 5.14. Gain versus output signal power for different input pump powers - Gain saturation

In addition to high optical gain and low noise figure (see Fig. 5.12), optical amplifiers are desired to provide high output power. Figure 5.14 shows the gain variation of an EDFA versus the delivered output signal power, for different pump powers. The pump and signal wavelengths are 980 nm and 1532 nm, respectively. It is shown that the amplifier gain remains almost constant for an extended range of output signal powers. However, the optical gain drops sharply, as the EDFA enters the self-saturation regime, without any significant change in the output power. This corresponds to a soft self-limiting action [38]. A parameter of great importance is the *saturation output power* ($P_{\text{out}}^{\text{sat}}$), defined as the amplifier output power corresponding to an optical gain reduced by 3 dB from the small-input-signal value. The saturation output power gives the output-power dynamic range over which the amplifier is providing near-maximum gain (within 3 dB from the maximum value). This parameter should not be confused with the intrinsic saturation power (P_k^{sat}) defined in (5.13), and discussed in reference to Fig. 5.8. Another parameter, frequently encountered in the literature (not to be confused with the previous two), is the *saturated output power*, which is defined as the maximum output signal power for a given input signal and pump power. This is always achieved at optimum length and is shown in Fig. 5.13a.

Due to the spectral variations of the emission and absorption cross-sections (see Fig. 5.5), the optical gain and noise figure of an EDFA are expected to be functions of the wavelength. However, what ultimately determines the gain and noise figure is the integrated population inversion along the EDFA length and, therefore, the choice of pump power and wavelength are of great importance too. Figures 5.15a,b show the variation of the gain and spontaneous emission factor as a function of wavelength for different input pump powers. The pump wavelength is 1480 nm, the input signal power is -30 dBm and the fibre length is 30 m. Pump and signal co-propagate.

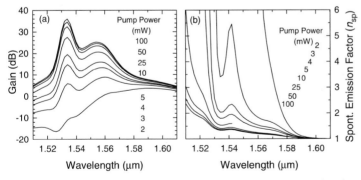

Fig. 5.15. (a) Gain and (b) spontaneous emission factor (n_{sp}) as a function of signal wavelength for various pump powers. The EDF length is 30 m and the pump wavelength is 1480 nm

It is shown that variation of the input pump power changes the population inversion along the fibre length and affects not only the maximum gain but also the gain spectrum (Fig. 5.15a). The spontaneous emission factor and, therefore, the EDFA noise performance, is also affected by the pump power variation, as shown in Fig. 5.15b. At low pump powers, gain is first attained at the long-wavelength tail (around 1600 nm) of the erbium transition. This is because this spectral region is characterised with sizeable emission and negligible absorption cross-sections (see Fig. 5.5). This is the basis for the implementation of long-wavelength EDFAs [39]. At high pump powers (> 25 mW), on the other hand, significant gain and low noise figure are achieved over a typical bandwidth of about 30 nm, centreed around 1550 nm. The gain spectrum is quite uneven across this bandwidth, peaking around 1536 nm, where the emission and absorption cross-section maxima occur. The spontaneous emission factor and, therefore, the optical noise figure remain close to the quantum limit. For intermediate pump powers (~ 5–10 mW), the EDFA spectrum becomes relatively smooth, at the expense, however, of reduced gain and deteriorated noise performance. The gain spectrum is flattened significantly with the incorporation of specially designed filters [40] and/or the use of alternative host materials (like alumino-germano-silicate [41], fluorozirconate [42] or tellurite [43] glasses). EDFAs with flattened gain spectrum and low noise figure are of paramount importance for the realization of future, high-capacity WDM optical links and networks.

The population inversion along the EDF length is also affected by the input signal power. It is, therefore, expected that the amplifier peak gain as well as the gain spectrum are dependent on the input signal power. Figure 5.16 shows the ASE power spectrum of an EDFA for different levels of saturation, determined by the input signal power. The pump and signal wavelengths are 1480 nm and 1555 nm, respectively. The input pump power is 50 mW the EDF length is 30 m. The input signal power is assumed to be -50 dBm, 10 dBm and 0 dBm. The corresponding gain is 25 dB, 21.3 dB and 14.7 dB, respectively. It is shown that saturation due to increased input signal power

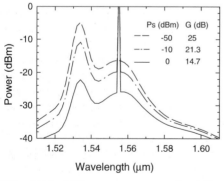

Fig. 5.16. ASE power spectrum for different degrees of gain saturation

not only reduces the signal gain but also changes significantly the entire ASE spectrum. This is a direct consequence of the fact that the erbium transition around 1.55 μm wavelength is predominantly homogeneously broadened.

Main EDFA Types. In the previous section, it has been clearly demonstrated that, under appropriate pumping conditions, standard silica-based EDFAs are capable of providing high small-signal gain (in excess of 45 dB), quantum-limited noise performance and high saturation output powers (in excess of 10 dBm) over an optical bandwidth of about 30 nm. These characteristics enable EDFAs to be used as power amplifiers or boosters, and line amplifiers, as well as preamplifiers. The optical characteristics of and requirements on optical amplifiers will vary with the particular application and optical-link topology.

A *power amplifier* or *booster* is placed just after the optical transmitter, at the start of the optical link. It is mainly used to boost the optical power, launched into the optical link, in order to either increase the propagation distance and/or compensate for the additional losses that occur during the multiple power splitting into an increased number of different routes and subscribers. Optical transmitters provide output powers of several mW by operating them in deep saturation. Power amplifiers are, therefore, designed to provide high saturated output powers. In that respect, the SNR is always maintained at a high level and the amplifier noise figure is of secondary importance. Power amplifiers are required to have high power-conversion efficiency [defined as PCE $= (P_s^{out} - P_s^{in})/P_p^{in}$]. Its maximum value is easily shown to be given by the ratio λ_p/λ_s. Therefore, pumping at 1480 nm [44] will result in higher PCE when compared with 980 nm pumping [45]. It is also shown that backward pumping results in the highest PCE [46].

An optical *preamplifier* is usually placed at the end of an optical link, just in front of the optical detector, and constitutes one of the most important components of a sensitive optical receiver. The input signal is usually at a very low level and, therefore, the preamplifier should exhibit both high small-signal gain and low quantum-limited noise performance. Incorporation of an high-performance optical preamplifier increases the receiver sensitivity significantly [47].

Finally, the *line amplifiers* are placed within an optical link and are used to compensate for propagation losses and extend the fibre link length. The input signal is usually quite small and high-gain, low-noise performance is required. On the other hand, in order to increase the unamplified fibre spans, high saturation output powers are desirable. In that respect, line amplifiers should combine the characteristics of power amplifiers and preamplifiers.

Composite EDFAs. As has already been demonstrated, standard EDFAs combine a number of optical properties that permit their use in different positions along an optical link. It has been realized, however, that their performance can be further improved by incorporating a number of optical components or by using different fibre designs. These improved amplifier designs are often referred to as composite EDFAs.

Standard EDFAs operated in the small-signal regime, although they can provide high gain (in excess of 45 dB), exhibit a noise figure that departs considerably from the quantum limit as the required gain is increased. This is primarily due to the strong backward-propagating ASE that reduces the population inversion at the amplifier front end. It has been demonstrated that the incorporation of an optical isolator can reduce significantly the effect of the ASE(-). Such a composite EDFA has been shown to provide gain in excess of 54 dB *and* near-quantum-limited noise performance ($NF_{opt} \approx 3.1$ dB), when pumped with ~ 50 mW at 980 nm [34,48]. Insertion of an optical isolator also reduces significantly the deleterious effects on gain efficiency and noise figure caused by internal Rayleigh back-scattering [49].

From Figs. 5.15a and 5.16, it is obvious that the EDFA gain spectrum is quite uneven, under all operating conditions. Such a spectral gain variation renders the use of the EDFA in WDM repeated systems problematic, resulting in gain peaking and a dramatic reduction of the effective bandwidth. The EDFA spectrum can be flattened efficiently by incorporating an appropriate optical filter in the amplifier design [40]. In long haul optical links, the gain flatness should be kept to within ± 0.5 dB.

In addition to gain flatness, composite optical amplifiers can exhibit optical limiting action, with an input-power dynamic range of ~ 40 dB, by simply introducing a lump differential loss between pump and signal [38]. Also, use of a twin-core erbium-doped fibre results in an optical amplifier that provides power equalization. Such a composite amplifier relies on the wavelength dependence of the coupling properties of the twin-core fibre to spatially separate different channels and induce spatial 'hole burning' [50].

5.1.2   Other Fibre Amplifiers

Erbium-doped silica fibre amplifiers are extremely efficient in providing gain around the 1.55 μm window. However, in order to use the entire low-loss window offered by the optical fibre, spanning the 1.3−1.6 μm bandwidth, different amplifiers are needed.

Pr^{3+}-Doped Fibre Amplifiers. Spectroscopic analysis has shown that Nd^{3+} and Pr^{3+} ions exhibit transitions in the 1.3 μm region and can be used to provide amplification in that spectral region. At present, the most promising approach for the realization of 1.3 μm amplifiers is through the use of Pr^{3+}-doped fluoride fibre [51]. These amplifiers, however, still suffer from low radiative quantum efficiency ($\sim 4\%$) and low gain efficiencies (~ 0.2 dB/mW). A newly developed PbF_2–InF_3-based fluoride fibre has resulted in an improved gain efficiency of 0.36 dB/mW and gain of 20 dB at

1.31 μm with 100 mW at 1017 nm pump power [52]. Fluoride fibres are not as robust as their silica counterparts and special care should be taken in assembling and packaging the amplifiers. Research into new low-phonon-energy materials, such as GLS glasses, could ultimately lead to the realization of 1.3 μm amplifiers with improved gain performance [53].

Fibre Raman Amplifiers. Fibre Raman amplifiers (FRAs) provide an alternative means for eliminating propagation loss in optical communications. FRAs rely on stimulated Raman scattering (SRS) as the physical mechanism for producing gain [54]. SRS is an inelastic process during which pump photons are scattered by the optical vibrational modes (optical phonons) of the material of propagation. During the scattering process energy is transferred from the pump photons (at frequency ν_p) into the optical phonons and another photon is created of reduced energy at a lower frequency ν_s (referred to as Stokes downshift). When a pump and signal at frequencies ν_p and ν_s, respectively, co- or counter-propagate into the fibre, energy is transferred from the pump into the signal through the SRS interaction. Due to the amorphous nature of silica, the optical phonon energy levels of silica molecules are merged together forming a continuum and resulting in a wide amplification band. For typical silica fibres, the Stokes shift is about 13.2 THz and the main amplification bandwidth is about 6 THz. The mean Stokes downshift and amplification bandwidth varies with the host composition [55].

Due to the non-resonant nature of the energy transfer (amplification) process, FRAs can in principle be operated at any wavelength over the entire low-loss communication window ($\sim 1.3-1.6$ μm). Recent advances in high pump power lasers have enabled successful demonstration of Raman amplifiers in the 1.3 μm [56], 1.4 μm [57] and 1.5 μm [58] windows. Raman amplification can also be combined with erbium-based amplifiers to produce composite gain units of low noise figure and a 3 dB bandwidth of about 76 nm [59].

5.2   Semiconductor Optical Amplifiers

5.2.1   Optical Gain in Compound Semiconductor Materials

Amplification in semiconductor materials is most well known for its application in semiconductor lasers. Pumping a semiconductor material with an electrical current can create a population inversion similar to the case of, for example, Er^{3+} ions in silica fibres when pumped with intense laser light. Passing a light signal through the semiconductor causes amplification through stimulated emission processes [60]. In a semiconductor laser, the amplifying material is combined with optical-feedback elements (for example, cleaved crystal facets acting as mirrors) to create a resonating cavity in which light propagates back and forth. In a semiconductor optical amplifier (SOA) the signal light passes through the amplifier only once and special care is taken to eliminate optical feedback.

In a semiconductor, for a certain range of electron energies, no energy states exist which electrons can occupy (e.g. see [61]). The size of the forbidden energy gap between the valence and conduction band, the bandgap, is determined by material properties such as composition and temperature. In thermal equilibrium, the states in the valence band are (nearly) all filled; only a small amount of electrons reside in the conduction band due to thermal excitation, leaving behind an equal amount of positively charged holes in the valence band. When passing an electrical current through the semiconductor, the electrons in the conduction band travel towards the high potential (positive) side, and the holes in the valence band travel towards the low potential (negative) side. The conduction properties of the semiconductor crystal can be strongly influenced by adding impurities as dopants. P-type doping enhances the hole concentration in the valence band; n-type doping enhances the electron concentration in the conduction band.

Consider the structure of semiconductor materials shown in Fig. 5.17, consisting of a sandwich of p-type material on top of an undoped (intrinsic) layer on top of n-type material. The material composition of the central or active layer is chosen differently from the surrounding layers in order to create a lower bandgap in the active layer. Applying a forward voltage over this heterostructure diode will cause electrons from the n-type material and holes from the p-type material to travel towards the active layer. Here, the electrical carriers accumulate as they are trapped in the low bandgap potential well. If large enough currents are applied, large concentrations of electrons and holes build up in the active layer, leading to population inversion. Photons passing through the active layer can stimulate radiative recombination of the carriers, resulting in amplification of the light. This is the basic effect enabling the structure to function as an optical amplifier. In addition, carriers can also decay spontaneously, leading to amplified spontaneous emission (ASE), or they can decay through non-radiative processes. The balance between the applied current and the decay of carriers determines the carrier concentration.

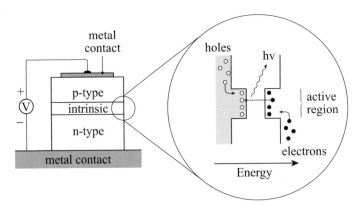

Fig. 5.17. Basic structure of a semiconductor active layer

5.2.2   Basic Heterojunction Device Structure

An optical amplifier based on semiconductor materials has to meet several requirements in order to function efficiently:

1. Electrical carriers (electrons and holes) must be transported to the active layer.
2. The carriers must be confined within the active layer in order to achieve a population inversion. This also implies that the carrier lifetime must be sufficiently long.
3. The light beam to be amplified needs to have sufficient overlap with the inverted active layer in order to achieve optical gain.
4. The light beam must be guided through the structure so that no light is lost along the length of the active layer.
5. Unwanted optical reflections must be avoided.
6. Signal light must be efficiently coupled into and out of the amplifier chip.
7. The amplification process shall be independent of the polarization of the incident light beam.

A typical SOA device structure is shown in Fig. 5.18 [62,63]. In this example, the device is based on a n-type InP substrate. The active layer consists of a stack of InGaAsP quantum wells separated by thin barrier layers (quantum wells will be discussed below). The active layer, and in particular the quantum wells, have a bandgap lower than that of InP in order to confine the electrical carriers. On top of the active layer, a p-type InP layer has been grown. The device is complemented with metal contacts. Note that the active layer is restricted in the lateral dimension as well. On either side, it is surrounded by isolating Fe-doped InP, ensuring that all the electrical current will flow through the active layer. This results in high current densities, which are necessary to achieve high carrier concentrations in the active layer. Lastly, in order to suppress optical feedback the active stripe has been placed at an angle with respect to the front and end facets [64,65]. In addition, the facets are anti-reflection coated, and so-called window regions have been included in which the signal light diverges before passing through the facets. All these measures are taken to reduce the effective optical reflections at the facets to less than 0.1–0.01%.

In addition to confining electrical carriers, the active layer also confines light passing through the structure. The lower bandgap with respect to the surrounding materials implies a larger refractive index for the active layer. This leads to waveguiding within the active region. For efficient amplification, the waveguided light beam in the amplifier must have sufficient overlap with the active layer. Also, it is of paramount importance that the waveguided light is coupled efficiently in and out of the amplifier chip, usually to single-mode optical fibre. This implies that the amplifier should operate as a single-mode waveguide, with a circular beam waist that can be matched to that of the single-mode fibre. In general, careful design of the material composition and

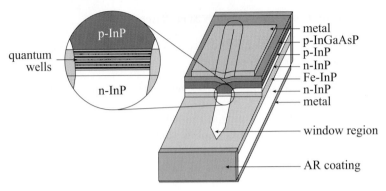

Fig. 5.18. Schematic view of a 1.3 μm semiconductor optical amplifier

layer thicknesses is necessary to obtain both the desired electrical properties and optical properties.

In order to fabricate the desired semiconductor optical amplifier structure, compound-semiconductor crystal growth techniques are used. One such technique is low-pressure Metal Organic Vapour Phase Epitaxy (MOVPE). A schematic of the growth technique is illustrated in Fig. 5.19. Substrates are placed on a heated susceptor (around 625 °C) within the reaction chamber. Group III alkyls ($Ga(CH_3)_3$ and $In(CH_3)_3$) and group V hydrides (AsH_3 and PH_3) are fed into the chamber with hydrogen carrier gas. From the laminar gas flow, gas molecules can diffuse towards the hot substrate where they decompose via different chemical stages to deliver the constituents In, Ga, As, and P. On the substrate these atoms can further diffuse to find lattice positions on the semiconductor crystal. The layers grown in this way can be doped by introducing small amounts of $Zn(C_2H_5)_2$ or H_2S for p- and n-type doping with Zn and S, respectively. Due to the high velocities of the reactants and a fast switching between gas flows, atomically flat interfaces between layers of different composition are readily achieved. Alternative crystal growth techniques, in particular Molecular Beam Epitaxy (MBE), may also be employed.

Using epitaxial growth of semiconductor layers, different types of active layer can be grown. The simplest type of active layer consists of a bulk semiconductor layer with a bandgap smaller than that of the surrounding layers [66–70]. Figure 5.20 shows the Fermi-Dirac distribution for electrons and holes in (a) a bulk active layer, and (b) a quantum well. The corresponding density-of-states functions are also shown. The parabolic band structure of the bulk layer together with the Fermi-Dirac distribution governing the occupation of states by electrons leads to the given carrier concentration versus energy. In turn this determines the light emission spectrum and parameters such as the noise figure. The second type of active layer is the quantum well [70,71]. In quantum wells the electrical carriers are restricted in one dimension, as the thickness of the layer approaches the de Broglie wavelength

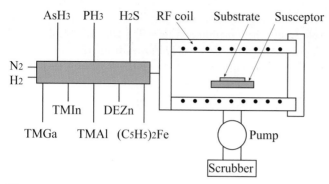

Fig. 5.19. Schematic of an MOVPE reactor chamber used for the growth of compound-semiconductor crystals

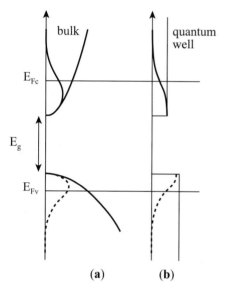

Fig. 5.20. Fermi-Dirac distributions (*dashed lines*) for electrons and holes in (**a**) a bulk semiconductor material and (**b**) semiconductor quantum wells. The Fermi energies are indicated. The density-of-states functions are shown as solid lines. Note that the graphs have been drawn to scale and the total amount of electrical carriers integrated over the energy is constant

of the electrons. This results in a square-shaped density-of-states function. Combined with the Fermi-Dirac distribution, this leads to a significantly different carrier concentration versus energy compared to bulk material.

5.2.3   Rate Equations, Saturation Behaviour, Noise Figure

There are three key performance parameters that describe the semiconductor optical amplifier: optical gain, saturation output power and noise figure.

Clearly, it is not possible to amplify an input signal of arbitrary power; if the input power is too high, the amplifier gain will saturate. Similarly, if the input signal is too low, it cannot be discriminated from the (amplified) spontaneous emission.

Optical Gain and Saturation Behaviour. Consider the schematic of the amplifier structure shown in Fig. 5.18. Following [72], the propagation of the optical signal power P through the amplifier is described by

$$\frac{\mathrm{d}P}{\mathrm{d}z} = gP - \alpha P \ . \tag{5.29}$$

Here, g is the gain coefficient and α is the absorption coefficient describing all power losses not related to stimulated emission and absorption. The gain coefficient can be expressed as

$$g = \Gamma A_{\mathrm{g}}(N - N_{\mathrm{tr}}) \tag{5.30}$$

with N the carrier concentration, N_{tr} the carrier concentration at transparency, A_{g} the differential gain coefficient (often expressed as $\mathrm{d}g/\mathrm{d}N$), and Γ the confinement factor.

The carrier concentration may be determined by solving the rate equation for electrical carriers

$$\frac{\mathrm{d}N}{\mathrm{d}t} = \frac{I}{qV_{\mathrm{act}}} - \frac{N}{\tau_{\mathrm{c}}} - \frac{gP}{h\nu A_{\mathrm{act}}} \ . \tag{5.31}$$

The first term on the right-hand side of (5.31) refers to the amount of carriers pumped into the active layer, and consists of the electrical current I, electron charge q and active layer volume V_{act}. The second term describes all non-radiative (i.e. carrier loss) processes through the carrier lifetime τ_{c}. The last term describes the loss of electrical carriers through stimulated emission processes, with $h\nu$ the photon energy, and A_{act} the cross-sectional area of the active layer.

In the steady state (5.31) is set to zero, and a solution may be obtained for N. Using this result and with (5.30) in (5.29) this leads to

$$\frac{\mathrm{d}P}{\mathrm{d}z} = \frac{g_0}{1 + P/P_{\mathrm{sat}}} P - \alpha P \ , \tag{5.32}$$

with $P_{\mathrm{sat}} = h\nu A_{\mathrm{act}}/\Gamma Ag\tau_{\mathrm{c}}$ and $g_0 = \Gamma A_{\mathrm{g}}(I\tau_{\mathrm{c}}/qV_{\mathrm{act}} - N_{\mathrm{tr}})$. Integrating (5.32) over the amplifier length L (and neglecting the waveguide losses α) leads to

$$\frac{P_{\mathrm{out}}}{P_{\mathrm{in}}} = G = \exp\left(g_0 L - \frac{P_{\mathrm{out}} - P_{\mathrm{in}}}{P_{\mathrm{sat}}}\right) \ , \tag{5.33}$$

$$\ln(G) = \ln(G_0) - \frac{P_{\mathrm{out}}}{P_{\mathrm{sat}}} \ , \ (P_{\mathrm{out}} \gg P_{\mathrm{in}}) \ , \tag{5.34}$$

where G is the amplifier gain, $G_0 = \exp(g_0 L)$ is the small-signal amplifier gain, and the subscripts 'in' and 'out' indicate the input and output, respectively. The 3-dB saturation power $P_{-3\,\mathrm{dB}}$, defined as the amplifier output power at which the gain has dropped to half of the small-signal gain, may be determined from (5.34) and is

$$P_{-3\,\mathrm{dB}} = \ln(2)P_{\mathrm{sat}} = \ln(2)h\nu A_{\mathrm{act}}/\Gamma A_{\mathrm{g}}\tau_{\mathrm{c}} . \tag{5.35}$$

This equation states that amplifier saturation is determined by the material properties A_{g} and τ_{c}. The effect of saturation in a SOA is illustrated in Fig. 5.21.

Dynamic Behaviour. Equation (5.35) was derived under the assumption of continuous wave (CW) operation. In practice, optical signals carrying information are modulated in time; usually intensity modulation is applied to the signal. According to (5.34), the gain will adjust itself to the changing signal intensity, but only as long as the carrier concentration can react to the time-varying signal. The time constant governing the carrier concentration, or gain recovery time, determines whether this is the case. For the SOA, the carrier lifetime is of the order of 0.1 ns, hence the gain recovery time is short with respect to typical data rates ($10^9\,\mathrm{s}^{-1}$). Therefore, different signal intensity levels (say the "ones" and "zeros" in a digital signal) experience different optical gains, leading to signal distortion. Note that this is only significant when operating close to saturation conditions. This is in contrast to the situation for the EDFA, which has a gain recovery time of the order of 1 ms. Here, the time-varying signal will 'see' a constant gain determined by the average signal power.

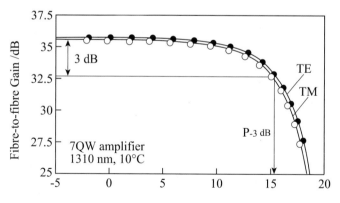

Fig. 5.21. Gain compression or saturation in a 7-quantum well semiconductor optical amplifier operated at 1310 nm. The data are shown for both TE and TM input signals. The 3 dB saturation power is indicated

The dynamic behaviour of the SOA leads to several application advantages as well as restrictions. Most importantly, for amplification at data frequencies lower than the gain recovery time, the SOA should not be operated in the saturation regime as this leads to signal distortion. This poses a limitation on the maximum amplifier output power. In addition, the gain modulation can lead to self-phase modulation (SPM), which in systems can cause frequency modulation or chirping. Notwithstanding these restrictions, SOAs have many useful applications as amplifiers, as will be discussed later in this chapter. For more demanding applications, gain-clamping of SOA structures by introducing optical feedback presents a solution to the issue of gain modulation (see Sect. 5.2.5). There are also applications which actually require amplifiers to react fast to changing signal powers or to switch. In addition, some applications such as wavelength conversion and signal regeneration utilize the amplifier's nonlinearity.

Noise. An active layer pumped with electrical carriers will also produce spontaneous emission noise through radiative recombination. In an amplifying structure the fraction of spontaneous emission coupled into the waveguiding mode will be amplified along the length of the amplifier resulting in amplified spontaneous emission (ASE). The beating of spontaneously emitted photons with signal photons leads to amplifier noise. The noise figure $F_{\text{sig-sp}}$, defined as the signal-to-noise ratio at the input over the signal-to-noise ratio at the output (assuming a shot-noise-limited input signal), for such signal-spontaneous beat noise can be expressed as follows [63]

$$F_{\text{sig-sp}} = \frac{S/N_{\text{input}}}{S/N_{\text{output}}} = \frac{2\rho_{\text{ase}\backslash\text{parallel}}}{gh\nu} = \frac{2n_{\text{sp}}K_{\text{x}}}{\eta_{\text{in}}}, \qquad (5.36)$$

with S/N the signal-to-noise ratio, $\rho_{\text{ase}\backslash\text{parallel}}$ the ASE power spectral density having the same polarization as the amplified signal, g the optical gain, n_{sp} the population inversion parameter, K_{x} the excess noise factor, and η_{in} the input coupling efficiency. Essentially, the noise figure is determined by the ratio between ASE power and gain, which is in turn determined by the level of population inversion. Clearly, the input coupling efficiency should be high in order not to degrade the signal intensity. The excess noise factor will be discussed later.

Polarization Dependence. Lastly, for many amplifier applications it is of importance that the gain be independent of the polarization of the input signal light. Semiconductor active layers are in general sensitive to the polarization for two reasons. First, the overlap of transverse electric (TE) light with the active layer is different from transverse magnetic (TM) light [73]. Second, crystal strain, which is commonly applied in semiconductor active layers, breaks the symmetry of the crystal leading to different optical gain for TE and TM light [71]. To obtain polarization-independent operation of the amplifier several techniques can be used. In bulk layers, with no crystal strain applied, the layer can be designed such as to have a square cross-section leading to equal overlap of TE and TM light [66,67]. Alternatively, the application of strain in a non-square bulk layer can be used to compensate for the differences in overlap [68,69]. In quantum wells, strain can be used to fabricate layers which produce only TE or TM gain. Combining such quantum wells in one active layer can also lead to polarization-insensitive operation [70].

To illustrate the parameters discussed above, Fig. 5.22 shows the gain and saturation power versus wavelength for a bulk semiconductor amplifier with tensile strain, and for a multi-quantum-well amplifier stack combining 'TE quantum wells' and 'TM quantum wells'.

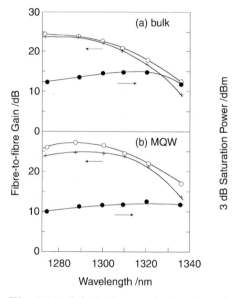

Fig. 5.22. Polarization-resolved gain and saturation power versus wavelength for (**a**) a bulk amplifier, and (**b**) a multi-quantum-well amplifier

5.2.4   Effect of Optical Reflections (Gain Ripple)

Optical reflections can severely influence the amplifier performance, especially in the case of high single-pass gains. Reflections from the amplifier end facets may cause feedback into the amplifier. If the single-pass gain exceeds the reflection coefficient, the condition for lasing is achieved, resulting in detrimental amplifier performance. (A more controlled manner of lasing in amplifiers will be discussed later). Even before lasing occurs, the reflections will create a Fabry–Perot type of etalon, leading to oscillations in the amplifier gain versus wavelength. This is more commonly known as gain ripple.

Assuming that the reflections show no wavelength dependence, the equation governing the transmitted optical intensity I_t relative to the input intensity I_i may be obtained by summing all the transmitted amplitudes and converting to intensity. This leads to a well-known equation for a Fabry–Perot resonator with optical gain [60]:

$$\frac{I_t}{I_i} = G = \frac{(1-R)^2 G_s}{(1-RG_s)^2 + 4RG_s \sin^2(\delta/2)} \,, \tag{5.37}$$

$$\delta = \frac{4\pi L n_{eff}}{\lambda} \,. \tag{5.38}$$

In this equation G is the real (i.e. measured) gain, G_s the single pass gain, R the reflection coefficient of the facets (it is assumed that each facet has the same reflection coefficient), and δ is the phase shift that the lightwave undergoes on traversing the length of the amplifier L twice (i.e. the round-trip phase shift). n_{eff} is the effective refractive index of the medium and λ the wavelength of the signal light.

Figure 5.23 shows a measurement of optical gain versus wavelength showing the effect of gain ripple. The data have been fitted according to (5.37).

Reflections in an optical amplifier also have an effect on the noise performance. Amplified spontaneous emission, which causes noise in optical amplifiers, is fed back into the amplifier and re-amplified enhancing the amount of spontaneous emission noise. A similar formula to (5.37) can be deduced for the spontaneous emission, which is generated along the whole length of the

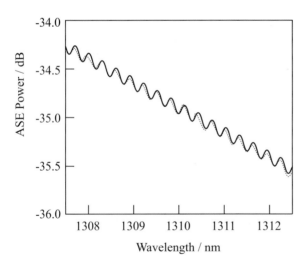

Fig. 5.23. Measurement of the gain ripple in a semiconductor optical amplifier. The data have been fitted according to (5.37)

amplifier and travels in both directions. The density of spontaneous emission, and so also the noise figure, is increased with the so-called excess noise factor K_x [72,74]

$$K_x = \frac{(1 + R_1 G_s)(G_s - 1)(g + \alpha_i)}{(1 - R_1)G_s G_s g} \, .$$
(5.39)

5.2.5   Gain-Clamping

One of the main disadvantages of semiconductor optical amplifiers is the effect of gain compression or saturation. This effect may be overcome by the use of gain-clamped semiconductor optical amplifiers (GCSOAs), in which the optical gain is constant over a large range of signal output power. Gain clamping can be achieved by introducing wavelength selective feedback, e.g. from a distributed Bragg reflector (DBR), to the amplifying medium. The device lases at the corresponding wavelength as soon as the round trip gain (material gain plus mirror losses) equals unity. At this point, the carrier concentration is fixed and hence the optical gain is clamped at a certain value.

Figure 5.24 shows a schematic of a GCSOA. The DBR mirrors are chosen at a wavelength outside of the signal band, so not to interfere with the signal. The optical spectrum of a GCSOA is shown in Fig. 5.25, and the gain versus output power diagram in Fig. 5.26 [75–78]. The behaviour in this diagram is distinctly different from that shown in Fig. 5.21. The gain of the GCSOA is flat up to a certain output power level, beyond which the gain droops steeply. At this power level the signal depletes the active layer of its carriers to such an extent that laser oscillation can no longer be sustained.

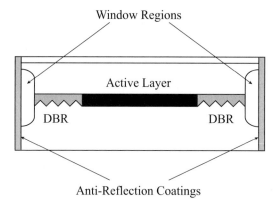

Fig. 5.24. Schematic diagram of a gain-clamped semiconductor optical amplifier

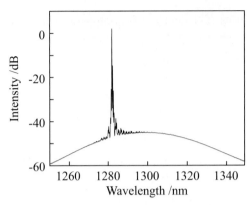

Fig. 5.25. Output emission spectrum of a gain-clamped semiconductor optical amplifier

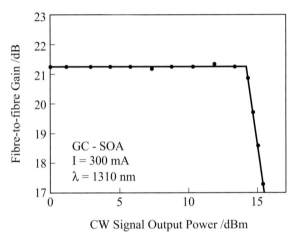

Fig. 5.26. Gain versus output power of a gain-clamped semiconductor optical amplifier

5.2.6   General Applications of Semiconductor Optical Amplifiers in Communication Systems

Basic performance parameters such as gain, noise figure and saturation power were introduced in the previous sections. The following sections will discuss the performance of SOAs in optical communication systems. In addition to the technical performance, the feasibility of optical amplifiers in different applications depends on factors such as costs, compactness and reliability. With respect to these issues, SOAs have distinct advantages over other optical amplifiers, as they are based on mature semiconductor laser technology, which is proven to be low-cost and reliable. In addition, SOA technology

Fig. 5.27. Packaged SOA module

can be applied in both the low-loss 1550 nm window and the zero-chromatic-dispersion 1310 nm window.

Figure 5.27 shows a fully-packaged hermetically-sealed 14-pin SOA butterfly module [64]. The input and output fibres are coupled efficiently to the amplifier chip using aspheric lenses. The amplifier chip is mounted on a Peltier cooler to maintain a constant temperature. The amplifier module can also be equipped with internal optical isolators. Such compact packaging allows for integration of the module on a printed circuit board.

5.2.7   Digital Transmission Systems

Figure 5.28 shows a diagram of a digital transmission system employing SOAs. Three amplifier types perform three basic functions required in such a system: power boosting [79], in-line amplification, and preamplification [80]. Power boosting is used to increase the laser transmitter power. In-line amplifiers are installed at regular intervals to compensate for optical losses encountered in the fibre. Preamplification is necessary to enhance the receiver sensitivity. The different functionalities lead to different requirements on the SOA performance.

One of the main performance limiting factors of SOA devices is gain compression. This is discussed in Sect. 5.2.3 and is illustrated in Fig. 5.21. As the gain recovery time in SOAs is of the order of, or shorter, than the bit length, the bits are amplified differently depending on their history. Figure 5.29 shows this for a 10 Gbit/s return-to-zero (RZ) bitstream at 1310 nm. Such distortions of the optical signal can severely degrade the signal-to-noise ratio, and eventually degrade the system performance. Nonetheless, if the

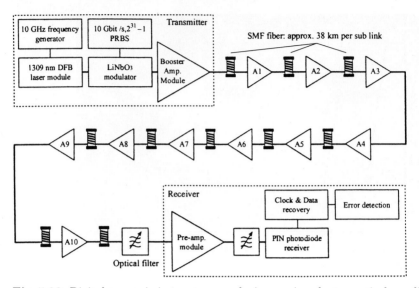

Fig. 5.28. Digital transmission system employing semiconductor optical amplifiers

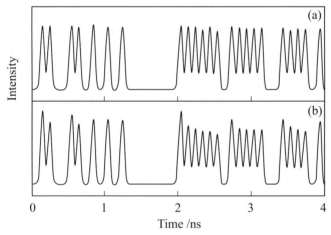

Fig. 5.29. Bit pattern of a 10 Gbit/s return-to-zero signal after a cascade of 6 SOAs. Bitstream was taken (**a**) under non-saturating power conditions; (**b**) under saturating conditions

distances are less than 1000 km, error-free transmission at high bit rates is possible using standard SOAs.

The feasibility of SOAs in digital transmission systems is demonstrated in Fig. 5.30, which shows the bit-error-rate for the system shown in Fig. 5.28. The system consists of a 10 Gbit/s RZ 1310 nm transmitter with a 5% extinction ratio connected to a SOA booster giving 0−2 dBm average output

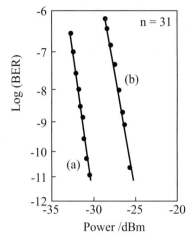

Fig. 5.30. Bit-error-rate results obtained at (**a**) 0 km and (**b**) after 420 km, plotted against the received power

power (7−9 dBm peak power). The signal is passed through 420 km of standard single-mode fibre with SOAs at 38 km intervals. At the receiver, the signal is amplified by a high-gain optical SOA preamplifier before detection. Details may be found in [80]. Similar experiments have been carried out elsewhere [81,82], including experiments in the field [83].

The performance of SOAs in digital systems is determined by the amplifier gain, saturation power and noise figure. These performance parameters may be translated to a figure-of-merit (FOM) that describes the signal quality degradation by a single amplifier, provided that an optimum signal power level is applied.

$$FOM = \frac{P_{-3\,\mathrm{dB}}}{F_{\mathrm{sig\text{-}sp}}G} \, . \tag{5.40}$$

This FOM may be understood as follows. $P_{-3\,\mathrm{dB}}$ defines the upper limit to the amplifier output power for a given distortion level; $F_{\mathrm{sig\text{-}sp}}$ defines the lower limit to the input power for a given signal-noise ratio degradation. Multiplying by G gives the margin between the two, and therefore the FOM defines the operating regime of the amplifier. When amplifiers are cascaded, the signal quality degradation effects increase with the number of amplifiers; i.e. the effective saturation power is $P_{-3\,\mathrm{dB}}/N_{\mathrm{amp}}$ and the effective noise figure is $N_{\mathrm{amp}}F_{\mathrm{sig\text{-}sp}}$. This implies that the remaining system margin M_{system} of the cascade is equal to the FOM divided by the square of the number of amplifiers N_{amp}. This relationship states that the number of amplifiers that can be cascaded depends on the performance of a single amplifier, described by the FOM, and the signal-quality budget necessary for error-free transmission (determined by the transmitter and receiver). Details may be found in [84,85]. Figure 5.31 illustrates how many amplifiers may be cascaded as a function of

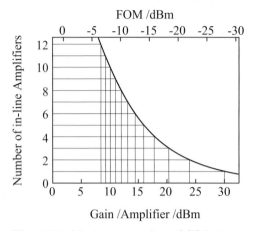

Fig. 5.31. Maximum number of SOAs in a cascade as a function of the gain per amplifier. The top axis shows the amplifier figure-of-merit *FOM*

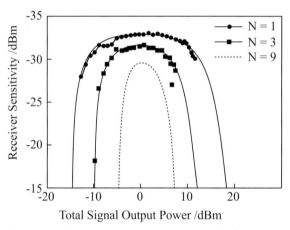

Fig. 5.32. System power budget versus SOA signal output power

the gain per amplifier and the *FOM* for a given (conservative) choice of the system margin.

Figure 5.32 plots the system power budget versus the SOA signal output power. Data has been measured for a cascade of 1, 3 and 9 amplifiers operated at 10 Gbit/s RZ 1310 nm transmission.

5.2.8   WDM Systems

Wavelength-division multiplexing (WDM) of signals in order to increase transmission capacity is very beneficial when applied in optically amplified systems. Instead of using a regenerator for each separate signal, all signal wavelengths are amplified in the same optical amplifier. This implies that crosstalk between the wavelength channels must be avoided.

In conventional SOAs, cross-gain modulation severely limits the performance in a WDM system. The power dependence of the optical gain leads to gain modulation in phase with the signal. As different signals pass through the amplifier simultaneously, each signal is subject to the gain modulation caused by the other signals. Hence, cross-gain modulation is detrimental in a WDM system. In gain-clamped amplifiers (Sect. 5.2.5), on the other hand, the gain is constant independent of signal power. Experiments have shown error-free amplification of 16×10 Gbit/s channels in a GCSOA device [86].

5.2.9   Analogue Transmission Systems

Analogue transmission of optical signals is primarily used for the transport and distribution of cable television (CATV). Each channel is amplitude modulated (AM) at a given carrier frequency; the carriers are frequency multiplexed to enable high signal-transmission capacity. Typically, analogue systems transport frequencies between 50 and 800 MHz. The AM nature of the signal places high demands on the linearity of components in the system. Composite second-order (CSO) distortion stems from the beating of two carrier frequencies f_1, f_2 leading to sum and difference intensities at $f_1 + f_2$ and $f_1 - f_2$; composite triple beat (CTB) is a result of mixing products at $2f_1 - f_2, 2f_1 + f_2, 2f_2 - f_1$ and $2f_2 + f_1$. Most analogue transmission systems operate at the zero chromatic-dispersion wavelength of 1310 nm.

Optical amplifiers operating in an analogue transmission system must meet the stringent requirements on linearity. Conventional SOAs cannot meet these needs when operated at high powers because of gain compression (Fig. 5.21). On the other hand, highly linear gain versus signal power is possible for the gain-clamped amplifiers discussed in Sect. 5.2.5. Figure 5.33 shows the CSO distortion and CNR as measured on a gain-clamped SOA employed in a 77 channel National Television System Committee (NTSC) system [87]. The CSO distortion of -55 dBc is close to the value of -65 dBc (electrical) required. The remaining nonlinearity in gain-clamped amplifiers is mainly due to spatial hole burning effects discussed in [78]. Such nonlinearity can be tackled by adjusting the ratio between the mirror reflectivities to achieve a laser light intensity distribution in the amplifier that matches the signal intensity.

5.2.10   Other Applications

One of the advantages of the SOA is its ability to switch at speeds comparable to typical data rates. This ability gives the SOA added functionality in systems which require switching between different optical paths or simple

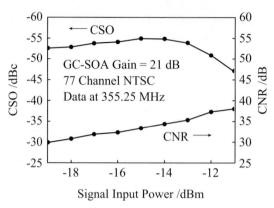

Fig. 5.33. CSO and CNR for a gain-clamped SOA

data switching. As optical communication systems evolve from simple point-to-point systems to intricate networks such functionality is becoming an important feature. Basic switching and signal modulation using SOAs has been demonstrated [88–92]. An important application of switches may be found in the all-optical cross-connect. Here N input channels can be configured to any of N output channels, requiring N^2 switching elements. Switching capabilities together with the possibility of large-scale integration makes SOAs ideally suited for such applications. An example of SOA application to a complex network is in access networks [93]. In upstream data transport, ASE noise from optical amplifiers accumulates. Figure 5.34 illustrates this process known as noise funnelling.

Although only one node is transmitted at a given time, all the amplifiers generate ASE noise leading to undetectable signals at the head end. By switching the optical amplifiers on only when they amplify a data cell, noise funnelling is overcome. This requires the amplifiers to be switchable on a cell-

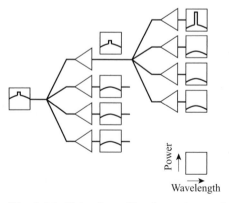

Fig. 5.34. Noise funnelling in upstream data transport in an access network

to-cell timescale of typically 10 ns. Other applications of SOAs may be found in the area of wavelength conversion and signal regeneration.

References

1. R. Heidemann, B. Wedding, and G. Veith, "10 Gb/s transmission and beyond," Proc. IEEE **81**, 1558 (1993)
2. A.R. Chraplyvy, "Limitations on lightwave communications imposed by optical-fibre nonlinearities," J. Lightwave Technol. **8**, 1548 (1990)
3. R.J. Mears, L. Reekie, I.M. Jauncey, and D.N. Payne, "Low-noise erbium-doped fibre amplifier operating at 1.54 μm," Electron. Lett. **23**, 1026 (1987)
4. E. Desurvire, J.R. Simpson, and P.C. Becker, "High-gain erbium-doped travelling-wave fibre amplifier," Opt. Lett. **12**, 888 (1987)
5. E. Snitzer, H. Po, F. Hakimi, R. Tumminelli, and B.C. McCollum, "Erbium fibre laser amplifier at 1.55 μm with pump at 1.49 μm and Yb sensitized Er oscillator," *Conf. Opt. Fibre Commun. (OFC '88)* (Opt. Soc. America, Washington, DC 1988), paper PD2
6. R.I. Laming, L. Reekie, P.R. Morkel, and D.N. Payne, "Multichannel crosstalk and pump noise characterisation of Er^{3+}-doped fibre amplifier pumped at 980 nm," Electron. Lett. **25**, 455 (1989)
7. W.J. Miniscalco, "Optical and electronic properties of rare earth ions in glasses," in *Rare Earth Doped Fibre Lasers and Amplifiers* (M.J.F. Digonnet, ed.), Chap. 2 (Marcel Dekker, Inc., New York 1993)
8. J.R. Armitage, "Introduction to glass fibre lasers and amplifiers," in *Optical Fibre Lasers and Amplifiers* (P.W. France, ed.), Chap. 2 (Blakie and Son Ltd., Glasgow and London 1991)
9. W.J. Miniscalco, "Erbium-doped glasses for fibre amplifiers at 1500 nm," J. Lightwave Technol. **9**, 234 (1991)
10. E. Desurvire, "Characteristics of erbium-doped fibres," in *Erbium-doped fibre amplifiers, Principles and Applications*, Chap. 4 (J. Wiley and Sons, Inc., New York 1994)
11. C.R. Giles and E. Desurvire, "Modeling erbium-doped fibre amplifiers," J. Lightwave Technol. **9**, 271 (1991)
12. B. Pedersen, A. Bjarklev, J.H. Povlsen, K. Dybdal, and C.C. Larsen, "The design of erbium-doped fibre amplifiers," J. Lightwave Technol. **9**, 1105 (1991)
13. P.F. Wysocki, "Accurate modeling of next generation rare-earth doped fibre devices," *OSA Trends in Optics and Photonics, Vol. 16, Optical Amplifiers and Their Applications*, (M.N. Zervas, A.E. Willner, and S. Sasaki, eds.) (Opt. Soc. America, Washington, DC 1997), p. 46
14. W.L. Barnes, R.I. Laming, E.J. Tarbox, and P.R. Morkel, "Absorption and emission cross section of Er^{3+} doped silica fibres," IEEE J. Quantum Electron. **27**, 1004 (1991)
15. W.J. Miniscalco and R.S. Quimby, "General procedure for the analysis of Er^{3+} cross sections," Opt. Lett. **16**, 258 (1991)
16. J.N. Sandoe, P.H. Sarkies, and S. Parke, "Variation of Er^{3+} cross section for stimulated emission with glass composition," J. Phys. D: Appl. Phys. **5**, 1788 (1972)

17. A.A.M. Saleh, R.M. Jopson, J.D. Evankow, and J. Aspell, "Modeling of gain in erbium-doped fibre amplifiers," IEEE Photon. Technol. Lett. **2**, 714 (1990)
18. C. Barnard, P. Myslinski, J. Chrostowski, and M. Kavenrad, "Analytical model for rare-earth-doped fibre amplifiers and lasers," IEEE J. Quantum Electron. **QE-30**, 1817 (1994)
19. E. Desurvire and J.R. Simpson, "Amplification of spontaneous emission in erbium-doped single-mode fibres," J. Lightwave Technol **7**, 835 (1989)
20. P.R. Morkel and R.I. Laming, "Theoretical modeling of erbium-doped fibre amplifiers with excited-state absorption," Opt. Lett. **14**, 1062 (1989)
21. C.R. Giles and E. Desurvire, "Propagation of signal and noise in concatenated erbium-doped fibre amplifiers," J. Lightwave Technol. **9**, 147 (1991)
22. B. Pedersen, "Small-signal erbium-doped fibre amplifiers pumped at 980 nm: a design study," Opt. Quantum Electron. **26**, 273 (1994)
23. H. Heffner, "The fundamental noise limit of linear amplifiers," Proc. Inst. Radio Eng. **50**, 1604 (1962)
24. H.A. Haus and J.A. Mullen, "Quantum noise in linear amplifiers," Phys. Rev. **128**, 2407 (1962)
25. K. Shimoda, H. Takahasi, and C.H. Townes, "Fluctuations and amplification of quanta with application to maser amplifiers," J. Phys. Soc. Japan **12**, 686 (1957)
26. Y. Yamamoto, "Noise and error rate performance of semiconductor laser amplifiers in PCM-IM optical transmission systems," IEEE J. Quantum Electron. **QE-16**, 1073 (1980)
27. E. Desurvire, *Fundamentals of noise in optical fibre amplifiers, Erbium-doped fibre amplifiers, Principles and Applications* (J. Wiley and Sons, Inc., New York 1994), Chap. 2
28. J.A. Arnaud, "Enhancement of optical receiver sensitivities by amplification of the carrier," IEEE J. Quantum Electron. **QE-4**, 893 (1968)
29. E. Desurvire, "Comment: Optical amplifier noise-figure reduction using a variable polarization beam splitter," Electron. Lett. **31**, 1743 (1995)
30. R. Olshansky, "Noise figure for erbium-doped optical fibre amplifiers," Electron. Lett. **24**, 1363 (1990)
31. H. Steinberg, "The use of a laser amplifier in a laser communication system," Proc. IEEE 51 (1963)
32. S.D. Personick, "Applications of quantum amplifiers in simple digital optical communication systems," Bell System Tech. J. **52**, 117 (1973)
33. A. Yariv, "Signal-to-noise considerations in fibre links with periodic or distributed optical amplification," Opt. Lett. **15**, 1064 (1990)
34. M.N. Zervas, R.I. Laming, and D.N. Payne, "Efficient erbium-doped fibre amplifiers incorporating an optical isolator," IEEE J. Quantum Electron. **QE-31**, 472 (1995)
35. M.N. Zervas, R.I. Laming, J.E. Townsend, and D.N. Payne, "Design and fabrication of high gain efficiency erbium-doped fibre amplifier," IEEE Photon. Technol. Lett. **4**, 1342 (1992)
36. M. Shimizu, M. Yamada, M. Horiguchi, T. Takeshita, and M. Oyasu, "Erbium-doped fibre amplifiers with an extremely high gain coefficient of 11.0 dB/mW," Electron. Lett. **26**, 1641 (1990)
37. T. Kashiwada, M. Shigematsu, T. Kougo, H. Kanamori, and M. Nishimura, "Erbium-doped fibre amplifier pumped at 1.48 µm with extremely high efficiency," IEEE Photon. Technol. Lett. **3**, 721 (1991)

38. O.C. Graydon, M.N. Zervas, and R.I. Laming, "Erbium-doped-fibre optical limiting amplifier," J. Lightwave Technol. **13**, 732 (1995)

39. J.F. Massicott, J.R. Armitage, R. Wyatt, B.J. Ainslie, and S.P. Craig-Ryan, "High gain, broadband 1.6 μm Er^{3+}-doped silica fibre amplifier," Electron. Lett. **26**, 1645 (1990)

40. M. Tachibana, R.I. Laming, P.R. Morkel, and D.N. Payne, "Erbium-doped fibre amplifier with flattened gain spectrum," IEEE Photon. Technol. Lett. **3**, 118 (1991)

41. B.J. Ainslie, "A review of the fabrication and properties of erbium-doped fibres for optical amplifiers," J. Lightwave Technol. **9**, 220 (1991)

42. D. Ronarc'h, M. Guibert, H. Hibrahim, M. Monerie, H. Poignant, and A. Tromeur, "30 dB optical net gain at 1.543 μm in Er^{3+} doped fluoride fibre pumped around 1.48 μm," Electron. Lett. **27**, 908 (1991)

43. M. Yamada et al., "Gain-flattened tellurite-based EDFA with a flat amplification bandwidth of 76 nm," Proc. *Optical Fibre Communications Conference (OFC '98)*, paper PD7

44. J.F. Massicott, R. Wyatt, B.J. Ainslie, and S.P. Craig-Ryan, "Efficient high-power, high gain Er^{3+}-doped silica fibre amplifier," Electron. Lett. **26**, 1038 (1990)

45. B. Pedersen, M.L. Dakss, B.A. Thompson, W.J. Miniscalco, T. Wei, and L.J. Andrews, "Experimental and theoretical analysis of efficient erbium-doped fibre power amplifiers," IEEE Photon. Technol. Lett. **3**, 1085 (1991)

46. R.I. Laming, J.E. Townsend, D.N. Payne, F. Meli, G. Grasso, and E.J. Tarbox, "High-power erbium-doped-fibre amplifiers operating in the saturated regime," IEEE Photon. Technol. Lett. **3**, 253 (1991)

47. R.I. Laming, A. H, Gnauck, C.R. Giles, M.N. Zervas, and D.N. Payne, "High-sensitivity two-stage erbium-doped fibre preamplifier at 10 Gbit/s," IEEE Photon. Technol. Lett. **4**, 1348 (1992)

48. R.I. Laming, M.N. Zervas, and D.N. Payne, "Erbium-doped fibre amplifier with 54 dB gain and 3.1 dB noise figure," IEEE Photon. Technol. Lett. **4**, 1345 (1991)

49. M.N. Zervas and R.I. Laming, "Rayleigh scattering effect on the gain efficiency and noise of erbium-doped fibre amplifiers," IEEE J. Quantum Electron. **QE-31**, 468 (1995)

50. M.N. Zervas and R.I. Laming, "Twin-core erbium-doped channel equaliser," J. Lightwave Technol. **13**, 721 (1995)

51. Y. Ohishi, T. Kanamori, T. Kitagawa, S. Takahashi, E. Snitzer, and G.H. Sigel Jr, "Pr^{3+}-doped fluoride fibre amplifier operating at 1.31 μm," Opt. Lett. **16**, 1747 (1991)

52. Y. Nishida, M. Yamada, T. Kanamori, K. Kobayashi, J. Temmyo, S. Sudo, and Y. Ohishi, "Development of an efficient praseodymium-doped fibre amplifier," IEEE J. Quantum Electron. **QE-34**, (8), 1332 (1998)

53. D.W. Hewak, R.C. Moore, T. Schweizer, J. Wang, B. Samson, W.S. Brocklesby, D.N. Payne, and E.J. Tarbox, "Gallium lanthanum sulfide optical fibre for active and passive applications," Electron. Lett. **32**, 384 (1996)

54. R.H. Stolen and E.P. Ippen, Appl. Phys. Lett. **22**, 276 (1973)

55. M. Nakazawa, "Raman optical amplification," *Optical Amplifiers and Their Applications* (S. Shimada and H. Ishio, eds.), Sect. 12.4.1 (John Wiley & Sons 1994)

56. S.G. Grubb, T. Erdogan, V. Mizrahi, T. Strasser, W.Y. Cheung, W.A. Reed et al., "1.3 µm cascaded Raman amplification in germano-silicate fibre," *Optical Amplifiers and Their Applications*, Techn. Digest Series (Opt. Soc. America, 1994), paper PD3

57. A.K. Srivastava et al., "High speed WDM transmission in AllWaveTM fibre in both 1.3 µm and 1.5 µm bands," *Optical Amplifiers and Their Applications*, Techn. Digest Series (Opt. Soc. America, 1998), paper PD2

58. L. Eskildsen, P.B. Hansen, S.G. Grubb et al., "Capacity upgrade of transmission systems by Raman amplification," OSA TOPS on Optical Amplifiers and Their Applications **5**, 306 (1996)

59. H. Masuda, S. Kawai, and K. Aida, "76 nm 3 dB gain-band hybrid fibre amplifier without gain equaliser," *Optical Amplifiers and Their Applications*, Techn. Digest Series (Opt. Soc. America, 1998), paper PD7

60. A. Yariv, *Optical Electronics* (Holt-Saunders, New York 1985)

61. C. Kittel, *Introduction to Solid State Physics* (Wiley, New York 1986)

62. L.F. Tiemeijer, P.J.A. Thijs, T. van Dongen, J.J.M. Binsma, E.J. Jansen, and S. Walczyk, "33 dB fibre to fibre gain, +13 dBm fibre saturation power polarization independent 1310 nm MQW laser amplifiers," *Proc. Optical Amplifiers and their Applications*, Techn. Digest 1995, Vol. **18**, PD1–5

63. L.F. Tiemeijer, P.J.A. Thijs, T. van Dongen, J.J.M. Binsma, and E.J. Jansen, "Polarization resolved, complete characterization of 1310 nm fibre pigtailed multiple-quantum-well optical amplifiers," J. Lightwave Technol. **14**, 1524 (1996)

64. L.F. Tiemeijer, "High performance MQW laser amplifiers for transmission systems operating in the 1310 nm window at bitrates of 10 Gbit/s and beyond," *Proc. 21st Europ. Conf. Opt. Commun.(ECOC '95)*, Brussels, Belgium, paper Tu.B. 2, **1**, 259 (1995)

65. A.E. Kelly, I.F. Lealman, L.J. Rivers, S.D. Perrin, and M. Silver, "Polarization insensitive, 25 dB gain semiconductor laser amplifier without antireflection coatings," Electron. Lett. **32**, 1835 (1996)

66. S. Kitamura, K. Komatsu, and M. Kitamura, "Very low power consumption semiconductor optical amplifier array," IEEE Photon. Technol. Lett. **7**, 147 (1995)

67. Ch. Holtmann, P.-A. Besse, T. Brenner, and H. Melchior, "Polarization independent bulk active region semiconductor optical amplifiers for 1.3 µm wavelength," IEEE Photon. Technol. Lett. **8**, 343 (1996)

68. P. Doussiere, "Recent advances in conventional and gain clamped semiconductor optical amplifiers," *Proc. Optical Amplifiers and their Applications*, Techn. Digest 1996, Vol. **11**, 220 (1996)

69. J-Y. Emery, P. Doussiere, L. Goldstein, F. Pommereau, C. Fortin, R. Ngo, N. Tscherptner, J-L. Lafragette, P. Aubert, F. Brillouet, G. Laube, and J. Barrau, "New process tolerant, high performance 1.55 µm polarization insensitive semiconductor optical amplifiers based on low tensile bulk GaInAsP," *Proc. 22nd Europ. Conf. Opt. Commun. 1996 (ECOC '96)*, Oslo, Norway, paper WeP. 2, p. 3.165 (1996)

70. L.F. Tiemeijer, P.J.A. Thijs, T. van Dongen, R.W.M. Slootweg, J.M.M. van der Heijden, J.J.M. Binsma, and M.P.C.M. Krijn, "Polarization insensitive multiple quantum well laser amplifiers for the 1300 nm window," Appl. Phys. Lett. **62**, 826 (1993)

71. P.J.A. Thijs, L.F. Tiemeijer, J.J.M. Binsma, and T. van Dongen, "Progress in long-wavelength strained-layer InGaAs(P) quantum-well semiconductor lasers and amplifiers," IEEE J. Quantum Electron. **30**, 477 (1994)

72. *Optical Amplifiers and their Applications* (S. Shimada and H. Ishio, eds.) (Wiley, Chichester 1994), Chap. 3

73. T.D. Visser, B. Demeulenaere, J. Haas, D. Lenstra, R. Baets, and H. Blok, "Confinement and modal gain in dielectric waveguides," J. Lightwave Technol. **14**, 885 (1996)

74. C.H. Henry, "Theory of spontaneous emission noise in open resonators and its application to lasers and optical amplifiers," J. Lightwave Technol. **LT-4**, 288 (1986)

75. Ch. Holtmann, P.-A. Besse, H. Melchior, and D.L. Williams, "Gain-clamped semiconductor optical amplifiers for 1.3 µm wavelength with 20 dB polarization independent fibre-to-fibre gain and significantly reduced pulse shape and intermodulation distortions," *Proc. 21st Europ. Conf. Opt. Commun.(ECOC '95)*, Brussels, paper Th.B. 3.7, (1995)

76. L.F. Tiemeijer, G.N. van den Hoven, P.J.A. Thijs, T. van Dongen, J.J.M. Binsma, and E.J. Jansen, "1310 nm DBR-type MQW gain-clamped semiconductor optical amplifiers with AM-CATV-grade linearity," IEEE Photon. Technol. Lett. **8**, 1453 (1996)

77. P. Doussiere, F. Pommereau, J-Y. Emery, R. Ngo, J-L. Lafragette, P. Aubert, L. Goldstein, G. Soulage, T. Ducellier, M. Bachmann, and G. Laube, "1550 nm polarization independent DBR gain clamped SOA with high dynamic input power range," *Proc. 22nd Europ. Conf. Opt. Commun. 1996 (ECOC '96)*, Oslo, Norway, paper WeD. 2.4, p. 3.169 (1996)

78. J.L. Pleumeekers, T. Hessler, S. Haacke, M.-A. Dupertuis, P.E. Selbmann, R.A. Taylor, B. Deveaud, T. Ducellier, P. Doussire, M. Bachmann, and J.-Y. Emery, "Relaxation oscillations in the gain recovery of gain-clamped semiconductor optical amplifiers: Simulation and experiments," *Proc. Optical Amplifiers and their Applications*, Victoria, BC, paper WB5, p. 224 (1997)

79. F. Tiemeijer, P.J.A. Thijs, T. van Dongen, J.J.M. Binsma, E.J. Jansen, P.I. Kuindersma, G.P.J.M. Cuijpers, and S. Walczyk, "High-output power (+15 dBm) unidirectional 1310 nm multiple-quantum-well booster amplifier module," IEEE Photon. Technol. Lett. **7**, 1519 (1995)

80. P.I. Kuindersma, G.P.J.M. Cuijpers, J.G.L. Jennen, J.J.E. Reid, L.F. Tiemeijer, H. de Waardt, and A.J. Boot, "10 Gbit/s RZ transmission at 1309 nm over 420 km using chain of multiple quantum well semiconductor optical amplifier modules at 38 km intervals," *Proc. 22nd Europ. Conf. Opt. Commun. 1996 (ECOC '96)*, Oslo, Norway, paper TuD. 2.1, p. 2.165 (1996)

81. A. Shipulin, G. Onishchukov, P. Riedel, D. Michaelis, U. Peschel, and F. Lederer, "10 Gbit/s signal transmission over 550 km in standard fibre at 1300 nm using semiconductor optical amplifiers," Electron. Lett. **33**, 507 (1997)

82. R.C.J. Smets, J.C.L. Jennen, H. de Waardt, B. Teichmann, C. Dorschky, R. Seitz, J.J.E. Reid, L.F. Tiemeijer, and P.I. Kuindersma, "114 km, repeaterless, 10 Gbit/s transmission at 1310 nm using an RZ data format," *Proc. Conf. Opt. Fiber Commun. (OFC '97)*, Dallas, USA, paper ThH2, 269 (1997)

83. J.J.E. Reid, P.I. Kuindersma, G.P.J.M. Cuijpers, G.N. van den Hoven, S. Walczyk, B. Teichmann, C. Dorschky, R. Seitz, C. Schulien, L. Cucala, H. Gruhl, R. Leppla, and A. Mattheus, "High bitrate 1.3 µm optical transmission in the

field using cascaded semiconductor optical amplifiers," *Proc. 23rd Europ. Conf. Opt. Commun. (ECOC '97)*, Edinburgh, UK, paper Mo3A. **5**, 83 (1997)

84. P.I. Kuindersma, G.P.J.M. Cuijpers, J.J.E. Reid, G.N. van den Hoven, and S. Walczyk, "An experimental analysis of the system performance of cascades of 1.3 μm semiconductor optical amplifiers," *Proc. 23rd Europ. Conf. Opt. Commun. (ECOC '97)*, Edinburgh, UK, paper Mo3A. **4**, 79 (1997)

85. G.N. van den Hoven, P.I. Kuindersma, G.P.J.M. Cuijpers, L.F. Tiemeijer, T. van Dongen, J.J.M. Binsma, E.J. Jansen, and S. Walczyk, "Optimizing semiconductor optical amplifiers for transmission systems," *Proc. 23rd Europ. Conf. Opt. Commun. (ECOC '97)*, Edinburgh, UK, paper Tu2B. **2**, 90 (1997)

86. M. Bachmann, P. Doussiere, J-Y. Emery, R. Ngo, F. Pommereau, L. Goldstein, G. Soulage, and A. Jourdan, "SOA with integrated spot-size convertor and DBR gratings for WDM applications at 1.55 μm wavelength," Electron. Lett. **32**, 2076 (1996)

87. V.G. Mutalik, G.N. van den Hoven, and L.F. Tiemeijer, "Analog performance of 1310 nm gain-clamped semiconductor optical amplifiers," *Proc. Conf. Opt. Fiber Commun. (OFC '97)*, Dallas, USA, paper ThG4, 266 (1997)

88. M.D. Feuer, J.M. Wiesenfeld, J.S. Perino, C.A. Burrus, G. Raybon, S.C. Shunk, and N.K. Dutta, "Single part laser amplifier modulators for local access," IEEE Photon. Technol. Lett. **8**, 1175 (1996)

89. C. Tai, W.I. Way et al., "Dynamic range and switching speed limitation of an $N \times N$ optical packet switch based on low gain semiconductor optical amplifiers," IEEE J. Lightwave Technol. **14**, 525 (1996)

90. F. Masetti et al., "Design and implementation for a fully reconfigurable all-optical crossconnect for high capacity multi wavelength transport networks," IEEE J. Lightwave Technol. **14**, 979 (1996)

91. J-Y. Emery, M. Di Maggio, M. Bachmann, J. Le Bris, F. Pommereau, R. Ngo, C. Fortin, F. Dorgeuille, E. Grard, M. Renaud, and G. Laube, "High performance 1.55 μm 4 clamped gain semiconductor optical amplifiers array module for photonic switching applications," *Proc. Optical Amplifiers and their Applications*, Victoria, BC, paper TuC2, 112 (1997)

92. S. Kitamura, H. Hatakeyama, T. Kato, M. Yamaguchi, and K. Komatsu, "Very-low-operating-current SOA-gate modules for optical matrix switches," *Proc. Optical Amplifiers and their Applications*, Victoria, BC, paper TuC3, 116 (1997)

93. I. Van de Voorde, C. Martin, H. Slabbinck, L. Gouwy, B. Stubbe, X.Z. Qiu, J. Vandewege, and P. Solina, "Evaluation of the super PON LAB-demonstrator," *Proc. 23rd Europ. Conf. Opt. Commun. (ECOC '97)*, Edinburgh, UK, paper We4C, 331 (1997)

6 Passive and Active Glass Integrated Optics Devices

Antoine Kévorkian

6.1 General Introduction

Optical integration technologies were uncovered early in the emergence of the optical telecommunication field. As early as 1973, a review reference such as [1] summarized some of the basic theoretical tools and device concepts later implemented in actual commercial components. Over the 25 years that ensued, the technology of planar devices on glass evolved, and they became available at the beginning of the 1990s, in the form of passive splitters.

Glass planar devices are made with collective fabrication techniques close to those used in microelectronics. For this reason they are also part of integrated optics technology. One particular characteristic of integrated optics glass devices is that they have both optical inputs and optical outputs, in contrast to the transmitters and receivers described in Chaps. 3 and 4. As such, they offer a wide possible array of applications, which extends far over that of the commercially available passive splitters. Two main integrated optics technologies present the level of performance and reliability required for long-distance applications. The first one, mostly studied in the present chapter, is based on transforming a glass chemical composition by local ion exchange. The second type or group of technologies relies on doping an otherwise pure layer and then etching it to create the appropriate patterns and is considered in Chaps. 7 and 11.

Before moving to the technical part, the reader should be aware of the rapid pace of change that optical techniques have undergone recently. The technology has now become market-pulled and its progress is governed by a combination of several factors:

- The main market segment has evolved: The primary motor for growth in the optical telecommunication market has long been the long-distance segment. In the past, this market has mostly relied on semi-custom, single-function components assembled in simple optical subsystems. The more recent evolution, brought about by the combination of optical amplification and dense wavelength division multiplexing (DWDM), has moved requirements towards more complex subsystems. These, in turn, justify more integrated functions.

- Other market segments are growing: The ongoing European deregulation is predicted to be very effective in the actual competition it will generate. Since many European cities are closely located, the metropolitan area network will be at the heart of the expansion. The successful technological solutions will call for a better performance-to-cost ratio favorable to integrated technologies. Future evolution of the Access Network market will broaden the market even more, this time calling for very rugged, mass-produced devices.

- The scope and technological performances of planar devices have improved: Ion-exchange planar or doped-silica devices present several areas of applications for passive as well as active functions. Table 6.1 shows how, in addition to the presently marketed devices, several new functions are now moving out of the research laboratories into a pre-commercial phase.

A given technology usually has intrinsic advantages in relation to the underlying physical process and is then developed further to fit more closely the customer technical requirements.

Table 6.1. Status of ion-exchange and doped-silica technologies for some DWDM applications

	Commercial devices	Prototypes/R&D
Passive $1 \times N, 2 \times N$	Ion-exchange Doped-silica	Yes
Passive WDM $1.55\,\mu m/1.3\,\mu m$ $1.55\,\mu m/0.98\,\mu m$	Doped-silica	Yes
Passive phasars	Doped-silica	Ion-exchange Doped-silica
Active amplifiers	Ion-exchange	Ion-exchange Doped-silica
Active/Passive $1 \times N$ gain splitters	Under development	Ion-exchange
Active DWDM lasers	Under development	Ion-exchange Doped-silica

The specific technical advantages of ion-exchange devices are twofold:

- from the user's point of view they present:
  - highly polarization-independent properties,
  - low insertion losses (coupling and on-chip) regardless of the confinement level,
  - both passive and active functions.
- from the manufacturer's point of view they present large flexibility due to:
  - low scattering,
  - low waveguide (WG) crossing losses,
  - 3-D controllable confinement.

Section 6.2 reviews the principles and performance of ion-exchange passive devices. Section 6.3 studies the main features of more advanced Yb/Er co-doped integrated amplifiers, from both theoretical and application standpoints. Section 6.4 addresses integrated optics glass lasers.

6.2   Passive Power Splitters

6.2.1   Splitters and Their Basic Functions

Waveguide splitters can be divided into different families based on the number of outputs and inputs, as outlined in Table 6.2. Application engineers expect such devices to present low insertion loss, polarization and wavelength dependence. Actual performance specifications depend on design, wafer processing and fibre-to-WG assembly. The following part of this section studies the various subelements of WG splitters and their technology.

$1 \times N$ divider/combiners were the first planar devices to become commercial. They generally exhibit very small wavelength dependence, so that their range of operation may span over the full range of wavelength between 1280 and 1680 nm. They also present the lowest excess loss and highest port-to-port uniformity. 2×2 and more generally $N \times N$ couplers are more wavelength dependent because they rely to some degree on mode interference to achieve the required functionality.

Two types of commercial technology are currently used to manufacture splitter devices: silica-on-silicon deposition and ion exchange in glass. Most of the theoretical and experimental description of circuit elements given in this section is relevant to the two technologies. However, manufacturing technology is only considered for ion exchange, since silica deposition is covered in other chapters.

As expected from the schematic designs in Table 6.2, all splitters are based on channel-WG structures, which provide the input and output (I/O) arms for attachment to fibres. In turn, these WGs can be bent, enlarged or branched out to provide the splitting function. I/O fibres are positioned in

Table 6.2. Families of $N \times P$ waveguide splitters

Schematic design	Function and practical number of ports
	1×2^n power divider/combiner
	$2^n \leq 128$
	$1 \times N$ power divider/combiner(N arbitrary)
	$N \leq 10$
	2×2 coupler
	$N \times N$ star coupler
	$N \leq 144$

front of the I/O waveguides of the chips with precision holders and permanently attached to the chips with adhesives. We will follow the simple order summarized in Fig. 6.1 to guide the study. Mode solvers are used for straight and bent portions of WGs and for abrupt discontinuities; coupled-mode theory is adapted to the modelling of couplers; finally, local-mode theory and beam propagation methods (BPM) can be used for general slowly varying structures.

Let us first start with a brief overview of a channel WG. It is essentially a local increase of the optical index across a section of the material, such as schematized in Fig. 6.2. In an ion-exchange structure, the index increase is graded, whereas for deposition methods, piecewise domains of constant index are juxtaposed, leading to higher index steps. Typical transverse dimensions of the cores range between 4 and 10 μm, so that the WG area is about 50 to 300 μm^2.

Mode solvers for straight and bent WGs are useful to derive precisely the intensity profile of the modes. Several books have dealt in depth with the description of dielectric WG modes [2,3]. Ion-exchange structures exhibit the specific property of presenting abrupt discontinuities at the air/glass interface, while presenting a graded ion distribution inside the glass. Taking into account the air/glass interface is important in correctly evaluating the

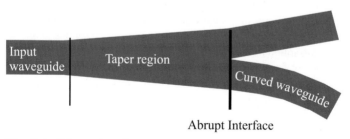

Fig. 6.1. Schematic subparts of a 1×2 splitter

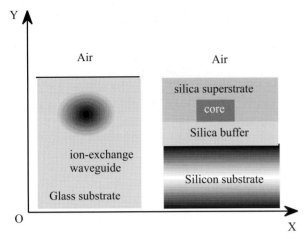

Fig. 6.2. Cross-sectional view of ion-exchange and silica-on-silicon waveguides

residual effect of dust particles or other perturbations, which could be present at the upper surface.

6.2.2   Computing Waveguide Modes

WGs, among which optic fibres make a particular subclass, exhibit particular "modes" of light propagation. Light may indeed be propagated in these structures with a transverse distribution of constant intensity along the propagation. In order to understand how the optical functions can be designed on planar circuits, more should be understood about WG modal properties.

We start with a general configuration and gradually show how various simplifications can make it tractable with a reasonable amount of computer power and programing time.

In the general case, one looks for a solution of Maxwell equations for a wave with pulsation ω propagated along an index structure invariant along z:

$$
\begin{aligned}
\nabla \times E &= -\mathrm{i}\omega\mu_0 H \\
\nabla \times H &= \mathrm{i}\omega\varepsilon H
\end{aligned}
\tag{6.1}
$$

where E is the electric field and H the magnetic field at angular frequency $\omega = 2\pi\nu$, μ_0 is the vacuum permeability and ϵ is the permittivity of the medium. ϵ is discontinuous across step index interfaces such as the air/glass interface and therefore (6.1) is true in the sense of distributions.

To solve for the guided modes of the WG structure is equivalent to:

(a) look for solutions of (6.1) where the field component dependence on x and y is separable from z, due to the physical invariance of the WG along z,
(b) summarize the dependence on z by a mostly harmonic exponential term,

(c) apply a so-called radiation condition, which states that no light comes from the lateral sides of the WG to sustain the WG mode when propagating along z.

Thus, any solution of the guided-wave problem represents a phenomenon originated by source terms located at $z = -\infty$ which are not explicitly described. The light remains in the WG and is therefore guided as it propagates without additional supply of energy.

Applying conditions (a) and (b), one looks for solutions of the form:

$$E = E(x,y) \exp(i\beta z)$$
$$H = H(x,y) \exp(i\beta z) \quad . \tag{6.2}$$

The constant β is called the effective propagation constant along z. It describes how the guided wave effectively progresses at a slower speed than if it were going along a straight path in the substrate or superstrate of the circuit. Condition (c) induces a particular property of guided modes: only a finite number of eigenmode $\{\beta, E_\beta(x,y)\}$ solutions of (6.2) are allowed (or in some cases none at all). The corresponding $H_\beta(x,y)$ can then be deduced through Maxwell's equations. In fact, one may classify the solutions by inspecting the range of allowed practical values for β in relation to condition (c).

Let us first note that in a homogeneous material, where the relative permittivity $\varepsilon/\varepsilon_0$ is constant in the entire space, β simply reduces to the propagation constant of a plane wave and E and H do not depend on (x, y). Now, returning to the WG structure, we also note that in order for the light to remain confined in the WG core the effective propagation constant should be higher than the propagation constants of the surrounding media. Indeed, if otherwise, possible coupling to plane waves would enable leakage from the core.

We therefore come to a first inequality:

$$\beta \geq \max\left(\frac{2\pi}{\lambda} n_{\text{superstrate}}, \frac{2\pi}{\lambda} n_{\text{substrate}}\right) \quad . \tag{6.3}$$

Similarly, if a mode is to propagate along z, there should be at least one medium, the core, with a propagation constant high enough to accommodate guided-wave propagation. In other words, if n_{max} is the maximum index of refraction found in the structure:

$$\beta < \frac{2\pi}{\lambda} n_{\text{max}} \quad . \tag{6.4}$$

If one of these conditions is not met, true guided behaviour is not possible: an exchange of energy takes place between the WG and the surrounding media. This happens, for example, when a plane wave crosses the WG or when energy leaks away from the core.

Solutions still exist to the general problem, but lie outside of the validity domain of condition (c). In this case, there exists a continuum of solutions.

Table 6.3 summarizes all the practical cases for a simple rectangular WG structure. The left column presents a schematic view of light intensities (shaded ellipses) and possible energy fluxes, the central column classifies the associated range of propagation constants and the right column the corresponding physical situation.

The finite number of guided modes and the continuum spectrum of radiation modes form a complete set of solutions, onto which *any* light propagating in or even crossing the WG can be projected. For a given WG structure, there is a wavelength beyond which only one single guided mode exists. Such a mode is called the fundamental mode of the WG. Most devices for telecom applications operate in the single-mode regime. At even longer wavelengths, the fundamental mode itself may experience cut-off, except in the case of structures with a particular geometrical symmetry.

One way to solve the general problem is to combine the set of equations (6.1) and project them to obtain the transverse vectorial Helmholtz equation:

Table 6.3. Classification of WG eigenmodes

Intensity	Domain of validity	Physical situation
	$\beta \geq \frac{2\pi}{\lambda} n_{\mathrm{max}}$	No physical propagation allowed, no solution
	$\beta < \frac{2\pi}{\lambda} n_{\mathrm{max}}$ and $\beta \geq \max\left(\frac{2\pi}{\lambda} n_{\mathrm{sup}}, \frac{2\pi}{\lambda} n_{\mathrm{sub}}\right)$	True guided wave regime. A finite number of solutions exist. Condition (c) applies
	$\beta < \max\left(\frac{2\pi}{\lambda} n_{\mathrm{sup}}, \frac{2\pi}{\lambda} n_{\mathrm{sub}}\right)$ and $\beta \geq \min\left(\frac{2\pi}{\lambda} n_{\mathrm{sup}}, \frac{2\pi}{\lambda} n_{\mathrm{sub}}\right)$	Energy radiates away or is reflected on one side of the WG. A continuum of solutions exists. Condition (c) does not apply
	$0 < \beta < \min\left(\frac{2\pi}{\lambda} n_{\mathrm{sup}}, \frac{2\pi}{\lambda} n_{\mathrm{sub}}\right)$	Energy propagates across the WG section. A continuum of solutions exists. Condition (c) does not apply
	$\beta \leq 0$	All modes evanesce, no propagation of radiation modes. Condition (c) does not apply

$$\Delta_T \overline{E}_T = \left(\beta^2 - k_0^2 \varepsilon\right) \overline{E}_T + \overline{\nabla}_T \left(\varepsilon \overline{E}_T \cdot \left\{\overline{\nabla}_T \varepsilon^{-1}\right\}\right)$$
$$+ \overline{\nabla}_T \left(\varepsilon \overline{E}_T \cdot \overline{u_S} \sigma \left(\varepsilon^{-1}\right) \delta_S\right) , \tag{6.5}$$

where $k_0 = \frac{2\pi}{\lambda}$ is the vacuum wave vector, $\{d\}$ denotes the regular part of the distribution d, $\sigma(d)$ the jump of d across an abrupt interface, δ_S is the Dirac distribution associated with the abrupt interface and $\overline{u_S}$ the unit vector normal to the interface.

All functional operators and fields are transverse, which is expressed with the x and y components only.

Equation (6.5) is quite complex to solve as it is fully vectorial and includes the jumps due to the discontinuities at interfaces. A close inspection of the E-field components and their partial derivatives shows that their jumps at an interface parallel to x may be described by the following set of relations:

$$\sigma\left(E_x\right) = \sigma\left(E_z\right) = 0 ,$$

$$\sigma\left(E_y\right) = \varepsilon E_y \sigma\left(\varepsilon^{-1}\right) ,$$

$$\sigma\left(\frac{\partial E_x}{\partial y}\right) = \sigma\left(\frac{\partial E_y}{\partial x}\right) = \frac{\partial \left(\varepsilon E_y \sigma\left(\varepsilon^{-1}\right)\right)}{\partial x} = \frac{\partial \sigma\left(E_y\right)}{\partial x} ,$$

$$\sigma\left(\frac{\partial E_y}{\partial y}\right) = E_x \sigma\left(\left\{\frac{\partial \log\left(\varepsilon^{-1}\right)}{\partial x}\right\}\right) + \varepsilon E_y \sigma\left(\left\{\frac{\partial \varepsilon^{-1}}{\partial y}\right\}\right) ,$$

$$\sigma\left(\frac{\partial E_x}{\partial x}\right) = 0 . \tag{6.6}$$

The first two relations describe the jump of the field's normal component across the interface, whereas the latter relations describe how some of the derivatives both jump and couple the field components at the same interface. In practice, however, these terms will only play a role if the derivatives of the dielectric constant are non-vanishing near the surface. Since the upper interface is usually far enough from the WG to reduce the impact of surface unevenness or impurities, one is justified in seeking approximations that are useful in reducing these general equations to several simpler ones.

Semi-vectorial and Scalar Approximations

The semi-vectorial approximation neglects the coupling between field components in the term $\nabla\left(\varepsilon E \cdot \left\{\nabla \varepsilon^{-1}\right\}\right)$ in (6.5) and the scalar-mode approximation neglects the term altogether. The semi-vectorial approximation retains jumps of E-field components normal to the dielectric interface as well as their gradient when they are continuous. This leads to a decoupling of the field components, and the solutions can be split into two independent families. The first is described by a nearly transverse electric field (here parallel to y) called the TE modes, and the other by a nearly transverse magnetic field called the TM modes.

The use of the scalar approximation is justified by the relatively slow variation of refractive index in the core for ion-exchange guides and the low core-cladding variation of the index for step-index guides. Efficient algorithms based on these approximations compute the E-field and the effective propagation constants, for example, with finite difference schemes [4]. They are, however, relatively complex to implement and require significant computer power.

Effective Index Methods

The effective index methods go one step further by separating the dependencies on x and y and splitting the channel WG problem into two consecutive planar WG problems. There are several versions of the effective index methods. In general, these methods offer an accurate and fast-to-compute approximation of the effective refractive index constant, but are less accurate at providing the transverse E-field components. They are, however, simple to implement and useful for evaluating residual superstrate losses as well as the residual birefringence created by the interface. We present here a version suited for graded-index structures and originally developed in [5,6].

Write the transverse field component as a product of two functions:

$$E\left(x, y\right) = E_x E_{xy} , \tag{6.7}$$

where E_{xy} is a function of y and a slowly varying function of x, and E_x a function of x only. Then,

$$\frac{\partial^2 E\left(x, y\right)}{\partial x^2} \approx \frac{\partial^2 E_x}{\partial x^2} E_{xy} . \tag{6.8}$$

Slice the structure along x as shown in Fig. 6.3. For a given x, define an effective index $n_{\text{eff}}\left(x\right)$ such that

$$\frac{\partial^2 E_{xy}}{\partial y^2} + k_0^2 \left[\varepsilon\left(x, y\right) - n_{\text{eff}}^2\left(x\right)\right] E_{xy} = 0 . \tag{6.9}$$

For each value of x, (6.9) is a unidimensional homogeneous differential equation in y with solutions $\{E_{xy}, n_{\text{eff}}\left(x\right)\}$. It is equivalent to finding the modes of a planar WG structure consisting of layers with refractive index equal to those of the slice. Far enough from the central slices, however, the core layer stops and the index increase relative to the substrate goes to zero. In other

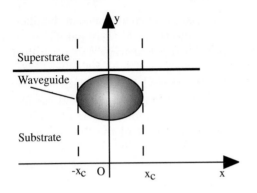

1: Actual index structure

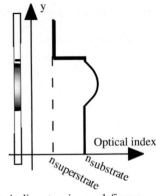

2: A slice at a given x defines a particular $n_{\text{eff}}(x)$

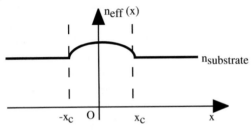

3: Compute the effective index of the equivalent planar waveguide with optical index $n_{\text{eff}}(x)$

Fig. 6.3. The effective index method

words, there is a value x_c beyond which no mode exists in the equivalent planar WG, and modelling may become a problem. An approximate solution is to notice that in this region of the WG, $n_{\text{eff}}(x)$ goes to $n_{\text{substrate}}$. One therefore chooses to replace $n_{\text{eff}}(x)$ with $n_{\text{substrate}}$ for $|x| \geq x_c$.

Once this first reduction is carried out, one then solves a new 1-D equation of a planar structure with layers assuming refractive indices equal to the effective indices computed as the solution of (6.9):

$$\frac{\partial^2 E_x}{\partial x^2} + \left[k_0^2 n_{\text{eff}}^2(x) - \beta^2 \right] E_x = 0 . \qquad (6.10)$$

Solving (6.10) yields the effective propagation constant β as well as the approximate E-field components. Solving 1-D equations such as (6.9)–(6.10) and their associated planar layer structure is straightforward and explained next.

1-D Waveguide Problem

Computing the modes of a graded-index or multilayer planar structure is computationally efficient because analytical forms exists for the fields in any homogeneous layer. In addition, part of the vectorial nature of the solution can be recovered by taking into account two families of solutions, the TE and TM modes.

In the present example, the WG structure is invariant in the (x, z) plane. Maxwell's equations can be decoupled into two families of independent variables:

- TE modes, linking the partial derivatives of E_x, H_y, H_z,
- TM modes, linking the partial derivatives of H_x, E_y, E_z.

One can then derive a mathematical propagation equation valid for both solutions:

$$\frac{\partial^2 \Omega}{\partial y^2} + \left[k_0^2 n^2(y) - \beta^2\right] \Omega = 0 , \tag{6.11}$$

with $\Omega = E_x$ for TE and $\Omega = H_x$ for TM.

Although direct integration algorithms may be applied, one can also approximate the planar WG structure with a stack of constant-index layers to represent the graded-index part when present, as shown in Fig. 6.4. Solutions to (6.11) can then be expressed in layer number j by:

$$\Omega = \left[A_j^- \exp\left(-ik_{yj}y\right) + A_j^+ \exp\left(+ik_{yj}y\right)\right] \exp\left(i\beta z\right) , \tag{6.12}$$

where $A_j^{\pm}$ is the amplitude of a plane wave going up (+) or down (−) with respect to y, $k_0^2 \varepsilon_j = k_{yj}^2 + \beta^2$ and k_{yj} is determined in the complex plane by $\mathrm{Re}\,(k_{yj}) + \mathrm{Im}\,(k_{yj}) > 0$.

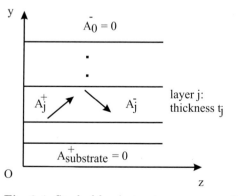

Fig. 6.4. Stacked-layer structure corresponding to the 1-D equation

For a given layer j, we define a reflection coefficient r_j, such that $r_j = \frac{A_j^+}{A_j^-}$, where r_j is the reflection coefficient of a wave propagating in the negative y direction at the top of layer j. Use the continuity (in-plane E-field component) or jumps (orthogonal E-field component) of the fields between layers j and $j+1$ and relate r_j to r_{j+1}:

$$r_j = \frac{N_j^+ r_{j+1} + N_j^-}{N_j^- r_{j+1} + N_j^+} \exp\left(-\mathrm{i}2k_{yj}t_j\right), \tag{6.13}$$

where $N_j^+ = 1 + \alpha_j \frac{k_{yj+1}}{k_{yj}}$ and $N_j^- = 1 - \alpha_j \frac{k_{yj+1}}{k_{yj}}$ and $\alpha_j = \begin{cases} 1 & \text{for TE} \\ \frac{\varepsilon_j}{\varepsilon_{j+1}} & \text{for TM} \end{cases}$.

Compute the reflection coefficient of the multilayer structure, for example, as viewed from the top, by iterating (6.13) from the substrate to the superstrate. The radiation condition is applied in the substrate as an initial condition for the iteration $r_{\mathrm{substrate}} = 0$, since $A_{\mathrm{substrate}}^+ = 0$. It is also applied to find the resonance condition for the multilayer structure: $A_0^- = 0$, so that the modes are defined by the values of β such that

$$N_0^- r_1 + N_0^+ = 0. \tag{6.14}$$

6.2.3   Tapers and Branches

The computation of WG modes is only valid for structures with no variation along the direction of propagation. In practice, optical functions always require changes in the shape of the guides. In some cases, the modal analysis may be adapted to study gradual modifications of WGs with the so-called local-normal-mode theory. In this approach, the WG structure is divided into thin segments of constant index distribution, *along the direction of propagation*, as represented in Fig. 6.5.

In a given slice, the total electric field is written as a linear combination of all the eigenmode fields, over the set of discrete guided modes and the

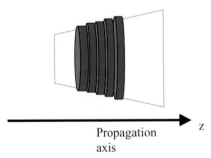

Propagation
axis

z

Fig. 6.5. Division of a slowly varying structure into constant profile slices

continuum of radiation modes:

$$E(x, y, z) = \sum_{g \text{ guided modes}} a_g E_g(x, y) + \int_{\beta \text{ rad. modes}} a(\beta) E_\beta(x, y) \, . \tag{6.15}$$

From one slice to the next, the local eigenmode family varies a little, so that a given field propagated on one slice will project over a slightly different linear combination of the local modes when entering the next slice. If the change in dimension is gradual enough, an input field consisting solely of the fundamental guided mode will see vanishingly small coupling to other local modes. One then speaks of an adiabatic process, since no energy is exchanged between the local fundamental mode and the other local modes during the propagation. The condition for an adiabatic regime (see Chap. 5 in [7]) can be written as a function of the linear change of the dielectric profile along the propagation axis:

$$\left| \iint_{\text{WG section}} E_g(x, y) E_p(x, y) \frac{\partial \varepsilon(x, y)}{\partial z} \right| \ll |\beta_g - \beta_p| \, , \tag{6.16}$$

where the subscript g denotes a given guided mode and p is any other mode locally supported by the WG. The z dependence of the fields and propagation constants is understood, since all eigenmodes are local. In practice, divergence angles of a few degrees lead to a satisfactory adiabatic regime for the fundamental mode in the taper region.

The taper-to-branching boundary, such as in Fig. 6.1, is usually gradual in the case of ion-exchange technology because the index structure resulting from the technology process shows gradual separation of the branches. This is due to ion diffusion, which tends to even out the region between the WGs when they are very close. In the case of deposition technologies, the change is abrupt and inevitable loss occurs. Typically, the power loss will occur through forward-travelling light remaining near the symmetry plane of the junction, instead of branching out into one WG or the other. In a similar fashion, the change in the branching angle between the WGs should be kept gradual to prevent subsequent losses.

6.2.4   Bends

In order to connect several Y junctions, S-shaped bends are used. The properties of bent WGs have been studied extensively in the literature (see Sect. 5.2 in [2]). Two types of phenomena need to be taken into account for proper design:

- when bending radii are too small, the light tends to escape out of the WG,

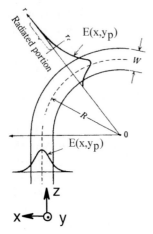

E(x,yp)

Fig. 6.6. Radiation of light in a curved dielectric WG, after [8]

- mode field profiles are deformed by bends.

The modal field is now expressed with a supplemental attenuation constant such that

$$E = E(x, y) \exp(i\beta z) \exp(-\alpha z) . \tag{6.17}$$

One way of looking at the first type of loss is to consider that, once far enough from the WG core, there is a point beyond which the optical power would propagate faster than the local speed of light, as outlined in Fig. 6.6. As a consequence, after this limit, optical power radiates away from the bent WG. Another way of describing the effect of bends is to replace the original bent WG with an equivalent straight WG [9]. The equivalent WG can then be defined by an equivalent dielectric constant distribution such that

$$\varepsilon_{eq}(x, y) = \varepsilon(x, y) \left(1 + 2\frac{x}{R}\right) . \tag{6.18}$$

With such an index distribution, the WG is not truly guiding, since the optical index increases with distance from the core. Attempts to compute the modes in such a structure will result in complex values for the effective propagation constant, which now includes an attenuation term.

For an ion-exchange WG, a detailed calculation leads to the following approximation for small values of the field attenuation constant:

$$\alpha = \frac{1}{4} \sqrt{\frac{\kappa}{\pi R}} \frac{\left| \int_{-\infty}^{+\infty} E\left(x_t, y_m\right) \mathrm{d}y \right|^2}{2\pi \int_x \int_y |E(x, y)|^2 \, \mathrm{d}x \, \mathrm{d}y} \exp\left[2\beta \left(x_t - \frac{\kappa^2}{3\beta^2} R\right)\right] , \tag{6.19}$$

where the subscript m refers to the point where the WG index increase is maximal along a given coordinate axis, and x_t, the effective turning point, is defined, for any value y_p, by

$$\beta^2 = k_0^2 \varepsilon (x_t, y_p) \ . \tag{6.20}$$

The effective transverse constant κ is defined such that

$$\kappa^2 = \beta^2 - k_0^2 \varepsilon_{\text{substrate}} \ . \tag{6.21}$$

The field deformation is essentially a lateral displacement along x, towards the outside of the bend, with an approximate value x_d such that:

$$x_d = \frac{k_0^2 \varepsilon (x_m, y_m) \, w_x^4}{4R} \ , \tag{6.22}$$

where w_x is the $1/e$ mode radius along x of the Gaussian fit to the straight waveguide E-field.

If a straight WG is directly connected to a bent WG section, mismatch loss will occur, resulting in radiation losses, usually after a series of lateral oscillations of the optical wave around the axis of the guide. In practice, one often chooses to continuously change the radius of curvature, making it minimally different from one section to the next one.

6.2.5   2 × 2 Splitters

Four-port devices such as 2×2 splitters are basic blocks for building more general $2 \times N$ functions, which may be used for redundancy applications in optical networks, or in an optical add–drop multiplexer (OADM). A 2×2 device must meet stringent criteria of splitting accuracy over a wide range of wavelengths. There have been many designs studied and published on wavelength-flattened 2×2 devices. They may be divided into the following three categories:

- resonant couplers with an extra-wavelength-flattening feature,
- purely adiabatic coupler devices,
- coupler–Mach–Zehnder combinations.

In the first category, one basic design is a 2×2 co-directional coupler modified by varying one of its governing parameters in order to compensate for its natural wavelength dispersion. Without entering a detailed analysis of co-directional couplers (see Chap. 6 in [7]), let us state some of their relevant properties. When two parallel WGs are sufficiently close, an exchange of energy takes place between the guides in a periodic fashion along the direction of propagation. We call I_1 and I_2 the total intensity propagated in WG 1 and 2, respectively, and assume that all the light is concentrated in WG 1 at the entrance of the structure. The following relationships link the light intensities in the two WGs:

$$I_1(z) = 1 - \left| \frac{\kappa_c}{\delta_{\text{eff}}} \right|^2 \sin^2 (\delta_{\text{eff}} z) \ , \tag{6.23}$$

$$I_2(z) = \left| \frac{\kappa_c}{\delta_{\text{eff}}} \right|^2 \cos^2 (\delta_{\text{eff}} z) \ , \tag{6.24}$$

where κ_c is the coupling constant of the coupler, a complex number characteristic of the structure geometry and of its optical mode fields, and

$$\delta_{\text{eff}} = \sqrt{(\beta_1 - \beta_2)^2 + |\kappa_c|^2} \ . \tag{6.25}$$

We observe immediately from (6.23)–(6.24) that optical power is fully transferred from one WG to the other only if they are identical, since in that case $\beta_1 - \beta_2 = 0$. In a 2×2 device, the incoming power must be split evenly between the two output arms. This is only possible if $(\beta_1 - \beta_2) \leq |\kappa_c|$.

Since κ_c and β_1, β_2 depend on the wavelength, identical WG coupler structures do not lead, in general, to an appropriate wavelength response. Structural elements offering counteracting wavelength dependence may be added to the basic coupler. Input and output arms may provide that function, or alternatively, a certain level of asymmetry between the guides can also be introduced, such as described in Fig. 6.7. An overall flatness of 0.5 dB has been achieved [10] across the entire 1300–1550 nm wavelength range. It should be nevertheless emphasized that such approaches are relatively dependent on the manufacturing process.

In the second category, devices are designed for a purely adiabatic transfer of energy between the WG modes. For example, the adiabatic 3 dB coupler in [11] has input asymmetric WGs that gradually change in width to become identical at the output side. A uniformity of ±0.6 dB has been reported over

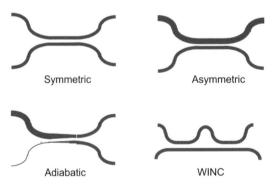

Fig. 6.7. Main types of 2×2 splitters (top view)

the wavelength range 1250–1650 nm. This type of device is less sensitive to technology, but is usually longer than the previous category.

Finally, wavelength insensitive couplers (WINCs) combine the wavelength properties of an asymmetric Mach–Zehnder interferometer with two coupler sections. The differential phase shift in the interferometer is designed to balance the wavelength dependence of the couplers. WINCs have been demonstrated to yield excellent wavelength flatness for unequal output ratios. A $20\% \pm 1.9\%$ output ratio has been demonstrated [12] for a device working over the wavelength range of 1250–1650 nm.

6.2.6 $P \times N$ Star Couplers

For higher port counts, radiative star couplers lead to more compact devices. They rely on the division and propagation of the guided signal in an intermediate planar WG joining the input and output WGs. The star couplers' planar geometry allows input signals to be distributed over any number of outputs. The minimum theoretical insertion loss in dB of $P \times N$ couplers is

$$L_{P \times N} = 10 \log_{10} \left(\max \{P, N\} \right) . \tag{6.26}$$

The theoretical loss is, therefore, the same for any $P \times N$ device with a given number of outputs N and any number of inputs P, with $P \leq N$. In practice, excess loss above this value will depend on the selected input port as well as on the wavelength. In an ideal situation, one hopes to achieve a uniform light distribution over the output WGs after propagation through the planar section. Tailoring this far field is possible by introducing co-directional coupling between the WG in the input section, before the planar lens. In this case, the Fourier transform of the input field is modified and tailored to achieve a better uniformity. Dummy WGs are added to the sides of the most peripheral inputs to maintain coupling to adjacent WGs as constant as possible. Further refinement is added by tapering inputs and outputs near the radiative planar region, which is also adapted to very high port counts by offsetting the centres of its composing circles. Low excess losses of 2.6 dB for an 8×8 device have been reported [13] and so has simultaneous operation at 1300 and 1550 nm, with a silica deposition process. In addition, star couplers with up to 144×144 ports have been demonstrated [14]. These devices were mainly realized with silica deposition technologies, whereby good control of guide-to-guide coupling is achieved. Other types of devices have been analysed but have not yet been turned into commercial products [15].

6.2.7   Ion Exchange in Glass

Theory

Ion exchange in glass is a generic term describing the exchange between ions per se and the field-assisted migration of ionic species in glass. One of the

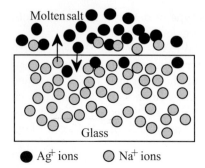

Molten salt

Glass

● Ag⁺ ions   ○ Na⁺ ions

Fig. 6.8. Exchange of ions between a glass and a molten salt

species is usually an alkali ion initially present in the glass, such as Na⁺ or K⁺. The other species, entering the glass, is another cation such as Ag⁺ or Tl⁺, as shown in Fig. 6.8. The substitution of one type of ion by another creates a change in the glass composition and an increase in optical index proportional to the local concentration of the new species. This is due to the difference in electronic polarizabilities between ionic species, itself giving rise to the change in the refractive index. The theory of ion exchange in silicate glass has been extensively studied [16], so only the most prominent characteristics will be described here.

If we call a and b the two ionic species involved in the exchange, the local change in concentration c_i of each species $i = \{a, b\}$ as a function of time can be written as

$$\frac{\partial c_i}{\partial t} = -\nabla j_i \, , \tag{6.27}$$

with j_i being the local current density of species i, such that

$$j_i = -D_i \nabla c_i + \mu_i c_i E_t \, , \tag{6.28}$$

where D_i is the diffusion coefficient, μ_i is the mobility, E_t is the total local electric field acting on the ions. E_t is due to the presence of an externally applied field and to the presence of a local space charge created during the exchange.

Mobility and diffusion coefficients are linked by the Nernst–Einstein relation:

$$\mu_i = \frac{eD_i}{h_r kT} \, , \tag{6.29}$$

where e is the charge of one electron, k is the Boltzmann constant, T is the temperature, and h_r is the Haven ratio; it describes possible correlation effects between successive diffusion steps in the glass and usually ranges between 0.1 and 1, depending on the type of glass host and ion.

The total electric field obeys Maxwell's equation:

$$\nabla \left(\varepsilon_0 \varepsilon_r E_t \right) = \rho , \tag{6.30}$$

where the local space-charge density can be written as a function of c_t the total local concentration of type-b ions initially present in the glass (and the same throughout the material) and the local concentrations of type-a and type-b ions observed as the exchange progresses,

$$\rho = e \left[c_t - (c_a + c_b) \right] . \tag{6.31}$$

It should be emphasized that during a field-assisted migration there are always areas of the material where a space charge exists. However, for total ion concentrations as low as 10^{20} m^{-3}, a small difference of relative concentration suffices to create an important space-charge field. In this case, although $\rho \neq 0$, the following approximate condition holds numerically:

$$c_t \cong c_a + c_b . \tag{6.32}$$

The total internal field can then be approximated by:

$$E_t = \frac{h_r kT}{e} \frac{1}{1 - \chi c} \left(\frac{J_t}{c_t D_b} - \chi \nabla c \right) , \tag{6.33}$$

where J_t represents the total current density, $c = c_a / c_t$ is the normalized concentration of type-a ions and $\chi = 1 - D_a / D_b$.

Finally, if we combine (6.27), (6.28), and (6.33), the concentration of incoming ions varies as

$$\frac{\partial c}{\partial t} = \nabla \left[\frac{D_a}{1 - \chi c} \left(\nabla c - \frac{J_t}{c_t D_b} c \right) \right] . \tag{6.34}$$

Since the variation of concentration in the glass is very large, diffusion coefficients and mobilities depend on the local ion concentration. In other words, the chemical composition of the glass is sufficiently modified so that the properties of diffusion and E-field-induced drift current become concentration dependent. The solution to (6.34) is usually numerical. Only in the specific case of constant diffusion coefficients and ion mobilities do analytical error function solutions apply.

Ion-Exchange Process and Device Manufacturing

A typical silver–sodium ion-exchange process for an embedded channel WG is presented in Fig. 6.9. It involves two main manufacturing steps. In the first step, a glass wafer covered by a mask is immersed in a molten salt containing silver ions and no sodium. Ion interdiffusion takes place between the sodium-rich glass and the melt. First-step exchange temperatures are in the $300-350\,^{\circ}$C range. This leads to the creation of a near-surface channel of

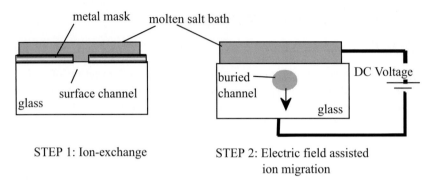

STEP 1: Ion-exchange   STEP 2: Electric field assisted
                                      ion migration

Fig. 6.9. The two main steps in an ion-exchange process

a modified glass with a high silver doping. Index changes as high as 0.1 can be reached near the glass surface. In the second step, the mask is removed and an electric field applied across the glass wafer to enable ion migration. Sodium-rich molten salt baths play the role of electrodes on the upper and lower side of the glass. In the $300-350\,°C$ temperature range, the silver ions in the glass move under the electric field, and simultaneously spread away from each other through thermal diffusion. From the top surface, sodium ions enter the glass and replace the silver ions, thus lowering the index of refraction. The end result is a channel WG with the shape shown in Fig. 6.10.

The entire device manufacturing process is outlined in Fig. 6.11. It is based on a 6-step procedure, dealing with wafers for steps 1 to 3 and circuits for steps 4 to 6. Besides the circuit processing itself, a critical step is the fibre-to-chip assembly. The most commonly used approach is to align the fibres in precision V-grooves etched on silicon or glass and to attach them to the glass circuit with adhesives. Thus, collective alignment of a fibre ribbon can be achieved by optimizing only a reduced proportion of the total number of channels. A combination of UV and thermally cured adhesives is usually

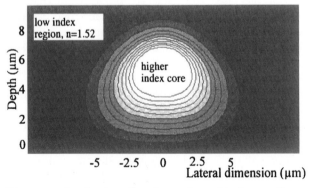

Fig. 6.10. Iso-density plots for computed index profile

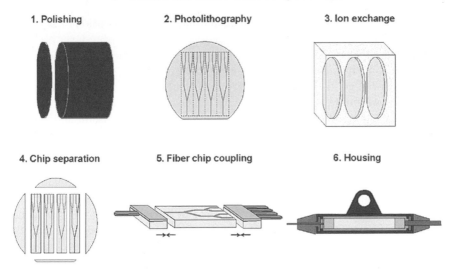

Fig. 6.11. Device manufacturing process (courtesy of IOT)

employed to enable fast and stable positioning of the fibre assembly. Even if a perfect matching of the WG and fibre modes is achieved, connection losses of 0.2 to 0.4 dB may appear because of imperfections in the V-grooves and residual eccentricity of the fibre cores, accounting for a misalignment of up to 1 µm. For more than 4 port connections, insertion losses tend to be lower for planar splitters than for fibre devices.

6.2.8   Characterization Methods

The manufacturing of integrated optics circuits requires the use of various levels of characterization methods in order to control the process as well as the performance of the circuits. This section focuses on methods for index profile measurement as well as basic on-chip performance assessment. Connected circuit characterization and equipment control methods are specific to each function and are not described here.

The basic features to be on-chip tested for passive WGs are propagation losses and mode intensity profiles. Losses and mode intensity profile variations may be traced to localized problems in the index distribution of the WGs. Thus, one has to be able to measure the relevant material properties or the index distribution.

Propagation Loss

On the whole, channel-WG losses tend to be higher than planar WG losses. Indeed, a larger part of the optical power is situated in boundary areas of a channel WG, where spurious phenomena may create absorption or scattering. In deposition technologies, edge roughness appears during layer etching and produces scattering loss. In ion-exchange technologies, absorption loss may appear if the silver ions are reduced to their metallic state. The propagation losses of good channel WGs are in the sub-0.1 dB/cm range. At these low values, attempts to directly measure the decrease in intensity by cutback or diffused intensity methods are too inaccurate. The most useful and simple approach is to compare the output power from a long length of WG with that from a shorter length reference arm. A Y-junction near the entrance of the test section splits an incident WG into the two WGs with different lengths. Since the longer WG often must be coiled on the wafer, care must be taken in the design to prevent spurious bend losses from interfering with the measurement. Cumulated lengths of more than 50 cm enable a measurement accuracy of better than 1 dB/m. Besides linear loss, possible localized defects on the WG may also contribute to the overall loss. These are detected by direct inspection, under visible and IR camera observation.

Mode Intensity Profiles

Mode intensity profiles can be measured in the following two different ways:

First, a near-field imaging approach, in which a lens with a suitable numerical aperture yields an image of the output light exiting the WG. Care is necessary to correctly analyse the images using Gaussian beam imaging theory. Measurements on highly confined WGs tend to be less accurate. However, this method is often used because it is easily implemented with commercially available equipment.

Second, a local-field imaging technique can be used to circumvent the problems brought about by very tight field confinements. The narrowed tip of a fibre is brought very near the output end of the WG, with positioning of an apparatus analogous to that employed in atomic force microscopy. The distance away from the exit end of the WG is usually of the order of 100 nm or less, so that the actual local-field intensity is recorded. The set-up may in-couple light into the WG, or alternatively may out-couple it. Supplemental information on the fibre-to-WG distance may be obtained by making the fibre vibrate perpendicularly to the WG end.

Refractive Index Measurements

Mode intensity measurements give quick estimates of the potential connection losses and thus help assess a part of the fabrication process. However, they fail to provide accurate information on the exact index distribution, mainly because different distributions can lead to almost identical mode intensities. Furthermore, in the case of ion-exchange WGs, index-distribution measurements of thick planar WGs are very helpful in the determination of the ion diffusion and mobility coefficients.

In addition to direct optical measurements, material methods such as the X-ray microprobe help to determine the local composition of the glass material and its relation to the index variations. It is, however, essential to directly measure the optical index distribution, since local stress in the material may create index perturbation through the elasto-optic effect. Differences of thermal expansion coefficients trigger this type of problem in deposition technologies, whereas uncontrolled chemical processes are more likely to be involved in ion-exchange technologies.

In the following, we present the m-lines technique and its associated inverse WKB index retrieval algorithm, useful for measuring planar WG index profiles, and the refracted near-field technique for characterizing more general channel-WG structures.

m-Lines Method

This method enables the measurement of planar WG propagation constants by decoupling the light propagating in the guide with a high-index prism brought in contact with the top surface of the glass. For each guided mode, a gradual loss of light takes place towards the prism; the emerging light rays propagate from the interface at an angle defined by the effective propagation constant and the prism index:

$$n_\gamma \cos \gamma = n_{\text{eff}} \ . \tag{6.35}$$

Forming an image with the emerging light rays gives rise to a series of lines in the focal plane of the imaging lens; one for each WG mode initially excited. A precise measurement of the propagation constants is therefore possible, since the perturbation in the WG due to the close contact with the prism can be shown to be very small.

Inverse WKB Algorithm

Once a set of effective index constants is known, this method allows an approximate reconstruction of the index profile. The starting point is the WKB (Wentzel, Kramers, and Brillouin) approximation, which enables the computation of the approximate eigensolution of the 1-D homogeneous equation

$$\frac{\partial^2 E}{\partial y^2} + \left[k_0^2 n^2 \left(y \right) - \beta^2 \right] E = 0 \ . \tag{6.36}$$

In particular, the approximate effective indices of a multimode surface WG with an index profile $n(y)$ are solutions of the integral equation

$$\int_0^{y_m} \left[n^2 \left(y \right) - n_{\text{eff}}^2 \left(m \right) \right]^{1/2} \mathrm{d}y = \lambda \frac{4m - 1}{8} \ , \tag{6.37}$$

where $n_{\text{eff}}(m) = \frac{\beta_m}{k_0}$ is the effective index of the mode number m, and $m = 1, ..., M$; in particular, the fundamental mode is labeled 1 here. $n_{\text{eff}}(m)$ is intentionally written here as a function of the mode number m and is not to be confused with $n(y)$, the refractive index, which is a function of the *space* variable y. The two functions coincide at each turning point y_m so that $n(y_m) = n_{\text{eff}}(m)$, the values of y_m are, however, initially unknown.

The Inverse WKB method, initially presented in [17] and then improved in [18,19], enables a reconstruction of the index profile from the otherwise measured values of the effective index of each mode. It is based on a simple but effective inversion scheme of the WKB method.

Let us assume that an experimental method such as the m-lines method has yielded a table of M values $n_{\text{eff}}(m)$, $m = 1, ..., M$. One can first evaluate the surface index $n(y = 0)$ by fitting an $M-1$ degree interpolating polynomial $N(m)$ to the function $n_{\text{eff}}(m)$. Consider (6.37) as a functional equation of the *real* variable m, with known values for positive integers. For $y_m \to 0$, $m \to \frac{1}{4}$ so that both sides of (6.37) go to 0. The surface value $n(y_0) = n(0)$ can therefore be approximated by $N(0.25)$. Since this scheme involves an extrapolation of the polynomial N, one should exercise caution, in particular if the measurement is quite noisy.

The method then gives an estimate of the values of y_m as follows:

For $m = 1$: $y_1 = \dfrac{9\lambda}{16}\left[\left(\dfrac{n(y_0) + 3n(y_1)}{2}\right)^{-1/2}(n(y_0) - n(y_1))^{-1/2}\right]$. (6.38)

For $m > 1$, the iterative formula is

$$y_m = y_{m-1} + \frac{3}{2}\left[\left(\frac{n^2(y_{m-1}) + 3n^2(y_m)}{2}\right)^{-1/2}[n^2(y_{m-1}) - n^2(y_m)]^{-1/2}\right]$$
$$\times \left[\frac{(4m-1)\lambda}{8} - \frac{2}{3}\sum_{i=1}^{m-1}\left(\frac{n(y_{i-1}) + n(y_i)}{2} + n(y_m)\right)^{1/2}\right.$$
$$\left.\times \left(\frac{y_m - y_{m-1}}{n(y_{m-1}) - n(y_m)}\right)F(k,m)\right] ,$$
(6.39)

with $F(k, m) = \{[n^2(y_{k-1}) - n^2(y_m)]^{3/2} - [n^2(y_k) - n^2(y_m)]^{3/2}\}$. (6.40)

Interpolation after this step-by-step inversion leads to the full reconstruction of the index profile of a planar WG as long as there are enough modes available to enable a sufficient accuracy. It is particularly useful for ion-diffused structures, as they exhibit a smooth, continuously varying index profile.

Refracted Near Field (RNF) Methods

RNF methods, initially adapted from fibre measurements to non-circular geometry, provide the only direct measurement of a channel-WG optical

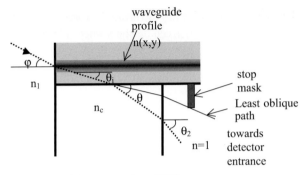

Fig. 6.12. Schematic of an RNF set-up

index profile. Different practical set-ups [20,21] have been used, but they are presented here in a unified manner. The generic scheme is presented in Fig. 6.12. Common to all designs, a probe light beam is focused on the WG entrance, where the optical index profile is to be recovered. A stopper mask, equally common to all designs, is located on the exit side of the probe light, partially masking a uniform photodetector. In specific cases, the reference material with optical index n_c may be extended to all the space up to the photodetector unit. In the general case depicted in Fig. 6.12, the equations describing the light path angles are the following:

$$n_1 \sin \varphi = n\,(x,y) \sin \theta_i \ ,$$
$$n_c \cos \theta = n\,(x,y) \cos \theta_i \ , \tag{6.41}$$
$$n_c \sin \theta = \sin \theta_2 \ .$$

In a first type of set-up the input, $n_1 = n_c = n_l$, as all these materials are replaced with an index-matching fluid. In this case, summing the squares of the first two equations in the set of (6.41) lead to the following result:

$$n_l^2 \left(\sin^2 \varphi + \cos^2 \theta \right) = n^2\,(x,y) \ . \tag{6.42}$$

Similarly, in a second type of set-up, $n_1 = 1$ and material c is a piece of bulk glass. Summing the squares of the three equations in the set of (6.41) leads to:

$$\sin^2 \varphi - \sin^2 \theta_2 + n_c^2 = n^2\,(x,y) \ . \tag{6.43}$$

A reference ray path is defined by the smallest angle $\varphi_{\min}$ not blocked by the stop mask and corresponding to the fixed output angles $\theta_{\min}$ or $\theta_{2\,\min}$, according to the actual set-up. This angle depends on the local refractive index $n\,(x,y)$ through (6.42) or (6.43). In order to produce an accurate measurement, the input light is focused on the input face in order to illuminate a very narrow area of glass over which the index of refraction is almost constant. The total detected light will be the angular integral of the input light from $\varphi_{\min}$ to a maximum angle $\varphi_{\max}$. This latter angle corresponds to certain characteristics of the light beam itself, given by the numerical aperture of the focusing lens:

$$I_p\left(n\left(x,y\right)\right) = \int_{\varphi_{\min}\left(n(x,y)\right)}^{\varphi_{\max}} I(\varphi)\,\mathrm{d}\varphi\,.\tag{6.44}$$

As is apparent in (6.44), the output signal of the detector will depend on the local index of refraction at input $n\left(x,y\right)$ since the integral *decreases* as $n\left(x,y\right)$ *increases*. The exact dependence of the signal is a function of the type of light source and also of the precise set-up being considered. In a practical situation, an accuracy of several 10^{-4} is possible on the refractive index final readout.

6.2.9   Performance and Reliability of Commercial Devices

Commercially available splitter devices have undergone continuous performance improvement over the past 10 years, to the point where they are near the fundamental physical performance limits. The striking features, as outlined in Fig. 6.13, include a 400 nm bandwidth window of low-loss operation; very low polarization-dependent loss, in the sub-0.1 dB range; and very good performance stability in fast-temperature cycling conditions. Excess losses of the order of 1 dB for 8-port and 16-port devices show this technology's excellent aptitude in offering a satisfactory response to market demand. It is anticipated that future devices will see enhancements, as more functions will be added to the initial power-splitting functions.

Passive splitter devices are commonly deployed in uncontrolled environments such as encountered in access network applications. For this reason, they must maintain an excellent level of performance while operating under harsh conditions. In the following, I will briefly review possible failure modes previously described [22,23]. The typical result of failure is a significant increase in transmission losses or reflection characteristics, driving them out of specification range. One should stress that present-day planar devices have achieved a very high level of reliability, with examples of maintained performance under Bellcore 1221 testing conditions.

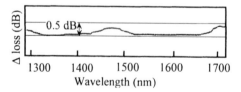

Fig. 6.13. Wavelength average dependence of a 1×8 splitter insertion loss (diagram Teem Photonics)

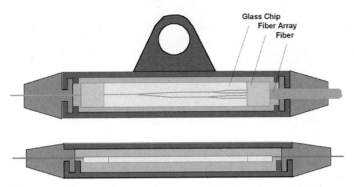

Fig. 6.14. Packaged component assembly (courtesy of IOT)

An example of a typical planar circuit assembly is presented in Fig. 6.14. The fibres are attached through a supporting precision V-groove holder. Various counter-plate schemes can be used to enhance adhesion and surface quality. Finally, the faces are angled to decrease reflection losses below −55 dB per port. The assembly is permanently bonded using UV or thermally cured adhesives. High alignment accuracy is thus obtained for the entire fibre block, and connection losses are mainly created by residual core eccentricity and mode mismatch.

Among these types of design, three main functional elements may generate failure problems. First and foremost is the fibre-to-WG interface, which usually requires the use of epoxy-based adhesives to insure proper index matching between the two guiding structures. Under the influence of humidity or certain contaminants, the adhesive may soften and the fibre-block move out of alignment with respect to the chip, in which case the transmission loss will increase. The adhesive may also detach from the chip surface, creating an air pocket and increasing the reflected light. Secondly, the fibres in the assembly block may fail under thermal cycling or because of improper stripping procedures. Finally, the chip itself may present cracks propagating from the facet edges; potassium-containing glasses have also proven to be sensitive to water contamination.

Among all possible conditions, the toughest tests appear to be the long-term high temperature and high humidity test ($85\,^\circ$C/85% rh) which combine temperature activation of any failure mechanism with exposure to high water concentrations. An Arrhenius–Lawson law helps in the estimation of the acceleration factor of the time to failure:

$$\mathrm{MTTF}\,(T)\exp\left(-\frac{E_\mathrm{a}}{kT}\right) + (\mathrm{rh})^2\,\eta = \text{constant}\;, \tag{6.45}$$

where $\mathrm{MTTF}(T)$ is the mean time to failure at temperature T, E_a is the activation energy of the failure, k is Boltzmann's constant, rh is the relative humidity in %, and η is the humidity factor.

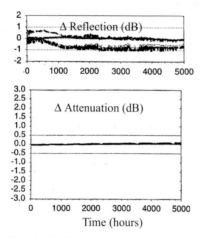

Fig. 6.15. Long-term reliability of optical splitters (after [24])

Reported values for these parameters are $0.8\,\text{eV} < E_a < 1.4\,\text{eV}$ and $\eta = -5 \times 10^{-4}$. Equation (6.45) enables us to easily estimate the reliability figures in actual field-deployment conditions. In practice, MTTF ranging from 10 to 125 failures in time (FIT) are estimated at 25 °C and 40% rh. The non-emergence of wear-out failures is also documented in Fig. 6.15, after [24].

Though very good at present, the reliability figures could be even further improved by the use of solder glass or on-chip fibre-maintaining U-grooves, using approaches already demonstrated at the laboratory level. At the same time, on-chip direct alignment approaches should lead to cost reductions because fibre-handling procedures would be simplified. They may, however, induce a drop in manufacturing yields as long as the production levels have not reached quantities large enough to stabilize the new process steps.

In conclusion, the road for larger-scale deployment of passive devices has thus been paved, with associated unrivalled levels of wide-band performance, small dimensions, ease-of-use features, and excellent reliability.

6.3   Integrated Optics Yb/Er Glass Amplifiers

6.3.1   Introduction

The optical amplifier, in the form of the Er-doped fibre amplifier (EDFA), represents a milestone in the recent development of optical telecommunications. Amplification enables network system designers to adopt an unrestricted approach where each new element in a transmission line is no longer perceived as a source of optical loss. As a result, new designs have been more rapidly implemented and have thus fostered the expansion of optical networks. A chief example of this is the simultaneous transmission capacity of dense wavelength division multiplexing (DWDM) multichannel systems, which have come to satisfy the ever-increasing needs of per-fibre data transmission rates. This new-found freedom, however, has entailed a rapid increase in system complexity as a large number of independent channels can now be transmitted through a single fibre. Since continuing demand will stretch the boundaries of technology, all the more attention is being devoted to simplifying amplification systems.

In the course of this development, integrating Er-doped amplifiers quite naturally combines the advantage of reduction of scale with the remarkable properties of glass amplification. Among its benefits are the following:

- it provides a more compact approach to multistage, complex devices,
- it requires fewer independent parts and therefore proves more reliable,
- it is suited to the fabrication of very stable multiple-wavelength laser sources, as well as very short pulse sources,
- finally, it is well suited to mass production at a reduced cost.

Despite all of these advantages, integrated amplifier development has been technically challenging, with the first products only recently becoming available [25,26]. The main reason for this delay lies in the problems inherent in integration. In order to obtain the same gain level from an integrated amplifier as from its fibre counterpart, the use of glass with an Er concentration more than 100 times higher is necessary. At these concentrations, the rare earth can no longer be considered a dopant. Its ability to chemically bond to the surrounding glass structure is critical, as the rare earth itself becomes a component of the glass.

The type of glass used in most chemical vapour deposition (CVD) processes, a rather pure silica with very small amounts of doping materials such as phosphorus, has proven incapable of hosting much more than 0.5% (wt) of rare-earth compounds without producing drastic reductions in efficiency. Working devices with a much higher Er concentration have been demonstrated in silicate soda-lime and phosphate glasses [27,28] deposited by sputtering, as well as in ion-exchanged phosphate glasses [29].

A general study of Er-based amplification can be found in Chap. 5. The present section is focused on the main theoretical elements and the particulars of Yb co-doping as well as on the constraints resulting from integration. An extension to integrated optics glass lasers is addressed in Sect. 6.4.

Let us first outline why co-doping a glass material with Yb^{3+} ions yields interesting features for integration. It provides an alternative and very efficient path for pumping the Er^{3+} population. At the same concentration, Yb becomes photo-excited by an optical pump at 980 nm wavelength more easily than Er, since its absorption cross-section is about 5 times bigger than that of Er in this wavelength range. As schematized in Fig. 6.16, once excited to its $^2F_{5/2}$ state, a Yb^{3+} ion can transfer its energy to an Er ion, because the energy levels of the $^4I_{11/2}$ manifold of Er^{3+} and the $^2F_{5/2}$ manifold of Yb^{3+} are nearly the same.

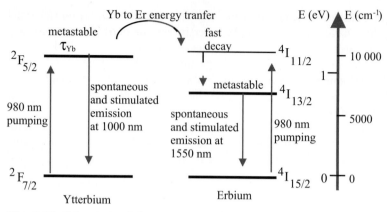

Fig. 6.16. Schematic of the energy transfer from Er to Yb

Yb^{3+} has another interesting feature: it helps alleviate some of the detrimental effects occurring at high rare-earth concentrations. The Yb energy diagram in Fig. 6.16 is a quasi-two-energy-level system, and presents fewer unwanted de-excitation mechanisms through higher or intermediate energy levels.

We shall thus begin our discussion by modelling the propagation of light in co-doped amplifiers. Two main types of equations describe how light is amplified as it propagates in an optically amplified medium:

- rate equations, which describe how the population of ions is locally excited or de-excited as a function of time, and therefore how energy is stored in matter;
- propagation equations, which describe how electromagnetic energy is exchanged between the ions and the various guided waves as they propagate.

Although the two sets of equations are coupled, we will show how to derive a general equation giving a global account of the power transfers in an amplifier. We will also independently examine each set of equations and derive important consequences for the properties of a co-doped amplifier.

6.3.2   Rate Equations for Yb/Er Co-doping

Rate equations describe how the various rare-earth energy level manifolds are locally populated or de-populated through interaction with the passing photons and the glass network. There is one rate equation per manifold for each rare-earth species. However, with straightforward reasoning, most of the information relevant to the computations can be summarized by studying the population of the metastable manifolds.

The first rate equation states that the variation per unit of time of the number of Yb ions in their metastable level $^2F_{5/2}$ is equal to the number of ions excited by the pump less the number of ions de-excited by other mechanisms. Three main de-excitation mechanisms are considered here:

- stimulated emission, which occurs when an incident 980 nm photon stimulates the emission of a second identical photon,
- spontaneous emission, which occurs with a decay constant τ_{Yb},
- energy transfer to the Er ions.

This last term is taken to be proportional to the population of Yb in the metastable level and to the population of Er ions at the fundamental level. The proportionality constant k_{tr} depends on the actual concentrations of both types of ions in a given glass.

We therefore write:

$$\frac{dN_m^{Yb}}{dt} = \underbrace{(W_p^{Yb} + W_{fm}^{Yb})N_f^{Yb}}_{\substack{\text{populating terms:} \\ \text{pump and 980 nm signal}}} - \underbrace{W_{mf}^{Yb}N_m^{Yb} - \frac{N_m^{Yb}}{\tau_{Yb}}}_{\substack{\text{de-excitation terms:} \\ \text{stimulated and spontaneous} \\ \text{emission from level m to level f}}}$$

$$- \underbrace{k_{tr}N_m^{Yb}N_f^{Er}}_{\substack{\text{energy transfer} \\ \text{to Er level 3}}} , \tag{6.46}$$

where N_m^{Yb}, N_f^{Yb} and N_f^{Er} are the population densities of the metastable Yb $^4F_{5/2}$ manifold and the fundamental Yb $^4F_{7/2}$ and Er $^4I_{15/2}$ manifolds, W_p^{Yb} is the 980 nm pump rate from the fundamental to the metastable manifold, W_{fm}^{Yb} is the 980 nm signal (if present) absorption rate, W_{mf}^{Yb} is the stimulated 980 nm signal emission rate, τ_{Yb} is the spontaneous decay rate for Yb, and k_{tr} is the transfer coefficient between Yb and Er.

The W terms above refer to transition mechanisms and are proportional to the local intensity of the relevant populating/depopulating light intensity. For example, in the case of the pump rate W_p^{Yb}, one writes

$$W_p^{Yb} = \frac{I_p}{h\nu_p}\sigma_{a,p}^{Yb} , \tag{6.47}$$

where $\sigma_{a,p}^{Yb}$ is the absorption cross-section for the pump and I_p is the local intensity of the pump. Similar expressions hold for the others terms, with the relevant parameters.

We assume here that the $^4I_{11/2}$ manifold of Er has a very short lifetime and therefore that any ion in this state is quickly de-excited to the metastable $^4I_{13/2}$ manifold. This ensures that no significant energy back-transfer occurs from Er to Yb. Back-transfer occurs in some silicates, but is nearly absent in phosphates.

With reasoning similar to that in (6.46), the rate equation for the Er metastable level can be written as follows:

$$\frac{dN_m^{Er}}{dt} = \underbrace{(W_p^{Er} + W_{fm}^{Er})N_f^{Er}}_{\substack{\text{populating terms:} \\ \text{pump and 1550 nm signal} \\ \text{absorption}}} - \underbrace{W_{mf}^{Er}N_m^{Er} - \frac{N_m^{Er}}{\tau_{Er}}}_{\substack{\text{de-excitation terms:} \\ \text{stimulated and spontaneous} \\ \text{emission from Er ions}}}$$

$$+ \underbrace{k_{tr}N_m^{Yb}N_f^{Er}}_{\substack{\text{populating term:} \\ \text{energy transfer from} \\ \text{Yb}}} , \tag{6.48}$$

where N_m^{Er} is the population density of the metastable Er $^4I_{13/2}$ manifold, W_p^{Er} is the direct pump rate from the $^4I_{15/2}$ to the $^4I_{11/2}$ Er manifold, W_{fm}^{Er} is the 1550 nm signal absorption rate, W_{mf}^{Er} is the stimulated 1550 nm signal emission rate, and τ_{Er} is the spontaneous decay rate for Er.

We again assume that no significant part of the Er population remains on that manifold, and that no significant diversion mechanism detracts energy from the $^4I_{11/2}$ to $^4I_{13/2}$ transition. Therefore, a direct pump term to the level $^4I_{11/2}$ is introduced in (6.48), as well as source terms originating from signal re-absorption and transfer from Yb.

Note that $k_{tr}N_m^{Yb}N_f^{Er}$ now acts as a source term to the excited Er population. More precisely, $k_{tr}N_m^{Yb} + W_p^{Er}$ describes how the pump is absorbed by both the Yb and Er ions to invert the Er population. Most of the optimization of an Yb/Er amplifier relies on an effective use of this process.

6.3.3 Propagation Equations

Propagation equations relate how the light intensity varies along the propagation path, as energy is exchanged with the rare-earth ions. In contrast to the rate equations, they focus on the build-up or absorption of the electromagnetic energy in the WG and therefore inherently describe a propagation process. There are, hence, as many propagation equations as there are guided waves. The guided waves in an optical amplifier are the 980 nm pumps (several pumps may be used), the 1550 nm signals (the DWDM channels), and finally the amplified spontaneous emission (ASE). Precise numerical models take all of the guided waves into account and split the ASE spectrum into subintervals, with as many propagation equations [30]. This analysis is restricted to the case of one pump and one signal, explicitly pointing out the main features of the co-doped integrated amplifiers. The influence of ASE is neglected, since the short lengths and overall gain of integrated amplifiers do not usually result in an ASE-limited regime.

The following differential (propagation) equations describe the local variation of the intensity profiles $I(x, y, z)$ for the pump and the signal guided waves:

$$\frac{\mathrm{d}I_\mathrm{p}(x,y,z)}{\mathrm{d}z} = I_\mathrm{p}(x,y,z)\left(\underbrace{\sigma_\mathrm{e,p}^{Yb}N_\mathrm{m}^{Yb} - \sigma_\mathrm{a,p}^{Yb}N_\mathrm{f}^{Yb}}_{\substack{\text{emission and absorption}\\\text{by Yb ions}}} - \underbrace{\sigma_\mathrm{a,p}^{Er}N_\mathrm{f}^{Er}}_{\substack{\text{absorption}\\\text{by Er ions}}} - \alpha_\mathrm{p} \right),$$

(6.49)

$$\frac{\mathrm{d}I_\mathrm{s}(x,y,z)}{\mathrm{d}z} = I_\mathrm{s}(x,y,z)\left(\underbrace{\sigma_\mathrm{e,s}^{Er}N_\mathrm{m}^{Er} - \sigma_\mathrm{a,s}^{Er}N_\mathrm{f}^{Er}}_{\substack{\text{emission and absorption}\\\text{by Er ions}}} - \alpha_\mathrm{s} \right),$$

(6.50)

where $\sigma_\mathrm{e,s}^{Er}$ and $\sigma_\mathrm{a,s}^{Er}$ are the signal emission and absorption cross-sections by Er ions, and α_s and α_p are the background losses for the signal and the pump. The energy transfer between Er and Yb ions does not appear explicitly in (6.49) and (6.50) as it results from a direct interaction between ions.

The solutions of (6.49) and (6.50) are the intensities of the guided waves. In order to link them to the usual observable parameters, they can be related to the optical power $P_w(z)$ carried at z by the pump ($w = \mathrm{p}$) or the signal ($w = \mathrm{s}$). We make the assumption that the changes brought by the amplification or absorption phenomena are so minute that they induce no major change in the guided-mode intensity profiles along the propagation axis. Transverse (x, y) variables can therefore be separated in the following manner [31]:

$$I_w(x,y,z) = P_w(z)\frac{\Psi_w(x,y)}{\int_{\mathfrak{R}^2}\int\Psi_w(x,y)\,\mathrm{d}x\,\mathrm{d}y},$$

(6.51)

where $\Psi_w(x,y)$ is the intensity envelope of the passive WG modes at pump or signal wavelengths and $\mathfrak{R}^2$ is the entire (x, y) plane. We further assume that

$$|\Psi_\mathrm{s,p}(x,y)|_\mathrm{max} = 1$$

(6.52)

and define the pump or signal mode power area by

$$S_w = \int_{\mathfrak{R}^2}\int\Psi_w(x,y)\,\mathrm{d}x\,\mathrm{d}y.$$

(6.53)

6.3.4   The Power-Transfer Equation

Within the previously stated limitations, the set of rate and propagation equations is sufficient to model an amplifier, provided a numerical integration scheme is used. While they describe local interactions, the coupled nature of these equations fails to offer a global view of the main energy transfers occurring along the propagation path. A more physical view is given by deriving an equation describing how energy is lost in an amplifying device. This approach is useful for both the physicist wanting to know the quantities which remain invariant and the development engineer looking for general rules governing the design of an integrated amplifier.

First, rewrite (6.49) and (6.50) by introducing the relevant pumping rates:

$$\frac{1}{h\nu_p}\frac{dI_p}{dz} = \left(W_{mf}^{Yb}N_m^{Yb} - W_p^{Yb}N_f^{Yb} - W_p^{Er}N_f^{Er}\right) - \frac{\alpha_p}{h\nu_p}I_p , \tag{6.54}$$

$$\frac{1}{h\nu_s}\frac{dI_s\left(x,y,z\right)}{dz} = \left(W_{mf}^{Er}N_m^{Er} - W_{fm}^{Er}N_f^{Er}\right) - \frac{\alpha_s}{h\nu_s}I_s . \tag{6.55}$$

Then, use rate equations (6.46) and (6.48) at steady state to replace the terms in parenthesis and derive

$$\frac{1}{h\nu_p}\frac{dI_p}{dz} = \left(-\frac{N_m^{Yb}}{\tau_{Yb}} - k_{tr}N_m^{Yb}N_f^{Er} - W_p^{Er}N_f^{Er}\right) - \frac{\alpha_p}{h\nu_p}I_p , \tag{6.56}$$

$$\frac{1}{h\nu_s}\frac{dI_s}{dz} = \left(W_p^{Er}N_f^{Er} - \frac{N_m^{Er}}{\tau_{Er}} + k_{tr}N_m^{Yb}N_f^{Er}\right) - \frac{\alpha_s}{h\nu_s}I_s . \tag{6.57}$$

Add together (6.56) and (6.57) and integrate the result over the section of the WG:

$$\frac{1}{h\nu_p}\frac{dP_p}{dz} + \frac{1}{h\nu_s}\frac{dP_s}{dz} = \left(-\frac{N_t^{Yb}}{\tau_{Yb}}K_{eff}^{Yb}S^{Yb} - \frac{N_t^{Er}}{\tau_{Er}}K_{eff}^{Er}S^{Er}\right)$$
$$- \frac{\alpha_p}{h\nu_p}P_p - \frac{\alpha_s}{h\nu_s}P_s , \tag{6.58}$$

where, for RE = Yb or Er, $N_t^{RE}K_{eff}^{RE}S^{RE} = \int_{\Re^2}\int N_m^{RE}\,dx\,dy = N_t^{RE}\int_{\Re^2}\int k^{RE}\,dx\,dy$; N_t^{RE} is the concentration of rare-earth ions and K_{eff}^{RE} is the effective inversion factor over the effective area S^{RE} for which inversion occurs. A more precise definition of the effective inversion parameters is given in Sect. 6.3.6. I note that in the case of uniform rare-earth doping S^{RE} can take any arbitrarily high value when the pump power is increased.

Equation (6.58) states that the variation of the total number of photons guided at the pump and signal wavelength (left-hand side) is equal to the sum of:

- the number of photons lost by spontaneous emission from Er and Yb ions (first two terms on the right-hand side),
- the number of photons absorbed by background losses (last two terms on the right-hand side).

Equation (6.58) is valid for co-directional pumping; otherwise one changes the sign of the terms involving P_p.

In order to integrate the differential equation (6.58), we assume identical background losses at the pump and signal wavelengths, with loss coefficient α, and we introduce the following loss-corrected effective volume of rare-earth ions that are in the metastable state:

$$V_{\text{eff}}^{\text{RE}} = \exp\left(-\alpha L\right) \int_0^L S^{\text{RE}} K_{\text{eff}}^{\text{RE}} \exp\left(\alpha z\right) \, \mathrm{d}z \, .$$

With these notations, we integrate (6.58) over the total length of the WG and obtain the following power-transfer equation:

$$\frac{P_{\text{p}}(L)}{h\nu_{\text{p}}} + \frac{P_{\text{s}}(L)}{h\nu_{\text{s}}} = \left(\frac{P_{\text{p}}(0)}{h\nu_{\text{p}}} + \frac{P_{\text{s}}(0)}{h\nu_{\text{s}}}\right) \exp\left(-\alpha L\right)$$
$$- \frac{N_{\text{t}}^{\text{Yb}}}{\tau_{\text{Yb}}} V_{\text{eff}}^{\text{Yb}} - \frac{N_{\text{t}}^{\text{Er}}}{\tau_{\text{Er}}} V_{\text{eff}}^{\text{Er}} \, . \tag{6.59a}$$

Again, let us identify the constitutive terms in (6.59a): $\frac{P_{\text{p}}(z)}{h\nu_{\text{p}}} + \frac{P_{\text{s}}(z)}{h\nu_{\text{s}}}$ represents the total photon flux per unit of time propagated by the pump and the signal guided waves at abscissa z along the propagation axis; $\frac{N_{\text{t}}^{\text{Yb}}}{\tau_{\text{Yb}}} V_{\text{eff}}^{\text{Yb}}$ and $\frac{N_{\text{t}}^{\text{Er}}}{\tau_{\text{Er}}} V_{\text{eff}}^{\text{Er}}$ represent the total number of photons emitted by the rare-earth ions per unit of time. With the notation: $\Phi_{\text{T}}(z) = \frac{P_{\text{p}}(z)}{h\nu_{\text{p}}} + \frac{P_{\text{s}}(z)}{h\nu_{\text{s}}}$, (6.59a) can thus be rewritten in the following way:

$$\underbrace{\Phi_{\text{T}}(L) - \Phi_{\text{T}}(0)}_{\substack{\text{variation of the total} \\ \text{guided photon flux}}} = \underbrace{-\Phi_{\text{T}}(0)[1 - \exp(-\alpha L)]}_{\text{guided photons lost by absorption}}$$

$$- \underbrace{\left(\frac{N_{\text{t}}^{\text{Yb}}}{\tau_{\text{Yb}}} V_{\text{eff}}^{\text{Yb}} + \frac{N_{\text{t}}^{\text{Er}}}{\tau_{\text{Er}}} V_{\text{eff}}^{\text{Er}}\right)}_{\substack{\text{spontaneous emission} \\ \text{photons lost by excited ions}}} \, . \tag{6.59b}$$

Equation (6.59b) describes an amplifier as a black box, disregarding the details of the power transfer between the pump and signal guided waves. What becomes apparent is that the cost of this transfer is the energy used to maintain effective volumes of Yb and Er ions in their metastable state as well as that used to compensate for background losses and the frequency difference between pump and signal wavelengths. No explicit reference to the details of the energy transfer between Er and Yb ions is visible in (6.59b), since we have assumed the mechanism to be free of losses.

Unsurprisingly, in the absence of any loss, the *total* guided wave photon flux would be invariant along the propagation. As will be seen in Sect. 6.3.8, if new loss mechanisms are taken into account, the right-hand side of (6.59a,b) is modified and the total guided photon flux at the output of the waveguide is even further reduced.

6.3.5   Yb/Er Co-doping Enhances the Inversion

An amplifier can be optimized very differently depending on its use. Whereas a power amplifier should provide as high an output power as possible, a preamplifier should offer a high gain, with a low noise figure, and require as little pumping as possible. In all cases, it is necessary to design a device that works as efficiently as possible. We shall now address in more detail how Yb co-doping can reach this end.

To study how energy is transferred from Yb to Er, rewrite (6.46) at steady state by replacing N_f^{Yb} with $N_t^{\mathrm{Yb}} - N_m^{\mathrm{Yb}}$ and group terms:

$$W_p^{\mathrm{Yb}} N_t^{\mathrm{Yb}} - \left(W_p^{\mathrm{Yb}} + W_{\mathrm{mf}}^{\mathrm{Yb}} + \frac{1}{\tau_{\mathrm{Yb}}} + k_{\mathrm{tr}} N_f^{\mathrm{Er}} \right) N_m^{\mathrm{Yb}} = 0 \; ; \tag{6.60}$$

transform it further by replacing the stimulated emission and pump rate terms:

$$N_m^{\mathrm{Yb}} = \frac{\dfrac{I_p}{h\nu_p} \sigma_{a,p}^{\mathrm{Yb}}}{\dfrac{I_p}{h\nu_p} \left(\sigma_{a,p}^{\mathrm{Yb}} + \sigma_{e,p}^{\mathrm{Yb}} \right) + \dfrac{1}{\tau_{\mathrm{Yb}}} + k_{\mathrm{tr}} N_f^{\mathrm{Er}}} N_t^{\mathrm{Yb}} \; . \tag{6.61}$$

We neglect (and in practice avoid) the ASE induced by the pump in the Yb absorption band by selecting a pump power wavelength for which the emission cross-section is equal to or larger than the absorption cross-section [30]. This is obtained for pump wavelengths greater than 970 nm.

In order to optimize the pump efficiency, a given pump intensity should lead to an energy transfer to Er ions that is as high as possible. Inspection of (6.61) shows that the spontaneous emission term $\frac{1}{\tau_{\mathrm{Yb}}}$ will compete with the energy transfer $k_{\mathrm{tr}} N_f^{\mathrm{Er}}$ in a proportion independent of the actual inversion level of Yb. As an obvious strategy, materials with a high value for the transfer coefficient k_{tr} should be sought. Once a material has been selected, a further improvement can be made, since k_{tr} also depends on the actual concentration of Yb and Er ions. For phosphates, it is reported [32] that k_{tr} exhibits a quadratic dependence on the Yb concentration when low and an almost constant value at concentrations above $10^{27} \, \mathrm{m}^{-3}$. To model this behaviour, assume a dependence following the simplified law

$$k_{\mathrm{tr}} = k_\infty \frac{\left(N_t^{\mathrm{Yb}} \right)^2}{\left(N_t^{\mathrm{Yb}} \right)^2 + \left(N_0^{\mathrm{Yb}} \right)^2} \; , \tag{6.62}$$

where N_0^{Yb} is a constant associated with the glass material. Such a law is represented in Fig. 6.17 with $k_\infty = 10^{-22} \, \mathrm{m}^{-3} \, \mathrm{s}^{-1}$ and $N_0^{\mathrm{Yb}} = 5 \times 10^{26} \, \mathrm{m}^{-3}$ for a phosphate. We observe how in a given glass matrix k_{tr} increases with higher Yb concentrations. This appears to be a general property of co-doped glass materials which is beneficial for the transfer of energy from Yb to Er.

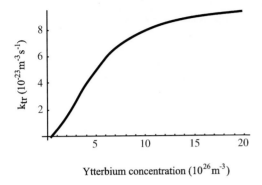

Fig. 6.17. Example of Yb concentration dependence of k_{tr} in a phosphate glass

Unfortunately, (6.61) tells us that, for high pump intensities, N_m^{Yb} is proportional to N_t^{Yb}, since we can then approximately write $N_m^{Yb} \cong \frac{\sigma_{a,p}^{Yb}}{\sigma_{a,p}^{Yb}+\sigma_{e,p}^{Yb}} N_t^{Yb}$. The spontaneous emission loss will therefore also increase with Yb concentration unless we simultaneously lower the ratio of absorption to emission cross-sections. This is made possible by adequately selecting the pump wavelength. Indeed, in Fig. 6.18, experimental data on phosphate glass show a rapid drop-off for the Yb absorption cross-section at longer wavelengths, whereas the emission fluorescence lies in the 980−1100 nm range. Optimized pump wavelengths thus move from 975 nm upward with increasing Yb concentrations. Since the present semiconductor pump wavelengths lie mainly in the range 975−985 nm, economic reasons allow only moderately high Yb concentrations to be used, roughly up to 5×10^{26} m^{-3}.

Experimental determination of k_{tr} can be done by measuring the reduction of the observed lifetime of Yb ions due to the energy transfer to Er and

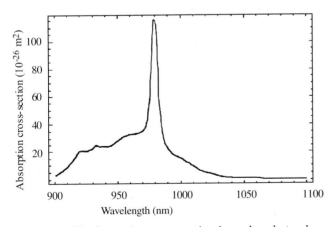

Fig. 6.18. Yb absorption cross-section in a phosphate glass

making use of the following relationship [33]:

$$\frac{1}{\tau_{\text{obs}}} = \frac{1}{\tau_{\text{Yb}}} + k_{\text{tr}} N_{\text{f}}^{\text{Er}} \; .$$
(6.63)

In order to simplify the following discussion, we now introduce reduced variables for the pump and signal:

$$i_{\text{p}} = \frac{I_{\text{p}}}{I_{\text{sat1}}} = \frac{I_{\text{p}}}{(h\nu_{\text{p}}/\sigma_{\text{a,p}}^{\text{Er}}\tau^{\text{Er}})} \; ,$$
(6.64)

$$i_{\text{s}} = \frac{I_{\text{s}}}{I_{\text{sat2}}} = \frac{I_{\text{s}}}{(h\nu_{\text{s}}/\sigma_{\text{a,s}}^{\text{Er}}\tau^{\text{Er}})} \; ,$$
(6.65)

where, with the usual notation, I_{sat} is the pump intensity which would locally bring transparency in the case of equal absorption and emission cross-section.

Equation (6.61) can be restated as

$$K^{\text{Yb}} = \frac{N_{\text{m}}^{\text{Yb}}}{N_{\text{t}}^{\text{Yb}}} = \frac{\sigma_{\text{a,p}}^{\text{Yb}}/\sigma_{\text{a,p}}^{\text{Er}} i_{\text{p}}}{(\sigma_{\text{a,p}}^{\text{Yb}} + \sigma_{\text{e,p}}^{\text{Yb}})/\sigma_{\text{a,p}}^{\text{Er}} i_{\text{p}} + \frac{\tau_{\text{Er}}}{\tau_{\text{Yb}}} + k_{\text{tr}}\tau_{\text{Er}}(1 - K^{\text{Er}})N_{\text{t}}^{\text{Er}}} \cdot$$
(6.66)

Similarly, (6.48) can be transformed with the same notations to obtain a relation linking the pump and signal intensities to the local inversion coefficient of Er and Yb:

$$K^{\text{Er}} = \frac{N_{\text{m}}^{\text{Er}}}{N_{\text{t}}^{\text{Er}}} = \frac{i_{\text{p}} + i_{\text{s}} + k_{\text{tr}}\tau_{\text{Er}}N_{\text{t}}^{\text{Yb}}K^{\text{Yb}}}{1 + i_{\text{p}} + i_{\text{s}}(1 + \sigma_{\text{e,s}}^{\text{Er}}/\sigma_{\text{a,s}}^{\text{Er}}) + k_{\text{tr}}\tau_{\text{Er}}N_{\text{t}}^{\text{Yb}}K^{\text{Yb}}} \cdot$$
(6.67)

In phosphate glass, even for relatively moderate Yb concentrations, the energy transfer from Yb to Er is efficient. In practice, the non-dimensional term $k_{\text{tr}}\tau_{\text{Er}}N_{\text{t}}^{\text{Yb}}K^{\text{Yb}}$ in (6.67) may take values of 100 or more. The local Er inversion is therefore much improved by the presence of Yb. From these first elements, we expect Yb co-doping to allow for a higher saturation power in boosters and an improved noise figure for preamplifiers. In the case of uniform rare-earth doping, it also leads to a better pumping of the annular region of the guided mode. In that region the local pump intensity quickly decreases and local inversion is weaker.

6.3.6   Effective Inversion Coefficients

The power-transfer equation (6.59a,b) clearly indicates that the effective volume of ions being pumped is a very significant source of loss. In a uniformly doped material, the effective volume is not limited; it is hence necessary to evaluate more precisely how the various design parameters can affect its value. We now define more precisely the effective inversion coefficient and effective areas used in (6.58).

For Yb ions, it appears natural to average the inversion over the pump intensity distribution. We therefore define this coefficient as

$$K_{\text{eff}}^{\text{Yb}} = \frac{1}{S_p p_p} \int_{\mathfrak{R}^2} K^{\text{Yb}} i_p(x, y) \, dx \, dy \,, \tag{6.68}$$

where $p_p = \frac{P_p}{I_{\text{sat}} S_p}$ is the reduced pump power, S_p being defined by (6.53). The corresponding effective area is then deduced from the relation

$$S^{\text{Yb}} K_{\text{eff}}^{\text{Yb}} = \int_{\mathfrak{R}^2} K^{\text{Yb}} \, dx \, dy \,. \tag{6.69}$$

For Er ions we proceed by reference to the signal intensity in a similar fashion to (6.68) and obtain the following set of relations:

$$K_{\text{eff}}^{\text{Er}} = \frac{1}{S_s p_s} \int_{\mathfrak{R}^2} K^{\text{Er}} i_s(x, y) \, dx \, dy \tag{6.70}$$

and

$$S^{\text{Er}} K_{\text{eff}}^{\text{Er}} = \int_{\mathfrak{R}^2} K^{\text{Er}} \, dx \, dy \,. \tag{6.71}$$

It should be emphasized that the different choice of references for Er and Yb does not matter when computing the integrals of (6.69) and (6.71). The usefulness of this choice will be made more apparent when actual gain computations are executed.

For uniform doping and small signals, both effective areas are always larger than S_p, from which they diverge as the pump power increases. In the examples below, we assume a circular Gaussian distribution for the pump and signal intensities, with a radial dependence given by

$$i_p = \frac{P_p}{P_{\text{sat}}} \exp\left[-\left(\frac{r}{2}\right)^2\right] \tag{6.72}$$

$$i_s = \frac{P_s}{P_{\text{sat}}} \exp\left[-\left(\frac{r}{2.35}\right)^2\right] \,. \tag{6.73}$$

With these hypotheses, Fig. 6.19 shows the effective inversion for several Yb concentrations, at small and large signals. The effective inversion is considerably improved for Yb concentrations of $3 \times 10^{26} \, \text{m}^{-3}$ or above. Similarly the effective inversion is improved for signal powers such that $\frac{P_s}{P_{\text{sat}}} = 3$. In the present case, $P_{\text{sat}} = I_{\text{sat}} S_p \cong 3.5 \, \text{mW}$.

The effective surfaces for Er and Yb ions are presented in Fig. 6.20. We can observe how the larger Yb concentration, favorable to a more complete inversion, also brings larger effective surfaces and will therefore induce higher losses by spontaneous emission from the rare-earth ions.

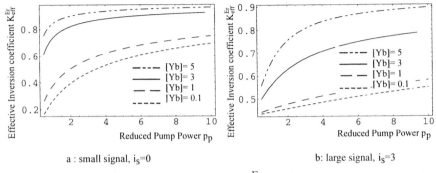

Fig. 6.19. Effective Er inversion coefficient $K_{\text{eff}}^{\text{Er}}$ versus reduced pump power for different Yb concentrations in factors of $10^{26}\,\text{m}^{-3}$. (**a**) Small signal, $i_{\text{s}} = 0$; (**b**) large signal, $i_{\text{s}} = 3$

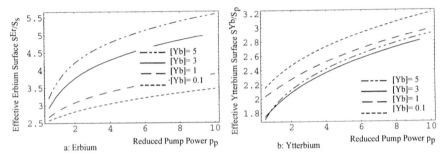

Fig. 6.20. Effective surfaces versus reduced pump power for different Yb concentrations in factors of $10^{26}\,\text{m}^{-3}$. (**a**) Er; (**b**) Yb

6.3.7   Gain of a Co-doped Waveguide Section

A complete computation of the gain can be performed with a direct numerical integration of (6.46)–(6.50). We can however use the previously introduced concepts to analyze the main factors acting on an amplifier's gain.

We first integrate (6.49) and (6.50) over the WG section and obtain the following set of differential equations:

$$\frac{dp_{\text{p}}(z)}{dz} = p_{\text{p}}(z)\left\{\sigma_{\text{e,p}}^{\text{Yb}} N_{\text{t}}^{\text{Yb}} K_{\text{eff}}^{\text{Yb}}(z) - \sigma_{\text{a,p}}^{\text{Yb}} N_{\text{t}}^{\text{Yb}}[1 - K_{\text{eff}}^{\text{Yb}}(z)]\right.$$
$$\left. - \sigma_{\text{a,p}}^{\text{Er}} \Gamma_{\text{Er}}^{\text{Yb}} N_{\text{t}}^{\text{Er}}[1 - K_{\text{eff}}^{\text{Er}}(z)] - \alpha_{\text{p}}\right\} , \tag{6.74}$$

$$\frac{dp_{\text{s}}(z)}{dz} = p_{\text{s}}(z)\left\{\sigma_{\text{e,s}}^{\text{Er}} N_{\text{t}}^{\text{Er}} K_{\text{eff}}^{\text{Er}}(z) - \sigma_{\text{a,s}}^{\text{Er}} N_{\text{t}}^{\text{Er}}[1 - K_{\text{eff}}^{\text{Er}}(z)] - \alpha_{\text{s}}\right\} , \tag{6.75}$$

where $\Gamma_{\text{Yb}}^{\text{Er}}$ accounts for the difference introduced by integrating $K^{\text{Er}} i_{\text{p}}(x, y)$ instead of $K^{\text{Er}} i_{\text{s}}(x, y)$ in (6.74). The definition of $K_{\text{eff}}^{\text{Er}}$ in (6.70) leads in a consistent manner to the local transparency for the signal when $K_{\text{eff}}^{\text{Er}} = 0.5$ if the Er emission and absorption cross-sections in (6.75) are equal (in the

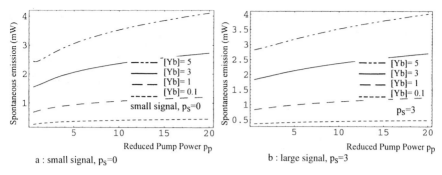

Fig. 6.21. Total spontaneous emission loss of a 1 mm long WG section for different Yb concentrations in factors of 10^{26} m^{-3}. (**a**) Small signal, $p_s = 0$; (**b**) large signal, $p_s = 3$

absence of background loss). The term $\sigma_{e,s}^{\mathrm{Er}} N_t^{\mathrm{Er}} K_{\mathrm{eff}}^{\mathrm{Er}}(z) - \sigma_{a,s}^{\mathrm{Er}} N_t^{\mathrm{Er}} \left[1 - K_{\mathrm{eff}}^{\mathrm{Er}}(z)\right]$ is the local gain of the amplifier in the absence of background loss. As the effective inversion is dependent on all parameters, the local gain will vary along the WG.

We integrate (6.75) over the length of the WG to obtain the maximum internal gain G_s:

$$P_s(L) = P_s(0) G_s \exp(-\alpha_s L) , \qquad (6.76)$$

where

$$G_s = \exp\left(\left(\sigma_{e,s}^{\mathrm{Er}} + \sigma_{a,s}^{\mathrm{Er}}\right) N_t^{\mathrm{Er}} \int_0^L K_{\mathrm{eff}}^{\mathrm{Er}}(z)\,\mathrm{d}z - \sigma_{a,s}^{\mathrm{Er}} N_t^{\mathrm{Er}} L \right) . \qquad (6.77)$$

Alternatively, the commonly used dB value for G_s is:

$$G_{\mathrm{dB}}(L) = \frac{10}{\ln(10)} \left(\left(\sigma_{e,s}^{\mathrm{Er}} + \sigma_{a,s}^{\mathrm{Er}}\right) N_t^{\mathrm{Er}} \int_0^L K_{\mathrm{eff}}^{\mathrm{Er}}(z)\,\mathrm{d}z - \sigma_{a,s}^{\mathrm{Er}} N_t^{\mathrm{Er}} L \right) . \qquad (6.78)$$

Although (6.78) can only be evaluated via numerical integration, useful insight into the characteristics of a co-doped amplifier can be obtained by studying the gain and spontaneous emission loss occurring over a short length of WG. We assume that the length under consideration is small enough so that the pump power, the effective surfaces and the inversion coefficients can be considered constant. Choosing a 1 mm length, Fig. 6.21a (small signal) and b (large signal) present the losses incurred by spontaneous emission computed from the power transfer (6.59a,b), and Fig. 6.22a and b present the corresponding net gain for various Yb concentrations.

For low Yb concentrations, one clearly observes the slow increase in the gain with the pump power, as expected from the lower Er absorption cross-sections. The power consumption remains equally low. As the Yb concentration is increased, the gain is sharply enhanced, at the cost of higher

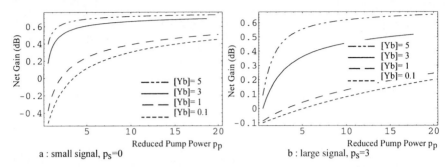

Fig. 6.22. Net gain of a 1 mm long WG section for different Yb concentrations in factors of $10^{26}\,\mathrm{m}^{-3}$. (**a**) Small signal, $p_s = 0$; (**b**) large signal, $p_s = 3$

power dissipation, until a limit is reached as the extra amount of power lost by spontaneous emission provides only a marginal gain increase. Therefore, a small signal optimization procedure will minimize the power consumption while preserving a high gain over a given length of WG. More confined guided waves would improve the overall performance, since the saturation power P_{sat} would be reduced.

For a higher signal power $P_s = 3$, the behaviour of the amplifier is exemplified in Fig. 6.22b. Here, high Yb doping clearly leads to an advantageous ability to withstand high signal power. This is due to the higher energy transfer from the pump allowed by the Yb ions.

The design of a co-doped amplifier therefore involves the weighing of the higher effective inversion generally brought by Yb against the extra consumption of pump power. For a given glass, the optimization critically depends on the law followed by the transfer coefficient. In general, very small amplifier lengths such as those required for a single-mode power laser will involve high Yb doping, in the range of $5 \times 10^{26}\,\mathrm{m}^{-3}$ or above. A small signal amplifier such as a preamplifier is likely to be longer, with much lower Yb doping, in the 1 to $3 \times 10^{26}\,\mathrm{m}^{-3}$ range, unless longer wavelength semiconductor laser sources become available.

Let us finally say a few words about rare-earth concentration profiling. Although the main equations are valid for any ion profiles, the results presented here mostly refer to a uniform rare-earth doping profile. It is well known that fibre devices exhibit better performances with rare-earth ions confined inside the core. One cannot, however, directly transfer this conclusion to integrated devices. This is because some essential optimization constraints to be applied in that case reduce the potential interest of core profiling. Indeed, integration induces the following constraints:

- the Er concentration is limited by high concentration effects (addressed in Sect. 6.3.8),
- the total length of a device is limited by integration requirements.

Consider, for example, a WG where only the core is doped with rare earth and compare it qualitatively with its uniformly doped counterpart. Write (6.78) as

$$G_{\mathrm{dB}}\left(L\right) = \frac{10}{\ln\left(10\right)} \left(\left(\sigma_{\mathrm{e,s}}^{\mathrm{Er}} + \sigma_{\mathrm{a,s}}^{\mathrm{Er}} \right) N_{\mathrm{t}}^{\mathrm{Er}} \int_{0}^{L} K_{\mathrm{eff}}^{\mathrm{Er}}(z)\,\mathrm{d}z - \Gamma_{\mathrm{s}} \sigma_{\mathrm{a,s}}^{\mathrm{Er}} N_{\mathrm{t}}^{\mathrm{Er}} L \right) ,$$

$$(6.79)$$

where $\Gamma_{\mathrm{s}} = \frac{1}{p_{\mathrm{s}}} \int_{\mathrm{doped\ core}} \int i_{\mathrm{s}}(x, y)\,\mathrm{d}x\,\mathrm{d}y$ is the overlap between the signal and the doped core.

In such a situation, the overlap term will be around 0.6, and the maximum effective inversion will be close to 0.5. In order to preserve the gain at high pump power, the selectively doped device should either be twice as long or twice as doped as the uniformly doped device. If possible, doubling the length can be beneficial. It will induce the effective volume of rare earth being pumped in the dopant-profiled WG to be approximately one half that in the uniformly doped device. Indeed, if the pump-inverted effective area is 0.8 for the dopant-profiled amplifier and 3 for a uniformly doped amplifier half as long, the ratio of the effective volumes of inverted ions is approximately $1/2$.

However, this favourable alternative is not always feasible as space is very precious on an integrated device. For example, an Er-doped integrated optic device such as reported in [34] required a 47 cm long WG, which used $9 \times 5\,\mathrm{cm}^2$ of substrate space. In some technologies, such a length could not even be used because the background losses would be too high. From a designer's point of view, there is a trade-off between pump efficiency and circuit integration, as discussed in [35].

In the case where a longer device is not a viable option, one must resort to increasing the Er concentration. This in turn may degrade the efficiency of the pumping process to the point of canceling the benefits of confining rare-earth doping to the core. Therefore, one should be cautious in directly applying to integrated devices some of the results that seem almost intuitive for EDFAs. In the following section, various detrimental effects of high rare-earth concentrations will be evaluated.

6.3.8   Adverse Effects of High Rare-Earth Concentrations

As the ions are brought sufficiently close together, stronger interaction effects begin to set in. Although such interactions are used beneficially in the cooperative energy transfer (CET) of the Yb level $^2F_{5/2}$ to the Er level $^4I_{11/2}$, as shown in Fig. 6.23a, they may also provide alternative paths for pump energy dissipation and thus decrease the device's overall efficiency.

These effects require two or more rare-earth ions to interact cooperatively. The exact nature of the interactions depends on the actual distribution of ions in the glass matrix as well as on the sites they occupy. They involve

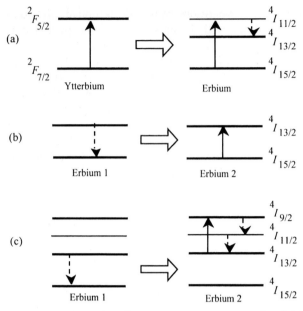

Fig. 6.23. Examples of cooperative energy transfer. (**a**) Yb to Er energy transfer; (**b**) energy migration; (**c**) homogeneous and inhomogeneous cross relaxation

multipolar coupling between ions with various dependencies on the distance between local ions.

When the ions are distributed evenly in the matrix, homogeneous CET leads to a reduction of the observed lifetime, sometimes accompanied by visible light emission. On the other hand, if the ions are unevenly grouped in clusters or pairs, inhomogeneous CET will then occur. One should therefore classify the various ion population subclasses according to their interaction distance. There would be a gradual change of properties between purely homogeneous and purely inhomogeneous CET. For the sake of simplicity, it is convenient in practice to group the effects and therefore the ion populations in only two categories, one with homogeneous CET only and the other with purely inhomogeneous CET.

A first type of CET is concentration quenching of the Er effective life-time. It involves the dual effect of energy migration, Fig. 6.23b, and loss to surrounding receptors, which lead to non-radiative deactivation of the metastable levels. In the case of OH^- receptors, second-harmonic absorption occurs around $1.5\,\mu m$, even more drastically reducing the effective lifetime of Er ions. Appropriate annealing process steps help reduce OH^- concentrations to values where there is no major effect on the lifetime. In practice, OH^- concentrations should be kept below $5 \times 10^{24}\,m^{-3}$.

In a given glass and for a given fabrication process, there is a concentration above which the observed lifetime falls significantly. In some phosphates, quenching starts for Er concentrations as high as $4 \times 10^{26}\,\mathrm{m^{-3}}$. This is the upper limit for efficient Er operation in the absence of other mechanisms. It is generally detrimental to work with glass with an Er concentration much greater than the quenching concentration. When present, the effect is summarized by a reduced value of the observed radiative lifetime of the rare earth under consideration.

Cooperative upconversion, a second type of homogeneous CET, is a cross-relaxation process (see Fig. 6.23c). It involves interactions between two ions in a metastable state and is modelled by a quadratic term in (6.48) as well as in the rate equations. In the case of (6.48) one obtains

$$\frac{dN_m^{Er}}{dt} = \left(W_p^{Er} + W_{fm}^{Er}\right) N_f^{Er} - W_{mf}^{Er} N_m^{Er} - N_m^{Er}\left(\frac{1}{\tau_{Er}} + C_{up} N_m^{Er}\right)$$
$$+ k_{tr} N_m^{Yb} N_f^{Er} , \tag{6.80}$$

where, for a dipole–dipole interaction [36],

$$C_{up} \cong \frac{16\pi^2 R_0^6}{9\tau_{Er}} N_t^{Er} \tag{6.81}$$

and $R_0 \cong 2.2\,\mathrm{nm}$ is a critical interaction distance between Er ions.

Experimentally, such a phenomenon will induce a reduction of the observed Er lifetime *as the metastable level becomes populated*. Once this level is fully populated (i.e. under sufficient pumping conditions) the phenomenon is self-limiting, with a lower value of the observed lifetime such that

$$\left(\frac{1}{\tau_{Er}}\right)_{(p_p=\infty)} = \left(\frac{1}{\tau_{Er}}\right)_{(p_p=0)} + C_{up} N_t^{Er} . \tag{6.82}$$

The main effect of this type of CET will thus be an increase in the pump power required to reach a given inversion level. Typical values have been measured for silicates [27] by direct lifetime measurement, revealing a halving of the lifetime under high pump power. In phosphates, similar upconversion values have been obtained [37] by fitting the upconversion coefficient as a loose parameter. In practice, upconversion will become significant when the Er/Yb ratio is sufficiently high so that a major part of the total loss comes from Er ion pumping. Other CET phenomena may exist, such as cooperative frequency doubling, in which 550 nm light is produced by cooperative interaction between two Er ions.

Another mechanism present at high rare-earth concentrations is the inhomogeneous CET called clustering and also described by Fig. 6.23c. Adopting a simplified approach, assume [38,39] that a fraction k_c of the Er ions are assembled in clusters. Assume further that interactions between ions within a cluster are so strong that at most one ion can be found in the metastable state, all the other ions in the cluster being subject to rapid decay to their ground state. Modelling such an effect does not call for quadratic terms as in homogeneous CET. Two populations of Er ions are taken into account separately. The ions not trapped by inhomogeneous CET obey rate equations (6.46) and (6.48), whereas ions in clusters are in the ground state.

The propagation equations are therefore modified in the following manner:

- N_t^{Er} is replaced by $(1 - k_c) N_t^{Er}$,
- supplemental (non-bleachable) absorption terms appear for pump and signal as $k_c N_t^{Er}$ ions are never inverted.

Rearranging the factors leads to the following modified propagation equations for the pump and signal:

$$
\begin{aligned}
\frac{dp_p(z)}{dz} =\ & p_p(z) \left\{ \left[\left(\sigma_{e,p}^{Yb} + \sigma_{a,p}^{Yb} \right) K_{eff}^{Yb}(z) - \sigma_{a,p}^{Yb} \right] N_t^{Yb} \right. \\
& \left. - \sigma_{a,p}^{Er} \Gamma_{Er}^{Yb} N_t^{Er} \left[1 - (1 - k_c) K_{eff}^{Er}(z) \right] - \alpha_p \right\},
\end{aligned}
\tag{6.83}
$$

$$
\frac{dp_s(z)}{dz} = p_s(z) \left[\left(\sigma_{e,s}^{Er} + \sigma_{a,s}^{Er} \right) (1 - k_c) N_t^{Er} K_{eff}^{Er}(z) - \sigma_{a,s}^{Er} N_t^{Er} - \alpha_s \right].
\tag{6.84}
$$

With these notations, the inversion factor K can still become equal to one but only for an Er subpopulation with ion density $(1 - k_c) N_t^{Er}$.

When compared to (6.78), the gain is reduced accordingly:

$$
G_{dB}(L) = \frac{10}{\ln(10)} N_t^{Er} \left(\left(\sigma_{e,s}^{Er} + \sigma_{a,s}^{Er} \right) (1 - k_c) \int_0^L K_{eff}^{Er}(z) \, dz - \sigma_{a,s}^{Er} L \right).
\tag{6.85}
$$

From these equations, we can deduce two detrimental effects of ion clustering: firstly, it creates extra pump and signal absorption, and secondly, it reduces the maximum available gain. Consequently, clustering has a very strong impact on the characteristics of an amplifier. For example, it is apparent in (6.85) that if $k_c \cong 0.5$, inversion will become *impossible* near the zero-phonon line because too many inactivated sites will prevent population inversion.

A more complete image of the limitations created by clustering can be obtained with the power-transfer equation. Following a derivation similar to the one leading to (6.59a,b), we come to a general equation including both homogeneous and inhomogeneous CET. However, we cannot assume the guided-photon losses to be equal for pump and signal. We therefore do not integrate the differential equation and write a difference equation across a small length δz of WG:

$$\delta\Phi_{\mathrm{T}} = -\Phi_{\mathrm{p}}(z)(\alpha_{\mathrm{p}} + k_{\mathrm{c}}\sigma_{\mathrm{a,p}}^{\mathrm{Er}} N_{\mathrm{t}}^{\mathrm{Er}})\delta z - \Phi_{\mathrm{s}}(z)(\alpha_{\mathrm{s}} + k_{\mathrm{c}}\sigma_{\mathrm{a,s}}^{\mathrm{Er}} N_{\mathrm{t}}^{\mathrm{Er}})\delta z$$
$$- \left(\frac{1}{\tau_{\mathrm{Yb}}} + k_{\mathrm{c}}k_{\mathrm{tr}}N_{\mathrm{t}}^{\mathrm{Er}}\right) N_{\mathrm{t}}^{\mathrm{Yb}}\delta V_{\mathrm{eff}}^{\mathrm{Yb}} - \frac{1-k_{\mathrm{c}}}{\tau_{\mathrm{Er}}^{\infty}} N_{\mathrm{t}}^{\mathrm{Er}}\delta V_{\mathrm{eff}}^{\mathrm{Er}} , \qquad (6.86)$$

where $\delta V_{\mathrm{eff}}^{\mathrm{Yb}} = K_{\mathrm{eff}}^{\mathrm{Yb}}(z)S^{\mathrm{Yb}}(z)\,\delta z$ and $\delta V_{\mathrm{eff}}^{\mathrm{Er}} = K_{\mathrm{eff}}^{\mathrm{Er}}(z)\,S^{\mathrm{Er}}(z)\delta z$ are incremental effective volume elements of excited rare-earth ions. We present (6.86) in Table 6.4, explaining the role of each term explicitly.

Table 6.4 highlights the opposite energy conversions which lead to undesirable losses: part of the guided-wave electromagnetic energy is dissipated through interaction with surrounding matter and converted into phonons; correspondingly, part of the energy stored in rare-earth ions is dissipated by electromagnetic radiation and by interaction with the neighbouring atoms or ions.

The same approach could be used to include other type of loss mechanisms, according to the material and technology being modelled.

By examining Table 6.4 in more detail, we observe that, in addition to decreasing the maximum available gain described in (6.85), the inhomogeneous CET also fulfills three other roles:

- It reduces the gross amount of pump and signal available for power transfer by adding to the background loss in the two first terms.

Table 6.4. Step-by-step account of the loss mechanisms in an amplifier of length δz

Equation (6.86)	Physical meaning
$\delta\Phi_{\mathrm{T}} =$	The variation of the total guided photon flux across a length δz is equal to the sum of:
	The energy lost by guided photon absorption:
$-\Phi_{\mathrm{p}}(z)\alpha_{\mathrm{p}}\delta z$	Pump photons absorbed by background loss
$-\Phi_{\mathrm{p}}(z)k_{\mathrm{c}}\sigma_{\mathrm{a,p}}^{\mathrm{Er}} N_{\mathrm{t}}^{\mathrm{Er}}\delta z$	Pump photons lost in cluster excitation
$-\Phi_{\mathrm{s}}(z)\alpha_{\mathrm{s}}\delta z$	Signal photons absorbed by background loss
$-\Phi_{\mathrm{s}}(z)k_{\mathrm{c}}\sigma_{\mathrm{a,s}}^{\mathrm{Er}} N_{\mathrm{t}}^{\mathrm{Er}}\delta z$	Signal photons lost in cluster excitation
	The energy lost by excited rare-earth ions:
$-\frac{1}{\tau_{\mathrm{Yb}}}N_{\mathrm{t}}^{\mathrm{Yb}}\delta V_{\mathrm{eff}}^{\mathrm{Yb}}$	Radiation loss by Yb spontaneous emission
$-k_{\mathrm{c}}k_{\mathrm{tr}}N_{\mathrm{t}}^{\mathrm{Er}} N_{\mathrm{t}}^{\mathrm{Yb}}\delta V_{\mathrm{eff}}^{\mathrm{Yb}}$	Energy lost by Yb to Er-cluster transfer
$-\frac{1-k_{\mathrm{c}}}{\tau_{\mathrm{Er}}^{\infty}}N_{\mathrm{t}}^{\mathrm{Er}}\delta V_{\mathrm{eff}}^{\mathrm{Er}}$	Radiation loss by non-clustered Er spontaneous emission (Er lifetime is reduced by homogeneous CET)

- Simultaneously, it increases (third term) the pump losses due to Yb spontaneous emission by channelling energy away to pump Er clusters.
- It reduces the energy lost to maintain non-clustered Er ions in the inverted state, since there is only a fraction of $1 - k_c$ of these ions available for inversion (fourth term); this advantage is, of course, more than compensated for by the extra absorption terms caused by clustered ions.

In (6.86), the effect of homogeneous CET is summarized by setting all along the WG an Er lifetime equal to its high pump value as described in (6.82). Additional homogeneous CET phenomena such as concentration quenching can equally be taken into account by selecting a lower observed lifetime and substituting it into the right-hand side of (6.82). This mechanism increases the pump losses due to Er inversion.

In conclusion, the effects of clustering are much more detrimental than those of homogeneous CET. In practice, this may lead us to select the rare-earth co-doping differently. Co-doping with Yb dilutes the Er in Yb clusters and is therefore beneficial. At the same time, the interaction between Er and Yb ions can actually be reinforced if they belong to the same cluster. A more complete study of the impact of such a dilution remains to be carried out. Finally, an additional rare earth such as lanthanum may be used to fine-tune the optimization of the doping, as it is optically inactive but chemically similar to Er or Yb. The actual optimization of an amplifier is very specific to a glass family and even to the fabrication process and is beyond the scope of this section. The main results published to date are presented in the next section.

6.3.9   Technologies and Devices

Technologies and Comparative Results

There are several glass WG technologies currently contending for the fabrication of integrated amplifiers. They include

- ion exchange in phosphates, silicates and germanates,
- flame hydrolysis deposition of phosphate-doped silica,
- silicate and phosphate sputtering,
- doped silica plasma-enhanced CVD (PECVD),
- sol-gel deposition of silica, germania, phosphate.

To date, the best results have been obtained with the first three technologies; their main process steps are summarized here. The reader will find a detailed review of the sol-gel technology in [40] and results on PECVD in [41].

Ion Exchange. In silicates and germanates, the process is basically the same as described in Sect. 6.2.7. In phosphates, additional steps must be taken in order to reduce potential damage to the glass during the exchange [37]. The salts are dehydrated by heating under moderate vacuum and the samples are exchanged under vacuum or argon-gas flow. Rare-earth compounds are added during glass fabrication and are therefore uniformly present in the material. Phosphates accept very high rare-earth oxide concentrations (up to 21% in weight). The process yields WGs of high quality in the materials where ion exchange is possible.

Flame-Hydrolysis Deposition (FHD). Originally developed for the fabrication of optical fibre preforms, FHD has been adapted to WG fabrication. It is based on a hydrolysis reaction in a H_2/O_2 glass burner: $SiCl_4 + 2H_2O \rightarrow SiO_2 + 4HCl$.

Multicomponent glass is obtained by the simultaneous flow of $TiCl_4$ and $GeCl_4$ to synthesize TiO_2 and GeO_2, respectively. Various compositions may therefore lead to the increase in optical index required for waveguiding.

A torch distributes the fine glass particles (otherwise known as soot) over the substrate. The rare-earth ions are then introduced by dipping the soot in an alcohol solution of $(RE)Cl_3$. A sintering step at temperatures above $1200\,°C$ then shrinks the soot into a film, reducing its thickness by as much as 90%.

Very low losses are obtainable in FHD deposited WGs, in the dB/m range, but the process is well known and controlled for a rather restricted range of materials. New glass matrices accepting much higher dopant concentration will require a substantial technological effort.

Sputtering. Sputtering can be successfully used to deposit a variety of glass materials. In a vacuum chamber with $10^{-2}-10^{-3}$ Torr, a plasma is formed by magnetron discharge in a rare gas such as argon. The plasma ions collide with the material to be deposited and gradually remove it. In turn, the particles thus created condense on the substrate to form a thin film. The process, purely mechanical, can also be enhanced by chemical reactions if a gas such as oxygen is added to the vacuum chamber. This process is very flexible since a wide variety of materials can be sputtered. Two drawbacks are (a) the material deposited may not have the same exact chemical composition as the original material, and (b) WGs in the as-deposited material tend to exhibit high scattering loss, which can be reduced by special annealing steps.

Both FHD and sputtering require the active core layer to be etched to pattern it into a WG. This leads to increased scattering loss by WG walls when highly confining, high-index structures are sought. On the other hand highly confining ion-exchange WGs exhibit very little loss, but rare-earth ions cannot be selectively located in the core.

Comparative Results. In an attempt to evaluate the achievements obtained with these technologies, we adopt two different approaches:

- a "black-box" approach, appropriate for system design;
- a technology approach which seeks intrinsic comparative parameters.

In the black-box approach, the system designer will be looking for the following figures:

- the net gain for a pigtailed device, here referenced at 120 mW pump for small-signal single-pass,
- the noise figure,
- the insertion loss at 1300 nm.

Other important selection criteria include the usable wavelength window at 1550 nm, the 3 dB drop-off saturation gain and the usable pump wavelength window. But the lack of available data did not allow for a comparative presentation of these parameters.

In the technology approach, we introduce a basis for evaluating the maturity of a given process with the gain to pump intensity (GPI) ratio:

$$\text{GPI} = \frac{G_{\text{dB}} S_{\text{p guide}}}{P_{\text{p fibre}}} ,$$

in $\text{dB}\,\mu\text{m}^2/\text{mW}$, where $P_{\text{p fibre}}$ is the *injected* pump power, *external* to the device, and $S_{\text{p guide}}$ is the *on-chip* pump-mode-power area, as defined by (6.53).

The gain is chosen to be 80% of the infinite pump net gain, and the corresponding injected pump value is then used. The GPI ratio is sensitive to the spectroscopic quality of the glass, which has a strong impact on the GPI ratio. It is also sensitive to the connection losses between fibre and WG because of the specific choice made with regard to the GPI ratio to measure the pump and the mode area. Connection losses are critical for the efficiency of an optical amplifier. By construction, the GPI ratio does not directly depend on the mode confinement on the chip.

The second figure of merit is the internal gain per unit length. It compares the maximum achievable gain of a technology relative to the WG length necessary and therefore to the space occupied on the wafer. This figure of merit also reflects the ability to proceed to high rare-earth doping concentrations.

Since the information available in the literature does not include all the necessary figures, some missing parameters have been deduced from the information provided. Despite the care devoted to this reconstitution, estimated values should be considered to have large error margins, of the order of ±25%. The results are presented Table 6.5.

Notes:
1. Various WG performances have been reported in [25] for the type of technology used in [27], with much lower loss, but also lower gains; the higher-gain/higher-loss type is indicated here.

Table 6.5. Comparison between the main integrated optic technologies (best results for each category in bold face)

Technology	System design performance figures			Technology figures of merit	
	Net gain at 120 mW pump power (dB)	Noise figure (dB)	Insertion loss (estim.) at 1300 nm (dB)	GPI (estim.) (dB μm^2/mW)	Internal gain/length (dB/cm)
FHD [34]	**18.9** Double-side pumping	3.8	− 5.5	**6.9**	0.68
Ion-exchange phosphate [29]	16	4	**− 0.8**	2.5	2.9
Ion-exchange silicate [42]	10.6	Not available	− 2.7	0.97	2.3
Sputtering silicate	8 [27]	**3.1** [27]	−6	0.75 [27] 0.6 [43]	**4.2**
Sputtering phosphate [28]	4.1 (25 mW pump power)	Not available	< −8	0.3	4.1

2. A large distortion may arise from the double-side pumping used in [33], which tends to enhance the low pump gain. The 120 mW gain and the GPI index would be lower in a single-side pumping scheme.

Table 6.5 reveals that the FHD technology, closest to more standard EDFA technology, displays the highest GPI ratio, indicative of its maturity and optical efficiency. At the same time, FHD presents a low ability for compact integration. At the other end of the spectrum, sputtering techniques, while not yet pump-photon efficient, yield the highest gain per unit length. Finally, ion-exchange data exhibit balanced qualities, giving ion exchange an advantage for immediate commercial applications.

As a concluding remark, we observe that the problems incurred by high doping concentrations are being gradually overcome. The performances of some of these integrated devices make them marketable. Likewise, attention is now focused on obtaining a better gain flatness over a larger wavelength window. Since a large variety of glass materials are currently under investigation, some of these future devices should exhibit a wider gain window than the current EDFA technology.

Amplifiers: Present Results and Trends

Sputtering and ion-exchange technologies have led to very short integrated components. The device shown in Fig. 6.24 is less than 7 cm long, including the fibre-to-chip connections. Adding a WDM (Wavelength Division Multiplexer) injection coupler such as demonstrated in [44] does not lead to a much larger size component.

The 1535 nm gain of ion-exchanged amplifiers with various WG lengths is shown in Fig. 6.25. The highest net gain (fibre-to-fibre) was 24 dB for a 9-cm-long WG at 180 mW of 980 nm pump power. The flat slope region, more apparent for longer WGs, is attributed to significant pump wavelength fluctuations at low power. Figure 6.26 shows the value of the gain along the length of the same WG. The measurements were made using a cut-

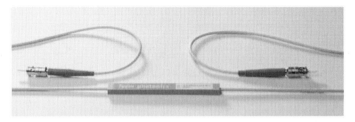

Fig. 6.24. An integrated optic glass amplifier (photo Teem Photonics)

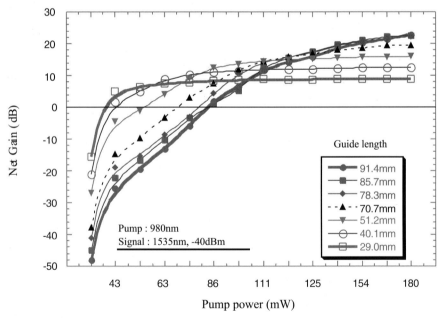

Fig. 6.25. Measurements of small-signal net gain versus injected pump power for various WG lengths

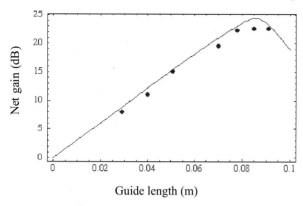

Fig. 6.26. Theoretical and experimental values of the small signal gain versus WG length for 180 mW (signal wavelength 1535 nm)

back technique, whereas the theoretical curve was computed with a model including the effect of clustering and ASE.

The highest large-signal gains reported so far have shown best values of deep saturation around 14 dBm for a double-pump amplifier. This is a general consequence of the devices still having modest efficiencies which, with 0.12 dB/mW, do not leave enough pump power for large signal gain [see (6.59a,b)]. It is not, however, a fundamental physical limit and future devices are expected to fare significantly better. In particular, it is expected that Yb/Er-co-doped devices will lead to a very high saturated output power, because of the high pumping rate permitted by Yb ions. This has been shown in microchip lasers based on co-doped glasses, where several hundreds of mW can be extracted, indicating an ability to provide gain at high signal powers.

The noise figure in an integrated optics glass device can be very close to the fundamental (3 dB) limit. An excess noise factor very close to 1, corresponding to a near-ideal noise figure of 3.1 dB, has been reported for an Er-doped device [45]. Yb/Er ion-exchange amplifiers also exhibit a sub-4-dB small-signal noise figure. This trend should be confirmed in the next generation of devices, which will offer even higher confinement and therefore an even more complete inversion of the rare-earth ion population. Such low-noise figures have led to advantageous results in system performance. Indeed, no significant degradation has been reported for 10 Gb/s long-distance transmission experiments based on planar amplifiers [46], with no floor or degradation down to bit-error-rates of 10^{-12} over distances of 70 to 186 km of propagation in fibre.

The usable spectral bandwidth over which rare-earth amplifiers can be operated is dependent on the glass matrix. Whereas EDFA technology has moved away from very pure component fibre-technology, planar amplifiers came from a background of multicomponent glasses. Because of this natu-

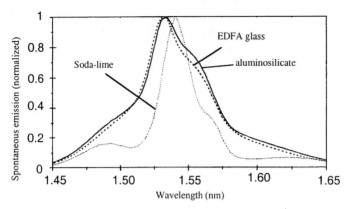

Fig. 6.27. Emission spectrum of Er in different glass compositions compared to EDFA glass (after [25])

rally wider material diversity, it is anticipated that future integrated optics devices will exhibit wider wavelength windows. Recent results along that road have begun to appear. Figure 6.27 shows an example of a highly Er-doped aluminosilicate glass with a large emission spectrum used for sputtered glass WG amplifiers.

Active/Passive Functions

A prominent feature of integrated technologies is their ability to group several functions in a small amount of space. For example, a more evolved amplifier will include WDM (Wavelength Division Multiplexer) pump injection, gain flatness correction and tap control channels. Different strategies are envisaged to meet that goal.

For small-to-medium series of specialized devices such as required for transmission applications, it appears cheaper and offers a better yield to make passive and active functions on different glass and to hybridize them. One such approach has led to a loss-compensated amplifier/splitter, which could be used for multiplexing 8 DWDM channels. The spectral transfer function of the device is shown in Fig. 6.28. Although the amplifying section of the device has no gain-flattening filter, the usable spectral span is about 15 nm.

For larger series, on-chip co-integration of active and passive functions may become commercially attractive. With this alternative, different sections of a given circuit will be selectively doped with rare-earth ions. Several laboratories have demonstrated very low transition loss between doped and undoped WG sections. The challenge that they are presently facing is how to readily obtain satisfactory specifications for the co-integrated functions.

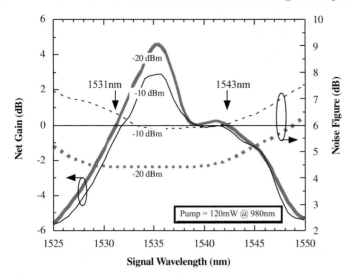

Fig. 6.28. Gain spectrum and noise figure versus signal wavelength of a loss-compensated amplifier/splitter, operated as a 1 × 8 combiner, for two input power levels

6.4   Integrated Optics Er/Yb Laser Oscillators

Lasers in co-doped glass systems [47] are a very interesting area of development because they simultaneously combine a narrow emission linewidth with good emission-wavelength stability. They are suited to DWDM systems using external modulation. Similar in principle to the fibre lasers studied in [48], the first demonstration of an integrated optics glass laser was achieved in Nd-doped glass [49] in 1974; Er-doped WG lasers followed 17 years later [50]. This in turn led to Yb/Er-co-doped devices [51,52] as the potential applications to the telecommunications industry came into consideration.

As for amplifier devices, the very high concentrations of rare earth possible in some silicates and phosphates have given them an edge in the creation of short, high-quality lasers.

The most important technical details of a glass WG laser in continuous wave operation are its excellent wavelength stability, a high signal-to-noise ratio and a very narrow linewidth. This section only reports on specific prospects and experimental results of integrated optics glass lasers. Readers interested in a broader study of lasers may consult [53].

6.4.1   Continuous Wave (CW) Operation

In order to quickly assess the domain of CW oscillation, direct use of the previous study on optical amplification leads to a simple prediction model. We start from the output-power/pump-power characteristics of the amplifying

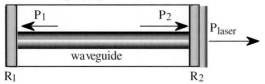

Fig. 6.29. Schematic of a WG laser cavity

WG section studied in Sect. 6.3.9. It is assumed, as indicated in Fig. 6.29, that the laser is an amplifier fitted with mirror reflectors. We call the effective mirror reflection coefficients R_1 and R_2, assuming they include some loss, so that for each mirror $R+T < 1$, where T is the effective transmission coefficient including possible loss.

Define G as the total net gain per pass at a given wavelength. As such, G includes the possible propagation loss incurred in the amplifier. The laser oscillation will occur when the gain within the cavity overcomes the losses introduced by the reflections on the mirrors. Equating gain and loss, the condition for oscillation is

$$G^2 R_1 R_2 = 1 . \tag{6.87}$$

The total light power incident on the mirrors (from the inner sides) is

$$P_T = P_1 + P_2 = P_2 \left(1 + R_2 G\right) . \tag{6.88}$$

Substituting G from (6.87) into (6.88) leads to

$$P_T = P_2 \left(1 + \sqrt{\frac{R_2}{R_1}}\right) . \tag{6.89}$$

Therefore, the laser output power exiting from side 2 is related to the total power by

$$P_{\text{laser}} = P_2 T_2 = P_T T_2 \left(1 + \sqrt{\frac{R_2}{R_1}}\right)^{-1} , \tag{6.90}$$

where T_2 is the effective transmission coefficient of mirror 2, including possible losses due to mirror imperfections.

In order to directly use the experimental and theoretical data on the sections in amplifiers, the output power of a unidirectional amplification stage is set equal to P_T. It is clearly an approximation, since we should in fact use the exact bidirectional signal output power characteristics as the defining elements. Nevertheless, comparisons with a full propagation model reveal only a small difference in computed output power. In addition, this approximation has no impact on the calculation of the lasing threshold.

The theoretical (experimentally validated) curves of the amplifier output power characteristics presented in Fig. 6.30 are used as input to compute the lasing characteristics of an integrated laser with various mirror reflection coefficients, presented in Fig. 6.31. On the density plot, we observe a flat maximum for high output powers, which can be used to optimize a rugged design of a laser source.

For the telecommunication and cable TV application domains, a good spectral purity as well as a high signal-to-noise ratio are necessary. For short WG lengths, the Yb/Er lasers have an emission linewidth in the Schawlow–Townes regime, as detailed below. The Schawlow–Townes formula

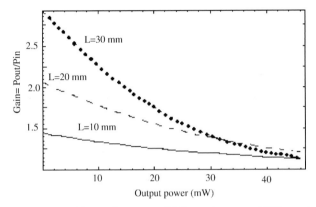

Fig. 6.30. Gain of various WG sections at 1535 nm (pump power 100 mW at 980 nm)

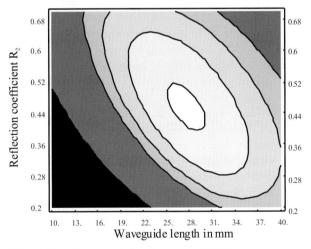

Fig. 6.31. Computed density plot of 1535 nm laser output power for various WG lengths. Contours are for 0.99, 0.9, 0.8 and 0.7 (and threshold) of the maximum 12 mW output power

describes [54, p. 580] how the spectral profile of a laser oscillator is widened by the effect of spontaneous emission within the laser cavity. The full width at half maximum of the laser frequency spectrum is

$$\frac{\Delta\nu_{\text{laser}}}{\nu_{\text{laser}}} = \frac{2\pi h \left(\Delta\nu_{\text{c}}\right)^2}{P_{\text{laser}}} \frac{K_{\text{av}}^{\text{Er}}}{2K_{\text{av}}^{\text{Er}} - 1} , \tag{6.91}$$

where we have used an average inversion of the Er ions

$$K_{\text{av}}^{\text{Er}} = \frac{1}{N_{\text{t}}^{\text{Er}}} \int\limits_{\text{length, } L} \left(\int_{\Re^2} \int N_m^{\text{Er}} \, \mathrm{d}x \, \mathrm{d}y \right) \mathrm{d}z . \tag{6.92}$$

The width $\Delta\nu_{\text{c}}$ of the cavity resonance is related to the decay lifetime of a passive-cavity mode by the relation

$$\Delta\nu_{\text{c}} = \frac{1}{2\pi\tau_{\text{c}}} . \tag{6.93}$$

τ_{c} is called the photon lifetime. It is the time necessary to empty the passive cavity of its electromagnetic energy. For a cavity of length L and with our notations [54, p. 148],

$$\tau_{\text{c}} = \frac{2n_{\text{eff}}L}{c[2\alpha_{\text{s}}L - \ln(R_1 R_2)]} . \tag{6.94}$$

For a numerical example, we choose a laser with $L = 1\,\text{cm}$, $R_1 R_2 = 0.5$, a linear background loss constant $\alpha = 2.3\,\text{m}^{-1}$ (corresponding to $0.1\,\text{dB/cm}$) and an effective index of 1.55. We in turn obtain a photon lifetime $\tau_{\text{c}} = 140\,\text{ps}$ and, correspondingly, $\Delta\nu_{\text{c}} = 1.1\,\text{GHz}$. We then assume an emitted power of $1\,\text{mW}$ and an average effective inversion of 0.75 and obtain $\Delta\nu_{\text{laser}} = 1.6\,\text{kHz}$. For cavity lengths around or shorter than $1\,\text{cm}$, the relative effect of the background loss is negligible in (6.93), and $\Delta\nu_{\text{laser}}$ will scale as L^{-2}. Other phenomena, such as mechanical vibrations, usually broaden the linewidth to values around $10\,\text{kHz}$. Glass cavities shorter than $4\,\text{mm}$ will thus exhibit laser linewidths limited by the Schawlow–Townes limit [see (6.91)].

To date, actual single longitudinal-mode operation of an Er/Yb integrated glass laser has not been reported. The experimental examples shown here were obtained with a few longitudinal modes. However, recent results obtained in an Er/Yb fibre device [55] with a geometry very similar to an integrated component confirm the excellent prospects afforded by this approach. The reported values for the optical linewidth are as low as $18\,\text{kHz}$. Overall, the transmission characteristics show no degradation of the bit error rate when compared to a semiconductor laser. The higher optical signal-to-noise ratio ($65\,\text{dB}$) and low relative intensity noise (RIN) level of $-168\,\text{dB/Hz}$ above $10\,\text{MHz}$ make it highly attractive for CATV applications.

At present, the highest performance for a CW integrated optics glass laser presented in [56] shows a slope efficiency of 27%, a signal-to-noise ratio around

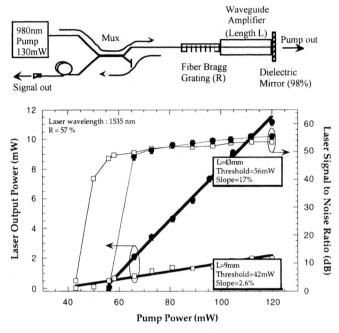

Fig. 6.32. Measured output power and signal-to-noise ratio of 1535 nm lasers with 57% Bragg grating reflectors

55 dB and a maximum output power of 16 mW for 120 mW of pump power. The output power and signal-to-noise ratio of a similar device are shown in Fig. 6.32. A schematic of the experimental device is also shown.

The lowest threshold of 14.8 mW of pump power was achieved in an 8.6 mm long silicate glass cavity [57], although propagation losses were at a high 2.5 dB/cm in that case. The device delivered around 1 mW of laser radiation at 1536.6 nm, for a total 33 mW of pump power at 966.6 nm.

Yb co-doping is very instrumental in suppressing self-pulsing in these devices. The first reason is the higher ability to effectively pump the Er ions and therefore to move farther away from threshold, a condition known to increase the damping of relaxation oscillations in lasers. The Er pairs or clusters appear to be the primary source of instabilities, behaving as a saturable absorber [58]. The second reason for improvement comes from the dissolution of Er in Yb clusters, which reduces the relative amount of saturable absorption centers. The experimental results shown in [55] confirm that trend with a maximum RIN level of −133 dB/Hz at the relaxation oscillation frequency of 2.8 MHz. Using the previous notations, the relaxation oscillation frequency for a 3-level laser can be written approximately as

$$\nu_{\text{relax}} = \frac{1}{2\pi} \left[\frac{K_{\text{av}}^{\text{Er}}}{2K_{\text{av}}^{\text{Er}} - 1} \left(\frac{x-1}{\tau_{\text{Er}}\tau_{\text{c}}} \right) \right]^{1/2} , \qquad (6.95)$$

where $x = \frac{P_{p\,\text{in operation}}}{P_{p\,\text{at threshold}}}$.

Depending on the exact device characteristics, (6.95) gives oscillation frequencies in the MHz range for Er-doped structures.

In all of these devices, the wavelength domain of tunability usually covers most of the Er window but depends on the specific material being used. With their unique combination of integration, environmental stability, high output power, and very low noise, integrated optics glass lasers offer an interesting solution to multiple-wavelength laser heads for high-bit-rate DWDM applications ($> 2.5\,\text{Gbit/s}$) where external modulators become necessary.

6.4.2   Experimental Soliton and Q-Switch Operation

In addition to CW operation, integrated optics glass lasers have been demonstrated to yield efficient operation in pulsed mode. One such example, the soliton laser source, is considered to be a promising approach to future long-distance communications. Soliton waves have the unique property of sustaining long-distance propagation without suffering from the usual dispersion spreading, because the effects of nonlinear self-phase modulation and negative dispersion compensate each other. Sub-picosecond solitons fit time-domain multiplexing transmission requirements for data bit rates in the 100 Gbit/s range; they could also be used for DWDM spectrally sliced sources because of their broad wavelength spectrum.

A ring laser based on an integrated optics glass gain section working in the femtosecond soliton regime has been demonstrated [59] by the use of additive pulse mode (APM)-locking [60]. The laser produced 116 fs pulses at a repetition rate of 130 MHz, with a pulse energy of 160 pJ. The ring-laser configuration helps to lower the self-starting of passive-mode-locked

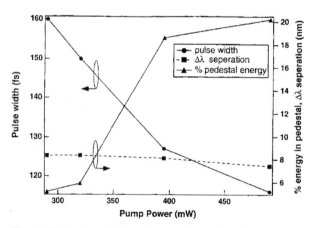

Fig. 6.33. Pulse width and energy in pedestal versus pump power for a mode-locked laser operating at a repetition rate of 130 MHz (after [59])

operation. As can be seen in Fig. 6.33, the pulse duration decreases from 160 fs at a pump power of 290 mW (the onset of the pulsed mode) down to 116 fs at 490 mW. The integrated optics amplifying section brings several improvements, all related to the shortening of the cavity. Firstly, since a single pulse is circulated within the ring, the repetition rate is increased. In the present experiment, replacing an EDFA section with a planar device led to a shorter total optical path of 2.25 m. If this path was further reduced by a factor of 20, it would enable repetition rates in the GHz regime. A second major advantage of the reduced length is the improvement in the pulse shape. The solitons have a shorter duration because the (parasitic) resonant sidebands are generated over a shorter length; their peak power is higher because the reduced non-linearity enables the use of higher pump powers without entering into an unstable multipulse regime.

Other results for mode-locked integrated optics lasers include an actively mode-locked, Er-doped silica amplifier with electro-optic modulation. The device emitted 5.2 ps pulses at a 6.3 GHz modulation frequency [61] with time–bandwidth products of 0.51. The resulting peak power for each peak is evaluated to be around 20 mW.

A Q-switched linear cavity laser has also been demonstrated [62]. Q-switching was produced with an acousto-optic modulator at a repetition rate of 500 Hz. The results presented in Fig. 6.34 show 200 ns pulses with a peak power of 3.2 W. Since important cavity losses resulted in a limitation of the performance in the CW mode, significant improvements in the characteristics are anticipated if full use of the WG amplification properties is achieved.

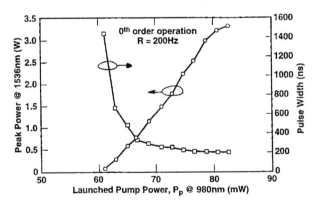

Fig. 6.34. Dependence of peak power and pulse width on launched power for operation of a Q-switched WG laser at a repetition rate of 200 Hz (after [62])

References

1. D. Marcuse (ed.), *Integrated Optics* (IEEE Press, New York 1973)
2. C. Vassallo, *Optical waveguide concepts, Optical Wave Science and Technology* (Elsevier, Amsterdam 1991)
3. A.W. Snyder and J.D. Love, *Optical Waveguide Theory* (Chapman and Hall, London 1983)
4. D. Heatley, G. Vitrant, and A. Kévorkian, "Simple finite-difference algorithm for calculating waveguide modes," Opt. Quantum Electron. **26**, 151 (1994)
5. K. Van de Velde, H. Thienpont, and R. Van Geen, "Extending the effective index method for arbitrarily shaped inhomogeneous optical waveguides," IEEE J. Lightwave Technol. **6**, 1153 (1988)
6. P.K. Tien, S. Riva-Sanseverino, R.J. Martin, A.A. Ballman, and H. Brown, "Optical waveguide modes in single-crystalline $LiNbO_3$-$LiTaO_3$ solid-solution films," Appl. Phys. Lett. **24**, 503 (1974)
7. R. März, *Integrated Optics, Design and Modeling* (Artech House, Boston, London 1995)
8. H. Nishihara, M. Haruna, and T. Suhara, *Optical Integrated Circuits*, McGraw-Hill Optical and Electro-Optical Engineering Series, R.E. Fisher and W.J. Smith, Series eds. (Mc Graw-Hill, New York 1989)
9. D. Marcuse, "Field-deformation and loss caused by curvature of optical fibres," J. Opt. Soc. Am. **66**, 311 (1978)
10. A. Takagi, K. Jinguji, and M. Kawachi, "Wavelength characteristics of 2 × 2 optical channel-type directional couplers with symmetric or non-symmetric coupling structures," IEEE J. Lightwave Technol. **10**, 735 (1992)
11. Y. Shani, C.H. Henry, R.C. Kistker, R.F. Kazarinov, and K.J. Orlowski, "Integrated optic adiabatic devices on silicon," IEEE J. Quantum Electron. **27**, 556 (1991)
12. K. Jingui, N. Takato, A. Sugita, and M. Kawachi, "Mach–Zehnder interferometer type optical waveguide coupler with wavelength-flattened coupling ratio," Electron. Lett. **26**, 1326 (1990)
13. K. Okamoto, H. Takahashi, M. Yasu, and H. Hibino, "Fabrication of wavelength insensitive 8 × 8 star coupler," IEEE Photon. Technol. Lett. **5**, 61 (1992)
14. K. Okamoto, H. Okazaka, Y. Ohmori, and K. Kato, "Fabrication of large-scale integrated optic $N \times N$ star couplers," IEEE Photon. Technol. Lett. **4**, 1032 (1992)
15. S.I. Najafi, "Glass integrated optics and optical fibre devices," *Critical reviews of optical science and technology, CR 53, Conf. Proceedings*, 24–25 July 1994, San Diego, USA, (SPIE Press 1996) p. 61
16. A. Lupascu, A. Kévorkian, T. Boudet, F. Saint-André, D. Persegol, and M. Levy, "Modeling ion-exchange in glass with concentration-dependent diffusion coefficients and mobilities," Opt. Eng. **35**, 1603 (1996)
17. J.M. White and P.F. Heidrich, "Optical waveguide refractive index profiles from measurement of mode indices, a simple analysis," Appl. Opt. **15**, 151 (1976)
18. K.S. Chiang, "Construction of refractive-index profiles of planar dielectric waveguides from the distribution of effective indices," IEEE J. Lightwave Technol. LT-**3**, 385 (1985)

19. P. Hertel and H.P. Mentzel, "Improved inverse WKB procedure to reconstruct refractive index profiles of dielectric planar waveguides," Appl. Phys. Lett. B **44**, 75 (1987)
20. R. Göring and M. Rothardt, "Application of the refracted near-field technique to multimode planar and channel waveguides in glass," J. Opt. Commun. **7**, 82 (1986)
21. P. Oberson, B. Gisin, B. Huttner, and N. Gisin, "Refracted near-field measurements of refractive index and geometry of silica-on-silicon integrated optical waveguides," Appl. Opt. **37**, 7268 (1998)
22. H. Hanafusa, F. Hanawa, Y. Hibino, and T. Nozawa, "Reliability estimation for PLC-type optical splitters," Electron. Lett. **33**, 238 (1997)
23. H. Eckstein, N. Fabricius, and J. Ingenhoff, "Reliability of passive integrated optical splitters," *Proc. Workshop on Fibre and Optical Passive Components*, Pavia, Italy (1998)
24. G. Gallo, A. Piccirillo, T. Tambosso, and G. Zaffiro, "Reliability of planar optical branching devices", *Proc. 10th Europ. Conf. Integr. Optics (ECIO'99)*, Torino, Italy, 441 (1999)
25. J. Shmulovich, "Er-doped glass waveguide amplifers on silicon," in *Rare-Earth-Doped Devices* (S. Honkanen, ed.), SPIE **2996**, 143 (1997)
26. M. Hempstead, "Ion-Exchanged Glass Waveguide Lasers and Amplifiers," in *Rare-Earth-Doped Devices* (S. Honkanen, ed.), SPIE **2996**, 135 (1997)
27. J. Shmulovich, Y.H. Wong, G. Nykolak, P.C. Becker, R. Adar, A.J. Bruce, D.J. Muehlner, G. Adams, and M. Fishteyn, "15 dB net gain demonstration in Er^{3+} glass waveguide amplifier on silicon," *Conf. Opt. Fiber Commun. (OFC '93)*, vol. 4, OSA Techn. Digest Series (Opt. Soc. America, Washington, DC 1993), PD-3
28. Y.C. Yan, A.J. Faber, H. de Waal, P.G. Kik, and A. Polman, "Net optical gain at 1.53 µm in an Er-doped phosphate glass waveguide on silicon," Appl. Phys. Lett. **71**, 2922 (1997)
29. D. Barbier, P. Bruno, C. Cassagnettes, M. Trouillon, R.L. Hyde, A. Kévorkian, and J.-M.P. Delavaux, "Net gain of 27 dB with a 8.6 cm long Er/Yb doped glass planar amplifier," *Conf. Opt. Fibre Commun. (OFC'98)*, vol. 2, OSA Techn. Digest Series (Opt. Soc. America, Washinton, DC 1998), p. 421
30. C. Lester, A. Bjarklev, T. Rasmussen, and P. Geltzer Dinesen, "Modelling of Yb^{3+}-sensitized Er^{3+}-doped silica waveguide amplifiers," IEEE J. Lightwave Technol. **13**, 740 (1995)
31. E. Desurvire, *Er-doped fibre amplifiers, principles and applications* (John Wiley and Sons, New York 1994), p. 13
32. V.P. Gapontsev, S.M. Matitsin, A.A. Isineev, and V.B. Kravtchenko, "Er glass lasers and their applications," Optics and Laser Technol. **14**, 189 (1982)
33. L.A. Riseberg and M.J. Weber, "Relaxation phenomena in rare-earth luminescence," in *Progress in Optics, vol. XIV* (E. Wolf, ed.), (North-Holland, Amsterdam 1976)
34. T. Kitagawa, K. Hattori, K. Shuto, M. Oguma, J. Temmyo, S. Suzuki, and M. Horiguchi, "Er-doped silica-based planar amplifier module pumped by laser diodes," *Proc. 19th Europ. Conf. Opt. Commun. (ECOC'93)*, Montreux, Switzerland, **3**, 41 (1993)
35. O. Lumholt, A. Bjarklev, T. Rasmussen, and C. Lester, "Rare earth-doped integrated glass components, modeling and optimization," IEEE J. Lightwave Technol. **13**, 275 (1995)

36. M. Federighi and F. Di Pasquale, "The effect of pair-induced energy transfer on the performance of silica waveguide amplifiers with high Er^{3+}/Yb^{3+} concentrations," IEEE Photon. Technol. Lett. **7**, 303 (1995)
37. A. Shoostari, P. Meshkinfam, T. Touam, M.P. Andrews, and S. I Najafi, "Ion-exchanged Er/Yb phosphate glass waveguide amplifiers and lasers," in *Rare-Earth-Doped Devices II* (S. Honkanen and S. Jiang, eds.), Proc. SPIE **3280**, 67 (1998)
38. E. Delevaque, T. George, M. Monerie, P. Lamouler, and J.-F. Bayon, "Modeling of pair induced quenching in erbium-doped silicate fibres," IEEE Photon. Technol. Lett. **5**, 73 (1993)
39. P. Myslinski, D. Nguyen, and J. Chrostowski, "Effects of concentration on the performance of erbium-doped fibre amplifiers," IEEE J. Lightwave Technol. **15**, 112 (1997)
40. G.C. Righini, M.A. Forestiere, M. Guglielmi, and A. Martucci, "Rare-earth-doped sol-gel waveguides: a review," in *Rare-Earth-Doped Devices II* (S. Honkanen and S. Jiang, eds.), Proc. SPIE **3280**, 57 (1998)
41. K. Shuto, K. Hattori, T. Kitagawa, Y. Ohmori, and M. Horiguchi, "Er-doped phosphosilicate glass waveguide amplifier fabricated by PECVD," Electron. Lett. **29**, 139 (1993)
42. P. Camy, J.E. Roman, M. Hempstead, P. Laborde, and C. Lerminiaux, "Ion-exchanged waveguide amplifier in erbium doped glass for broadband communications," *Proc. Optical Amplifiers and Applications (OAA'95)*, Davos, Switzerland, 181 (1995)
43. C.C. Li, H.K. Kim, and M. Migliuolo, "Er-doped glass ridge-waveguide amplifiers fabricated with a collimated sputter deposition technique," IEEE Photon. Technol. Lett. **9**, 1223 (1997)
44. K. Hattori, T. Kittagawa, M. Oguma, Y Ohmori, and M. Horiguchi, "Er-doped silica-based waveguide amplifier integrated with a 980/1530 nm WDM coupler," Electron. Lett. **30**, 856 (1994)
45. G. Nykolak, M. Haner, P.C. Becker, J. Shmulovich, and Y.H. Wong, "Systems evaluation of an Er^{3+}-doped planar waveguide amplifier," IEEE Photon. Technol. Lett. **5**, 1185 (1993)
46. J.-M.P. Delavaux, S. Granlund, O. Mizuhara, L.D. Tzeng, D. Barbier, M. Rattay, F. Saint-Andre, and A. Kévorkian, "Integrated optics Er/Yb amplifer system in 10 Gb/s fibre transmission experiment," IEEE Photon. Technol. Lett. **9**, 247 (1997)
47. K.A. Winick, "Rare-earth-doped waveguide lasers in glass and $LiNbO_3$: a review," in *Rare-Earth-Doped Devices II*, (S. Honkanen and S. Jiang, eds.), Proc. SPIE **3280**, 88 (1998)
48. M.J.F. Digonnet, *Rare-earth doped fibre lasers and amplifiers* (Marcel Dekker Inc., New York, Basel 1993)
49. M. Saruwatari and T. Izawa, "Nd-glass with three dimensional optical waveguide," Appl. Phys. Lett. **24**, 603 (1974)
50. T. Kitagawa, K. Hattori, M. Shimizu, Y. Ohmori, and M. Kobayashi, "Guided wave laser based on erbium-doped silica planar lightwave circuit," Electron. Lett. **27**, 334 (1991)
51. D. Barbier, J.-M.P. Delavaux, R.L. Hyde, J.M. Jouanno, A. Kévorkian, and P. Gastaldo, "Tunability of Er-Yb intergrated optic laser in phosphate glass," *Proc. Optical Amplifiers and Applications (OAA'95)*, Davos, Switzerland, PD-3 (1995)

52. J. Roman, P. Camy, M. Hempstead, W. Brocklesby, S. Nouh, A. Béguin, C. Ler-
    miniaux, and J. Wilkinson, "Ion exchange Er/Yb waveguide laser at 1.5 μm
    pumped by a laser diode," Electron. Lett. **31**, 1345 (1995)
53. O. Svelto, *Principles of lasers, 3rd ed.* (Plenum Press, New York 1989)
54. A. Yariv, *Quantum Electronics 3rd ed.* (John Wiley & Sons, New York 1988)
55. W.H. Loh, B.N. Samson, L. Dong, G.J. Cowle, and K. Hsu, "High Performance
    Single Frequency Fibre Grating-Based Er: Yb-Codoped Fibre Lasers," IEEE
    J. Lightwave Technol. **16**, 114 (1998)
56. D. Barbier, J.-M.P. Delavaux, T.A. Strasser, M. Rattay, R.L. Hyde, P. Gastaldo,
    and A. Kévorkian, "Sub-centimeter length ion-exchanged waveguide lasers in
    Er/Yb doped phosphate glass," *Proc. 23rd Europ. Conf. Opt. Commun.
    (ECOC'97)*, Edinburgh, UK, **4**, 41 (1997)
57. G.L. Vossler, C.J. Brooks, and K.A. Winick, "Planar Er,Yb glass ion exchanged
    waveguide laser," Electron. Lett. **31**, 1162 (1995)
58. M. Ding and P.K. Cheo, "Effects of Yb,Er-codoping on suppressing self-pulsing
    in Er-doped fibre lasers," IEEE Photon. Technol. Lett. **9**, 324 (1997)
59. D.J. Jones, S. Namiki, D. Barbier, E.P. Ippen, and H.A. Haus, "116-fs soli-
    ton source based on an Er-Yb codoped waveguide amplifier," IEEE Photon.
    Technol. Lett. **10**, 666 (1998)
60. L.E. Nelson, D.J. Jones, K. Tamura, H.A. Haus, and E.P. Ippen, "Ultrashort-
    pulse fibre ring lasers," Appl. Phys. Lett. B **65**, 277 (1997)
61. S. Kawanishi, K. Hattori, H. Takara, M. Oguma, O. Kamatani, and Y. Hibino,
    "Actively modelocked ring laser using Er-doped silica-based planar waveguide
    amplifier," Electron. Lett. **31**, 363 (1995)
62. J.-M.P. Delavaux, A. Yeniay, J. Toulouse, T.A. Strasser, R. Pedrazzanni, and
    D. Barbier, "Q-switched Yb co-doped channel waveguide laser," *Proc. 8th Eu-
    rop. Conf. Integr. Optics (ECIO'97)*, Stockholm, Sweden, 173 (1997)

7   Wavelength-Selective Devices

Meint K. Smit, Ton Koonen, Harald Herrmann, and Wolfgang Sohler

7.1   Introduction

A single optical fibre is capable of transporting multiple signals, each at its own wavelength. At the fibre entrance, the signals fed by a number of transmitters, each operating at a specific wavelength, are combined by a wavelength multiplexer and launched into the transmission fibre. At the fibre exit, a wavelength demultiplexer routes these signals again according to their wavelengths to one or more receivers. If the demultiplexer selects only one wavelength channel, it is usually called a wavelength filter. When the fibres at the input and at the output of the device are identical, a wavelength multiplexer can typically be used as a wavelength demultiplexer in the opposite direction also, and vice versa.

In a point-to-point link, each wavelength constitutes an independent communication path in the fibre, as illustrated in Fig. 7.1. Thus the data transport capacity of the fibre is increased by this same number of wavelengths. Besides this capacity enhancement, in an optically transparent multipoint-to-multipoint network the wavelength dimension can also be used as an extra degree of freedom to route signals. If the network itself does not perform any wavelength-selective routing but leaves the wavelength selection to the terminals, the signals arrive from source to destination in a single hop. Wavelength-

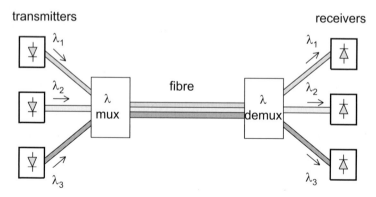

Fig. 7.1. Point-to-point multi-wavelength communication link

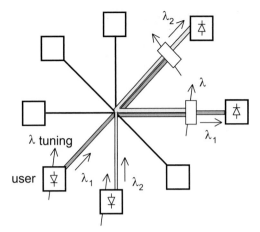

Fig. 7.2. Single-hop multi-wavelength multipoint-to-multipoint network

routeing inside the network guides the signals from node to node in a multi-hop fashion.

In a single-hop network, as shown in Fig. 7.2, the transmitter of each user may be set to a specific wavelength position. Also, the receiver of each user may be tuned to the wavelength desired. When different sets of users make their wavelength choices adequately, multiple communication paths may be established simultaneously in the same network without data-packet collisions. At the transmitting end as well as at the receiving end, the wavelength-selective mechanism can be fixed or tunable. When the transmitter wavelengths are fixed and the receiver wavelengths are tunable, the network operates in the 'broadcast-and-select' mode. Each user has access to the wavelength he is interested in, thus reducing the privacy of information transmitted if no extra measures are taken. On the other hand, broadcasting information from one user to all other users is inherently easy. When the transmitter wavelengths are tunable and the receiver wavelengths are fixed, the network operates in the 'wavelength addressing' mode. Provided that the user cannot manipulate the wavelength setting of his receiver, privacy is secured. Broadcasting information, however, is hampered. Making some of the transmitters as well as some of the receivers tunable yields a network operation mode in between these two extreme modes.

In a multi-hop network, as illustrated in Fig. 7.3, the optical signals transverse several nodes before arriving at their destination. Each node may route the signal to the next part of the network depending on its wavelength, thus performing a wavelength-crossconnect function. A so-called 'optical path' is established between two users when they are interconnected by a communication path on a single wavelength. The example in Fig. 7.3 shows that multiple optical paths at the same wavelength may exist in the same network provided that they do not overlap. This so-called wavelength re-use limits the precious resource of wavelengths needed. By introducing wavelength con-

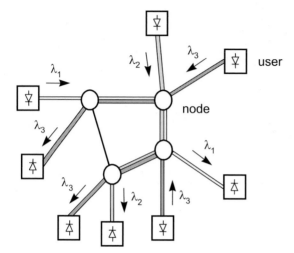

Fig. 7.3. Multi-hop multi-point-to-multipoint network

version at the nodes, the information carried on a specific wavelength can be transferred to another wavelength. For example, if two communication paths having the same wavelength arrive at the same node and are to be routed to another node via the same fibre, one of these paths can be transposed to another wavelength, thus avoiding the overlap. Including the wavelength conversion, the path between the users is called a "virtual wavelength path". Wavelength conversion increases the possibility of establishing collision-free communication paths in a network without adding new wavelengths, and thus enhances the network's data throughput. In Fig. 7.3, for instance, only two wavelengths (λ_2 and λ_3) are needed to provide the required connections, when the communication path originally starting at λ_1 now starts at λ_2, is converted to λ_3 in the second node passed, and converted back to λ_2 in the third node for the final link to its destination.

Optical local area networks may use multiple wavelengths in order to set up multiple collision-free data paths simultaneously. In, for example, a ring-shaped network topology, at each access node a user terminal can extract or inject information on a particular wavelength by means of a so-called 'wavelength add-drop node'. This wavelength is selected from the available set and assigned to the user by the network management system, which supervises the demand of other users and optimises the network throughput.

The core transport network typically carries high-speed data at wavelength carriers generated by sophisticated transmitters capable of bridging long lengths of fibre with low pulse dispersion (and mostly aided by optical fibre amplifiers with limited wavelength operating range). As illustrated by Fig. 7.4, wavelength add-drop nodes in the core transport link provide access to local networks. The core transport wavelengths may differ from the ones used in the local area networks, where low-cost transmitters operating at medium data rates are predominant. When coupling these local networks to

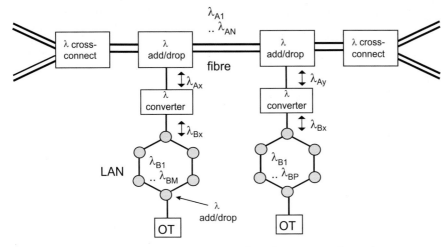

Fig. 7.4. Multi-wavelength interconnection of local area networks to the core transport network by means of wavelength add-drop nodes and wavelength converters (OT = Optical Terminal)

the core transport network, wavelength converters are needed to translate the wavelength channel of the local area network to the one selected of the core transport network, and vice versa. They also enable the same wavelengths to be used in neighbouring local area networks, and still making access by these networks possible to different core network wavelengths. This wavelength re-use thus advantageously limits the variety of wavelengths needed. Another issue is the bit rate adaptation between the core network wavelength and the local area network wavelength. By performing optical time-division multiplexing in combination with wavelength conversion, the gateway between the core network and the local area network may optimally match the moderate number of high-speed wavelength channels in the core network to the high number of moderate-speed wavelength channels in the local area networks.

7.2 Device Specifications

The characteristics of wavelength-selective passive optical devices may be fixed, or tunable. In point-to-point multiwavelength links, as shown in Fig. 7.1, the wavelength multiplexers and demultiplexers are usually fixed; at most, some slight wavelength tuning is used to align to small wavelength deviations of the transmitters. In the broadcast-and-select network shown in Fig. 7.2, however, active tuning of the wavelength filter at the receiving side is needed to select the appropriate wavelength channel.

A wavelength demultiplexer of which the output ports are equipped with the same type of fibre as the input port, can generally also be used in the reverse direction as a wavelength multiplexer.

The main specifications of a wavelength demultiplexer are:

- Number of output ports; at each output port a specific part of the spectrum of the input signal is available. This part corresponds usually to one of the wavelength channels. The power transfer characteristic of the input port to the output port has usually a bandpass shape.
- Insertion loss; i.e., the loss which the signal at an output port has suffered with respect to the same spectral part of the input signal.
- Bandwidth; e.g., the spectral range over which the insertion loss increases with 1 dB with respect to the insertion loss in the centre of the bandpass characteristic of the output port.
- Crosstalk attenuation; i.e., the attenuation of the non-desired signals (usually having a wavelength positioned at the centre of the passband of the other output ports) at an output port with respect to the desired signal.
- Polarization dependent loss; i.e., the variation in the insertion loss when the state of polarization of the input signal is varied across all possible states.
- Stability; the characteristics of the device should vary as little as possible during the lifetime of the device, and during ambient variations (such as temperature, humidity, and vibration).
- Dimensions; preferably as small as possible.
- Cost (last but not least important in, for example, widespread applications in subscriber-access networks).

For a tunable demultiplexer, the following specifications may be used as well:

- Tuning range; i.e., over which wavelength range can the device characteristics be deliberately varied?
- Tuning speed; i.e., how fast can this variation of the characteristics be accomplished? For packet-switched networks, speed requirements are significantly higher than for circuit-switched ones.
- Resolution; how close to each other may the wavelength channels be positioned while still allowing the non-selected channels to be sufficiently attenuated with respect to the desired channel? This is closely related to the crosstalk performance.
- Maximum number of discernible channels; this equals the tuning range divided by the resolution.

7.3   Fabry-Perot Interferometer Filters

A Fabry-Perot interferometer filter has a single input port and a single output port. It is capable of selecting one wavelength channel out of a multitude. When more channels need to be selected, the equivalent number of Fabry-Perot interferometer filters is needed.

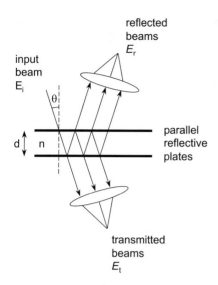

reflected
beams
E_r

input
beam
E_i

θ

d   n

parallel
reflective
plates

transmitted
beams
E_t

Fig. 7.5. Fabry-Perot interferometer

A Fabry-Perot interferometer filter basically employs two highly reflecting plates which are positioned in parallel. A light beam entering between these plates is reflected multiple times, and with each reflection a small part of its power escapes through the plate. The beams which thus emerge subsequently from one of the plates can be brought into the focal point of a lens, as illustrated in Fig. 7.5. By means of interference, the phase differences of the beams yield a wavelength-dependent light intensity in this focal point. With sufficiently high reflectivity of the plates, this interference process can be very wavelength-selective. Mathematically, the amplitude E_t of the electric field after focusing by the lens is described by [1].

$$E_t = at^2 E_i + at^2 a^2 r^2 e^{-i\delta} E_i + at^2 a^4 r^4 e^{-i \cdot 2\delta} E_i + at^2 a^{2n} r^{2n} e^{-i \cdot n\delta} E_i$$

$$= E_i \cdot t^2 \cdot a \cdot \sum_{n=0}^{\infty} (a\, r)^{2n} e^{-i \cdot n\delta} = E_i \cdot \frac{a \cdot t^2}{1 - a^2 \cdot r^2 \cdot e^{-i\delta}}, \tag{7.1}$$

where a is the factor with which the electric field amplitude is attenuated when travelling across the medium between the plates, and t and r are the factors by which the electric field is attenuated when transmitted and reflected, respectively, at a plate.

The phase shift δ of the light beam during one round-trip in the cavity (consisting of the two parallel reflecting plates spaced by d, and refractive index of the medium between the plates n) is given by

$$\delta = 2 \cdot 2\pi \cdot d \cdot \cos\theta \cdot \frac{n}{\lambda_0} = \frac{4 \cdot \pi \cdot n \cdot d \cdot \nu}{c_0} \cdot \cos\theta , \tag{7.2}$$

where ν is the light frequency, coupled to the speed of light in vacuum c_0 and wavelength in vacuum λ_0 by $\nu = c_0/\lambda_0$. To achieve the steepest possible filtering characteristic, the input light beam will usually be launched with nearly normal incidence to the plates, i.e. $\theta \approx 0$. The intensity transmission factor T_{FP} of the Fabry-Perot filter, given by the quotient of the intensity I_t of the transmitted light and the intensity I_i of the incident light, depends on the wavelength λ_0 via the phase shift δ according to

$$T_{\text{FP}}(\nu) = \frac{I_t}{I_i} = \frac{E_t \cdot E_t^*}{E_i \cdot E_i^*} = \frac{T_0}{1 + \left[\frac{2}{\pi} F_{\text{R}} \cdot \sin\left(\frac{\delta}{2}\right)\right]^2} \tag{7.3}$$

with the so-called reflectivity finesse factor F_{R} given by

$$F_{\text{R}} = \pi \cdot \frac{\sqrt{A \cdot R}}{1 - A \cdot R} \tag{7.4}$$

and the maximum transmission factor

$$T_0 = \frac{A \cdot (1 - R)^2}{(1 - A \cdot R)^2} , \tag{7.5}$$

where $R = |r|^2$ is the intensity reflection coefficient of the plates, the reflective plates are assumed to be loss-free (i.e., $R + T = 1$), and $A = |a|^2$ is the attenuation factor of the light intensity for half a cavity round-trip.

Similarly, the intensity reflection factor R_{FP} of the Fabry-Perot filter, given by the quotient of the intensity I_r of the transmitted light and the intensity I_i of the incident light, is

$$R_{\text{FP}}(\nu) = \frac{I_r}{I_i} = \frac{E_r \cdot E_r^*}{E_i \cdot E_i^*} = \frac{R_0 + \left[\frac{2}{\pi} \cdot F_{\text{R}} \cdot \sin\left(\frac{\delta}{2}\right)\right]^2}{1 + \left[\frac{2}{\pi} \cdot F_{\text{R}} \cdot \sin\left(\frac{\delta}{2}\right)\right]^2} , \tag{7.6}$$

where the minimum reflection factor R_0 is

$$R_0 = R \cdot \left(\frac{1 - A}{1 - A \cdot R}\right)^2 . \tag{7.7}$$

If the medium between the plates is loss-less, then obviously

$$T_{\text{FP}}(\nu) + R_{\text{FP}}(\nu) = 1. \tag{7.8}$$

Figure 7.6 shows T_{FP} versus the light frequency ν for such a lossless device.

The characteristics of both T_{FP} and R_{FP} versus the light frequency ν are periodic. The period is called the Free Spectral Range (FSR), and is given in the frequency domain by

$$FSR_\nu = \frac{c_0}{2 \cdot n \cdot d} \quad [\text{Hz}] \tag{7.9}$$

and in the wavelength domain by

$$FSR_\lambda = \frac{\lambda_0^2}{2 \cdot n \cdot d} \quad [\text{m}] . \tag{7.10}$$

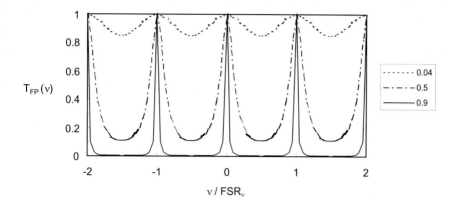

Fig. 7.6. Intensity transmission factor T_{FP} of Fabry-Perot filter versus the light frequency ν normalized to the frequency free spectral range FSR_ν (parameter: reflectivity R of the plates)

For example, for a plate distance $d = 120\,\mu m$, $\lambda_0 = 1.55\,\mu m$ and $n = 1$, $FSR_\lambda = 10\,nm$.

A useful performance parameter is the contrast factor C, which is the ratio of the maximum and the minimum of the transmission factor T_{FP},

$$C = \frac{T_{max}}{T_{min}} = \left(\frac{1 + A \cdot R}{1 - A \cdot R}\right)^2 . \tag{7.11}$$

The contrast factor C therefore determines the crosstalk attenuation of the Fabry-Perot device. For example, for $A = 1$ (i.e. a loss-less medium between the plates, which is a good approximation in the case of air) and $R = 0.9$, $C = 361$, which corresponds to a crosstalk attenuation of 25.6 dB. Figure 7.7 shows how the reflectivity finesse F_R and the contrast factor C depend on the reflectivity R of the plates.

The $-3\,dB$ bandwidth $\Delta\nu_{FWHM}$ (FWHM, full width at half maximum) of the Fabry-Perot bandpass curves is

$$\Delta\nu_{FWHM} = \frac{c_0}{2\pi \cdot n \cdot d} \cdot \frac{1 - A \cdot R}{\sqrt{A \cdot R}} \quad [Hz] \tag{7.12}$$

and is related to the free spectral range by

$$\Delta\nu_{FWHM} = \frac{FSR_\nu}{F_R}, \tag{7.13}$$

which indicates that the ratio of the FSR and the $-3\,dB$ bandwidth is given only by the reflectivity R of the plates and the loss factor A of the medium between them. If the plates are not perfectly flat, the finesse factor is reduced.

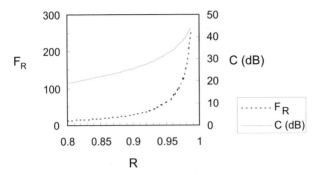

Fig. 7.7. Reflectivity finesse F_R and contrast factor C of a lossless Fabry-Perot filter versus plate reflectivity R

For instance, if the plates exhibit a spherical curvature such that the distance between the plates measured in the centre and measured at the edges differs by λ_0/M (where M is a positive real number), the so-called flatness finesse factor is given by $F_F \approx M/2$. The resulting finesse F is given by

$$F^{-2} = F_R^{-2} + F_F^{-2}. \tag{7.14}$$

Commercially available Fabry-Perot devices have finesse factors F ranging from 30 to more than 200. F is a measure of the wavelength filtering resolution of the device, and thus determines the maximum number of wavelength channels which can be discerned by a Fabry-Perot device. For instance, when a crosstalk attenuation of better than 10 dB is required (implying a crosstalk penalty of less than 0.5 dB), this number N_{max} is given by

$$N_{max} < \frac{F}{3}. \tag{7.15}$$

An $F = 150$ for a lossless device with perfectly flat and parallel plates thus requires a reflectivity $R = 98\%$, which enables us to resolve maximally about 50 wavelength channels.

The Fabry-Perot filter may be tuned by varying the spacing d between the plates. Alternatively, the refractive index n of the medium between the plates may be varied by, for example, changing the pressure of the gas in the cavity. The fastest tuning is achieved by varying the spacing d by means of piezo-electric transducers. The maximum transmission is obtained if the plate distance equals an integer number times half the wavelength of the light in the cavity; i.e., if the resonance condition $d = m\lambda_0/2n$ is fulfilled where the integer m is called the order of interference. Thus the light frequencies at which transmission maxima occur are given by the resonance frequencies

$$\nu_{res} = \frac{c_0}{\lambda_0} = c_0 \cdot \frac{m}{2\,n\,d} \tag{7.16}$$

and are therefore separated by the free spectral range $FSR_\nu = c_0/2\,n\,d$.

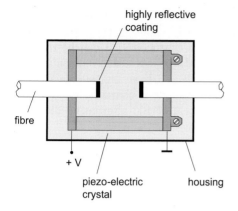

Fig. 7.8. Fibre Fabry-Perot filter, tunable by means of piezo-electric transducers

A typical commercially available implementation of a tunable fibre Fabry-Perot filter (FFP) is shown in Fig. 7.8. The reflective plates are created by depositing a highly reflective coating on two carefully aligned parallel fibre endfaces. The medium between the endfaces is air ($n \approx 1$), and the distance d is varied by applying a voltage V at the piezo-electric crystals. The voltages required may be fairly high (of the order of 100 Volt). When tuning the device, some backlash effects in the piezo-electric transducers have to be accounted for. The tuning time is of the order of milliseconds, which is adequate for circuit-switched networks but too slow for packet-switched ones. The polarization dependency is negligible, as standard type fibres are used which have a nearly perfect rotation symmetry. Finesses in excess of 150 can be achieved, and typical minimum transmission losses are 1 to 2 dB. As the devices are fibre-based, they can be inserted with low losses in fibre communication systems, e.g. directly in front of an optical receiver. Higher finesse factors may be obtained by cascading two Fabry-Perots, with an optical isolator in between to avoid parasitic resonance cavities; thus total finesses exceeding 1000 may be reached [2]. By choosing slightly different FSRs for the two Fabry-Perots, and tuning one device while keeping the other one fixed, the two comb-shaped transmission curves are shifted with respect to each other. Due to the difference in FSR, a vernier effect occurs, which yields a step-wise tuning behaviour [3].

7.4   Dielectric Interference Filters

Dielectric interference filters for application in Wavelength Division Multiplexing (WDM) devices are usually structured as complex multi-cavity Fabry-Perot filters. The main difference with normal Fabry-Perot filters is that the cavity space is formed by a thick dielectric layer, rather than an air gap, and that the mirrors are formed as a stack of alternating quarter-wave layers of 'high' and 'low' refractive index materials and are, therefore, very wavelength selective. The overall filter can be conceived of as a series of cascaded Fabry-Perot cavities. By properly choosing the thicknesses of the different layers many filter transmission characteristics can be realized. Interference filters for Dense WDM (DWDM) devices ideally transmit a flat passband surrounding the signal wavelength and have high attenuation at the adjacent signal wavelengths. In practice such a characteristic requires interference-filter designs comprising in excess of 100 alternating dielectric layers.

When used for WDM applications, the dielectric interference filter usually separates wavelength bands by transmitting one wavelength band to be captured in one optical fibre and reflecting the remaining light (which comprises the other wavelength bands) to be captured in a second optical fibre, as shown in Fig. 7.9. Such a mounting can be realised with a low back-reflection into the input port. As the filter response is angle dependent the filter design should be optimized for the applied angle. A device like this can be used as an add or a drop filter as shown in the Fig. 7.9. By adding a fourth fibre-collimator a combined add-drop filter can be obtained. Because of the large difference in signal level between the add and the drop port, this mounting imposes high requirements on the contrast factor (7.11).

A four-channel demultiplexer can be realised by cascading four of these drop filters. Figure 7.10 shows the configuration and the response of a commercial device with filters. Typical insertion loss is below 5 dB, and crosstalk attenuation is better than 20 dB. The fourth filter can be omitted in principle. In practice it is needed to suppress residual reflections at the 'dropped' wavelengths. In cascades like these the reflection loss of the drop filter should be low to prevent the last channel having much more loss than the first one. Therefore, these interference filters are formed from non-absorbing materials, usually oxides, and the filter deposition process is tailored to produce low-loss optical properties.

Figure 7.11 shows another type of demultiplexer based on dielectric interference filters. Here the input signal is transformed into a collimated beam which is coupled into a block of transparent material. At the place where it

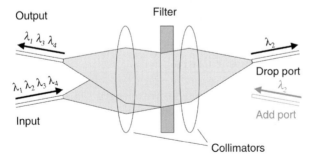

Fig. 7.9. Add-drop multiplexer based on a dielectric interference filter

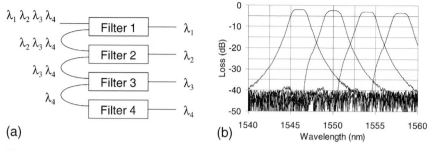

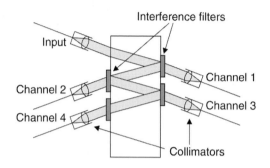

Fig. 7.10. Circuit scheme **(a)** and wavelength response **(b)** of a 4-channel demultiplexer based on cascaded interference filters

Fig. 7.11. Micro-optic 4-channel wavelength demultiplexer based on interference filters

first hits the side wall the first filter is mounted, followed by a fibre collimator which couples the dropped signal into output port 1. All other signals are reflected to the second filter, where wavelength 2 is dropped, and so on. The insertion loss claimed for these devices is low (a few dBs).

7.5   Fibre Gratings

Starting in the early nineties a novel class of filters has been developed based on phase gratings inside a fibre [4,5]. Many silica-based doped glassed exhibit photosensitivity, i.e. after exposure to an intensive UV light beam the index of the glass is slightly changed. The physical background of this index change is not yet fully understood. For some materials, e.g. Ge-doped silica glass, it appears to be very stable, so that devices created in this way are suitable for long-term operation.

Using the UV-illumination technique a phase grating can be created in the fibre core by exposing the core to the holographic fringe pattern of two interfering UV-beams coupled directly into the fibre or generated by a phase plate positioned close to the fibre. Figure 7.12a schematically illustrates the structure of the resulting device. By adjusting the angle of the UV-beams (or the period of the phase plate) the period of the grating can be controlled. For a reflection grating the grating period should equal half the signal wavelength

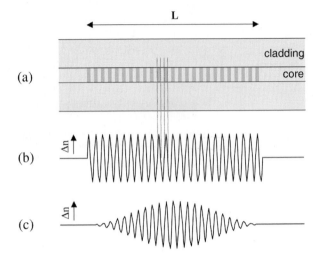

Fig. 7.12. (a) Structure of a fibre grating, (b) Uniform grating and (c) Apodized grating

inside the fibre. In such a grating a large coupling between the forward and backward propagating mode will occur and by proper design most of the power can be effectively reflected, whereas signals with other wavelengths are transmitted. An important advantage of fibre gratings is their natural fibre-compatibility; thus losses in connecting them to other fibres are very low and the total insertion loss for these devices is usually well below 1 dB, both for the reflected and the transmitted signals.

Figure 7.13 gives an example of the spectral reflectivity of a uniform fibre grating, i.e. a grating in which the index variation is uniform along the grating as depicted in Fig. 7.12b. From the figure it is seen that such a grating exhibits an inconveniently high side-lobe level. The shape of the spectral response is uniquely determined by the product κL of the coupling coefficient κ between the counter propagating waves and the fibre grating length L. The coupling coefficient is mainly determined by the magnitude of the index variation; the larger the index variation the higher the coupling. From Fig. 7.13 it is seen

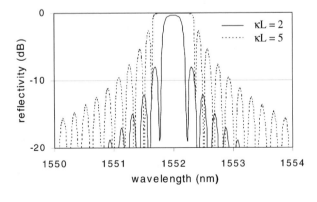

Fig. 7.13. Calculated spectral reflectivity of uniform fibre gratings for two different values of the product κL, grating length L = 4 mm

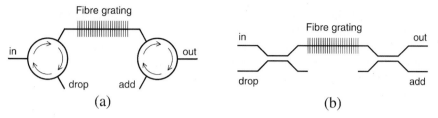

Fig. 7.14. Add-drop multiplexer based on fibre gratings: **(a)** with circulators, **(b)** with 3-dB couplers

that with increasing κL the response becomes more rectangular, which is favourable from a systems point-of-view, but it is also seen that the side-lobe level increases. The width of the reflectivity curve scales inversely with the grating length L if κL is kept constant.

The high side-lobe level observed in Fig. 7.13 can be reduced by "apodization" of the grating, i.e. changing the coupling coefficient (i.e. the index variation) more gradually as shown in Fig. 7.12c. Gaussian or square cosine profiles give much lower side-lobe levels. Side-lobe suppression levels in excess of 30 dB have been reported.

Another advantage of fibre gratings is that they can be produced relatively cheaply. A disadvantage is that the reflected beam is in the input fibre, so we need a circulator to couple it effectively to a drop port, as shown in Fig. 7.14a, and another circulator is needed for adding a signal. As the circulators are relatively expensive components, part of the low-cost advantage of fibre gratings is lost in this solution. If we can accept an additional loss of 6 dB, cheaper 3 dB couplers might be used instead of the circulators (Fig. 7.14b). Now a large reflection will occur in the input fibre too, and an isolator might be needed to eliminate it. Obviously, fibre gratings will not be the solution to all WDM-problems, but they may play an important role in future networks.

7.6   Grating-based Demultiplexers

A wavelength demultiplexer employing a diffraction grating usually has a single input port and a number of output ports. At each output port, one of the wavelength channels from the input port is available. A grating-based device thus performs the wavelength selection of a number of channels in parallel, and is therefore optimally suited for processing multiple channels.

A grating diffracts light at an angle which depends on its wavelength and angle of incidence. There are two basic types: transmission gratings and reflection gratings.

A transmission grating contains a large number of slits of identical width, which are closely spaced in parallel. At each slit, diffraction of the incident light takes place. By constructive interference of the diffracted beams, light maxima occur in certain angular directions. When projected with a lens in a plane behind the grating, the spots of maximum intensity are formed at locations determined by the wavelength, the slit distance, and the angle of incidence of the light on the grating. At other locations in the plane, the beams destructively interfere, resulting in light extinction.

A reflective grating has on its surface a large number of parallel closely spaced grooves, which are provided with a highly reflective coating (such as gold or silver for infrared light). For a ruled grating, these grooves have a specified angle (the blaze angle) with the grating surface, which optimizes the performance for a specified wavelength range. Similarly to the transmission grating, light maxima are formed in certain angular directions by means of diffraction by the grooves and interference of the diffracted light beams. These angular directions θ_r are coupled to the angle of incidence θ_i on the grating, the groove spacing d, the light wavelength in vacuum λ_0, and the refractive index n of the medium, according to the grating equation

$$\sin(\theta_i) + \sin(\theta_r) = \frac{m \cdot \lambda_0}{n \cdot d} , \tag{7.17}$$

where the integer m is the order of interference. The groove shape of the grating is usually chosen such that its efficiency is maximum for $m = 1$. After projection by the lens in a plane positioned in front of the grating, the spots of maximum intensity can again be found at locations determined by λ_0, d and θ_i. The maximum efficiency of the grating (i.e., the lowest diffraction losses) is reached when λ_0 equals the blaze wavelength λ_B, which is related to the blaze angle θ_B by

$$\lambda_B = 2 \cdot n \cdot d \cdot \sin(\theta_B). \tag{7.18}$$

For a ruled grating, the blaze angle equals the angle which the groove facets make with the plane in which the grating has been ruled. In the so-called Littrow mount, the angle of incidence θ_i is chosen such that at the wavelength of interest λ_0 the diffracted light beam follows nearly the same path as the incident one. This set-up is shown in Fig. 7.15. Only one lens is needed to collimate the incident light into a parallel beam towards the grating, as well as to project the diffracted beams into a specific plane. Thus, lens aberrations (in particular astigmatism) are reduced, which improves the system efficiency. The wavelength demultiplexing losses are lowest when the incidence angle θ_i as defined in Fig. 7.15 is small (i.e. normal incidence to the grating facets). The grating efficiency depends also on the polarization state of the incident light. At the blaze angle, however, this dependency is minimal. This is of advantage in a fibre communication system, where in general at the receiving end of a long link the polarization state may fluctuate considerably.

With a reflection grating demultiplexer, several closely spaced wavelength channels can be separated in one step. Figure 7.16 exemplifies the configuration of such a device, employing the Littrow mount. The wavelength channel

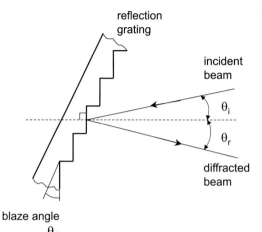

Fig. 7.15. Reflection grating in Littrow mount

spacing is given by

$$\Delta\lambda = \frac{D_s}{f} \cdot \frac{d\lambda}{d\theta_r} = \frac{D_s}{f} \cdot \sqrt{d^2 - (\lambda_c/2)^2}, \tag{7.19}$$

where the central wavelength $\lambda_c = 2\,n\,d\,\sin\theta_i$ is the wavelength for which at the specified angle of incidence θ_i ($= 7.5$ degrees in Fig. 7.16) the Littrow

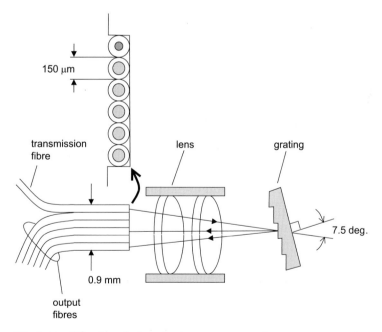

Fig. 7.16. Wavelength demultiplexer, using a reflection grating in Littrow mount [6]

condition is met, f is the focal length of the lens, and D_s is the spacing of the core centres of the output fibres.

By using step-index multimode output fibres with a core diameter which is significantly larger than the diameter of the core of the input fibre, the bandpass curves become broader and flatter. The projection of the core of the input fibre onto the array of output fibres moves across the array when the wavelength is varied (see Fig. 7.17). Due to the difference in core diameter, the coupling loss remains constant over a certain wavelength range, corresponding to the flat part in the bandpass curve A typical demultiplexer performance obtained with oversized-core output fibres is shown in Fig. 7.18 [6]. Neglecting lens aberrations, the $-1\,\mathrm{dB}$ bandwidth $\Delta\lambda_{\mathrm{BW}}$ of the flat parts of the characteristics is given by

$$\Delta\lambda_{\mathrm{BW}} = \Delta\lambda \cdot \Delta\Phi/D_{\mathrm{s}}, \qquad (7.20)$$

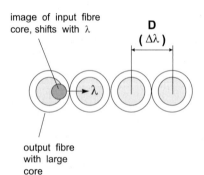

image of input fibre core, shifts with λ

output fibre with large core

Fig. 7.17. Obtaining flat passbands in a grating-based wavelength demultiplexer by oversized cores of the output fibres

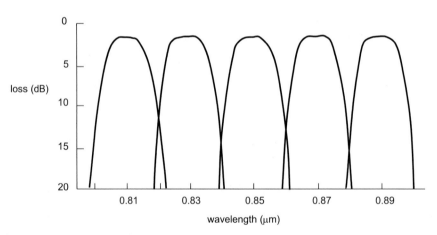

Fig. 7.18. Characteristics of a wavelength demultiplexer deploying a reflection grating and oversized-core output fibres [6]

where $\Delta\Phi$ denotes the difference in core diameter between the input fibre and the output fibres. The broadening of the bandpass characteristics allows a larger tolerance in the wavelength of the laser diodes in the communication system. Thus selection and stabilization of the laser diodes becomes simpler, and therefore cheaper.

Another form of grating-based demultiplexer, which uses a graded-index rod lens and is very compact, is shown in Fig. 7.19. This set-up has no glass-air interfaces, and thus does not need antireflection coatings on the lens surfaces in order to eliminate ghost reflections. The device is, however, not tunable.

Due to its ability to process multiple wavelength channels in one step, a grating-based wavelength demultiplexer is most suited for separating a large number of closely spaced wavelength channels. On the other hand, a demultiplexer using interference filters (e.g., Fabry-Perot cavities) needs as many filter elements as there are wavelength channels, and therefore is best suited for separating a small number of not too closely spaced wavelength channels. When some closely spaced channels as well as a more distant channel need to be resolved, a combination of a reflective grating and an interference filter may be attractive, as exemplified in Fig. 7.20. Using a $50/125\,\mu$m core/cladding diameter graded-index fibre at the input port and the bidirectional $1.3\,\mu$m port, $100/140\,\mu$m core/cladding diameter step-index output fibres at the 820 nm and 845 nm output ports, achromatic doublet lenses (with focal length $f = 10$ mm), a long-wavelength-pass interference filter placed at $45°$ angle of incidence, and a gold-coated reflection grating with 600 grooves per mm and a blaze wavelength of 750 nm, this device exhibited the characteristics given in Fig. 7.21 [7]. The losses in the passbands centred around 821 nm and 846 nm were 2.8 and 3.0 dB, respectively. The $1.3\,\mu$m wavelength channel has a loss of 1.4 dB (of which 0.9 dB is due to the lens coupling between the two graded-index fibres). The edge of the $1.3\,\mu$m long-wavelength pass-filter characteristic is less steep, because the $45°$ angle of incidence gives rise to polarization splitting of the filter curves. As the fibres at the input port and

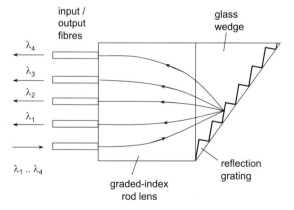

Fig. 7.19. Compact wavelength demultiplexer using a graded-index rod lens and a reflection grating replicated on a glass wedge

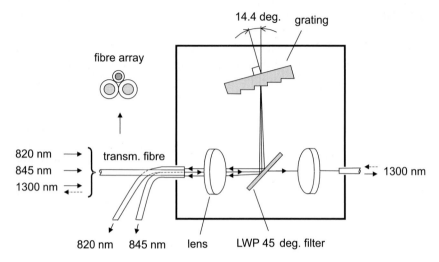

Fig. 7.20. Wavelength demultiplexer separating a number of closely spaced wavelength channels as well as a more distant channel [7]

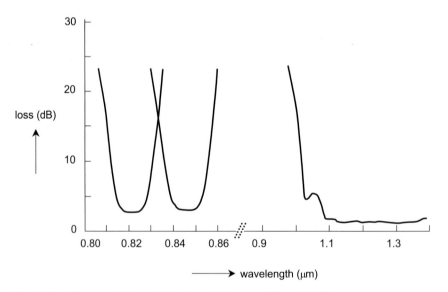

Fig. 7.21. Characteristics of the demultiplexer of Fig. 7.20 [7]

at the 1.3 μm port are both of the same type, the 1.3 μm port can be used as an input port as well as an output port; the 1.3 μm channel can therefore be used for bidirectional traffic. The crosstalk attenuation between all channels exceeds 28 dB. The flat part of the 820 nm and the 845 nm channel is more than 10 nm wide. The angle at which the grating is mounted can be adjusted

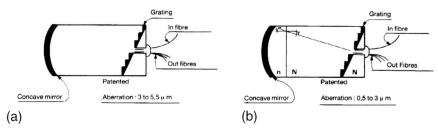

Fig. 7.22. Wavelength demultiplexer employing a concave mirror and planar grating (STIMAX) configuration **(a)** basic configuration, for multimode output fibres; **(b)** configuration with correction of the spherical aberration caused by the concave mirror, by means of another medium with specific refractive index, for single-mode output fibres

(by means of fine-threaded screws), and thus the passbands of these channels can be shifted easily.

Using a solid glass block with a concave mirror at one end and a planar grating at the opposite end, a rugged and compact grating (de)multiplexer design can be achieved as shown in Fig. 7.22. This setup has no glass-air interfaces, and thus does not need antireflection coatings on the lens surfaces in order to eliminate ghost reflections. It features good thermal stability and low sensitivity to environmental conditions; however, it does not provide means for tuning. When single-mode output fibres are to be deployed, the spherical aberration of the concave mirror needs to be corrected by shaping it parabolically, or by introducing another medium with specific refractive index [8]. For a device capable of demultiplexing or multiplexing 20 channels separated by 1.6 nm in the 1515 nm to 1546 nm region, the losses per channel are 3−6 dB with single-mode fibres at both input port and output ports [9]. When using multimode 50 μm core-diameter graded-index fibre at the output ports, 6 channels spaced at 20 nm can be demultiplexed in the 1490 nm to 1590 nm range, with losses of 1.5−2.5 dB [9].

7.7 PHASAR-based Devices

7.7.1 Introduction

Phased-array wavelength demultiplexers are planar devices which are based on an array of waveguides with both imaging and dispersive properties. They image the field in an input waveguide onto an array of output waveguides in such a way that the different wavelength signals present in the input waveguide are imaged onto different output waveguides [10–13]. They are known under different names: Arrayed Waveguide Gratings (AWGs), Waveguide Grating Routers (WGRs) and Phased Arrays (PHASARs); in this chapter we will use the latter name. Since the early nineties their popularity has been rapidly increasing. Starting in 1993 an increasing number of system experiments involving PHASAR-demultiplexers have been reported [14–18]. In 1994 the first phased-array (de)multiplexers became commercially available in silica-on-silicon technology. Phased-array devices have the advantage that they can be realised in a single-mask planar waveguide technology, which makes them robust and fabrication-tolerant and potentially low-cost. Their main potential lies in applications with moderate crosstalk requirements and in monolithic or hybrid integration in more complex devices like multi-wavelength receivers and transmitters, add-drop filters and optical crossconnects. In this section the properties of PHASAR-demultiplexers and their application in a number of advanced multi-wavelength devices will be discussed.

7.7.2   Principle of Operation

Figure 7.23 shows the schematic layout of a phased-array demultiplexer. The operation is understood as follows. When the beam propagating through the transmitter waveguide enters the Free Propagation Region (FPR) it is no longer laterally confined and becomes divergent. On arriving at the input aperture the beam is coupled into the waveguide array and propagates through the individual waveguides to the output aperture. The length of the array waveguides is chosen such that the optical path length difference ΔL between adjacent waveguides equals an integer multiple m of the central wavelength of the demultiplexer:

$$\Delta L = m \cdot \frac{\lambda_c}{n_g} \, . \tag{7.21}$$

The integer m is called the order of the array, λ_c is the central wavelength of the demultiplexer, λ_c/n_g is the wavelength inside the array waveguides, and

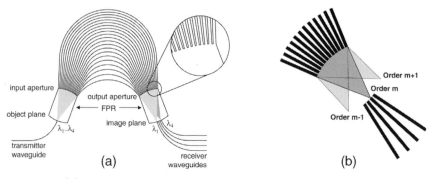

Fig. 7.23. (a) Geometry of the PHASAR demultiplexer, **(b)** beam focusing geometry in the free propagation region

n_g is the group index of the guided mode. For this wavelength the signals in the individual waveguides will arrive at the output aperture with equal phase (apart from an integer multiple of 2π), so that the field distribution at the input aperture will be reproduced at the output aperture. The divergent beam at the input aperture is thus transformed into a convergent one with equal amplitude and phase distribution, and an image of the input field in the object plane will be formed at the centre of the image plane.

The dispersion of the PHASAR is due to the length increment ΔL of the array waveguides, which will cause the phase difference $\Delta \Phi = \beta \Delta L$ between adjacent array waveguides, and thus the tilting of the outgoing wavefront, to vary linearly with the signal frequency (β is the propagation constant of the waveguide mode). As a consequence the focal point will shift along the image plane. The shift $ds/d\nu$ per unit frequency change is called the spatial dispersion D of the PHASAR; it is easily inferred that (for details see [13]):

$$D = \frac{1}{\nu_c} \cdot \frac{\Delta L}{\Delta \alpha} \tag{7.22}$$

in which ν_c is the central frequency of the PHASAR and $\Delta \alpha$ is the divergence angle between adjacent array waveguides near the input and output aperture. From the formulas for ΔL and D ((7.21) and (7.22)) it is seen that the dispersion is fully determined by the order m and the divergence angle $\Delta \alpha$ between adjacent array waveguides. By placing receiver waveguides at proper positions along the image plane, spatial separation of the different wavelength channels is obtained.

7.7.3   Technologies

Devices reported so far can be divided into two main classes: fibre-matched waveguide devices and compact waveguide devices. Fibre-matched waveguide structures have a waveguide core and an index contrast comparable to those of a single-mode fibre so that their modal field matches well with that of a fibre; thus, it is relatively easy to couple them to fibres. They combine low propagation loss with a high fibre-coupling efficiency. Most important technology today is silica-on-silicon with propagation losses below 0.1 dB/cm and fibre-to-chip coupling losses well below 0.5 dB. Silica-on-silicon was the first technology to be commercialized. More recently, silicon-based polymer devices [19,20] and lithium niobate devices [21] were reported which also employ fibre-matched waveguide structures. The potential of silica for integration of active functions is limited due to its passive character. The feasibility of integration with switches, however, was demonstrated by Okamoto who reported successful integration of thermo-optical switches with phased arrays in a 16-channel photonic integrated add-drop multiplexer [22,23].

Semiconductor-based devices have the potential to integrate a wide variety of functions on a single chip; they are suitable for integration of passive devices, electro-optical switches and modulators, optical amplifiers and also nonlinear devices such as wavelength converters. Another advantage is that they can be very compact due to the compact waveguide structure used, which makes it possible to integrate complex circuits on a single chip. A serious drawback of the compact waveguide structure are the higher fibre-coupling losses due to the small size of the waveguide cross-section (typically of the order of 1 μm as compared to almost 10 μm for the fibre core). Also, the propagation losses due to edge roughness are considerably higher than for silica waveguides (typically of the order of 1 dB / cm). Despite these relatively high on-chip propagation losses, the total on-chip loss of semiconductor devices can be kept low due to the small component size which is possible in semiconductor-based photonic integrated circuits. The high fibre-coupling loss and the problems of alignment to the small waveguides make semiconductor devices less competitive for passive functions with a low complexity; their main potential is for active functions. Integration of spot-size converters for obtaining a better mode-match to cleaved fibres is considered as a key issue for broadening their application. Indium phosphide-based devices are the most important class of semiconductor devices because of their suitability for operation in the long-wavelength window (both 1300 and 1550 nm). Monolithic integration of passive devices with active components such as detectors [24–28], optical amplifiers and modulators [29–32], and switches [33–35] is making rapid progress.

7.7.4 Device Characteristics

As explained in Sect. 7.7.2, PHASARs are lens-like imaging devices; they form an image of the field in the object plane at the image plane. Because of the linear length increment of the array waveguides the lens behaves dispersively: if the wavelength changes the image moves along the image plane without changing shape, in principle. Most of the PHASAR properties can be understood by considering the coupling behaviour of the focal field in the image plane to the receiver waveguide(s). This coupling is described by the overlap integral of the normalized receiver waveguide mode $U_r(s)$ and the normalized focal field $U_f(s)$ in the image plane, as illustrated in Fig. 7.24:

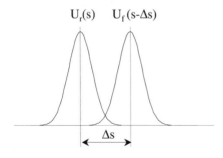

$U_r(s)$ $U_f(s-\Delta s)$

Δs

Fig. 7.24. The receiver waveguide mode profile $U_r(s)$ and focal field $U_f(s)$

$$\eta(\Delta s) = \left| \int U_{\mathrm{f}}(s - \Delta s) U_{\mathrm{r}}(s) ds \right|^2 , \tag{7.23}$$

in which Δs is the displacement of the focal field relative to the receiver waveguide centre. If $U_{\mathrm{r}}(s)$ and the image field $U_{\mathrm{f}}(s)$ have the same shape, which will be the case if identical waveguides are used for the transmitter and receiver waveguides, then the coupling efficiency can be close to 100% for proper design of the PHASAR. The power transfer function $T_i(\nu)$ for the i-th receiver waveguide is found by substituting $\Delta s = D \cdot (\nu - \nu_i)$:

$$T_i(\nu) = T_{i_0} \cdot \eta \{ D \cdot (\nu - \nu_i) \}, \tag{7.24}$$

in which D is the spatial dispersion of the PHASAR (7.22). T_{i_0} is the power transmission coefficient from the input waveguide to the image plane (which will be smaller than 1 due to transmission losses in the PHASAR), ν_i is the central frequency for channel i. Figure 7.25 shows an example of a demultiplexer response curve. The different device characteristics are indicated in the figure and will be explained below.

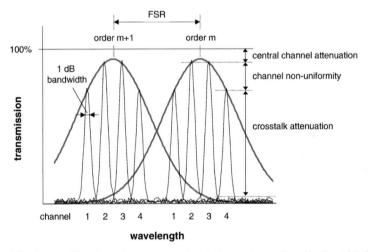

Fig. 7.25. Wavelength response of a 4-channel wavelength demultiplexer

Crosstalk Attenuation. One of the most important characteristics of the device is the crosstalk attenuation A_{x}. It is the attenuation of unwanted signals, e.g. the attenuation of the signal at wavelength ν_i in channel $i + 1$. The theoretical crosstalk attenuation follows from the overlap of the focal field with the mode of the unwanted receiver (7.23) as $A_{\mathrm{x}} = \eta(d)$, with d the distance between adjacent receiver waveguides. From this formula it is seen that an arbitrary large crosstalk attenuation is possible by spacing the receiver waveguides sufficiently wide. Usually a gap of 1–2 times the waveguide width is sufficient for more than 40 dB crosstalk attenuation. In practice other mechanisms appear to limit the crosstalk attenuation, however. The most important ones are errors in the phase transfer of the array waveguides. They are due to non-uniformities in layer thickness, waveguide width and refractive index and cause a rather noisy 'crosstalk floor' which is below −30 dB for good devices. Semiconductor-based devices exhibit higher crosstalk figures than silica-based devices. Crosstalk levels below −35 dB are difficult to achieve with today's technology.

Crosstalk figures provided for experimental and commercial devices usually refer to single-channel crosstalk levels, i.e. the crosstalk resulting from a single channel. In an operational environment crosstalk contributions from all active channels will increase the crosstalk level as compared to the single-channel crosstalk attenuation. This effect is discussed in [36].

Bandwidth. Without special measures the channel response has a parabolic shape. The 1 dB bandwidth is usually 30–40% of the channel spacing; it follows from the overlap of the receiver waveguide mode profile with itself according to (7.23) and (7.24).

The parabolic shape of the channel response may cause problems if a number of devices with similar bandpass characteristics have to be cascaded, e.g. a number of add-drop multiplexers or crossconnects. The overall wavelength response of the whole system consists of the product of the response curves of all individual devices, which will become very narrow in the case of a parabolic shape. The bandwidth can be increased using band-flattening techniques. If we succeed in giving the focal field of the PHASAR a 'camel-like' shape as indicated in Fig. 7.26a then the wavelength response, which follows from the overlap of this field with the mode of the receiver waveguide, will get a flat region as shown in Fig. 7.26b. There are several ways to realise a 'camel-like' field in the image plane. The most straightforward ones modify the modal field of the transmitter waveguides by applying a 1×2 multimode-interference (MMI) coupler with overlapping images [37], a parabolic horn [38] or a Y-junction with closely spaced output branches. Other approaches proposed are manipulation of the phase-transfer of the array in order to get a flattened image [39], or the use of two interspersed arrays with slightly shifted, but partially overlapping focal images [40].

Due to the mismatch between the 'camel-like' focal field and the waveguide mode the approach described above introduces a loss penalty. For moderate flattening (1 dB bandwidth of 50–60% of the channel spacing) this loss penalty may be restricted to 2–3 dB, for 80% band-flattening it may exceed 5 dB. At the receiver side this loss penalty can be avoided by applying wide (multimode) receiver waveguides. This approach is similar to the one described in Sect. 7.6 for grating-based multiplexers. Over a certain wavelength range the focal field will be fully captured by the multimode

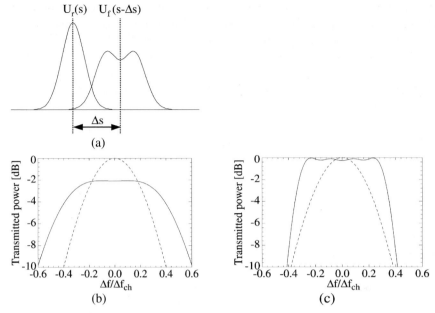

Fig. 7.26. Wavelength response flattening: **(a)** shape of the focal field U_f required for obtaining a flat region in the overlap with the receiver mode U_r; **(b)** wavelength response obtained by applying a camel-shaped focal field (the dashed curve indicates the non-flattened response obtained by applying a non-modified focal field, i.e. $U_f = U_r$); **(c)** Wavelength response obtained by applying a non-modified focal field and a wide multimode receiver waveguide

receiver waveguide. In this region the wavelength response will be almost flat, as shown in Figs. 7.18 and 7.26c, and the coupling loss will be very small because almost all of the signal power is coupled into the multimode receiver waveguide. As the signal power in the receiver waveguide is carried by several modes the signal should be directly applied to a detector, it can no longer be coupled to a single-mode fibre.

Central Channel Attenuation. The attenuation A_0 of the central channel is found from the total transmission loss T_{PH} of the PHASAR as $A_0 = -10 \log T_{PH}$. The most important contribution to this loss comes from the junctions between the free propagation regions (FPR) and the waveguide array. For low loss the fan-in and fan-out sections should operate adiabatically, i.e there should be a fluent transition from the guided propagation in the array to the free-space propagation in the FPRs and vice versa. This will occur if the divergence angle Δa between the array waveguides is sufficiently small and the vertex between the waveguides is sufficiently sharp. Due to the finite resolution of the lithographical process, blunting of the vertex will occur. Junction losses for practical devices are between 1 and 1.5 dB per junction (i.e. between 2–3 dB for the total device). Propagation loss in the PHASAR and coupling losses due to a mismatch between the imaged field $U_f(s)$ and the receiver waveguide mode $U_r(s)$ are usually much smaller.

Free Spectral Range. The dispersion of the phased array is obtained because, due to the length difference ΔL, the phase difference $\Delta \Phi = \beta \Delta L$ between adjacent waveguides increases with increasing frequency, which will cause the outgoing beam to be tilted. If the frequency change is such that $\Delta \Phi$ has increased by 2π the transfer will be the same as before, so the response of the PHASAR is periodical, as shown in Fig. 7.25. The period $\Delta \nu$ in the frequency domain is called the free spectral range, it follows directly from the condition $\Delta \beta \Delta L = 2\pi$ and (7.21) as

$$\Delta \nu_{FSR} = \frac{c_o}{n_g \Delta L} = \frac{\nu_c}{m} .   (7.25)$$

The free spectral range should be larger than the whole frequency range spanned by the channels in order to avoid crosstalk problems with adjacent orders. For a demux with 8 channels with 200 GHz channel spacing the FSR should thus be at least 1600 GHz. If the channels are centered around 1550 nm, this would require an array with an order of at least 121 according to (7.25).

If the device is to be used in combination with erbium-doped fibre amplifiers (EDFAs), the FSR should be chosen such that adjacent orders do not coincide with the peak of the EDFA gain spectrum in order to avoid cumulation of Amplified Spontaneous Emission (ASE).

Channel Non-Uniformity. From Fig. 7.25 it is seen that the outer channels have more loss than the central ones. This reduction is caused by the fact that the far field of the individual array waveguides drops in directions different from the main optical axis. The envelope, as indicated in the figure, is mainly determined by the far-field radiation pattern of the individual array waveguides. The non-uniformity ΔA is defined as the attenuation difference between the central channel(s) and the outermost channels (No. 1 and N):

$$\Delta A = -10 \log \frac{T_{1,N}(\nu_{1,N})}{T_0} ,   (7.26)$$

in which T_0 is the transmission of the central channels.

The power which is lost from the main lobe will appear in adjacent orders, as shown in Fig. 7.23b. If the free spectral range is chosen close to N times the channel spacing then the spacing between the outermost channels (No. 1 and N) will be close to the FSR. From Fig. 7.25 it is seen that in this case the outermost channels will experience almost 3 dB loss, the power is almost equally divided over two different orders, and the non-uniformity for such a PHASAR will be close to 3 dB.

From the above it is clear that in order to reduce the non-uniformity the FSR has to be increased; this increases the device-size.

Polarization Dependence. Phased arrays are polarization independent if the array waveguides are polarization independent, i.e. if the propagation constants for the fundamental TE- and TM-mode are equal. Waveguide birefringence, i.e. a difference in propagation constants, will result in a shift $\Delta\nu_{\rm pol}$ of the spectral responses with respect to each other, which is called the polarization dispersion. For semiconductor-based double-hetero waveguides typical polarization dispersion values are of the order of a few hundred GHz. In fibre-matched waveguide structures the birefringence should be zero, in principle. Due to mechanical stress resulting from the difference in thermal expansion coefficients between waveguide layers and substrate, polarization dispersions of more than 20 GHz may still occur in practice, which is too much in WDM systems.

A very elegant method applied in silica-based devices is the insertion of a $\lambda/2$-plate in the middle of the phased array [41,42]. Light entering the array in a TE-polarized state will be converted by the $\lambda/2$-plate and travel through the second half of the array in a TM-polarized state, and TM-polarized light will similarly traverse half the array in a TE-state. As a consequence, TE- and TM-polarized input signals will experience the same phase transfer regardless of the birefringence properties of the waveguides applied.

For the half-wave plate a strongly birefringent thin polyimide film is used, which is mounted in a trench diced in the wafer transversely through the waveguides in the centre of the phased array. The remaining space is refilled with an index-matched polymer. As the polyimide half-wave plates have a thickness of more than $10\,\mu$m they are only applicable to waveguide structures with a small numerical aperture (NA) which can bridge this distance with small diffraction losses.

In semiconductor waveguides the method is not practical due to the large NA of these waveguides. Several methods have been reported to arrive at polarization-independent designs. The most obvious way is by eliminating the birefringence of the waveguide. This can be done by making the waveguide cross-section square if the index contrast is the same in the vertical and lateral direction as, for example, in buried waveguide structures. Small deviations of the square shape, for example due to non-perfect control of the waveguide width, will disturb the polarization independence. If the index contrast between core and cladding is high, the tolerance requirements on waveguide-width control become impractically tight. Tolerant design requires, therefore, low-contrast waveguides with a relatively large waveguide core (which is also advantageous for achieving low fibre-coupling loss) [43–45]. Another solution is found in compensation of the polarization dispersion by inserting a waveguide section with a different birefringence in the phased array [46,47].

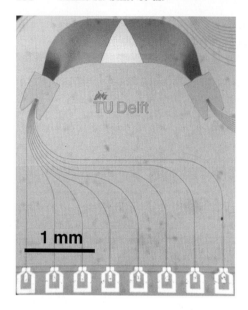

Fig. 7.27. WDM-receiver chip containing a PHASAR-demultiplexer integrated with 8 detector diodes (bottom) [26]

Figure 7.27 shows a multi-wavelength receiver with a demultiplexer based on this principle. In the triangular section the waveguide structure has been modified in order to create a local increase in birefringence. Through a proper choice of the polarity of the triangle this increased birefringence can be made to compensate the birefrigence in the rest of the array.

7.7.5   Wavelength Routeing Properties

An interesting device is obtained if the PHASAR is designed with N input and N output waveguides and a free spectral range equaling N times the channel spacing. With such an arrangement the device behaves cyclically: a signal disappearing from output N will reappear at output 1 if the frequency is increased by an amount equal to the channel spacing. Such a device is called a cyclical wavelength router [12]. It provides an important additional functionality as compared to multiplexers and demultiplexers and

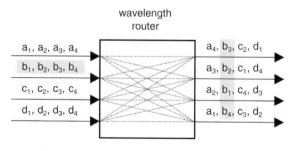

Fig. 7.28. Schematic diagram illustrating the operation of a wavelength router

plays a key role in more complex devices such as add-drop multiplexers and wavelength switches. Figure 7.28 illustrates its functionality. Each of the N input ports can carry N different frequencies. The N frequencies carried by input channel 1 (signals $a_1 - a_4$ in Fig. 7.28) are distributed among output channels 1 to N, in such a way that output channel 1 carries frequency N (signal a_4) and channel N frequency 1 (signal a_1). The N frequencies carried by input 2 (signals $b_1 - b_4$) are distributed in the same way, but cyclically rotated by 1 channel in such a way that frequencies $1 - 3$ are coupled to ports $3 - 1$ and frequency 4 to port 4. In this way each output channel receives N different frequencies, one from each input channel. To realize such an interconnectivity scheme in a strictly non-blocking way using a single frequency a large number of switches would be required. Using a wavelength router, this functionality can be achieved using only one single component. Wavelength routers are key components also in multi-wavelength add-drop multiplexers and crossconnects (see Sect. 7.7.7).

7.7.6 Multiwavelength Transmitters and Receivers

Multi-wavelength Lasers. Today's WDM systems use wavelength-selected or tunable lasers as sources. Multiplexing a number of wavelengths into one fibre is done using a power combiner or a wavelength multiplexer. A disadvantage of this solution is the large number of lasers required, each of which has to be wavelength-controlled individually. Integrated multi-wavelength lasers have been realized by combining a DFB-laser array (with a linear frequency spacing) with a power combiner on a single chip [48,49], (see also Chap. 11, Sect. 11.7).

Using a power combiner for multiplexing the different wavelengths in a single fibre is a tolerant method, but it introduces a loss of at least $10 \log N$ dB, N being the number of wavelength channels. The combination loss can be reduced by applying a wavelength multiplexer at the cost of more stringent requirements on the control of the laser wavelengths.

An elegant solution to this problem is integration of a broadband optical amplifier array with a multiplexer into a Fabry-Perot cavity as depicted

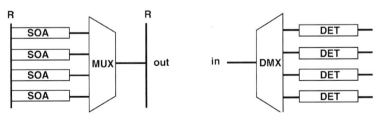

Fig. 7.29. Multi-wavelength laser consising of four optical amplifiers integrated with a multiplexer in a Fabry-Perot cavity (*left*), and a multi-wavelength receiver (*right*)

in Fig. 7.29 (left) [29–32]. If one of the Semiconductor Optical Amplifiers (SOAs) is excited the device will start lasing at the passband maximum of the multiplexer channel to which the SOA is connected. All SOAs can be operated and (intensity) modulated simultaneously, in principle. An important advantage of this component is that the wavelength channels are automatically tuned to the passbands of the multiplexer and that they are coupled to a single output port with low loss. The main disadvantage of this type of laser is its large cavity length; the associated large round-trip time makes it less suitable for direct modulation at high data rates. Despite their long cavity length, these lasers show single-mode operation over a wide range of operating conditions [30]. Direct modulation speeds up to around 1 Gb /s have been reported. Recently error-free transmission at 2.5 Gb /s data rate of 16 channels over 627 km has been demonstrated using two interspersed 8-channel multi-wavelength lasers [50].

The main potential of this type of laser might be in applications at short or medium distances at medium bit rates (622 Mb / s) or as discretely tunable CW-laser sources.

Multi-wavelength Receivers. The most straightforward way to realise a multi-wavelength receiver is by connecting N detectors to the output ports of a wavelength demultiplexer. Integration of a PHASAR-based demultiplexer with detectors on a single chip has been reported by several authors [24–26], one of these devices is shown in Fig. 7.27. A device like this has been hybridly integrated with 8 front-end amplifiers and packaged in an industry-standard 14-pin butterfly package [27] which illustrates the volume reduction that can be achieved using monolithic integration. In InP the receiver amplifiers can also be integrated; this was first reported by Chandrasekhar et al., who realized an 8×2.5 Gb /s MW-receiver with integrated heterojunction bipolar transistor (HBT) preamplifiers [28]. Monolithic integration has the potential to bring down the cost and increase the reliability of MW-receiver modules. Further it leads to a drastic volume reduction. The main challenge in exploiting the advantages of integration are the problems related to electrical crosstalk inside the package. With conventional packaging techniques it is difficult to realise acceptable crosstalk figures at data rates higher than 1 Gb/ s.

7.7.7   Multiwavelength Add-Drop Multiplexers and Crossconnects

Add-drop multiplexers (ADMs) are key components in WDM-networks. They are used for coupling one or more wavelengths out of a link or a ring (drop function) and coupling input signals at the same wavelengths to the link or ring (add function). Two classes can be distinguished: fixed and (re)configurable add-drops.

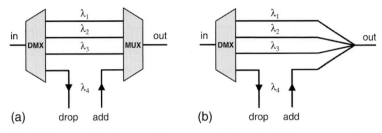

Fig. 7.30a,b. Add-drop multiplexers with fixed add-drop channel

In Sects. 7.4–7.6 some examples of fixed add-drops using dielectric filters, bulk and fibre gratings have been explained. Figure 7.30 shows how the same function can be realised using a demultiplexer and a power combiner or a multiplexer. Main performance requirements for add-drop multiplexers are a good isolation between the add and the drop part (necessary because the signal power level in the add port is usually much higher than in the drop port) and a good suppression of the drop signal in the output port. The configuration of Fig. 7.30a,ba performs well in both respects: there is an almost perfect isolation between the add and the drop port, and the crosstalk of the dropped signal ending up in the output is suppressed again by the multiplexer so that the total crosstalk attenuation is doubled. This is not the case for the cheaper solution of Fig. 7.30a,bb using the power combiner.

An add-drop multiplexer as shown in Fig. 7.30a,b can be made configurable, i.e. the added and dropped wavelengths can be selected with an external control signal by combining the (de)multiplexers with switches as shown in Fig. 7.31. Again the multiplexers may be replaced by power combiners at the price of an increased loss and a reduced drop-signal suppression. Although both schemes perform the same function, they differ strongly from a crosstalk point-of-view. In the configuration of Fig. 7.31a the add-signal may end up in the drop-port at an undesirably high level through crosstalk in the switch. In the configuration of Fig. 7.31b, which requires two 1×2 switches per channel

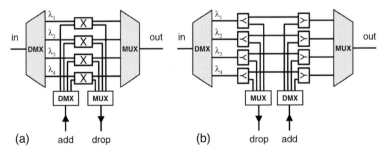

Fig. 7.31. Two reconfigurable add-drop multiplexers, based on one 2×2 switch (a) and two 1×2 switches (b)

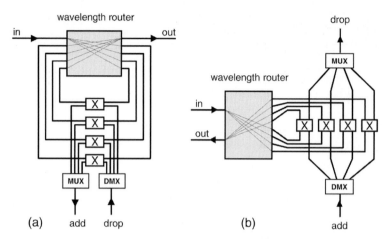

Fig. 7.32. Two add-drop multiplexer configurations based on a single wavelength router in a loop-back configuration **(a)** and in a fold-back configuration **(b)**

instead of one 2×2, the isolation is almost perfect. A 16-channel add-drop multiplexer according to the configuration of Fig. 7.31a has been realised in integrated form in silica-on-silicon technology [22,23]. Crosstalk was $-30\,\mathrm{dB}$ and loss between fibres was $10\,\mathrm{dB}$.

A disadvantage of the configurations shown in Fig. 7.30a,b and Fig. 7.31 is that for low insertion loss operation they require two (de)multiplexers. Figure 7.32 shows an add-drop multiplexer realised with a single $(N + 1) \times (N + 1)$ wavelength router [14,34]. In the first pass through the wavelength router the four wavelengths are demultiplexed. They are fed to four different switches with which they can be switched to the drop port or be looped back to the wavelength router which multiplexes them to the output port. In the drop state a signal applied to the add-port will be multiplexed to the output port. A disadvantage of this loop-back configuration is that the crosstalk of the PHASAR is coupled directly into the main output port, where it competes with the transmitted signal which is attenuated in the loop. This problem can be reduced by applying the PHASAR in a fold-back configuration as shown in Fig. 7.32 [15,35]; this, however, requires a larger PHASAR.

Another key device for advanced multi-wavelength networks is an optical crossconnect. Figure 7.33 shows an example of an experimental crossconnect which has been investigated in a European research project (ACTS OPEN). A mesh of high-capacity optical fibre links can be cross-connected through such transparent photonic nodes. The crossconnect has four input and four output ports carrying four wavelengths each and is able to transport bit rates up to $40\,\mathrm{Gb/s}$ per wavelength. The incoming signals are distributed over four switch matrices, one for each output, without filtering; each switch matrix routes the desired signals to four tunable filters which pick the signals to be routed out of the whole spectrum. Next the signals are multiplexed

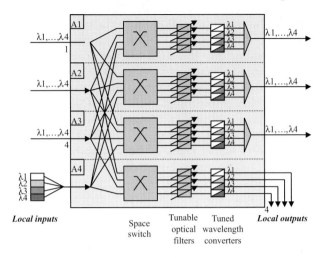

Local inputs   Space   Tunable   Tuned   Local outputs
switch   optical wavelength
filters converters

Fig. 7.33. Architecture of the ACTS-OPEN cross-connect

to the output fibre (except for ports 4, which are local input and output ports). Because the signals applied to the four multiplexer ports may have the wrong wavelength to end up in the multiplexer output port, four wavelength converters are needed to provide them with the right wavelength.

The costs presently involved in crossconnects like these are prohibitive for large-scale application of optical cross-connected networks. Application will only be economical at the highest levels of the network, where the costs can

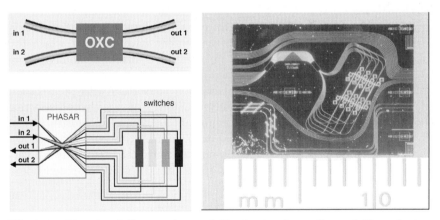

Fig. 7.34. Functional diagram (*upper left*), circuit scheme (*lower left*), and microscope photograph (*right*) of a compact photonic integrated crossconnect in a foldback configuration. The PHASAR wavelength router is seen at the upper left corner, the triangular structure makes it polarization independent. The four switches on the right are positioned at a specific angle to the crystal axis to make them polarization independent

be shared by many users. For broader applications both the cost and volume of devices like this will have to be reduced. Present progress in photonic integration holds a great promise that this goal can be achieved. Figure 7.34 shows the function, the circuit scheme and a photograph of a photonic integrated circuit (PIC) which may be used to crossconnect two fibre links. The device, which was monolithically integrated in InP-based semiconductor material, measures only $9 \times 12 \, \text{mm}^2$. At present the performance of PICs like this is not yet adequate for operational applications: crosstalk levels of the first experimental prototype were higher than $-20 \, \text{dB}$. Silica-based devices show better performance, but have a much larger device size. Technological progress will lead to a steady improvement of the performance of these devices, which may give rise to a new generation of highly sophisticated multi-wavelength networks.

7.8   Integrated Acousto-Optical Devices in LiNbO$_3$

7.8.1   Introduction

During the last few years a variety of integrated acousto-optical LiNbO$_3$ devices has been developed for applications in WDM communication systems [51,52]. They take advantage of a wavelength-selective polarization conversion induced by a *collinear* acousto-optical interaction.

The basic building blocks of these devices are acousto-optical polarization converters and polarization splitters (Fig. 7.35). Details of their operational characteristics will be discussed in the next subsections. To understand the device principles it is sufficient to know that in the acousto-optical converter a *wavelength selective polarization conversion* is induced by a surface acoustical wave. Its frequency determines the optical wavelength at which the polarization conversion occurs. The polarization splitters/combiners separate/combine the TE and TM polarized guided-optical waves.

By combining acousto-optical polarization converters with polarization splitters (or polarizers) a whole family of integrated wavelength-selective devices can be realized. The most important ones are wavelength filters, wavelength-selective switches and add-drop multiplexers, which are shown schematically in Fig. 7.36. An acousto-optical converter between two polarization splitters forms a single-stage bandpass filter. Polarization-independent

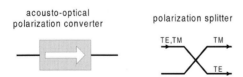

Fig. 7.35. Schematic structure of the integrated acousto-optical polarization converter and the polarization splitter used as the basic building blocks of integrated acousto-optical devices

Polarization independent double-stage wavelength filter

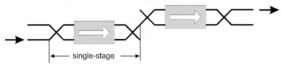

Wavelength-selective 2x2 switch

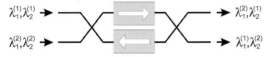

Add-drop multiplexer

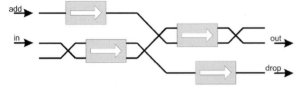

Fig. 7.36. Schematic drawing of the most important integrated acousto-optical devices. The basic structure of a wavelength filter, a wavelength-selective switch and an add-drop multiplexer is shown. All devices can be realized by combining the basic building blocks of acousto-optical polarization converter and polarization splitter

operation is achieved by applying the principle of polarization diversity, i.e. the two polarization components are converted separately and recombined by the rear polarization splitter. To improve the performance characteristics one can cascade two filters forming a double-stage tunable wavelength filter. A wavelength selective 2×2 switch is realized using two converters and two splitters. The device allows each incoming wavelength channel to be switched independently of the switching state of the other channels. The add-drop multiplexer consists of four converters and four splitters. Wavelength channels can be inserted and extracted from the transmission line.

Such integrated acousto-optical devices are of particular interest for WDM systems as they offer some unique features such as a broad tuning range with electronic control, fast tuning speed and, especially, simultaneous multi-wavelength operation. In the following the basic building blocks of these devices are briefly discussed. Subsequently, a description of the state-of-the art of integrated acousto-optical devices for WDM systems as well as some applications is given.

7.8.2   Basic Building Blocks

Acousto-Optical Polarization Converters. The central building block of integrated acousto-optical devices is the acousto-optical polarization converter. Due to the interaction of a Surface Acoustic Wave (SAW) with optical waves guided in Ti-indiffused stripe waveguides (fabricated in X-cut, Y-propagating LiNbO$_3$) a wavelength-selective polarization conversion, i.e. $TE \rightarrow TM$ or $TM \rightarrow TE$, is performed.

A propagating surface acoustic wave induces a periodic perturbation of the dielectric tensor, which results in a coupling of orthogonally polarized optical modes. To achieve an efficient polarization conversion the interaction process must be phase-matched. The difference between the wave numbers of the optical modes must be compensated by the wave number of the SAW, i.e. one obtains the phase-matching relation

$$|n_{\text{eff}}^{\text{TE}} - n_{\text{eff}}^{\text{TM}}|/\lambda_0 = f_{\text{a}}/v_{\text{a}}, \tag{7.27}$$

where f_{a} and v_{a} are the frequency and the velocity of the SAW, respectively, and $n_{\text{eff}}^{\text{TE}}$ and $n_{\text{eff}}^{\text{TM}}$ are the effective indices of the TE and TM polarized optical modes. The phase-matching condition makes the conversion process wavelength-selective. Via the frequency of the SAW the optical wavelength λ_0 of the modes to be phase-matched can be adjusted. For wavelengths in the third communication window around $\lambda_0 = 1.55\,\mu\text{m}$ the SAW frequency for phase-matching is around 170 MHz with a tuning slope of about 8 nm /MHz; these properties are determined by the birefringence of LiNbO$_3$.

Nowadays, most acousto-optical devices take advantage of integrated acoustical waveguides to confine the SAWs into localized regions, yielding large acoustical power densities even at low or moderate overall acoustic power levels (Fig. 7.37). Such acoustical waveguides are fabricated by a Ti-indiffusion into the cladding region of the guide which stiffens the material and, hence, increases the acoustic velocity. The SAW is guided in the un-doped region between the Ti-diffused claddings. Moreover, besides straight acoustical waveguides even more complex guiding structures, e.g. acoustical directional couplers, can be realized. Such directional couplers have been used to improve the spectral conversion characteristics as discussed below.

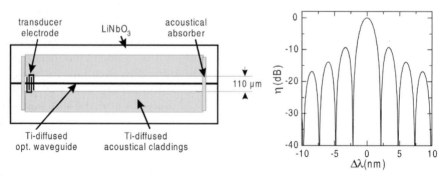

Fig. 7.37. Acousto-optical polarization converter with straight acoustical waveguide (*left*) and the corresponding calculated conversion characteristic, assuming a 12 mm long acousto-optical interaction length (*right*)

In a simple (unweighted) acousto-optical converter an optical waveguide is embedded in a straight acoustical guide (Fig. 7.37) [53,54]. The SAW, excited by applying an RF-signal to the interdigital transducer electrodes, propagates co- or contra-directionally to the optical waves. The theoretical conversion characteristic, i.e. the converted power as function of optical wavelength, is a sinc^2-function as shown in the right-hand diagram of Fig. 7.37. The spectral half-width of the curve is proportional to $1/L$ with L being the interaction length. Severe disadvantages of such devices are the high side-lobes of about -10 dB.

The spectral conversion characteristic is approximately given by the Fourier transform of the interaction strength. Therefore, to suppress the side-lobes one can apply a weighted coupling scheme (apodization) [55–57]. Instead of an abrupt change of the interaction strength a soft onset and a soft cut-off is achieved using an acoustical directional coupler. The optical waveguide is embedded in one arm of the acoustical directional coupler (Fig. 7.38). The SAW is excited in the other arm and couples into the adjacent guide and back again (Fig. 7.39). Therefore, side-lobes of the conversion characteristics are strongly suppressed (> 20 dB) as shown in the right-hand diagram of Fig. 7.38.

At ideal phase-matching the conversion efficiency$+\eta$, i.e. the ratio of the optical power converted into the orthogonal polarization state to the input power, is given by

$$\eta = \sin^2(\gamma\sqrt{P_a L}), \tag{7.28}$$

where γ is a constant determined by the overlap integral between the normalized optical mode fields of both polarizations and the acoustical mode, L is the interaction length and P_a the power of the acoustical wave. By adjusting the acoustical power the conversion efficiency can be controlled.

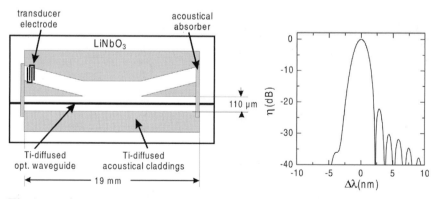

Fig. 7.38. Acousto-optical polarization converter with acoustical directional coupler for weighted coupling (*left*) and the corresponding calculated conversion characteristic (*right*)

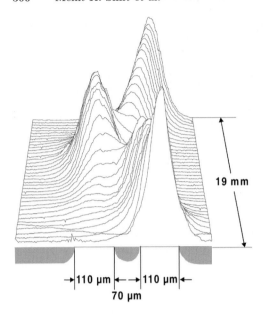

19 mm

→|110 μm|←   →|110 μm|←
       70 μm

Fig. 7.39. Measured intensity profile of a surface acoustic wave (SAW) propagating in the acoustical directional coupler shown in Fig. 7.38

The acousto-optical polarization conversion is accompanied by a frequency shift. The frequency of the converted optical wave is shifted by the SAW frequency. The direction of the shift depends on the direction of conversion, i.e. $TE \rightarrow TM$ or $TM \rightarrow TE$, and on the propagation direction of the SAW relative to the propagation direction of the optical waves.

A unique feature of acousto-optical mode converters is their multi-wavelength capability. By simultaneously exciting several acoustical waves at different frequencies in the polarization converter, simultaneous conversion at different optical wavelengths can occur [58]. However, by multi-wavelength operation additional crosstalk can be induced [59–61].

Polarization Splitters and Polarizers. Polarization splitters are used to separate the TE and TM components of an incoming wave and route them to different optical waveguides. Two schemes have been used to realize such polarization splitters for integrated acousto-optical devices (Fig. 7.40). Both schemes yield polarization splitters with splitting ratios exceeding 20 dB and additional insertion losses below 1 dB.

The first scheme uses a passive directional-coupler structure fabricated by applying solely the Ti-indiffusion technique [62,63]. Taking advantage of the polarization-dependent refractive-index profiles, the couplers have been designed to route TE-polarized waves to the cross-state output and TM-polarized waves to the bar-state output of the structure.

The second scheme is based on a hybrid Ti-indiffusion/proton-exchange technology [64]. In the proton-exchanged regions the extraordinary refractive index is increased whereas the ordinary is reduced. Therefore, TE-polarized

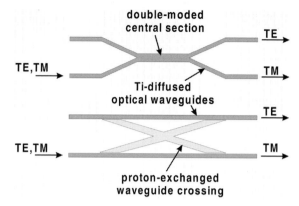

Fig. 7.40. Schematics of passive integrated polarization splitters used for acousto-optical devices. The upper structure is a zero-gap directional coupler fabricated by Ti-indiffusion technology, whereas in the lower structure the waveguide crossings are abricated using proton-exchange technology

waves are coupled into the proton-exchanged regions and routed towards the adjacent Ti-diffused channel guide whereas the TM-polarized waves remain in the initial Ti-diffused guide.

Several types of integrated polarizers have been developed [62,65]. However, as their fabrication requires further technological processes, the fabrication of the whole device becomes more complex. Therefore, nowadays polarization splitters are used most often as polarizers.

7.8.3 Tunable Wavelength Filters

The first polarization-dependent filters consisted of an integrated acousto-optical polarization converter between crossed polarizers (either external or integrated) [53,66]. Polarization insensitivity has been obtained by applying the principle of polarization diversity, i.e. by a separate processing of the polarization components [67–69]. Improved filters use cascaded structures to provide a double-stage filtering with the advantage of strongly suppressed baseline levels and reduced side-lobes [62,70].

Recently, a polarization-independent double-stage wavelength filter with weighted coupling in each stage has been realized [70]. The design of the circuit is shown in Fig. 7.41. The incoming wave is split into its polarization components in the first polarization splitter. They are routed to separate optical waveguides which are both embedded in a common acoustical waveguide, being one branch of an acoustical directional coupler. After passing this polarization converter the signals are recombined by the second polarization splitter. Because the state of polarization of the phase-matched waves has been changed, they are separated from the unconverted ones. The converted waves are routed to the second filter stage, whereas the unconverted ones

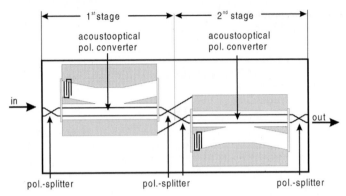

Fig. 7.41. Integrated optical, acoustically tunable double-stage wavelength filter realized by cascading two single-stage filters on a common substrate

are fed to a waveguide which is terminated on the substrate outside of the interaction area.

Such double-stage filters offer several advantages. First, the filter characteristics of the overall device are the product of the filter characteristics of the individual stages. This yields a strong suppression of the side-lobes and a narrowing of the spectral filter response. Even if one stage has a non-ideal performance, e.g., large side-lobes or bad splitting ratios of the polarization splitters, cascading with the other stage still results in good overall device performance. Second, due to the double-stage design there is no net frequency shift imposed on the waveguide modes. The opposite frequency shifts of the polarization components in the first stage are compensated by reverse frequency shifts in the second stage. There is a third advantage of this particular double-stage filter structure: The device only uses acousto-optical polarization converters and polarization splitters as basic building blocks. No polarizers are required, simplifying strongly the fabrication technology. As only Ti-indiffusion techniques are applied, this filter offers a great potential for future mass production.

In Fig. 7.42 the transmission of a pigtailed and packaged wavelength filter (Fig. 7.43) is shown. Two curves are drawn corresponding to an input polarization with minimum and maximum insertion loss at the peak transmission, respectively. (The state of polarization at the input of the device cannot be determined after pigtailing. Therefore, the minimum and maximum insertion loss has been used as a criterion for adjusting the input polarization.) The bandwidth (full-width at half maximum) is 1.6 nm. There are no pronounced side-lobes in the filter characteristics and the baseline, i.e. the residual transmission at a wavelength far away from the filter peak, is about 35 dB below the transmission maximum. The polarization dependence is quite small: only a small shift of about 0.07 nm occurs for the peaks of maximum transmission. The overall insertion loss (fibre-to-fibre) is less than

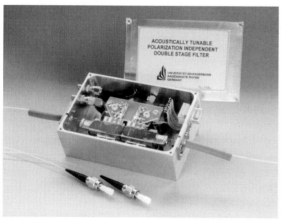

Fig. 7.42. Measured spectral bandpass characteristic of the double-stage wavelength filter. The two curves correspond to an input state of polarization with minimum and maximum insertion loss at the peak wavelength, respectively

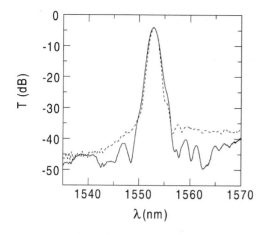

Fig. 7.43. Photograph of a pigtailed and packaged integrated acousto-optical double-stage wavelength filter

4.2 dB, with a polarization dependence smaller than 0.1 dB. The tuning range of the filter exceeds the spectral range for typical WDM applications. This filter could be tuned from 1530 nm to 1570 nm without readjusting the drive power of about 100 mW for both stages together.

7.8.4   Wavelength-selective Switches and Add-Drop Multiplexers

A wavelength-selective switch is a 2×2 switch matrix allowing the individual routeing of the wavelength channels of a WDM-transmission line to the cross- and bar-state outputs of the device. In Fig. 7.44 the design of a single-stage device is shown [63]; it consists of two acousto-optical polarization converters and two polarization splitters. The light entering, for example, input port 'in-1' is divided into its TE- and TM- polarized components by the first splitter; the second splitter acts as combiner. The optical power is routed to output port 'out-1' ('bar-state') if no mode conversion is performed. For

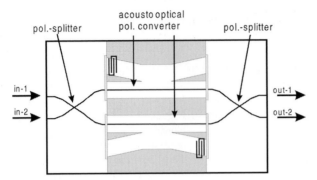

Fig. 7.44. Single-stage integrated acousto-optical wavelength-selective 2×2 switch

converted waves the state of polarization changes and, therefore, these signals are routed to output port 'out-2' ('cross-state'). The SAWs in the mode converters propagate in opposite directions to achieve identical frequency shifts for TE and TM components. This avoids beating effects that would disturb the performance of the switch seriously.

Typical spectral switching characteristics are shown in Fig. 7.45. The spectral halfwidth is about 2 nm. Fibre-to-fibre insertion losses are typically in the range of 3–5 dB. The side-lobe suppression is about -15 dB to -20 dB. The extinction of the notch curve is typically larger than 25 dB if the input light is either exactly TE- or TM-polarized. However, as the notch curves for the two polarizations do not exactly coincide, extinction of the notch curve for unpolarized input (as shown in the diagram) is typically limited to -15 dB to -20 dB. The tuning range of such devices exceeds 70 nm. The

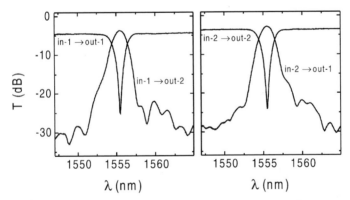

Fig. 7.45. Measured switching characteristics of a single-stage 2×2 switch. In the diagrams the transmission characteristics for unpolarized input waves are shown. The switch has been operated with an acoustical wave to route optical signals at $\lambda = 1556$ nm to the cross-state

switching time is determined by the build-up time of the acoustical wave in the converter structure. For 20 mm-long converters it is about 5 μs.

Crosstalk suppression is the most stringent requirement for components applied in WDM transmission systems. With single-stage switches the necessary crosstalk suppression cannot be achieved. Therefore, a double-stage design must be used leading to dilated switches [61,71,72]. Combining four single-stage switches yields a device with crosstalk suppressed to the order of ε^2, if ε is the crosstalk of a single switch. Up to now, no monolithically integrated dilated acousto-optical switch has been realized.

A first step towards a monolithically integrated dilated switch is the double-stage add-drop multiplexer shown in Fig. 7.46 [73]. Add-drop multiplexers do not require the full functionality of a 2×2 switch as no direct routing from the add- to the drop port is required. Therefore, a partially dilated switch can be used to form an improved add-drop multiplexer. The circuit consists of two switches in series and two frequency shifters. In the first switch the drop-function is performed and in the second switch the add-

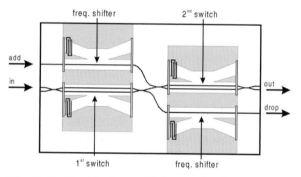

Fig. 7.46. Double-stage add-drop multiplexer

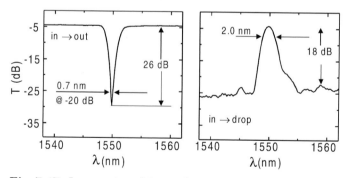

Fig. 7.47. In → out and in → drop spectral characteristics of the double-stage add-drop multiplexer. (The add → out characteristic is similar to the in → drop curve). The measured transmission curves have been obtained using an unpolarized broadband input source. The multiplexer has been operated to add and drop signals at $\lambda = 1550$ nm

function. Due to this local separation, i.e. spatial dilation, of the add- and drop functions, crosstalk between add- and drop-ports is strongly suppressed ($< 50\,\mathrm{dB}$). Moreover, the device acts as double-stage notch filter for the transmission from 'in' to 'out' resulting in a suppression of crosstalk due to incomplete dropping of about $26\,\mathrm{dB}$ (Fig. 7.47). The frequency shifters in the add- and in the drop-arms are required to compensate the frequency shift induced by the acousto-optical polarization conversion in the switches.

7.8.5   Applications in WDM Systems

It is beyond the scope of this article to discuss details of WDM systems using acousto-optical devices. However, a brief overview demonstrating their potential will be given.

The simplest application of acousto-optical wavelength filters is their use as channel selectors in front of a receiver. Another filter application using specific acousto-optical properties is the power equalization of different wavelength channels [74]. Taking advantage of the multi-wavelength capabilities of such a device the transmission of each channel can be adjusted by applying a set of RF signals with appropriate frequencies and power levels.

A further interesting filter application is the WDM channel analysis [75]. To monitor the state of the transmission system fast scanning over the spectral range of interest is required. This can be performed by electronic means using an acousto-optical filter without any moving mechanical parts.

Acousto-optical switches and add-drop multiplexers are used in dynamically reconfigurable WDM networks. The multiplexers can be used in network nodes to insert or extract information at certain wavelengths to/from the transmission line. Due to the tunability of the devices the configuration is flexible: the devices can be applied, for example, in systems with dynamic wavelength reuse. Within several microseconds, i.e. the build-up time of the acoustic wave in a polarization converter, the wavelength for polarization conversion can be switched.

2×2 wavelength-selective switches form the basic building blocks of more complex switching nodes as required for optical crossconnects. By combining several such switches $n \times n$-switches can be realized. The architecture is strongly simplified in comparison to WDM nodes without wavelength-selective switches. As a first example a 4×4 acousto-optical switching node

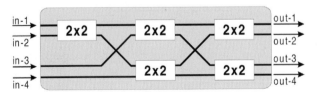

Fig. 7.48. 4×4 cross-connect using five acousto-optical 2×2 switches

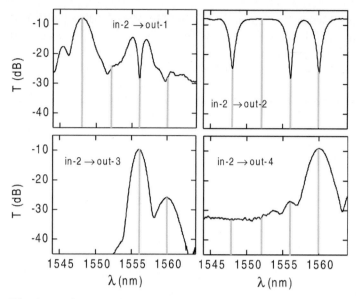

Fig. 7.49. Measured transmission characteristics of the 4 × 4 cross-connect for a specific configuration (see also text)

consisting of five 2 × 2 switches (Fig. 7.48) has been demonstrated [52]. To demonstrate the crossconnect performance broadband radiation has been coupled into input port 'in-2' (Fig. 7.49). The crossconnect has been driven to demultiplex four wavelength channels (grey bars in Fig. 7.49) separated by 4 nm to the output ports 'out-1' to 'out-4'. The worst-case crosstalk for this routeing configuration is 16 dB. Although the performance of this first prototype suffers from the limited crosstalk suppression of the individual components, the basic principle could be successfully demonstrated. By replacing the single-stage switches by dilated switches a significant improvement can be expected.

7.8.6   Outlook

Within the last few years a variety of integrated acousto-optical devices in LiNbO$_3$ has been developed taking advantage of acousto-optical polarization conversion and polarization filtering. Prototype wavelength filters and wavelength-selective switches/add-drop multiplexers have now reached a state of maturity such that commercial exploitation becomes possible.

However, there exist further challenges in improving the devices. It is of special interest to reach a higher state of integration. For instance, the formation of monolithically integrated dilated switches should provide an ideal basis for the construction of complex WDM cross-connects. Another task for future research is to develop devices with narrower spectral filtering to meet the requirements of dense WDM systems with 0.8 nm (100 GHz) channel spacing.

References

1. M. Born and E. Wolf, *Principles of Optics, 6th ed.* (Pergamon, Cambridge 1997)
2. I.P. Kaminow, P.P. Iannone, J. Stone, and L.W. Stulz, "A tunable vernier filter Fabry-Perot filter for FDM demultiplexing and detection," IEEE Photon. Technol. Lett. **1**, 24 (1989)
3. L.H. Laarhuis and A.M.J. Koonen, "Analysis and design of a high speed, wide range stepwise tunable optical filter/amplifier," *Proc. 19th Europ. Conf. Opt. Commun. (ECOC'93)*, Montreux, Switzerland, Vol. **2**, 261 (1993)
4. C.R. Giles, "Lightwave applications of fiber Bragg gratings," IEEE J. Lightwave Technol. **13**, 1391 (1997)
5. IEEE J. Lightwave Technol., Special issue on fibre gratings, **15**(8), August 1997
6. K. Aoyama and J. Minowa, "Low-loss optical demultiplexer for WDM systems in the 0.8 μm wavelength region," Appl. Opt. **18**, 2834 (1979)
7. A.M.J. Koonen, "A compact wavelength demultiplexer using both interference filters and a diffraction grating," *Proc. 7th Europ. Conf. Opt. Commun. (ECOC'81)*, Copenhagen, Denmark, paper 8.5 (1981)
8. J.P. Laude, J. Flamand, J.C. Gautherin, D. Lepere, P. Gacoin, F. Bos, and J. Lerner, "STIMAX, a grating multiplexer for monomode or multimode fibers," *Proc. 9th Europ. Conf. Opt. Commun. (ECOC'83)*, Geneva, Switzerland, 417 (1983)
9. Product sheet on WDM STIMAX wavelength multiplexers and demultiplexers, made by Jobin Yvon Instruments S.A.
10. M.K. Smit, "New focusing and dispersive planar component based on an optical phased array," Electron. Lett. **24**, 385 (1988)
11. H. Takahashi, S. Suzuki, K. Kato, and I. Nishi, "Arrayed-waveguide grating for wavelength division multi/demultiplexer with nanometre resolution," Electron. Lett. **26**, 87 (1990)
12. C. Dragone, "An $N \times N$ optical multiplexer using a planar arrangement of two star couplers," IEEE Photon. Technol. Lett. **3**, 812 (1991)
13. M.K. Smit and C. van Dam, "PHASAR-based WDM devices: principles, design and applications," IEEE J. Select. Topics Quantum Electron. **2**, 236 (1996)
14. Y. Tachikawa, Y. Inoue, M. Kawachi, H. Takahashi, and K. Inoue, "Arrayed-waveguide grating add-drop multiplexer with loop-back optical paths," Electron. Lett. **29**, 2133 (1993)
15. O. Ishida, H. Takahashi, S. Suzuki, and Y. Innoue, "Multichannel frequency-selective switch employing an arrayed-waveguide grating multiplexer with fold-back optical paths," IEEE Photon. Technol. Lett. **6**, 1219 (1994)
16. B.R. Hemenway, M.L. Stevens, D.M. Castagozzi, D. Marquis, S.A. Parikh, J.J. Carney, S.G. Finn, E.A. Swanson, I.P. Kaminow, C. Dragone, U. Koren, T.L. Koch, R. Thomas, C. Özveren, and E. Grella, "A 20-channel wavelength-routed all-optical network deployed in the Boston Metro area," *Conf. Opt. Fiber Commun. (OFC '95)*, OSA Techn. Digest Series, Postdeadline Papers, San Diego, USA, PD8-1, PD8-5 (1995)

17. O. Ishida, T. Hasegawa, M. Ishii, S. Suzuki, and K. Iwashita, "4 × 4, 7-FDM-channel reconfigurable network hub emplying arrayed-waveguide-grating (AWG) multiplexers," *Conf. Opt. Fiber Commun. (OFC '95)* Techn. Digest Series, Postdeadline Papers, San Diego, USA, PD9-1, PD9-5 (1995)

18. M. Fukui, K. Oda, H. Toba, K. Okamoto, and M. Ishii, "10 channel × 10 Gbit/s WDM add/drop multiplexing/transmission experiment over 240 km of dispersion-shifted fibre employing unequally-spaced arrayed-waveguide-grating ADM filter with fold-back configuration," Electron. Lett. **31**, 1757 (1995)

19. Y. Hida, Y. Innoue, and S. Imamura, "Polymeric arrayed-waveguide grating multiplexer operating around 1.3 μm," Electron. Lett. **30**, 959 (1994)

20. M.B.J. Diemeer, L.H. Spiekman, R. Ramsamoedj, and M.K. Smit, "Polymeric phased array wavelength multiplexer operating around 1550 nm," Electron. Lett. **32**, 1132 (1996)

21. H. Okayama and M. Kawahara, "Waveguide array grating demultiplexer on LiNbO₃," *Conf. Integrated Photonics Research (IPR'95)*, vol. **7**, OSA Techn. Digest Series (Opt. Soc. America, Washington, DC 1995), p. 296

22. K. Okamoto, K. Takiguchi, and Y. Ohmori, "16-channel add/drop multiplexer using silica-based arrayed-waveguide gratings," Electron. Lett. **31**, 723 (1995)

23. K. Okamoto, M. Okuno, A. Himeno, and Y. Ohmori, "16-channel optical add/drop multiplexer consisting of arrayed-waveguide gratings and double-gate switches," Electron. Lett. **32**, 1471 (1996)

24. M.R. Amersfoort, J.B.D. Soole, H.P. Leblanc, N.C. Andreakakis, A. Rajhel, and C. Caneau, "8 × 2 nm polarization-independent WDM detector based on compact arrayed waveguide demultiplexer," *Conf. Integrated Photonics Research (IPR'95)*, vol. **7**, OSA Techn. Digest Series (Opt. Soc. America, Washington, DC 1995), Postdeadline Papers PD3-1, PD3-2

25. M. Zirngibl, C.H. Joyner, and L.W. Stulz, "WDM receiver by monolithic integration of an optical preamplifier, waveguide grating router and photodiode array," Electron. Lett. **31**, 581 (1995)

26. C.A.M. Steenbergen, C. van Dam, A. Looijen, C.G.P. Herben, M. de Kok, M.K. Smit, J.W. Pedersen, I. Moerman, R.G.F. Baets, and B.H. Verbeek, "Compact low loss 8 × 10 GHz polarization independent WDM receiver," *Proc. 22nd Europ. Conf. Opt. Commun. (ECOC'96)*, Oslo, Norway, vol. **1**, 129 (1996)

27. A.A.M. Staring, C. van Dam, J.J.M. Binsma, E.J. Jansen, A.J.M. Verboven, L.J.C. Vroomen, J.F. de Vries, M.K. Smit, and B.H. Verbeek, "Packaged PHASAR-based wavelength demultiplexer with integrated detectors," *Proc. 11th Intern. Conf. Integr. Optics and Opt. Fibre Commun./23rd Europ. Conf. Opt. Commun. (IOOC-ECOC'97)*, Edinburgh, UK, **3**, 75 (1997)

28. S. Chandrasekhar, M. Zirngibl, A.G. Dentai, C.H. Joyner, F. Storz, C.A. Burrus, and L.M. Lunardi, "Monolithic eight-wavelength demultiplexed receiver for dense WDM applications," IEEE Photon. Technol. Lett. **7**, 1342 (1995)

29. M. Zirngibl and C.H. Joyner, "12 frequency WDM laser based on a transmissive waveguide grating router," Electron. Lett. **30**, 701 (1994)

30. M. Zirngibl, B. Glance, L.W. Stulz, C.H. Joyner, G. Raybon, and I.P. Kaminow, "Characterization of a multiwavelength waveguide grating router laser," IEEE Photon. Technol. Lett. **6**, 1082 (1994)

31. C.H. Joyner, M. Zirngibl, and J.C. Centanni, "An 8-channel digitally tunable transmitter with electroabsorption modulated output by selective-area epitaxy," IEEE Photon. Technol. Lett. **7**, 1013 (1995)

32. A.A.M. Staring, L.H. Spiekman, J.J.M. Binsma, E.J. Jansen, T. van Dongen, P.J.A. Thijs, M.K. Smit, and B.H. Verbeek, "A compact 9 channel multi-wavelength laser," IEEE Photon. Technol. Lett. **8**, 1139 (1996)

33. C.G.M. Vreeburg, T. Uitterdijk, Y.S. Oei, M.K. Smit, F.H. Groen, E.G. Metaal, P. Demeester, and H.J. Frankena, "First InP-based reconfigurable integrated add-drop multiplexer," IEEE Photon. Technol. Lett. **9**, 188 (1997)

34. C.G.P. Herben, C.G.M. Vreeburg, D.H.P. Maat, X.J.M. Leijtens, M.K. Smit, F.H. Groen, J.J.G.M. van der Tol, and P. Demeester, "A compact integrated InP-based single-PHASAR optical crossconnect," IEEE Photon. Technol. Lett. **10**, 678 (1998)

35. C.G.P. Herben, D.H.P. Maat, X.J.M. Leijtens, Y.S. Oei, F.H. Groen, P. Demeester, and M.K. Smit, "Compact integrated polarization-independent optical crossconnect," *Proc. 24th Europ. Conf. Opt. Commun. (ECOC'98)*, Madrid, Spain, **1**, 257 (1998)

36. H. Takahashi, K. Oda, H. Toba, and Y. Inoue, "Transmission characteristics of arrayed waveguide $N \times N$ wavelength multiplexer," IEEE J. Lightwave Technol. **13**, 447 (1995)

37. M.R. Amersfoort, J.B.D. Soole, H.P. Leblanc, N.C. Andreadakis, A. Rajhel, and C. Caneau, "Passband broadening of integrated arrayed waveguide filters using multimode interference couplers," Electron. Lett. **32**, 449 (1996)

38. K. Okamoto and A. Sugita, "Flat spectral response arrayed-waveguide grating multiplexer with parabolic horns," Electron. Lett. **32**, 1661 (1996)

39. K. Okamoto and H. Yamada, "Arrayed-waveguide grating multiplexer with flat spectral response," Opt. Lett. **20**, 43 (1995)

40. A. Rigny, A. Bruno, and H. Sik, "Multigrating method for flattened spectral response wavelength multi-/demultiplexer," Electron. Lett. **33**, 1701 (1997)

41. H. Takahashi, Y. Hibino, and I. Nishi, "Polarization-insensitive arrayed-waveguide grating wavelength multiplexer on silicon," Opt. Lett. **17**, 499 (1992)

42. Y. Innoue, Y. Ohmori, M. Kawachi, S. Ando, T. Sawada, and H. Takahashi, "Polarization mode converter with polyimide half waveplate in silica-based planar lightwave circuits," IEEE Photon. Technol. Lett. **6**, 626 (1994)

43. B.H. Verbeek, A.A.M. Staring, E.J. Jansen, R. van Roijen, J.J.M. Binsma, T. van Dongen, M.R. Amersfoort, C. van Dam, and M.K. Smit, "Large bandwidth polarization independent and compact 8 channel PHASAR demultiplexer/filter," *Conf. Opt. Fiber Commun. (OFC '94)*, Techn. Digest Series, (Opt. Soc. America, Washington, DC 1994), Postdeadline Papers, p. 63

44. H. Bissessur, F. Gaborit, B. Martin, and G. Ripoche, "Polarization-independent phased-array demultiplexer on InP with high fabrication tolerance," Electron. Lett. **31**, 1372 (1995)

45. J.B.D. Soole, M.R. Amersfoort, H.P. Leblanc, N.C. Andreadakis, A. Rajhel, C. Caneau, M.A. Koza, R. Bhat, C. Youtsey, and I. Adesida, "Polarization-independent InP arrayed waveguide filter using square cross-section waveguides," Electron. Lett. **32**, 323 (1996)

46. M. Zirngibl, C.H. Joyner, and P.C. Chou, "Polarization compensated waveguide grating router on InP," Electron. Lett. **31**, 1662 (1995)

47. C.G.M. Vreeburg, C.G. P Herben, X.J.M. Leijtens, F.H. Groen, J.J.G.M. van der Tol, and P. Demeester, "An improved technology for eliminating polarization dispersion in integrated PHASAR demultiplexers," *Proc. 11th Intern. Conf. Integr. Optics and Opt. Fibre Commun./ Proc. 23rd Europ. Conf. Opt. Commun. (IOOC-ECOC'97)*, Edinburgh, UK, **3**, 83 (1997)

48. M.G. Young, U. Koren, B.I. Miller, M. Chien, T.L. Koch, D.M. Tennant, K. Feder, K. Dreyer, and G. Raybon, "Six wavelength laser array with integrated amplifier and modulator," Electron. Lett. **31**, 1835 (1995)

49. C.E. Zah, F.J. Favire, B. Pathak, R. Bhat, C. Caneau, P.S.D. Lin, A.S. Gozdz, N.C. Andreakakis, M.A. Koza, and T.P. Lee, "Monolithic integration of multiwavelength compressive-strained multiquantum-well distributed-feedback laser array with star coupler and optical amplifiers,"Electron. Lett. **28**, 2361 (1992)

50. R. Monnard, A.K. Srivastava, C.R. Doerr, C.H. Joyner, L.W. Stulz, M. Zirngibl, Y. Sun, J.W. Sulhoff, J.L. Zyskind, and C. Wolf, "16-channel 50 GHz channel spacing long-haul transmitter for DWDM systems," Electron. Lett. **34**, 765 (1998)

51. D.A. Smith, R.S. Chakravarthy, Z. Bao, J.E. Baran, J.L. Jackel, A. d'Alessandro, D.J. Fritz, S.H. Huang, X.Y. Zou, S.M. Hwang, A.E. Willner, and K.D. Li, "Evolution of the acoustooptic wavelength routing switch," IEEE J. Lightwave Technol. **14**, 1005 (1996)

52. H. Herrmann, A. Modlich, Th. Müller, W. Sohler, and F. Wehrmann, "Advanced integrated, acousto-optical switches, add-drop multiplexers and WDM cross-connects in LiNbO₃," *Proc. 8th Europ. Conf. Integr. Optics (ECIO '97)*, Stockholm, Sweden, 578 (1997)

53. J. Frangen, H. Herrmann, R. Ricken, H. Seibert, W. Sohler, and E. Strake, "Integrated optical, acoustically tunable wavelength filter," Electron. Lett. **25**, 1583 (1989)

54. D.A. Smith and J.J. Johnson, "Low drive power integrated acousto-optic filter on X-cut, Y-propagating LiNbO₃," IEEE Photon. Technol. Lett. **3**, 923 (1991)

55. H. Herrmann and S. Schmid, "Integrated acousto-optical mode convertors with weighted coupling using surface acoustic wave directional couplers," Electron. Lett. **28**, 979 (1992)

56. D.A. Smith and J.J. Johnson, "Sidelobe suppression in an acoustooptic filter with a raised cosine interaction strength," Appl. Phys. Lett. **61**, 1025 (1992)

57. H. Herrmann, U. Rust, and K. Schäfer, "Tapered acoustical directional couplers for integrated acousto-optical mode converters with weighted coupling," IEEE J. Lightwave Technol. **13**, 364 (1995)

58. K.W. Cheung, S.C. Liew, C.N. Lo, D.A. Smith, J.E. Baran, and J.J. Johnson, "Simultaneous five-wavelength filtering at 2.2 nm wavelength separation using integrated optic acousto-optic tunable filter with subcarrier detection," Electron. Lett. **25**, 636 (1989)

59. F. Tian and H. Herrmann, "Interchannel interference in multiwavelength operation of integrated acousto-optical filters, switches," IEEE J. Lightwave Technol. **13**, 1146 (1995)

60. M. Fukutoku and K. Oda, "Optical beat-induced crosstalk of an acoustooptic tunable filter for WDM network application," IEEE J. Lightwave Technol. **13**, 2224 (1995)

61. J.L. Jackel, M.S. Goodmann, J.E. Baran, W.J. Tomlinson, G.K. Chang, M.Z. Iqbal, G.H. Song, K. Bala, C.A. Brackett, D.A. Smith, R.S. Chakravarthy, R.H. Hobbs, D.J. Fritz, R.W. Ade, and K.M. Kissa, "Acousto-optic tunable filters (AOTF's) for multi-wavelength optical cross-connects: crosstalk considerations," IEEE J. Lightwave Technol. **14**, 1056 (1996)

62. F. Tian, Ch. Harizi, H. Herrmann, V. Reimann, R. Ricken, U. Rust, W. Sohler, F. Wehrmann, and S. Westenhöfer, "Polarization independent integrated op-

tical, acoustically tunable double stage wavelength filter in LiNbO$_3$," IEEE J. Lightwave Technol. **12**, 1192 (1994)

63. F. Wehrmann, Ch. Harizi, H. Herrmann, U. Rust, W. Sohler, and S. Westenhöfer, "Integrated optical, wavelength selective, acoustically tunable 2 × 2 switches (add-drop multiplexers) in LiNbO$_3$," IEEE J. Select. Topics Quantum Electron. **2**, 263 (1996)

64. J.E. Baran and D.A. Smith, "Adiabatic 2 × 2 Polarization Splitter in LiNbO$_3$," IEEE Photon. Technol. Lett. **4**, 39 (1992)

65. U. Hempelmann, H. Herrmann, G. Mrozynski, V. Reimann, and W. Sohler, "Integrated optical proton exchange in LiNbO$_3$: modelling and experimental performance," IEEE J. Lightwave Technol. **13**, 1750 (1995)

66. B.I. Heffner, D.A. Smith, J.E. Baran, A. Yi-Yan, and K.W. Cheung, "Integrated-optic, acoustically tunable infrared optical filter," Electron. Lett. **24**, 1562 (1988)

67. D.A. Smith, J.E. Baran, K.W. Cheung, and J.J. Johnson, "Polarization-independent acoustically tunable optical filter," Appl. Phys. Lett. **56**, 209 (1990)

68. K.W. Cheung, D.A. Smith, J.E. Baran, and J.J. Johnson, "1 Gb/s system performance of an integrated, polarization-independent, acoustically tunable optical filter," IEEE Photon. Technol. Lett. **2**, 271 (1990)

69. T. Pohlmann, A. Neyer, and E. Voges, "Polarization independent Ti : LiNbO$_3$ switches and filters," IEEE J. Quantum Electron. **27**, 602 (1991)

70. H. Herrmann, K. Schäfer, and Ch. Schmidt, "Low-loss tunable integrated acousto-optical wavelength filter in LiNbO$_3$ with strong sidelobe suppression," IEEE Photon. Technol. Lett. **10**, 120 (1998)

71. D.A. Smith, A. d'Alessandro, J.E. Baran, D.J. Fritz, and R.H. Hobbs, "Reduction of crosstalk in an acoustooptic switch by means of dilation," Opt. Lett. **19**, 99 (1994)

72. J.L. Jackel, M.S. Goodmann, J. Gamelin, W.J. Tomlinson, J.E. Baran, C.A. Brackett, D.J. Fritz, R. Hobbs, K. Kissa, R. Ade, and D.A. Smith, "Simultaneous and independent switching of 8 wavelength channels with 2 nm spacing using a wavelength-dilated acoustooptic switch," IEEE Photon. Technol. Lett. **8**, 1531 (1996)

73. H. Herrmann, A. Modlich, Th. Müller, and W. Sohler, "Double-stage, integrated, acousto-optical add-drop multiplexers with improved crosstalk performance," *Proc. 11th Intern. Conf. Integr. Optics and Opt. Fibre Commun./23rd Europ. Conf. Opt. Commun. (IOOC-ECOC'97)*, Edinburgh, UK, **3**, 10 (1997)

74. S.F. Su, R. Olshansky, G. Joyce, D.A. Smith, and J.E. Baran, "Gain equalization in multiwavelength lightwave systems using acousto-optic tunable filters", IEEE Photon. Technol. Lett. **4**, 269 (1992)

75. S. Schmid, S. Morasca, D. Scarano, and H. Herrmann, "High-performance integrated acousto-optic channel analyzer", *Conf. Opt. Fiber Commun. (OFC '97)*, Techn. Digest Series, (Opt. Soc. America, Washington, DC 1997), Vol. **6**, 7 (1997)

8 Optical Switching

R. Ian MacDonald, Ken Garrett, Philip Garel-Jones,
Winfried H.G. Horsthuis, and Edmond J. Murphy

8.1 Introduction

Controlling the route light takes through a set of alternative fibre pathways
is a basic photonic function. Optical switching is required in all aspects of
photonics, from the manufacture of basic components to the operation of
large-scale telecommunications networks. Recent developments are placing
a new importance on switching.

In the first optical communications systems a fibre carried one optical
signal. To route signals it was only necessary to determine the path taken by
all the optical power carried on each fibre. In the last few years wavelength-
division multiplex (WDM) transmission has become very important. With
WDM a single fibre carries many optical signals on different wavelengths
that can be separated by passive optical filters. The wavelength channels of
a single fibre are almost equivalent to separate fibres running in parallel.
The connectivity of wavelength channels in WDM networks is more complex
than the connectivity afforded by the physical layout of fibres, because the
wavelength channels carried toward a switch on one fibre may leave it on
several others. Since many tens of WDM channels may be carried on a single
fibre, there is a large increase in the number of possible optical routes and
a corresponding requirement for optical routeing functions.

Optical switching is now widely viewed as a very important area for
development. Very high optical performance has already been established
in switches, setting major hurdles for the introduction of new technologies.
New technologies, on the other hand, may be needed for the applications now
emerging. We review here the status of optical switching, focusing on technologies and applications that have acquired or show promise of commercial
acceptance. Because applications determine requirements, we commence with
a review of applications in which optical switches have achieved commercial
success or the near-term promise of it.

8.2 Applications

8.2.1 Optical Component Characterization and Testing

Optical fibre switches currently find a wide variety of applications in the
testing and characterization of components and systems, both for manufac-

turers and customers. Optical systems are becoming increasingly complex and sophisticated; hence the components used in these systems need to meet more demanding performance and reliability specifications than ever before. Manufacturers need to be able to demonstrate the acceptability of their products under increasingly severe conditions and considerable time and money must be invested to accomplish this. Many customers also choose to perform their own product performance tests before placing large orders, and have routine incoming inspection programs in place to ensure the ongoing quality of devices they are purchasing.

Guidelines for the kind of performance and reliability testing that need to be met for particular classes of components are published in documents such as those from Telcordia in the USA. These include performance requirements and objectives for various parameters of an optical device that should be met under a wide variety of environmental conditions. Some of the required tests last as long as 5000 hours, (approximately 6 months), therefore it is important to ensure that the testing procedure and test equipment itself is reliable so that the data obtained are as accurate as possible.

Switches are frequently used to connect many devices to a single test set, thus minimising the capital-equipment cost and allowing for a much more efficient and controllable method of testing multiple devices. A fairly typical basic test system as recommended in Telcordia GR-326-CORE is shown in Fig. 8.1. The importance of optical switches in such apparatus is evident.

A more versatile test set is shown in Fig. 8.2. This is the Optical Component Evaluation Test System (OCETS) available from JDS Uniphase Corp. Such a test set allows measurements to be made on attenuation, return loss, isolation, directivity and polarization-dependent loss (PDL), and can be expanded to include other measurements such as wavelength dependence. In this example the main routeing elements are the two large multiport switches, between which over 100 devices under test (DUTs) can be inserted. The switches allow individual input or output ports of a multiport DUT to be accessed as required for any given test. Other switches can be used to

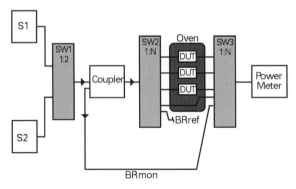

Fig. 8.1. Switching for device test

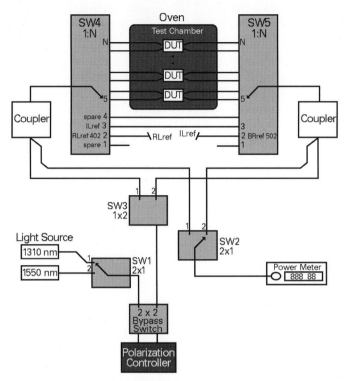

Fig. 8.2. Optical Component Evaluation Test System

select from a number of different optical sources and to route this light along a chosen path and direction to the device to be tested. Switches used in such test systems are required to have excellent optical characteristics because poor performance on loss, return loss, crosstalk or polarization dependent loss (PDL) in the switch, or indeed any other component in line with the DUT, will limit the dynamic range of the test system for these parameters. Repeatability, particularly for the 1xN switches is also very important because any change in the connection to the DUT from one measurement cycle to another can be mistakenly attributed to a change in the DUT itself. Control software with these test sets allows complex test sequences to be carried out on large numbers of devices over long periods of time with capture and storage of the data for later display and analysis. System reliability is obviously vital because of the time and expense involved in the procurement and preparation of test samples, setting up the test programme, and tying up the equipment. It is disastrous to have a 5000 hour test fail after 4500 hours and have to be repeated. The performance and reliability demands on switches used in test systems such as OCETS are met only by electromechanical switch technology at present.

8.2.2   Test Access

Optical fibre-based communication systems have become ever more reliable over the years, due to improved component reliability and also to the system designs which use various protection and self-healing techniques to minimise the impact of failures in any part of the system (see also Chap. 1). Also over the years the information-carrying capacity of individual fibres has been pushed higher and higher, making the impact of a non-recoverable failure even more disastrous. One of the keys to providing reliable service in any industry is good maintenance and regular monitoring of the condition of the physical plant. In the case of fibre optic systems one of the causes of failure can be degradation in the optical transmission of the fibre itself. Modern optical fibres are virtually inert and transmission changes in the fibre would only occur under the most severe conditions of environmental stress on the fibre installation. In most cases transmission degradation is caused by inadvertent physical damage to the fibre or fibre cable, more often than not as a result of construction work.

Various monitoring schemes have been used to provide an indication of the condition of installed fibre. Early systems had the benefit of being able to transmit monitoring signals down unused, ('dark'), fibres in a cable. Whilst this technique was not able to monitor fibres that were being used for active transmission, some peace of mind could be gained by knowing that one or two fibres in a cable were behaving themselves and that this provided at least an indication about the others. Early monitoring was performed by manually connecting directly to the fibres to be tested. Both continuous wave and subsequently optical time-domain reflectometry (OTDR) were used. Soon multiport optical switches were employed to route the OTDR to the fibre under test allowing more efficient measuring to be done.

Active fibre monitoring was made possible by using an out-of-band wavelength from the OTDR to inspect the fibre without interfering with the normal traffic being carried by the fibre. Typically a 1550 nm OTDR is used for 1300 nm traffic and vice versa. These systems require that a WDM coupler is installed as an access point for the OTDR signal and also a blocking filter is used to protect the receiver from the OTDR monitoring pulses.

In the last few years monitoring systems have progressed to the point where, by using a single OTDR and multiple routeing switches to direct the OTDR signal to the desired fibre link, monitoring can now automatically provide frequent updates on the state of a vast number of fibre lines. The out-of-band monitoring wavelength has now been pushed out to the 1625 nm region, deemed less usable for traffic, at least for the present time. These systems store a reference profile for the particular fibre being tested and continuously compare its present state to that original signature. Changes detected outside of pre-set limits will trigger more detailed investigation or cause alarms to be sent to appropriate systems or to personnel who can act immediately to remedy the problem depending on its severity. Knowing the exact location of the fault in the fibre and the installed route followed by the fibre permits a map reference to be automatically sent to field personnel, who can use the Global Positioning System (GPS) to pinpoint the location in the field.

Multiport switches used in these recent systems must meet lower insertion loss and higher return loss specifications, as they are frequently cascaded together to allow access to a larger number of fibres and the reach of the OTDR has to be maintained as much as possible. There are also more stringent requirements on the physical size of the switches as increasing numbers of switches occupy premium space. The shift of the operating wavelength to the 1625 nm region also means that more care has to be taken in routeing fibre within the switch to avoid small bend radii, since the mode is more weakly bound to the fibre core at the longer wavelength.

8.2.3   Telecommunications

Optical switches are now used to route telecommunications signals, performing the same physical function as the circuit switches in early telephone networks but on much more expensive traffic. It is foreseeable that many hundreds of optical signals, each bearing data at 10 Gb/s or even more, will be handled by an optical router. Switch reliability and performance are critical, and many issues remain unresolved. Optical circuit switches make physical connections and there is also something of a mismatch between this analogue nature of optical switches and the basic digital nature of the network. The importance of this mismatch is indicated by the considerable effort now being expended to establish the possibility of optical packet switching [1]. The functions required are very difficult to achieve with light signals. Because optical packet switching is unlikely to be practical in the foreseeable future this review considers only optical circuit switches.

Optical circuit switches, 'switching matrices', are constructed by assembling component optical switches that perform relatively simple functions. Most component switches fall into one of two classes: those that can deliver one input signal to N outputs ($1 \times N$ switch), and those that can accept two inputs and either carry them through in parallel or exchange their routes (2×2, or 'exchange-bypass' switch). For the cost of the insertion loss of a single device, the $1 \times N$ switch impresses $\log_2 N$ bits of routing information. To get the same result with 2×2 switches, the signal undergoes a loss (in dB) of ($\log_2 N$) times the loss of a single switch. Since the loss of a single 2×2 switch is not intrinsically smaller than that of a $1 \times N$ switch, a considerable benefit can be gained with the $1 \times N$ switching function. Figure 8.3 shows how switching matrices can be assembled from these two types of element.

The matrix based on $1 \times N$ switches shown in Fig. 8.3a is of the 'switched distribution, switched recombination' (SDSR) type. This form is commonly used with switches based on mechanical elements such as stepping motors. The $1 \times N$ switching function required in the SDSR type of matrix can be

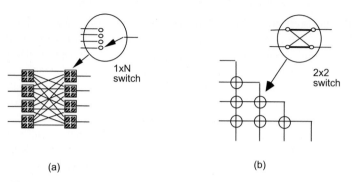

(a)   (b)

Fig. 8.3. Basic architectures for optical switching matrices: (**a**) switched distribution, switched recombination type, (**b**) exchange-bypass switching matrix ('crossbar switch')

obtained with trees of 1×2 elements (often 2×2 with one port unused). Where loss is not an important issue the $1 \times N$ switches in the column at the input or output of the SDSR can be replaced by passive $1 \times N$ power splitters. These matrices (passive distribution, switched recombination (PDSR) or vice versa) can offer broadcasting or incasting, as well as lower cost.

The matrix of exchange-bypass elements (Fig. 8.3b) is often called a 'crossbar switch', although it is not the same as an electronic crossbar switch. This architecure is usual for matrices assembled from integrated optical components since the 2×2 function is naturally obtained with directional couplers or Mach-Zehnder interferometers. Fast solenoid-driven optomechanical switches also are often of this type.

Both matrices shown in Fig. 8.3 are 'strictly non-blocking', which means that a signal arriving at any input port can be delivered to any free output port, irrespective of what other connections exist. Other designs may allow blocking or may be of the 'rearrangeably non-blocking' type, in which any connection can always be made, but could require the rearrangement of existing connections. Rearrangeable switches are probably no more useful than blocking switches for the purposes of optical networks, because the switching times of most optical elements are very slow compared to the bit times of the signals they carry. It is very difficult to synchronize optical switching with data and thus signals on existing connections would be significantly interrupted during a reconfiguration.

Initially, optical switches in telecommunications were used primarily in 'bench top' applications for research and development purposes. As data communications and telecommunications providers began to adopt fibre as the preferred choice for a high bandwidth transport medium requirements developed for optical switching in networks. Initial implementations focused on network testing and monitoring; optical route switching developed later.

Optical switches were first introduced in data communications networks for protection, to provide route diversity in the event of a network failure (see Chap. 1, Sect. 1.5). These initial networks used 1×2 and 2×2 switches to bypass potential problem areas and keep mission critical networks working. Telecommunications usage of fibre optics remained largely in point-to-point networks through the early 1990s and requirements for optical network switches were minimal. The emergence of WDM technologies and the ever-increasing need for bandwidth has seen fibre emerge as the primary transmission medium in both local and long-haul telecommunications networks, raising new networking requirements and new applications for optical-switching applications. In addition to test access as discussed before and protection switching (see below), these also include traffic management and direct routeing of broadband circuits.

Protection Switching. Fibre optic networks are susceptible to two types of failure: problems with the equipment transmitting or receiving the light; and path interruptions, generally as a result of a fibre or cable break. Equipment or individual fibre failures can be addressed by a second, unused system to which an optical switch can direct traffic if a problem is encountered. This type of solution, 1:1 protection, requires installing redundant transmission hardware for each fibre and becomes unnecessarily expensive as the numbers of fibres increase (see Chap. 1, Sect. 1.5).

Optical matrix switches can be used to construct cross-connect capability at the optical level. Optical cross-connects have recently been introduced to protect against multiple fibre or overall cable breaks. These systems offer service providers the capability to map out several contingency routes depending on the location and the magnitude of the cable break. An optical cross-connect can share redundant systems on a 1:N basis, thereby reducing additional equipment costs associated with network protection. They may also provide a back-up for existing recovery mechanisms in ring or Synchronous Optical Networks (SONET) by offering the ability to reconstruct the network in the event that more than one failure or break occurs within the ring. Interference between SONET recovery mechanisms and restoration at the optical level, however, raises many issues. Primary among these is the speed of the optical switch. The electromechanical optical switching technologies currently deployed usually do not recover the physical network fast enough to be invisible to the SONET layer. On the other hand, other switch technologies may not presently provide adequate optical performance.

All of the above applications focus on the management of the physical facilities of networks rather than management of fibre optic transmission. Optical switching may, however, find its most important application in the routine routeing of transmissions in networks.

Optical Network Switching. It was noted that WDM greatly increases the number of physical optical signals carried by a single fibre. With current technology more than 100 wavelengths having individual routeing requirements

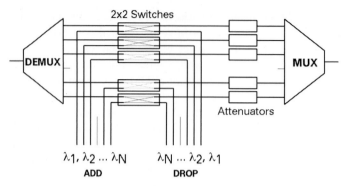

Fig. 8.4. Configurable add/drop multiplexer

may be carried. In such systems it is necessary that wavelength channels be added to the multiplexed optical signal or dropped from it at a particular location, creating a need for flexible optical add-drop multiplexers. Flexible WDM add/drop units using wavelength multiplexer/demultiplexer pairs with a 2×2 switch for each wavelength, such as that shown in Fig. 8.4 are now available. Usually only a few channels need to be accessed at a particular location, and it is advantageous only to demultiplex those that may be required.

The scenario becomes even more complex when multiple network nodes are interconnected by multiple fibre-optic cables with each fibre containing multiple wavelengths. It will be advantageous to allocate transmission capacity where it is needed. Traffic in optical networks with wavelength-division multiplexing can be managed by an optical cross-connect with access to an overall network management control system. It should be noted that optical cross-connects provide a universal optical management point that is not limited by bandwidth or dependent upon a particular transmission protocol.

As the demand for bandwidth increases, it is expected that individual customers will develop requirements for the entire capacity of a single optical channel. This capacity is likely to be at the OC-48 (2.488 Gb/s) level, the OC-192 (9.952 Gb/s) level, or even higher levels of the optical transmission hierarchy. In such a situation optical routers will be required to deliver fully loaded optical channels directly to their destinations. Networks with the capability to do optical switching in this sense become completely independent of the signal format. A customer who has purchased a physical channel would have the freedom to transmit non-standard signals. At the same time, the network operator retains control over the physical deployment of connections through the optical switching system.

Optical Routeing Switches. It is not clear at present how large an optical matrix switch will be required to meet the routeing demands of WDM systems. However, with individual fibre capacities exceeding 100 wavelength channels it is not unreasonable to speculate that matrix dimensions of 1000×1000 could be required. To obtain such large routers it will be necessary to assemble multiple matrix stages. Multistage switching is not uncommon in optics. Indeed the exchange bypass matrix shown in Fig. 8.3b may be considered itself to be a multistage matrix formed from individual 2×2 submatrices. Blocking often becomes a difficult architectural issue in multistage matrices. For example, many common arrangements of 2×2 switches are blocking to some degree. Strictly non-blocking multistage architectures such as the Clos design may become increasingly important as the requirement for large optical switches develops.

An advantage of WDM is that optical signals can travel in opposite directions on the same fibre. Optical routers provide continuous optical paths and therefore establish bidirectional paths for such signals. However it cannot be inferred that non-blocking, transparent optical switch matrices always provide the connectivity required in bidirectional optical systems. For example, the evolution of network traffic tends toward greater asymmetry. WDM provides a possible remedy in that some of the wavelength channels on a fibre could be assigned in either direction in response to asymmetric demand. (Not all optical transmission systems will allow this.) The result at the switches is that input and output ports become indistinguishable. The switching requirement is to connect pairs of ports together bidirectionally in any combination. This requirement is different from the switching function normally performed by optical switches. The point is illustrated in Fig. 8.5. The customary optical switch matrix has two sets of ports which are situated 'in line' in the transmission path. Ports from one set can be bidirectionally connected to ports from the other, but the ports in each set cannot be interconnected among themselves. The function required in the asymmetric bidirectional system, on the other hand, is to provide any interconnection among all paths that terminate on it. Since there is no grouping of ports into two disjunct sets corresponding to opposite sides of the switch, we refer to this type of switch as 'one-sided'.

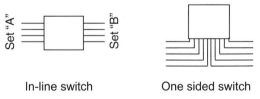

In-line switch   One sided switch

Fig. 8.5. Bidirectional optical switches

8.3   Technologies

Many different optical switches based on a wide variety of principles have appeared in the engineering literature. Given the ingenuity applied to the invention of optical switches it is remarkable that most of them in current commercial use are mechanical and employ principles not very different from those of the electromechanical switches of early telephone systems. Optomechanical switches have prevailed because light is a precious commodity. They exhibit lower loss, lower polarization dependence and lower spectral dependence than any other type of optical switch. In recent times alternative technologies have appeared specifically to address the perceived shortcomings of optomechanical switches, i.e. their size, weight, and perhaps reliability, without relinquishing their advantages.

Most optical matrix switches attempt to deliver all the light arriving on an incoming path to a selected outgoing path. Alternatively it is possible to deliver an optical signal to its selected output by distributing the incoming light to all possible outputs and discarding it at all but the one required. In such 'gate' switches only a portion of the optical power is delivered. Since loss is very significant in optical systems, amplifiers must usually be included. There has been extensive research into the use of semiconductor amplifiers themselves as the gates, since these amplifiers, when not forward biased to provide a population inversion, absorb the light they would amplify. While interesting results have been obtained with gate switches (e.g. the 8×8 matrices developed by Optivision Inc.), costs are high and they are not in widespread commercial use as routers. We do not consider them further here.

Very broadly, the switches that make up optical routers fall into two types: interferometric and non-interferometric. Interferometric switches rely on phase relationships among optical paths. Phase control is typically achieved through the Pockels effect or by thermal means. Such devices are very sensitive to their environment, particularly to ambient temperature, and they have a cyclic response to the control signal which generally necessitates monitoring the optical output to maintain a desired state. Directional couplers are typical interferometric switches.

Non-interferometric routers can be made in a wide variety of ways. In general they have a single-valued response to the control signal and usually do not require feedback to establish a state. They may offer the further advantage of latching. The term 'latching' has two meanings. From the point of view of the switch designer it is useful to have a switch action that is stable, i.e. the switching function saturates into the desired states. There is no need to provide feedback to hold a particular switch state, and ideally no need to bias the switch. From the point of view of the network administrator, however, a stronger form of latching is required: the switches must hold their states in the event of interruption of power. Devices with states that are determined, for example, entirely by electromagnets would not latch under this definition, if a power failure interrupted the magnetic field. The latter use of the term 'latching' is the common one.

Non-interferometric devices tend to be less sensitive than interferometers to polarization, wavelength, temperature and other influences that may be hard to control. Insensitivity also manifests as a disadvantage. Many non-interferometric switches require more power and time to activate than do interferometers.

The dynamic range of the switching function (or switch contrast) can be very high for non-interferometric switches in which, for example, a beam may be physically interrupted. In interferometers, on the other hand, dynamic range depends on accurate balance in the optical power of interfering beams and is generally lower and harder to maintain.

We can generalize that non-interferometric switches are useful primarily in applications that require good optical performance and latching, but not speed, whereas interferometric switches are better adapted to high-speed applications. The near-term prospective markets for optical switching in such areas as cross-connect, restoration, LAN reconfiguration, and traffic engineering are circuit-based rather than time-based. It is generally accepted that switching times of the order of ms are acceptable, but latching is often a requirement. Requirements for submillisecond speed in optical switching relate primarily to longer term developments such as optical packet routeing. It seems that non-interferometric switches should be the major area of commercial development and this observation is borne out by the commercial offerings, which are primarily mechanical. In the research literature, however, there is an emphasis on speed, integration, and other issues that point to the future of optical switching in telecommunications.

8.3.1 Non-interferometric Switches

Non-interferometric routers can be made in a wide variety of ways. In general they latch in both senses described above, and usually do not require feedback to establish a state. Optomechanical and some thermo-optic switches are of this type.

Optomechanical Switches. Optomechanical switches route light through physical displacement of the path of a light beam. These switches rely on mechanical actuators such as electronic relays, stepper motors and piezoelectric elements. Optomechanical switches can offer unsurpassed performance on optical parameters such as loss, crosstalk, dispersion, and other spectral and polarization dependence. Other technologies compete only on issues of size, switching speed, and reliability.

The most common approach to optomechanical switching uses a lens, typically of the gradient-index type, which collimates the light arriving on an input fibre to form a beam propagating in free space. The collimator may be moved or the collimated beam may be intersected by a mirror or prism to direct the light to a second lens that focuses the beam into an output fibre. Mirror-based systems are normally used in switches that require 1×1, 1×2 or 2×2 routeing. (The 1×1 designation refers to an interrupter or 'gate'.) Prism motion and physical motion of collimators are typical of $1 \times N$ switches of larger dimension. Prism motion is used for switches up to about 1×8, whereas motion of the collimator is normally used for slower switches of dimension up to 1×100.

When the mirror is not present the beam couples to one of the output ports. When the relay is activated the mirror is positioned very precisely to reflect the collimated beam to a second output port adjacent to the input port. Very high precision is required, because the angular tolerances for coupling collimated beams into monomode fibres with low loss are very tight. Switching is normally executed in $5-10\,\mathrm{ms}$, which includes the optical transition time and the time to energize the actuators and insert the device into the light path. Loss of the order of $0.5\,\mathrm{dB}$ is typical in single-mode systems. The approach is simple and offers a reliable way to perform the 1×2 switching operation. If the mirror is two-sided, a second input port can also be included to provide the 2×2 operation; however, the thickness of the mirror becomes an important issue in the alignment of the ports.

The performance of the actuator is a very important issue. The engineering effort that has been put into enhancing the reliability of relays over many years makes these the natural choice. When appropriate relays are used, these switches have the ability to latch in the last position without the need for constant powering. The switch position can be monitored by means of contacts that are closed when the mirror moves. A 2×2 solenoid-operated mirror switch can be made small enough that its electrical connections mount in a DIP socket ($4.8 \times 1.8 \times 0.9\,\mathrm{cm}^3$, JDS Uniphase Series SN).

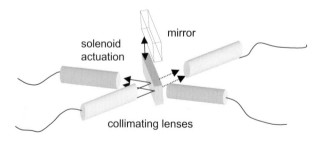

Fig. 8.6. Mirror-based optomechanical switch

A related type of switch uses a prism or series of prisms to route collimated beams of light. These offer many of the advantages of the mirror-based
switches, but can address applications requiring larger configurations, up to
about 1×16, because the prisms can be used in combination. Prism-based
switches may be slightly slower than mirror-based switches because the mass
of the prism is larger than the mass of a mirror.

In larger $1 \times N$ optomechanical switches, stepping motors are used to align
the collimated signal beam to a lens focusing into the desired output fibre.
Usually the stepping motors are rotary. A moving 'arm' carries a collimator
gradient-index lens attached to the input fibre around a disc in which multiple
gradient-index focusing lenses are fixed in an arc.

The stepping motor moves the arm to align the selected lenses. This
method depends on precise alignment of the arm with the disc to insure
optical performance. A schematic diagram is shown in Fig. 8.7.

Rotary optomechanical switches are commonly used in $1 \times N$ switching
applications where N is larger than 8, and can be up to 100 ports. Since

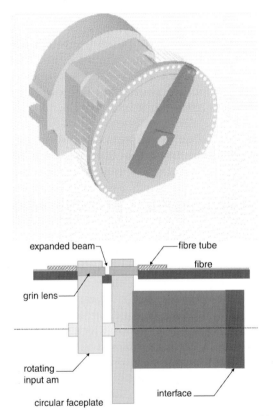

Fig. 8.7. Optomechanical rotary fibre switch. *Above*: configuration, *below*: mechanism

the optical system for a connection consists only of two lenses, without even a mirror or prism, the optical performance is normally high: low insertion loss, low polarization-dependent loss and high return loss are available for single-mode operation. As an example, performance data of a commercial device (JDS Uniphase SP series) for up to 1×100 capability are $0.7\,\mathrm{dB}$ insertion loss, $0.03\,\mathrm{dB}$ PDL and return loss exceeding $50\,\mathrm{dB}$ ($14 \times 2.8 \times 14\,\mathrm{cm}^3$ package size).

Resetting the position of a stepper motor switch requires stepping through all locations between the original position and the new one. Since step times between positions are about $15\,\mathrm{ms}$, the resulting delays can be hundreds of ms in large switches. As a consequence, switches based on stepping motors are not suitable for protection applications that require switching in time scales of $10\,\mathrm{ms}$, such as SONET-invisible restoration. However, because of their excellent optical performance, they can be the best choices for optical test applications and for network applications such as facility management and installation that do not require high speed.

Optical switches performing the $1 \times N$ selection function have been developed that use two position actuators rather than a stepping motor. These devices combine the speed of a relay-based switch with output port numbers more typical of stepper motor-based switches. Typically such switches use multiple mirrors or prisms to deflect a collimated beam into a focusing lens corresponding to each output. The optical path may be designed so that a single deflecting element is introduced for each port, giving low loss, or so that combinations of deflecting elements are used, reducing the number of mechanical components.

Extending the concept of combined deflecting elements leads to the 'binary switch' concept, in which light passes through a combination of M switching devices to obtain a 1×2^M switching action. Binary switches are among the most recent commercial offerings for $1 \times N$ switching. One such device, shown in Fig. 8.8a, uses transmission through glass plates to deflect a beam to the left or right, depending on the angular position of the plate (AMP Inc). Each successive plate has twice the thickness of the previous one, so that binary combinations of lateral deflection can be formed. In another type, shown in Fig. 8.8b, the combination of switches leads to a binary summation of deflection angles. In this device a glass block is moved into close proximity with a glass surface to actuate the switch. A beam is incident on the surface from within the glass. When the block is present, light transits the interface and reflects from the rear surface of the block, which is polished at an angle to its contacting surface. When the block is moved out of contact, the light reflects from the interface by total internal reflection (Optical Switch Corp).

The largest optical switching matrices currently available for single-mode fibre (for example the 32×32 JDS Uniphase SG 2000) use stepping motor switches and the SDSR architecture. Current $1 \times N$ component switch sizes

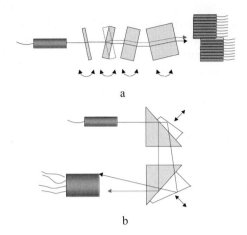

a

b

Fig. 8.8. Binary switches: **(a)** transmissive plate, **(b)** total internal reflection

would allow for matrices of dimension up to 100×100. Since an $N \times N$ switch requires N^2 fibre connections between the switches, matrices of this dimension are not currently available. The practical limit for the dimension of SDSR matrices that use $1 \times N$ stepper motor switches is probably nearer 32×32 than 100×100. However, with switches of high optical perfomance it is feasible to pass the signal through several stages of switching in tandem. A 32×32 submatrix suffices to make a non-blocking 3-stage switch of dimension 512×512 in the Clos architecture. No such matrix is currently available. The practical limit is the overall cost and the physical size of the multistage matrix.

Another type of optomechanical switch technology directly provides a matrix switch as the basic component. These systems use piezoelectronic actuators in a chamber to move fibres in both 'X' and 'Y' planes, directing the light from an array of N input ports to one of M output ports. Each moving fibre is positioned in front of a stationary lens. By moving fibres individually, the collimated light beam emerging from any lens of the input array can be directed to a particular point in a second array of lenses that refocus the beam into one of the output fibres. Both the input and output ports are monitored through a servo-control mechanism which continually aligns all the beams to ensure maximum throughput. This active alignment and management of the light beams allows for a modular switch that can be expanded simply by adding more collimators, actuators and lenses. As long as the required beam deflections remain within the capability of the system it is only necessary to update the software to add ports.

The 'beam steering' technique eliminates the N^2 scaling of interconnection costs associated with fibre-based SDSR switches because the corresponding connections in the beam-steering switch are in free space. Because of the need for an active servo-control system to maintain the optical connection, this type of switch does not latch in the event of a power failure. Beam-steering matrices are available for dimension up to 72×72 (Astarte 7200 series). Difficulties in keeping the beams aligned, and sensitivity to vibration restricts these switches to multimode operation at present.

Over the past fifteen years, optomechanical switches have been the primary method of actively routeing optical signals in fibre networks. Although these devices exhibit excellent optical performance their long-term reliability has been questioned. By their very nature, they rely on the physical movement of an element to route the light. These elements and their associated actuating devices can be subject to environmental stresses as well as wear resulting from prolonged use and high number of switching cycles. The most difficult question to switch suppliers regarding the use of optomechanical switches in networks, is: 'How can you ensure that if the switch remains in one position for several years it will switch when required ?' To respond to this challenge, suppliers are now offering switches that are hermetically sealed and built to withstand shock, vibration, and temperature extremes, as well as using improved actuator elements which have a proven track record of working under prolonged and intensive use. The actual reliability of optomechanical switches in the field has so far been excellent. In a field deployment of more than 5000 mechanical switches (JDS Uniphase) there have been no failures up to the present, three years into the project.

Micromechanical Optical Switches. Microelectromechanical switching techniques (MEMS) have become a very promising approach to the future development of switching components for optical fibre systems. MEMS devices are mechanical elements that are fabricated by a variety of photolithographic techniques that stem from microelectronics. The intention is to reduce the cost of mechanical devices by parallel fabrication or integration, and to raise their speed and reliability. Many micromechanical optical switches have been reported. There are two basic approaches. Light may be carried through the micromechanical device on a waveguide or a fibre that is attached to the micromechanical device, or the micromechanical element may serve to direct optical beams propagating in free space. Micromechanical optical switches are currently at the research or development stage. Commercial offerings of some types are probably not far off.

An $N \times N$ crosspoint switch matrix can be implemented very directly using collimated beams and micromechanical movable mirrors at the crosspoints that normally lie on the substrate, out of the beams, but can be are flipped up to intercept a beam and redirect it. Using micromechanical foundry processes, such a switch has been implemented recently, as illustrated in Fig. 8.9 [2] and exhibited switch times in the $100\,\mu s$ regime. Such switches are not restricted to coupling one set of beams to another set propagating at right angles. For example three sets of beams arranged at 60 degree intervals can be switched by the device. It is also possible to use both the front and the back side

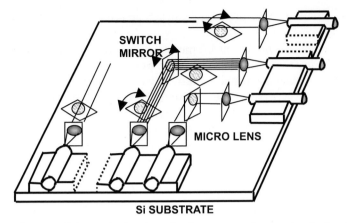

Fig. 8.9. Schematic drawing of a free-space micromechanical optical switching matrix using movable mirrors [2]

of the mirrors to deflect beams. In all cases the individual devices act as $N/2(1 \times 2)$ or $N/2(2 \times 2)$ switches, where N is the number of sets of beams to be coupled. While such an architecture leads to a blocking switch, methods have been discovered to assemble such switches into a multistage switch that is non-blocking [3].

Micromechanical switches can also make use of free space propagation in a manner similar to the beam-steering switch described above. Such switches would be based on planar arrays of micromirrors such as those made by Texas Instruments Inc. that perform a $1 \times N$ switching function. The micromachined mirrors normally lie parallel to the substrate but can be tilted out of this plane to reflect an incident beam to one of multiple output positions.

Various waveguide and fibre-based micromechanical switches have also been built. Fibre-based micromechanical devices use the micromechanics to create an actuator that moves a fibre. These require a more complex assembly than is needed with purely integrated optical devices, but generally have good optical performance that reflects the fibre properties. A $1 \times N$ fibre switch has been reported using two v-groove arrays laterally translated on silicon microstages [4].

By forming a cantilever that carries an integrated optical waveguide, and flexing the cantilever in order to interrupt or line up the waveguide with other waveguides carried on the same substrate, a very robust and inexpensive switch can be made. The cantilever is formed by depositing silicon oxide on silicon, photolithographically forming an opening, and etching the silicon away with an etch that does not attack the oxide. Undercutting by the etch frees the protrusion to form a cantilever projecting over a pit etched in the silicon. A typical problem is residual stress in the oxide which causes the cantilever to bend out of the substrate plane. The growth of low-stress oxides and inclusion of a stiffening structure with the cantilever are methods that have been used to eliminate this problem, which now seems to have been solved.

The cantilever switch can be driven by an integrated electrostatic linear motor (comb-drive). Electromechanical instability forces the comb-drives to fixed alignment stops, achieving rapid switching and a low drive voltage, (± 30 V). (As the cantilever moves, the force on it increases; this process proceeds until it hits the stop.) Relatively high losses are still characteristic of such switches due to the difficulty of maintaining a small gap between the end of the cantilever and the pick-up guides.

Waveguide-based micromechanical switches have the potential (in most cases) for integrating several switches together in the microfabrication step, thus minimising fibre pigtailing. Waveguide-based micromechanical switches are, however, at an earlier stage than fibre-based micromechanical switches.

Non-mechanical, Non-interferometric Switches

Adiabatic Switches. At a Y-branch in a single-mode optical waveguide, incident optical power ideally divides equally into the two outgoing branches. If the indices of the waveguides in the outgoing branches can be strongly controlled so that the refractive index of the waveguide in one branch can be made significantly higher than that in the other branch, then the mode field in the branching region is made asymmetric. The power in the mode can be pulled toward the high-index side. If the Y-branch is properly designed, then essentially all the power can be forced into one of its arms by adiabatic mode evolution.

The requisite control of refractive index is achievable through carrier injection or Pockels effect in semiconductors or Pockels effect in LiNbO$_3$, but the length of the devices is large because of the restriction of the switching process to very small branching angles. It is advantageous to make the device in a material that permits long devices without a cost penalty. Thermal control of refractive index potentially provides a way of using non-crystalline materials, but the very large thermal coefficient required precludes the use of glass. Waveguides based on polymers are ideally suited for this thermo-optic application due to their strong dependence of refractive index on temperature. However, standard available polymers can not be used due to high optical losses in the wavelength windows used in telecommunications. Specialty optical polymers with very low intrinsic optical absorption losses, well below 0.1 dB/cm for both the 1.3 and 1.5 μm windows, have been developed for this purpose by companies such as Akzo Nobel. Slab waveguides based on these materials have been demonstrated with losses around 0.08 dB/cm.

These low-loss polymers have been used to develop several optical devices, including 1×2, 2×2, 1×4 and 1×8 switches. The strong dependence of refractive index on temperature of the polymer waveguide leads to low drive powers, of the order of $50-80$ mW per switching stage, with response times of around 1 ms. These drive power levels are at least 10 times lower than those required for a similar design in glass or silica waveguides.

Despite the low intrinsic optical losses and the low losses recorded in slab waveguides, insertion losses for present polymer waveguide devices are still relatively high, most likely induced by the processing steps to create single mode waveguides. There are indications from various research efforts that these additional losses can be reduced significantly, which could render solid-state optical switches with more acceptable insertion losses (e.g. < 2 dB at 1550 nm for an 1 × 8 switch).

Polymer waveguide switches based on adiabatic mode evolution are commercially available (JDS Uniphase) in configurations of 1 × 2 to 1 × 8 and in integrated arrays such as four 1 × 2s or four 2 × 2s in a single hermetic package. Typically, these adiabatic polymer switches have ms switching times and generally good optical properties. The most important advantages over interferometric designs probably come from the 'digital' nature of the switching curve, leading to low polarization, wavelength, drive voltage (power) and temperature dependencies. These adiabatic switches can, for instance, be simultaneously used at optical signals in the 1.3 and 1.5 μm windows with identical crosstalk and isolation values. Also, more complex devices such as integrated 8×8 switches in adiabatic switch technology, require only one generic drive voltage to drive each individual switching unit, unlike interferometric switch assemblies, where often all switch elements need to have a precisely set voltage setting. This becomes a major advantage for integrated router–selector matrices, since an 8 × 8 switch has 112 basic 1 × 2 switch elements, which means 224 electrodes to be driven.

The packaging has been well developed for polymer waveguide switches, protecting the polymer from such influences as humidity through hermetic sealing, which has a positive impact on the long-term reliability of the devices. Major areas for product improvements could be identified as the insertion loss and crosstalk. The latter specification, which for commercially available devices is around 25−30 dB, will provide sufficient isolation in dilated applications such as cross-connects, but may prove troublesome in applications such as ring protection.

Liquid Crystal Polarization Switches. By applying an electric field to a suitably prepared layer of a liquid crystal, the transmission of the liquid crystal can be altered. Since liquid crystal technology is highly optimized for displays, the adaptation of existing display devices for switches has been held to be cost effective. One of the difficulties, however, is that the liquid crystal material is strongly birefringent and therefore may introduce polarization dependence, depending on the design of the rest of the switch.

Recently a 1 × 2 liquid crystal switch product with fairly good properties has been announced for optical-fibre switching (MacroVision LLC). Insertion loss of < 2.0 dB and polarization dependence of 0.3 dB are specified.

8.3.2   Interferometric Switches

Interferometric switches operate by splitting input light into two or more waveguide paths, or 'arms', that recombine at the output. Because devices are small and the coherence length of optical signals in fibre is often large, interference fringes are formed upon the recombination. An output guide may receive a dark fringe corresponding to switch state 'off' or a bright fringe corresponding to 'on', thus yielding a gate function. Control of the device requires the ability to control the optical phase difference for propagation in the different arms.

Phase-controlled devices are very sensitive to environmental influences. Furthermore, devices operating on relative phase shifts have an intrinsically cyclic response and therefore some form of feedback may be required to establish that the switch remains in the desired state. It will be difficult to design stable latching switches based on interference principles. On the other hand, they can be very fast, and this may be their chief advantage. Such switches will be necessary for such applications as packet-by-packet routeing in the optical domain.

Interferometers can be controlled by very subtle changes in relative optical phase delay between the paths: a path-length change of one half optical wavelength suffices to interchange bright and dark interference fringes and thus can switch light from one output to another. Interferometric switches can therefore be made in materials offering only weak control of refractive index; thermal control of interferometric switches works well in glass, whereas it is not practical to make adiabatic switches in glass, since these require large refractive index changes.

Modal interference devices, in which the operating principle is interference between multiple modes in a single waveguide, and switched directional couplers that modify the coupling between modes of adjacent waveguides have similarities to interferometers. They also rely on the control of optical phase and have cyclic response functions.

The phase shifts needed to operate interferometric switches have been obtained through the control of refractive index in many ways. The second-order (Pockels) and third-order (Kerr) optical nonlinearities in materials can be used to control the optical phase via refractive index. Pockels-effect devices use DC electric field to influence the refractive index. They must be made in non-centrosymmetric materials: $LiNbO_3$ and GaInAs are common examples. Unfortunately, electro-optic control via the Pockels effect is not available in amorphous materials such as glass.

Kerr-effect devices, on the other hand, use a light signal to control the refractive index and can be made in glass. Interesting research results in very high speed routeing of optical pulses have been obtained through Kerr-effect switching, but such devices have not become commercially important for switching up to the present. The principal difficulty of using the third-order optical nonlinearity for routeing purposes is that large optical intensities are required, a situation often incompatible with communications signals.

Pockels Effect (Electro-Optic) Switches

Directional Couplers. Directional couplers are optical waveguides that come in close enough proximity that the evanescent fields of their modes interpenetrate, as sketched in Fig. 8.10. All power is transferred from each guide to the other over a characteristic length measured in wavelengths. For a particular wavelength this length is determined by waveguide parameters such as core and cladding refractive index. If the refractive index can be altered, for example by the Pockels effect, then the coupling length can be altered and the crossover of optical power can be induced or prevented at will.

Switched directional couplers in $LiNbO_3$ and semiconductors have been the most intensively investigated optical switches over the past twenty years. They offer high speed, good optical performance on most measures, and the possibility of integration.

Nevertheless, progress has been slow by comparison with other developments in optical telecommunication and at present directional couplers are used primarily as external modulators for high-speed transmitters. The state of the art for fully engineered switch matrices based on integrated $LiNbO_3$ directional-coupler switch matrices is at a dimension of 8×8 [5]. In this device the following performance was achieved: loss (all 256 paths): -7.9 to -11.0 dB (93% within 1 dB); crosstalk: < 22 dB worst case; bias voltage: -6.7 ± 0.36 V; switching voltage: 9.4 ± 0.13 V.

While speed is very high and optical integration gives a packaging advantage, the optical performance in these fast switches is generally not as good as with mechanical switches. (8–11 dB typical loss and about -25 dB crosstalk in the 8×8 matrix). Moreover, a difficulty of $LiNbO_3$ devices in general is the need to adjust the drive voltages for slight variations in the performances of the individual crosspoints, and in some cases also to account for ageing. Bias and switching voltages of 5 to 10 V with a variation of ± 0.2 V are typical.

Mach–Zehnder Interferometers. The Mach-Zehnder interferometer is sketched in Fig. 8.11. The light is split, delayed by propagation in the two arms,

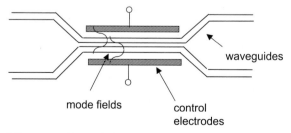

waveguides

mode fields

control electrodes

Fig. 8.10. Optical directional-coupler switch

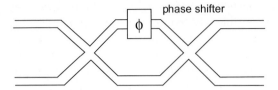

Fig. 8.11. Mach–Zehnder interferometer switch

then recombined. The relative phase on recombination determines whether constructive or destructive interference occurs at the output, and hence the device can be a gate switch. In this form it is often used as a modulator. It is wavelength independent if both arms are the same length.

In fact, the modulator device is fundamentally a four-port switch, with two ports unused. There are two outputs, and the phase relationship causes the light to exit via one of them rather than the other. (In some cases the second output may consist of radiation.) The device can act as a 1×2 switch, and when both inputs are also explicitly included, as shown in Fig. 8.11, as a 2×2 switch. Mach Zehnder routeing switches in the 1×2 and 2×2 configuration are commercially available (JDS Uniphase).

The Mach-Zehnder interferometer can be generalized to multiport switching. Such a switch would consist of a multiport splitter connected by multiple waveguides to a multiport combiner as shown in Fig. 8.12. By controlling the phase delays in the multiple interconnecting arms all the input power can be sent to a single output port, and a limited multi-input switching function can also be obtained. Such switches can be thought of as steerable phased-array antennas. The phase control in the arms allows the wavefront generated in the second coupler to be tailored to converge on the correct port. (In fact the tailoring may not be straightforward, since suitable optical splitters and combiners, known as multimode interference (MMI) couplers, have complicated relationships among the phases of the arms.) Such switches are at the research stage.

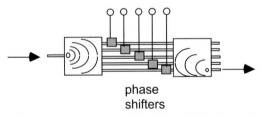

phase
shifters

Fig. 8.12. Generalized Mach-Zehnder interferometer switch using multimode-interference couplers

Polarization-independent Lithium Niobate Switch Arrays. There have been significant advances in the last few years on both device and system demonstration fronts. Low-crosstalk polarization-independent switch elements have been demonstrated. So too has the ability to produce large switch arrays with good yields. The most recent demonstrations have been with switch modules that provide 12×6 connectivity and contain 240 individual switch elements. These modules also contain other enhancements necessary for system performance including signal disconnect, crosstalk reduction, polarization mode dispersion and PDL compensation, and integrated variable attenuators. Thirty-two of these modules were fabricated with a typical insertion loss of 15 dB and crosstalk isolation of 43 dB. These devices are being deployed in a field trial employing a total of 48 lithium niobate switch modules with nearly 10 000 switch elements [6].

High-Speed Mach-Zehnder Modulators. High-speed interferometric modulators fabricated in lithium niobate [7] are commercially available. Device voltage stability issues have been addressed either by tailoring the electrical properties of the dielectric buffer layer or by tuning the devices to operate with no DC bias. These devices are used as external modulators in optical communication systems at 2.5 and 10 Gbits/s. Research devices operating at 40 Gbits/s have been demonstrated. The devices have several advantages over direct or electroabsorption modulation. First, external modulators separate light generation from modulation, enhancing the wavelength stability of the source. This advantage may become increasingly important as tunable lasers are deployed in communication systems. Secondly, external modulators can be designed with specific values of chirp or with variable chirp to compensate for fibre dispersion. Typically, the devices have an insertion loss of 3 to 4 dB, modulation voltages of 5 V, and extinction ratios of 15 to 20 dB. In many devices, other functions such as phase modulators or variable attenuators are integrated with the modulators to enhance system functionality.

Kerr-Effect (All-optical) Switches. The Kerr effect, i.e. the direct control of refractive index by optical intensity through the third-order optical nonlinearity in materials, provides direct control of light signals by light signals. It underlies a basic technology for very high speed optical logic that is at present a subject of research. The very short (fs) and very intense optical signals that can now be easily generated can interact directly in fibre or semiconductor waveguides. Optical digital operations at hundreds of Gb/s have been achieved. However, because of the intensities involved Kerr-effect switches do not have a role in routeing optical signals at the bandwidths and intensities currently used.

Thermally Driven Interferometer Switches. Thermally driven interferometer switch arrays have reached capabilities comparable to LiNbO$_3$

technology in matrix size, crosstalk and loss. Furthermore, these devices are relatively polarization insensitive. On the other hand they are slow, take a lot of power, and require very careful fabrication An 8×8 full crosspoint matrix based on the 'Planar Lightwave Circuit' (PLC) technology (pyrolithically deposited SiO_2) has demonstrated loss of -14 dB and crosstalk of -18 dB worst case, and a reconfguration time as low as 1.3 ms (NTT). It is interesting to compare the performance of a recent 8×8 adiabatic switch in polymer waveguide (Akzo-Nobel)), which has -10.7 dB loss, -30 dB crosstalk worst case, and a reconfiguration time of 1 ms.

8.4 Summary

Optical switching embraces a wide variety of technologies, some in development, some at a stage of mature commercial availability. At present, electromechanical optical switches dominate the commercial offerings and are widely and satisfactorily deployed in manufacturing, testing, sensing and telecommunications applications. The advent of wavelength-division multiplex transmission is widely seen as the harbinger of a new stage of development for optical switches, where new requirements on speed, size, reliability, capacity and economies in manufacture will come into play. A second-generation technology for optical switching is now being sought in many research organisations, but there is no clear evidence at present of what this technology will entail. At the beginning of the new millennium optical switching will very likely present commercial and scientific opportunities similar to those now seen in the worldwide development of wavelength-division multiplex.

References

1. C. Guillemot, M. Renaud, P. Gambini, C. Janz, I. Andonovic, R. Bauknecht, B. Bostica, M. Burzio, F. Callegati, M. Casoni, D. Chiaroni, F. Clérot, S.L. Danielsen, F. Dorgeuille, A. Dupas, A. Franzen, P.B. Hansen, D.K. Hunter, A. Kloch, R. Krähenbühl, B. Lavigne, A. Le Corre, C. Raffaelli, M. Schilling, J.-C. Simon, and L. Zucchelli, "Transparent optical packet switching: The European ACTS KEOPS project approach," IEEE J. Lightwave Technol. **12**, 2117 (1998)
2. L.Y. Lin, E.L. Goldstein, and R.W. Tkach, "Free-space micromachined optical switches with submillisecond switching time for large-scale optical crossconnects," IEEE Photon. Technol. Lett. **10**, 525 (1998)
3. J.M. Simmons, A.A.M. Saleh, E.L. Goldstein, and L.Y. Lin, "Optical crossconnects of reduced complexity for WDM networks with bidirectional symmetry," IEEE Photon. Technol. Lett. **10**, 819 (1998)
4. S.S. Lee, L.-Y. Lin, and C. Ming, "Realisation of FDDI optical bypass switches using surface micromachining technology," Proc. SPIE **2641**, 41 (1995)
5. E.J. Murphy, C.T. Kemmerer, D.T. Moser, M.R. Serbin, J.E. Watson, and P.L. Stoddard, "Uniform 8×8 lithium niobate switch arrays," IEEE J. Lightwave Technol. **13**, 967 (1995)

6. This work was done under the DARPA-sponsored MONET program. Reported in: E.J. Murphy, "Lithium Niobate Switch Arrays for Optical Crossconnects," *Optical Fiber Communication Conference (OFC '99)*, Techn. Digest Series (Opt. Soc. America, Washington, DC 1999), p. 28
7. F. Heismann, S.K. Korotky, and J.J. Veselka, "Lithium Niobate Integrated Optics," in *Optical Fiber Telecommunications III B* (I.P. Kaminow and T.L. Koch, eds.), Chap. 9 (Academic, San Diego 1997) p. 377

9 All-Optical Time-Division Multiplexing Technology

Masatoshi Saruwatari

9.1 Role of All-Optical TDM Technology

During the past two decades, great progress has been achieved in optical fibre communications technologies toward higher capacity and longer repeater-span transmission, leading to very economical trunk transmission systems for public communications networks. In particular, newly evolved technologies including erbium-doped fibre amplifiers have made very-high-speed 10 Gbit/s optical transmission systems of practical use. However, because these transmission techniques still rely on electronics for processing the high-speed signals, it is anticipated that the operational speed and transmission bit rate will hit the upper limit in the near future. Moreover, to accommodate the coming broadband network (B-ISDN) era, very-high-speed technologies must be developed not only for transmission lines, but also transmission nodes. The goal is to handle signal rates of over 1 Tbit/s, so that vast amounts of information, including data and pictures, can be provided to many subscribers through optical-fibre cables. To this end, novel all-optical signal-processing technologies capable of superseding conventional electron-based technologies are urgently required. Figure 9.1 illustrates a future ultrafast optical time-division-multiplexing (OTDM) transmission system that fully utilises all-optical high-speed signal processing technology. Key techniques include, optical pulse generation and modulation, all-optical multiplexing, optical linear or nonlinear (soliton) transmission, all-optical repeating, all-optical regenerating, all-optical demultiplexing, optical timing extraction, optical waveform measurement, etc.. This article reviews some of the most essential photonic technologies required for very-high-speed all-optical TDM (time-division multiplexing) transmission, such as ultrashort optical pulse generation, all-optical multiplexing/demultiplexing, optical timing extraction and optical pulse waveform measurement techniques. Recent all-optical TDM transmission experiments up to 400 Gbit/s and all-optical TDM/WDM (wavelength division multiplexing) transmission experiments of 400 Gbit/s–3 Tbit/s are also introduced together with major issues and future prospects.

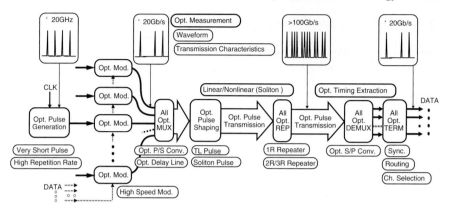

Fig. 9.1. Future ultrafast optical fibre transmission and related technologies

9.2 Key Technologies for All-Optical TDM Systems

9.2.1 Ultrashort Optical Pulse Generation Technology

In order to achieve very high-speed optical TDM transmission, it is essential for optical sources to generate transform-limited (TL) picosecond pulses, namely, $< 5\,\mathrm{ps}$ chirpless pulses at repetition-rates ranging from 5 to 20 GHz. In addition, tunable and controllable repetition rates are inevitably required to permit synchronization with other signals. Also, wavelength tunability is valuable for optimizing transmission characteristics. Table 9.1 compares various optical pulse generation techniques which have been applied to high-speed (over 10 Gbit/s) optical transmission experiments. These techniques include gain-switching of distributed-feedback laser diodes (DFB-LDs), gating of continuous-wave (CW) light with an electro-absorption (EA) modulator, mode-locking of laser diodes (LDs), harmonic mode-locking of erbium(Er)-doped fibre (EDF) laser, and supercontinuum (SC) generation by a low-dispersion single-mode fibre pumped with an intense picosecond pulse. In the following, these techniques will be discussed from the above-mentioned standpoint.

Gain-switching of Laser Diodes. Figure 9.2 shows the experimental set-up of gain-switching of DFB-LDs [1]. When a DFB-LD is driven by a sinusoidal current of several GHz, $25-30\,\mathrm{ps}$ optical pulses with red-shift chirping can be generated. When the gain-switched pulses transit through a dispersion-shifted fibre with normal dispersion of 46 ps/nm, their red-shift chirping can be compensated for, resulting in nearly TL compressed $5-7\,\mathrm{ps}$ pulses. Pulse waveforms and time-resolved spectra both before and after chirp-compensation are also shown in Fig. 9.2. The gain-switching technique is quite promising because it can generate $5-7\,\mathrm{ps}$ nearly-TL optical pulses at an arbitrary frequency up to 20 GHz, which makes it possible to

Table 9.1. Optical short pulse generation

Method	Repetition freq. [Limitation]	$\Delta\tau$ (ps)	$\Delta\tau \cdot \Delta\nu$	Comment
Gain Switching	Arbitrary [RC const.]	~ 20 (6) (0.8)	>TL ($\sim$TL) ($\sim$TL)	Red-shift chirp (Chirp compensation: CC) (CC+pulse compression)
CW+EA-Modulator	Arbitrary [Modulator]	~ 15	$\sim$TL	Large pulse width
Mode-Locking (LD)	Fixed	~ 10 (< 1)	>TL ($\sim$TL)	Conventional (CPM-type or EA modulator)
Harmonic Mode-Locking (EDF Laser)	Tunable [Mode locker]	~ 3	TL	λ-tunable, $\sim 20\,$nm
SC Generation	Tunable [Pump frequ.]	< 1	$\sim$TL	λ-tunable, > 200 nm

EA: electro-absorption; TL: transform-limited; SC: supercontinuum

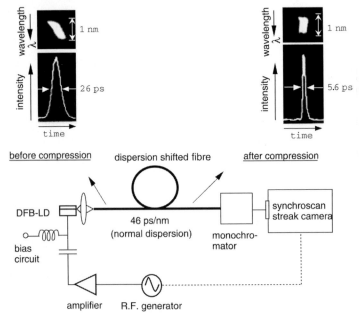

Fig. 9.2. Experimental configuration of gain-switched DFB-laser diode

easily synchronize with all other signals. Gain-switching has been applied for early high-speed experiments, including soliton transmission, all-optical demultiplexing, optical timing extraction and optical sampling. To further reduce gain-switched pulses to $< 4\,\mathrm{ps}$ duration, however, nonlinear pulse-compression techniques such as an adiabatic soliton compression technique should be incorporated together with the chirp compensation.

Gating of CW Light with Electro-absorption Modulator. The second pulse-generation technique is gating of CW light with sinusoidally driven EA modulators [2,3]. This is based on the nonlinear transmittance of InGaAsP-EA modulators with respect to applied voltage, and nearly-transform-limited 20 ps soliton pulses have been obtained at 10 Gbit/s. Generated soliton pulses were transmitted over a recirculating distance greater than 6000 km. However, this technique can hardly be applied to higher-speed TDM systems such as 100 Gbit/s due to its relatively large pulse width ($> 10\,\mathrm{ps}$). To improve this, a pulse-compression technique using a dispersion-decreasing fibre as a soliton adiabatic compressor has been demonstrated to obtain 2.5 ps TL pulses [3].

Mode-Locking of Laser Diodes. Mode-locking of LDs [4–10] is another way of producing high-repetition-rate ultrashort optical pulses. So far, a new type of LD capable of colliding-pulse mode-locking (CPM) has been demonstrated [4]. This can output ultra-high-speed, purely TL pulses with less than 1 ps duration at 40 and 350 GHz repetition frequency through active and passive CPM operation, respectively. Unfortunately, at present there is no way to utilise this extremely high operation speed. In order to mode-lock at moderate speed, a long-cavity LD integrated with a passive waveguide and a Bragg reflector was developed [5]. The cavity length is monolithically extended to 5.5 mm, leading to 8 GHz mode-locking operation. This was used for 8 Gbit/s 4000 km soliton transmission experiments. The main drawback of this method is no tunability of repetition frequency and relatively larger spectral width as compared with the TL condition.

To improve the operation characteristics, various mode-locked LDs have been reported, including pulse-width tunable subpicosecond pulse generation from an active mode-locked monolithic multiquantum well (MQW) LDs integrated with MQW-EA modulators [6], repetition rate tunable lasers using passive mode-locking of micromechanically-tunable LDs [7] or active mode-locking of external cavity LDs [8], and very high-speed or femtosecond optical pulse generation from an active mode-locked LD [9,10]. Figure 9.3 shows the structure and output pulse waveform and spectrum of a mode-locked monolithic MQW LD integrated with an MQW-EA modulator.

Harmonic Mode-Locking of EDF Lasers. Harmonic mode-locking of EDF lasers [11–17] is promising, because it can generate a purely TL pulse

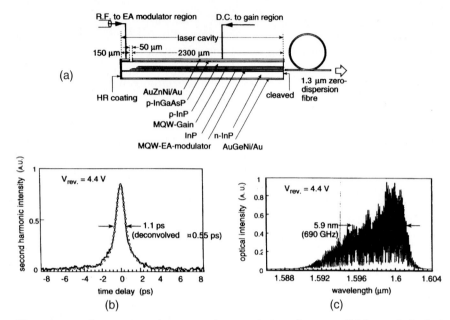

Fig. 9.3a–c. Structure and output characteristics of a monolithic mode-locked laser-diode integrated with an electro-absorption modulator

train in the 10 GHz region without requiring any chirp compensation or pulse compression. However, since Er fibre lasers are normally tens of metres long, they have to be mode-locked at very high harmonics of the cavity mode spacing, for example, harmonic orders of the order of 1000. This leads to the issue of how to stabilize such systems, which are characterized by the competition between many sets of supermodes. Moreover, mode-competition is enhanced by fluctuations in cavity lengths and polarization states. To stabilize mode-locked Er lasers, several techniques have been developed. The first method is inserting a high-finesse Fabry-Perot etalon in the cavity to eliminate all the unwanted laser cavity modes [12]. However, this straightforward method requires the strict condition that the free-spectral range of the etalon should coincide with the mode-locking frequency. The second method is dithering the cavity length at a kHz rate [13] to wash out spatial hole burning. Although this makes one set of supermodes dominant, polarization fluctuations still remain, resulting in unstable operation. The third method uses a single-polarization cavity to eliminate the mode-competition generated by polarization fluctuations [14–17]. Figure 9.4 depicts a wavelength-tunable mode-locked Er fibre ring-laser based on this principle.

All the fibres, including the Er-doped fibre, are polarization-maintaining (PM) single-mode fibres called PANDA fibres, and a pigtailed polarizer is inserted to ensure single-polarization oscillation. A tunable optical filter with 3 nm bandwidth and an optical delay line are inserted into this cavity to

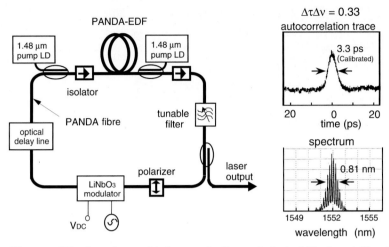

Fig. 9.4. Wavelength-tunable harmonically mode-locked Er-doped fibre laser

obtain wavelength tunability. With this configuration, 3.0−3.5 ps TL pulses having a time-bandwidth product of 0.33 and corresponding to $\text{sech}^2(t)$ waveform, were obtained at up to 20 GHz, as shown in Fig. 9.4. Stable oscillation characteristics without any pulse-to-pulse fluctuations were also verified by bit-error-rate (BER) measurements using external modulation [15]. Furthermore, wavelength tunable operation over 7 nm was confirmed and was applied to optimize the optical demultiplexer operation and fibre transmission characteristics at 100 Gbit/s, as described in Sect. 9.3.1.

Since EDF lasers have relatively long cavities, temperature change brings about considerable cavity length variations, resulting in unstable mode-locking. Figure 9.5 shows the configuration of a stabilized mode-locked EDF laser [16]. The cavity length is actively controlled by an optical delay line in the feedback loop. Since the onset of relaxation oscillation was found to be a good measure of the detuning, as shown on the right-hand side of Fig. 9.5, its RF power was used for the feedback signal. With this method, a very stable error-free operation throughout more than 10 hours has been confirmed [16]. The multi-frequency modulation technique was also applied to the mode-locked EDF laser to yield shortest pulses at a relatively slow repetition rate [17]. With this method, short pulses having constant pulse width independent of the repetition frequency can be obtained.

Supercontinuum (SC) Generation. When an intense optical pulse transits a low-dispersion silica fibre, an SC pulse, whose spectrum is spread over a wide spectral range, is induced by the combining effects of various optical nonlinear processes [18–22]. Figure 9.6 shows the experimental configuration together with the generated SC spectrum and pulse waveform. Here, a harmonically mode-locked 3 ps, 6.3 GHz Er fibre laser followed by an Er-doped

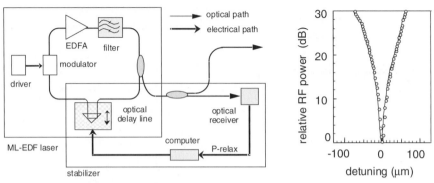

Fig. 9.5. Stabilized mode-locked EDF laser

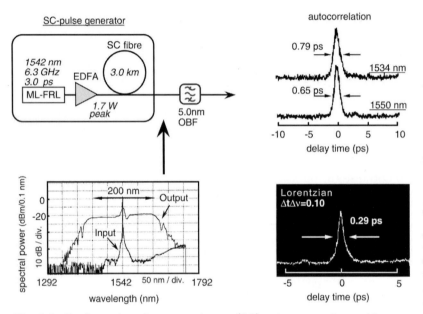

Fig. 9.6. Configuration of supercontinuum (SC) pulse generation and its properties

fibre amplifier (EDFA) is used as the pump source at 1542 nm wavelength, and a 'white' pulse with a 200 nm-wide spectrum ranging from 1450 nm to 1650 nm is generated from a 3 km dispersion-shifted fibre. The SC spectrum does not only exhibit 200 nm continuous, flat-top broadening, but has also coherency to some extent. Recently, it has been clarified that the desired flatly broadened SC spectra can be generated from a dispersion-decreasing fibre with a convex function having two zero-dispersion wavelengths at the input part [22]. With the use of these characteristics, penalty-free 3 ps pulse generation for WDM systems has been demonstrated, where many WDM

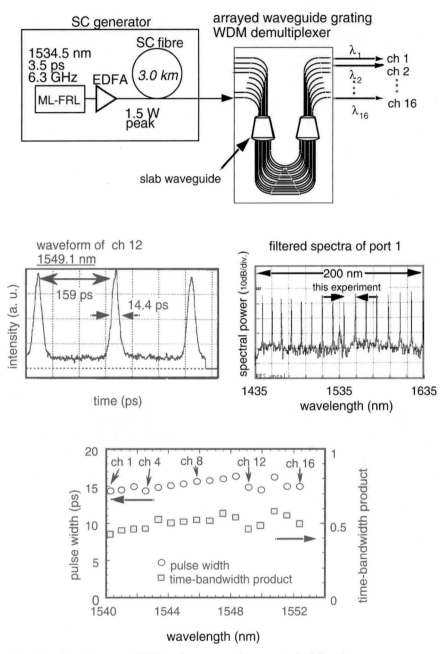

Fig. 9.7. Simultaneous WDM pulse output from a single SC pulse

channels can be produced at arbitrary wavelengths by merely filtering out the desired wavelengths [19]. Furthermore, by using a 5 nm bandpass filter and chirp compensation using a 1.3 μm zero-dispersion fibre, nearly-TL pulses shorter than 1 ps can be obtained within a wide spectral range [20]. The minimum achieved pulse width to date is 0.17 ps. By filtering a super-broadened SC spectrum with an arrayed-waveguide grating (AWG)-WDM demultiplexer, a multi-wavelength picosecond pulse source has been constructed for dense WDM/TDM systems [21]. Figure 9.7 depicts the proposed multi-wavelength pulse source using an SC generator and a 16 × 16 AWG WDM-demultiplexer with 100 GHz frequency spacing and 33 GHz bandwidth. A total of 16 different-wavelength 6.3 GHz pulses with a duration of nearly 15 ps and a time-bandwidth product of 0.5 can be simultaneously generated from a single SC pulse source. The output WDM pulses feature a very low jitter (< 0.3 ps) and high frequency stability (1.26 GHz/°C) determined by the pump source and AWG characteristics, respectively. SC pulse generation has been applied for picosecond signal sources used for the latest large-capacity transmission experiments such as $200-400$ Gbit/s OTDM transmission and $1-3$ Tbit/s OTDM/WDM transmission, and have also led to various kinds of all-optical signal processing, including 500 Gbit/s or multiple-output all-optical demultiplexing, 200-400 GHz optical timing extraction, and < 1 ps optical sampling as described in Sects. 9.2.2, 9.2.3 and 9.2.4.

9.2.2   All-Optical MUX/DEMUX Technology

Time-division-multiplexing/demultiplexing (MUX/DEMUX), where the channels of the lower digital hierarchy signals are multiplexed to those of the higher hierarchy signals (MUX) and vice versa (DEMUX) in the time-domain, is one of the key functions necessary to develop very high-speed optical transmission systems. The operation speed of conventional MUX/DEMUX circuits based on electronics is currently limited to around 40 Gbit/s at most. To circumvent such electronics bottlenecks, all-optical MUX/DEMUXs utilising ultra-fast third-order optical nonlinearities, e.g. nonlinear refractive index and parametric frequency-conversion processes, are promising.

All-optical MUX. In order to develop a very high-speed optical TDM system, a stable optical time-domain multiplexer is required. Although much attention has not been paid to all-optical MUX as compared with all-optical DEMUX, there have been some reports demonstrating very high-speed optical signal generation up to 500 Gbit/s. Here, some of prospective techniques used for all-optical TDM transmission experiments will be described. Figure 9.8a,ba shows a planar lightwave circuit (PLC)-based MUX where 2-by-2 couplers, denoted c_i, and optical delay lines are fabricated on a silicon substrate by a chemical-vapour-deposition method. The input signal is divided into two branches and then combined with a delay three or four times to yield

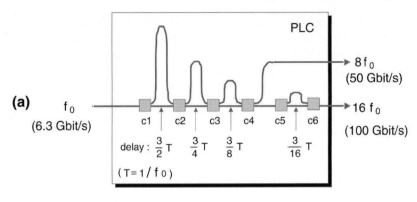

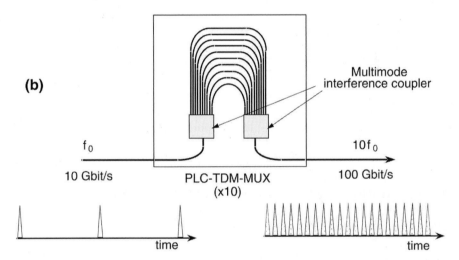

Fig. 9.8a,b. Planar lightwave circuit TDM multiplexer

2^3 or 2^4 times multiplexed signals. The delay of the nth stage ($3T/2^n$, where T is the original timeslot) corresponds to 1.5 times the pulse interval of the pulse train incident on the nth stage, to make the adjacent bit marks different. With this configuration, the original 6.3 Gbit/s signal is multiplexed to a stable 50 or 100 Gbit/s TDM test signal. Figure 9.8a,bb shows another kind of PLC-MUX using a couple of multimode waveguide regions as a branching or a combining circuit. Since both circuits are connected with an integer number of delay lines with the same length difference from one channel to the next, an arbitrary number of channels can be multiplexed with this circuit. This structure indicates ten-times multiplication to produce a 100 Gbit/s OTDM signal from a 10 Gbit/s baseline signal. Although these PLC-MUXs are not real MUXs combining different channels, they are very efficient for generating stable high-speed test signals up to 400–500 Gbit/s, in order to evaluate all-optical circuits and OTDM transmission systems as well. One example of an all-optical modulation technique is Four-Wave Mixing (FWM) in a Travelling-Wave Semiconductor-Optical Amplifier (TW-SOA) [23,24]. A higher repetition rate (100 GHz) optical pulse train is all-optically modulated by full sets of lower speed (10 Gbit/s) optical signals to make a fully multiplexed optical TDM signal. Presently, two channel error-free multiplexing of 10 Gbit/s individual signals on a 100 GHz optical clock was demonstrated using FWM in two sets of SOAs connected in series [24]. Recently, an intriguing and simple method to generate a high-repetition-rate optical pulse train from a lower one has been proposed and demonstrated [25]. This consists of an ultrashort optical pulse source and a dispersive medium like fibres to cause a linearly chirped pulse broadening. If the chirped pulse broadening exceeds the pulse interval, the chirped pulses partially overlap each other. When the frequency difference Δf between the overlapped pulses at the same temporal position satisfies $\Delta f = M f_0$, where f_0 is the original repetition rate and M is an integer, a harmonic pulse train with a repetition frequency M times higher than that of the original one will appear. Figure 9.9a,b shows the principle and experimental results of this clock multiplication method. Using a 10 GHz TL pulse train from a mode-locked EDF laser or SC generator, a 50 GHz or 400 GHz optical pulse train has been successfully generated by the proposed method. Since the generated pulse train is perfectly synchronized to the initial pulse train, an all-optical 100 Gbit/s MUX experiment, where the multiplied 100 GHz optical clock is all-optically modulated by three 10 Gbit/s optical channels with a nonlinear-optical loop mirror (NOLM) switch, has been successfully demonstrated [26]. Dispersion-shifted fibre transmission of this multiplexed 100 Gbit/s signal has also been realized without any degradation over a distance of 80 km.

All-optical Demultiplexer (DEMUX)

General. All-optical demultiplexers are key devices for developing ultrafast optical TDM transmission systems. To apply the DEMUXs to real communication systems, the following requirements must be satisfied; fast and stable bit-error-free operation, low control power suitability for LD or Er-laser pumping, polarization independent (PI) operation, synchronization to received high-speed signals, and cascadability or multi-output operation. Here, various demultiplexing techniques based on the third-order optical nonlinearity will be reviewed taking into account these requirements. Table 9.2 summarizes the various all-optical demultiplexing techniques reported so far. These are: optical Kerr switching [27–29], four-wave mixing using a fibre [30–33] or an SOA [34–36], cross-phase modulation (XPM)-based switching [37–39], and various fibre loop based switches, called NOLM [40–49], SLALOM [50,51] or

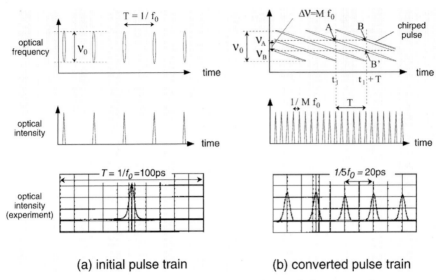

(a) initial pulse train   (b) converted pulse train

Fig. 9.9a,b. Principle and experimental result of optical clock multiplication

TOAD [52–54]. Polarization independence has been demonstrated in several cases as well as multiple output operation. The highest operation speeds reported so far were 500 Gbit/s using FWM in a fibre [33] and 640 Gbit/s using a NOLM switch [92]. In the following we will describe some of the demultiplexing techniques in more detail.

LD-pumped NOLM DEMUX. One example of an OTDM DEMUX is a fibre-based Sagnac interferometer, usually called the 'nonlinear optical loop mirror' (NOLM [40]). The operating principle of the NOLM is as follows. Incoming pulses are equally split into two counter-propagating pulses. They traverse the loop, recombine and are normally backreflected. If a control pulse propagates synchronously with one pulse, nonlinear effects introduce a phase shift to that particular pulse, and at the NOLM exit the phases of the corresponding pulse pair do no longer match to assure backreflection, but the combined pulse is partly directed to the NOLM output port. The back-reflected pulse is completely cancelled, if the phase shift introduced amounts exactly to π. In order to achieve error-free demultiplexing using LD sources, some kilometres of fibre length are generally required due to the small nonlinearity in fibres. The first successful demonstration of a NOLM used a 3−14 km-long fibre loop [43,44]. Recent NOLM DEMUX experiments include polarization-independent 32 Gbit/s operation [47], 64 Gbit/s-to-8 Gbit/s DEMUX with 8 GHz clock recovery, and 100 Gbit/s-to-6.3 Gbit/s PI-DEMUX using a po-larization maintaining PANDA fibre loop [48]. Multiple-output operation (1:5 DEMUX) has also been demonstrated using a chirped SC pulse as a signal to be demultiplexed in the NOLM switch [49].

Table 9.2. All-optical demultiplexing

Concept		Bit rate (Gbit/s)	BER	Author	Year	Ref.
Optical		2/60	–	Morioka (NTT)	1987/92	27, 28
Kerr		40	–	Patrick (BT)	1993	29
Switch						
FWM	Fibre	16	free	Andrekson (AT&T)	1991	30
Switch		100[a]	free	Morioka (NTT)	1994	31
		100[b] (4ch)	free	Morioka (NTT)	1994	32
		500	free	Morioka (NTT)	1996	33
	SOA	20	free	Ludwig (HHI)	1993	34
		100	free	Kawanishi (NTT)	1994	35
		200[a]	free	Morioka (NTT)	1996	36
XPM		60	–	Morioka (NTT)	1992	37
Switch		40	–	Patrick (BT)	1993	38
		100[b] (6ch)	free	Uchiyama (NTT)	1997	39
Loop	NOLM	1	–	Blow (BT),	1990	41
Mirror		5	–	Jinno (NTT)	1990	42
		40	–	Takada (NTT)	1991	43
		64	free	Andrekson (AT&T)	1992	44
		40	free	Patrick (BT)	1993	45
		20[a]	–	Bülow (Alcatel SEL)	1993	46
		32/100[a]	free	Uchiyama (NTT)	1993/94	47, 48
		100[b] (5ch)	free	Uchiyama (NTT)	1996	49
		640	free	Nakazawa (NTT)	1998	92
	SLALOM	9	–	Eiselt (HHI)	1993	50
		80	free	Diez (HHI)	1998	51
	TOAD	50/250	–	Sokoloff (Princeton University)	1993/94	52
		40	free	Ellis (BT)	1993	53
		160	free	Suzuki (NTT)	1994	54

BER: bit error rate; *free*: bit-error free operation
ch: number of channels
[a] polarization-independent operation; [b] multiple-output operation

Polarization-independent DEMUX. For applications to ultrafast optical TDM systems, polarization-independent (PI) DEMUXs have been developed with NOLM switches and FWM-based switches, both using polarization-maintaining PANDA fibres.

As shown in Fig. 9.10a,ba, the PI-NOLM DEMUX [47,48] utilises a PANDA fibre loop cross-spliced in the middle to cancel the overall polarization dispersion, resulting in no pulse walk-off when the signal pulse transits the 3 km-long loop. Since the control pulse equally excites two principal axes, two sets of single-polarization NOLMs can be formed corresponding to the orthogonal polarization states. This method yielded 32-to-6.3 Gbit/s and 100-

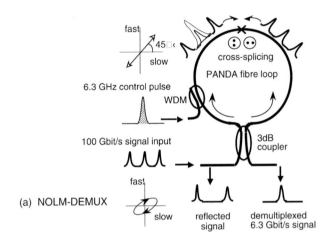

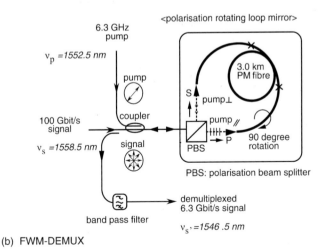

Fig. 9.10a,b. Polarization-independent all-optical demultiplexers

to-6.3 Gbit/s polarization-independent DEMUX operation. Figure 9.10a,bb shows the other PI-DEMUX. This uses FWM in a 3 km-long PANDA fibre [31]. The general principle of FWM-based demultiplexing is the following: Pump pulses from an EDF laser (at 6.3 GHz in the example shown in Fig. 9.10a,bb) propagate synchronously with pulses to be demultiplexed and sidebands are generated due to FWM. All outgoing signals are then passed through a bandpass filter, which lets the sideband wavelength pass, while the original data and the pump wavelength are blocked. The configuration shown in Fig. 9.10a,bb assures polarization independence, as orthogonal polarizations, namely S- and P-components, bidirectionally propagate the loop with their polarizations interchanged to each other. Since the polarization axis of the loop fibre is twisted by 90°, both components are lead to the input/output port through the polarization beam splitter. A polarization walk-off does not occur, because both components use the same principal axis of the loop. In order to assure the same FWM conversion efficiency for both components, the pump pulses are divided into S- and P-components with equal power. With this configuration, very stable PI-demultiplexing of 100 Gbit/s-to-6.3 Gbit/s has been achieved with less than 1 pJ pumping. Utilizing FWM in an SOA, polarization-insensitive DEMUX operation up to 200 Gbit/s was reported [36]. The laser amplifier used here exhibits small gain difference between TE and TM modes, resulting in small conversion efficiency difference between them. By depolarizing the input signal and pump pulses and compensating the output pulses, stable polarization-insensitive operation has been confirmed.

Multiple-output DEMUX. Figure 9.11 shows the principle of a multiple-output 100 Gbit/s demultiplexer based on multi-channel FWM employing a linearly-chirped square pump pulse [32]. The instantaneous frequency difference between signal and pump pulses depend on their relative temporal position, and as a consequence wavelength-converted pulses via FWM have different wavelengths according to the respective temporal position of the original 100 Gbit/s signal pulses. This experiment uses the SC-pulse source followed by a chirp generation fibre as a broad-square pump pulse generator. With this technique, 4-channel 10 Gbit/s signals were simultaneously demultiplexed from a 100 Gbit/s TDM signal. Another multiple-output DEMUX using a NOLM switch can be realized by using the chirped pulse, as shown in Fig. 9.12 [49].

The different colours are switched by the input TDM signal used as a pump. With this configuration, 5-channel 10 Gbit/s signals were demultiplexed from a 100 Gbit/s TDM signal at the same time. A novel multiple-output DEMUX has been proposed and demonstrated recently utilising XPM [39]. Its operation is based on temporal local chirp compensation of a down-chirped clock pulse through XPM induced by a signal pulse stream. Figure 9.13 shows the principle. A linearly down-chirped clock pulse train, and a randomly modulated 100 Gbit/s signal stream are coupled and launched

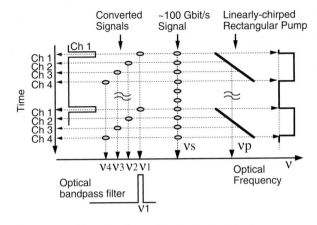

Fig. 9.11. Multiple-channel output demultiplexer

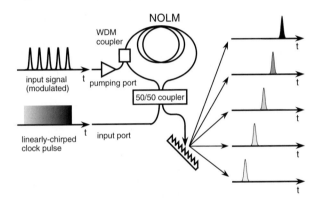

Fig. 9.12. Principle of multiple-output NOLM demultiplexer

into a nonlinear optical medium of positive nonlinear-index ($n_2 > 0$). Signal pulses at the time t_1 and t_2 induce a nonlinear optical frequency shift on the down-chirped clock pulse at the frequency ν_1 and ν_2 through XPM (Fig. 9.13a). Since the XPM-induced frequency shift is an up-chirp in the centre region (Fig. 9.13b), the down-chirp of the clock pulse is locally compensated in the time domain at t_1 and t_2, resulting in an increase in frequency components ν_1 and ν_2 according to the modulated signal pulses, as shown in Fig. 9.13c. Thus all TDM channels can be simultaneously converted into WDM signals with frequencies $\nu_1, \nu_2, \dots$, and a multiple-channel output DEMUX is achieved using a WDM demultiplexer. Figure 9.14 shows the experimental set-up. A 100 Gbit/s signal is obtained through time-division multiplexing 16 times a 6.3 Gbit/s baseline signal generated from a stabilized mode-locked EDF laser. The signal pulses have a pulse width of 3.7 ps and a wavelength of 1555 nm. The down-chirped clock pulses are generated by filtering out 6.3 GHz SC pulses at a centre wavelength of 1540 nm with

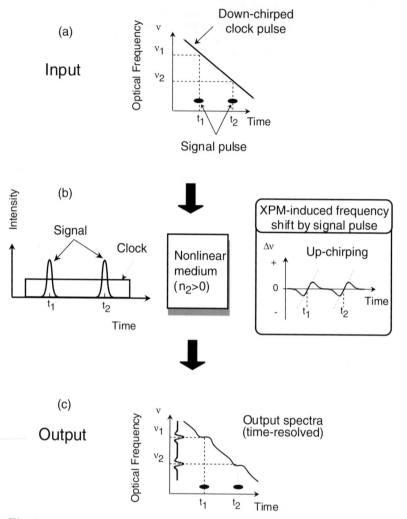

Fig. 9.13a–c. Principle of multi-output DEMUX using XPM-induced chirp compensation

a 20 nm bandpass filter and by providing linear down-chirp through 310 m standard fibre (SF) with 1.3 µm zero-dispersion wavelength. With this set-up, error-free, simultaneous 6-channel output, 100-to-6.3 Gbit/s demultiplexing has been successfully demonstrated. Since this method has a very simple set-up compared to the conventional methods using interferometers or parametric processes, it may offer a practical solution to all-optical TDM systems.

500 Gbit/s DEMUX. Very fast all-optical demultiplexing up to 500 Gbit/s has been accomplished using FWM in a polarization maintaining PANDA fibre [33]. Figure 9.15 shows the experimental set-up. Low-noise short SC pulses

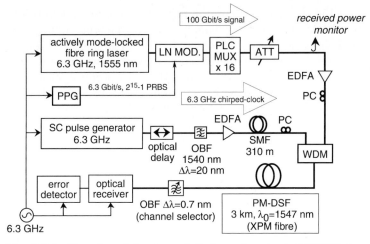

Fig. 9.14. Experimental setup of 6-channel output DEMUX based on XPM-induced frequency shift

were used for both the 500 Gbit/s signal and the 10 GHz pump pulses. The SC for the pump was generated by pumping a 3 km single-mode dispersion-shifted fibre with 10 GHz 3.5 ps TL pulses from an Er-doped fibre ring-laser (EDFRL), and that for the signal by pumping a 1 km PANDA dispersion-shifted fibre with 10 GHz 3.5 ps TL pulses from another EDFRL. SCs were filtered by 5 nm optical bandpass filters (OBFs) to produce 1562.6 nm 0.98 ps

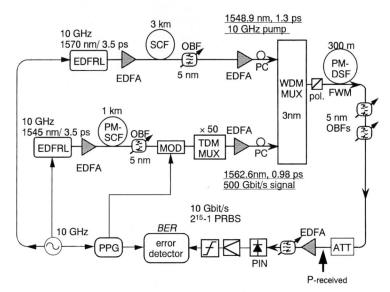

Fig. 9.15. Experimental set-up of 500 Gbit/s all-optical demultiplexing

signal pulses and 1548.9 nm 1.3 ps pump pulses, where the pump bandwidth is limited by the WDM multiplexer of 3 nm bandwidth. The rms timing jitter of the pump and signal pulses were measured to be 77 fs and 97 fs, respectively. The signal pulses were then modulated at 10 Gbit/s and time-division-multiplexed by a factor of 50 in a PLC multiplexer to produce 500 Gbit/s TDM signals and were combined with the pump pulses in a WDM-MUX after independent amplification by EDFAs. The nonlinear media for FWM was a 300 m PM-dispersion-shifted fibre (DSF) with a dispersion slope of $0.065\,\mathrm{ps\,nm^{-2}\,km^{-1}}$ and a zero-dispersion wavelength of 1549 nm. The combined pump and signal pulses were coupled into the PM-DSF with their polarization aligned along one of the principal axes. The generated FWM component was filtered out with two 5 nm OBFs in series having an overall bandwidth of 3.3 nm. The demultiplexed signal was preamplified by an EDFA and its BER performance was measured as a function of the received power before the EDFA. Figure 9.16 shows the output spectra from the PM-DSF together with temporal waveforms of the 500 Gbit/s signal stream, the 10 GHz pump pulse, and the demultiplexed 10 Gbit/s signal pulse. All the waveforms were measured with a newly developed optical sampling scope with a 0.5 ps temporal resolution described later (see Sect. 9.2.4). As shown here, the 10 Gbit/s signal pulses are clearly demultiplexed from the 500 Gbit/s signal, and the rms timing jitter was less than 0.1 ps.

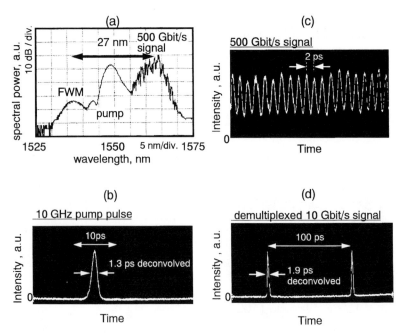

Fig. 9.16a–d. Spectrum and temporal waveform of 500 Gbit/s all-optical demultiplexing

The first demultiplexing experiments using cross-phase modulations in SOAs were performed in hybrid arrangements of fibre loop mirrors with SOAs [50,52]. The intention of these first experiments was a demonstration of the switching principle. Later, these hybrid switching devices were used in more advanced system experiments in order to demultiplex a low bit rate channel (e.g. 10 Gbit/s) from a multiplexed high bit rate data signal, up to 160 Gbit/s [48,49,51,53,54]. Recently, interferometric devices (Mach-Zehnder and Sagnac interferometers) with integrated SOAs were realized as switching elements for OTDM experiments. More details and ample reference to recent work in this field is given in [55].

9.2.3   Optical Timing Extraction Technology

Timing extraction, which extracts the timed clock from the received optical signals, is one of the key functions for constructing high-speed optical transmission systems. All transmission apparatuses such as repeaters and demultiplexers cannot operate without a clock. The requirements for optical timing extraction are as follows; ultra-fast operation, low phase noise, high sensitivity and polarization independence. In particular, ultra-low phase noise giving less than 1 ps rms jitter is required to develop a 100 Gbit/s system. The highest operating speed of currently used electrical timing extraction circuits is as high as 40 GHz, determined by the microwave mixer used as a phase detector in a phase-lock loop (PLL) circuit [56]. In order to raise the operating speed, many approaches based on photonics technology have been studied, as shown in Table 9.3. They are classified into three approaches: optical tank circuits using a Fabry-Perot resonator [57] or Brillouin gain in fibre [58], injection-locking of LDs [59–65] or EDF lasers [66–68], and phase-lock loop circuits using electrical voltage-controlled oscillators (VCOs) [69–75]. An optical tank circuit features ultra-high-speed operation and a simple configuration originating from its passive structure. The optical signal created by return-to-zero intensity modulation consists of a continuous component and line components in an optical frequency domain. All-optical timing extraction can be achieved by extracting the carrier spectral component f_c and line spectral components $f_c \pm f_0$, where f_0 denotes the data clock frequency. By using a Fabry-Perot optical tank circuit, an optical clock has been extracted from 2 Gbit/s pseudo-random optical data [57]. The injection locking utilises self-pulsating (SP) LDs such as multi-electrode LDs [59–62] or optical inverters which alternate between TE and TM modes [63], whose output repetition frequency is locked to that of an injected optical pulse train. Using these techniques, 5 GHz and 10 GHz clocks were generated all-optically. A wavelength- and polarization-insensitive clock recovery module based on an SP DFB-LD was also developed and tested in a 10 Gbit/s, 105 km transmission experiment [62]. These two approaches are intriguing, since the clock is generated all-optically. So far their performance, including rms jitter and relative phase error, is not good enough for speeds over 50 Gbit/s. An

Table 9.3. High-speed timing extraction

Principle		Speed (Gbit/s)	Author	Year	Ref.
Optical tank circuit	Fabry-Perot etalon	2	Jinno (NTT)	1992	57
	Brillouin gain in fibre	5	Miyamoto (NTT)	1993	58
Injection locking	Self-pulsating DFB-LD	0.2	Jinno (NTT)	1988	59
		5	Barnsley (BT)	1992	60
		18	As (HHI)	1993	61
		10	Sartorius (HHI)	1998/99	62
	Mode-locked LD	10	Takayama (NTT)	1990	63
		5	Ono (NTT)	1994	64
		10	Ono (NEC)	1995	65
	Mode-locked fibre laser	1	Smith (BT)	1992	66
		40	Ellis (BT)	1993	67
		20	Patrick (BT)	1994	68
Phase lock loop (PLL)	Electrical PLL	40	Ellis (BT)	1993	56
	LiNbO$_3$ mod.	14	Takayama (NTT)	1992	69
	Gain mod. in TW-SOA	1	Kawanishi (NTT)	1988	70
		50	Kawanishi (NTT)	1993	71
	FWM in TW-SOA	50	Kamatani (NTT)	1994	72
		100	Kawanishi (NTT)	1994	73
		200/400	Kamatani (NTT)	1994/95	74
	FWM in fibre	20	Saito (NEC)	1993	75

TW-SOA, travelling-wave semiconductor optical amplifier;

FWM, four-wave mixing

all-optical signal generator, which used an all-optical clock recovery scheme based on a mode-locked Er fibre laser [66,67] and a NOLM regenerator [76], has also been demonstrated. It restored the pulse timing and removed noise and intensity fluctuations [77], with, however, insufficient performance so far. Nevertheless, this technique could be applied to very long 1R repeatered transmission systems. A PLL shows no phase-error and complete retiming is, in principle, possible. The question is how fast the PLLs can operate. Operation speed is determined by the phase detector used to detect the phase difference between the input signal and a local clock. Figure 9.17 shows the principle of a PLL circuit, in which a TW-SOA is used as an all-optical phase correlator [70]. Because the SOA gain is instantaneously modulated by the intense optical clock, the cross-correlation (Δf component) between

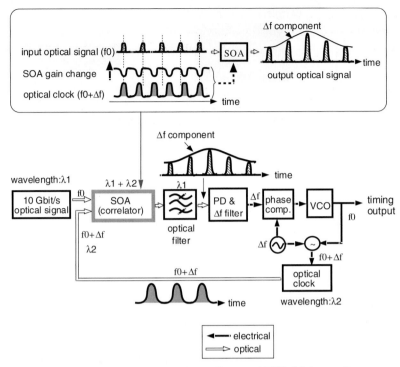

Fig. 9.17. Principle of PLL based on all-optical TW-SOA correlator

the 10 Gbit/s signal and the (10 GHz + Δf) clock driven by a VCO can
be obtained all-optically. The cross-correlated Δf signal, which contains the
phase difference between the optical signal and the optical clock, is compared
with the reference Δf signal in a conventional low-speed electrical phase comparator, the output of which is returned to the VCO to close the PLL. With
this PLL configuration, a 6.3 GHz retimed signal has been recovered using
a residual 6.3 GHz component in a 100 Gbit/s optical TDM signal. In the case
of completely multiplexed TDM signals, there is no low-frequency clock component capable of being extracted. To extract the retimed clock, the phase
detection must be processed at the multiplexed clock frequency, say 100 GHz
for a 100 Gbit/s signal. By applying a PLC multiplier, which generates an
N-times multiplied optical clock frequency to the above PLL, the original
6.3 GHz clock has been extracted from a completely multiplexed 50 Gbit/s
signal [71]. This indicates that the SOA phase detector has adequate response
to operate at 50 GHz or beyond. A prescaler PLL circuit, which extracts the
prescaled frequency clock of 6.3 GHz, 1/16th of 100 GHz, from a randomly
modulated 100 Gbit/s signal, has also been demonstrated [72]. Figure 9.18
shows the configuration.

  This utilises the harmonic frequency components that the short optical
clock possesses and the high-speed four-wave-mixing process in the TW-SOA

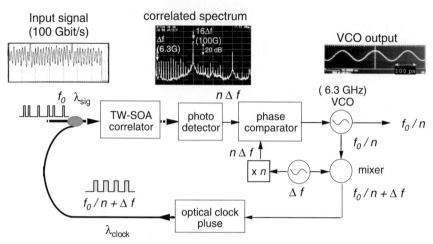

Fig. 9.18. Configuration of prescaled PLL

instead of gain-saturation. A prescaled 6.3 GHz clock was recovered from a 100 Gbit/s TDM optical signal using the 100 GHz phase comparison output obtained as the cross-correlation ($16 \times \Delta f$ component) between a 100 Gbit/s optical signal and a short (< 10 ps) 6.3 GHz $+ \Delta f$ optical clock pulse train in the FWM light; clock pulse multiplication is not required. The rms jitter is measured to be less than 0.3 ps. In order to increase the correlation signal level, pulses of less than 1 ps duration from the SC sources are used for the short optical clock pulse source and the optical TDM signal source as well. With this method, the prescaled frequency clock of 6.3 GHz, that is 1/64th of 400 GHz, can be extracted from a randomly modulated 400 Gbit/s signal [74]. This method has been applied to the latest 400 Gbit/s TDM and $1-3$ Tbit/s TDM/WDM transmission experiment described in Sect. 9.3.

9.2.4 High-Speed Optical Waveform Measurement

Optical waveform measurement with high temporal resolution is one of the key techniques in developing ultra-high-speed optical TDM transmission systems. In particular, eye diagram measurements, which measure variations or fluctuations in randomly modulated optical waveforms, are needed to evaluate very-high-speed optical transmission characteristics. Optical sampling, where optical waveforms to be measured are sampled (gated) by intense short optical pulses all-optically, provides a powerful method for measuring high-speed optical waveforms [78,79]. Several all-optical gates have been studied to perform the necessary cross-correlation between sampling and signal pulses. These include sum-frequency generation (SFG) in nonlinear crystals, XPM in fibres and FWM in fibres or SOAs. Here, recent SFG-based optical sampling having less than 1 ps resolution will be described together with its effective

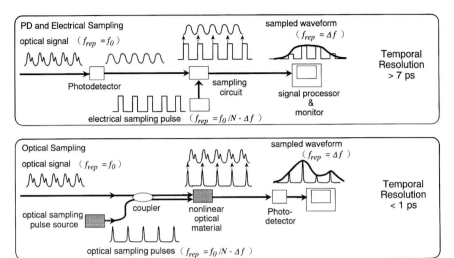

Fig. 9.19. Principle of optical sampling

data. Figure 9.19 compares the principles of optical sampling and conventional electrical sampling used for optical waveform measurements.

Temporal resolution of the conventional approach is limited by the photodetector bandwidth as well as that of the electrical sampler used. Optical sampling can circumvent the electronics limitation and its resolution is determined by the sampling pulse width, provided that the overall nonlinear response is sufficiently fast. This is the case for short SFG crystals. Also, the timing jitter of the sampling pulse should be minimized to achieve best temporal resolution of less than 0.6 ps [78]. Note that the timing jitter of the signal pulse stream should be evaluated itself; thus the conventional autocorrelation technique using second-harmonic generation (SHG) cannot be adopted in the case under consideration here. Figure 9.20 depicts the experimental set-up of optical sampling used for eye diagram measurements of a 100 Gbit/s optical signal [79]. As the signal source, 10 GHz 4.4 ps optical pulses from a stabilized mode-locked EDF laser were used. After modulation by a LiNbO$_3$ optical modulator, the 10 Gbit/s optical signal was multiplexed to a 100 Gbit/s optical TDM signal with a PLC-MUX inserted between two EDFAs. The 100 Gbit/s optical signal and the 10 GHz electrical clock f_0 were input into the optical sampling system. The timing clock generator generated electrical timing clocks of $f_0 - \Delta f$ and $(f_0 - \Delta f)/1024$.

These clocks were used for generating nearly 10 GHz optical pulses from the second EDF laser and for extracting every 1024th pulse from the 10 GHz pulses by the second modulator, respectively. The extracted 9.7 MHz optical pulses were compressed to 0.4 ps duration by the SC technique and amplified to 230 W peak by the EDFA. Both, the 100 Gbit/s signal and the optical sampling pulses were combined with a polarization beam splitter and cou-

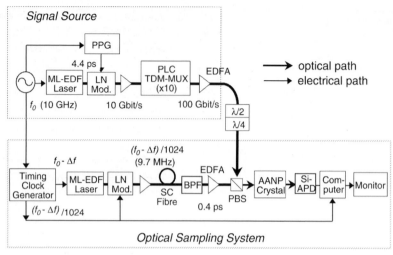

Fig. 9.20. Experimental set-up of optical sampling

pled into the organic AANP (2-adamantylamino-5-nitropyridine) crystal [80] for type II phase-matching SFG. The cross-correlation signals (SF light) at $(f_0 - \Delta f)/1024 \, (= 9.7 \, \text{MHz})$ were detected with an Si-APD and the obtained electrical signals were processed by a computer to be displayed on a monitor.

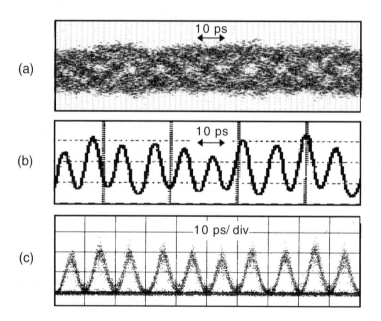

Fig. 9.21a–c. 100 Gbit/s eye-diagram

Figure 9.21a–c compares the waveforms of the 100 Gbit/s optical signal as measured by a 50 GHz electrical sampling oscilloscope (with a 50 GHz pin-photodiode), a streak camera, and the proposed optical sampling system, respectively.

It is apparent that the optical sampling can accurately measure the eye diagram of the 100 Gbit/s optical signal, whereas the streak camera shows the all mark rough pattern such as $\{1, 1, 1, 1\}$, because only the averaged waveforms could be measured at 1.5 μm due to its low sensitivity as well as limited response. This optical sampling system has been applied not only for evaluating ultra-fast optical devices such as mode-locked EDF lasers and all-optical DEMUXs, but also for monitoring ultra-fast optical TDM transmission characteristics as shown later.

9.3   Demonstration of OTDM and OTDM/WDM Transmission

All-optical time-domain signal processing technologies are being developed for realizing ultra-high bit rate OTDM transmission and OTDM-based WDM transmission as well. Here, recent progress on very-high-speed OTDM transmission experiments at up to 400 Gbit/s and very large capacity OTDM/WDM transmission experiments at up to 3 Tbit/s are introduced together with the essential technologies.

9.3.1   100–400 Gbit/s OTDM Transmission Experiment

In 1993, utilising newly developed key technologies including TL pulse sources, MUX/DEMUX circuits and timing extraction PLLs, the first successful 100 Gbit/s–50 km transmission was conducted by eightfold-OTDM together with twofold polarization multiplexing (2-PDM), where the actual switching speed was as high as 50 Gbit/s [81,82]. Since then, OTDM transmission technologies have been made a lot of progress toward much faster and longer optical transmission systems. Table 9.4 summarizes the development of OTDM transmission experiments together with the key technologies. As listed here, by improving these key technologies, including TL pulse sources, MUX/DEMUX circuits and timing extraction PLLs, 100 Gbit/s–35 km, 50 km, 200 km and 500 km transmission [73,83–85], 200 Gbit/s–100 km transmission [86], and lately 400 Gbit/s–40 km transmission [87] have been achieved only by the OTDM technique. Recently, using low-noise SC pulses as the WDM source, combined OTDM × WDM transmission experiments have been successfully demonstrated at 400 Gbit/s (100 Gbit/s × 4 channels) [88], 1 Tbit/s (100 Gbit/s × 10 channels) [89], 1.4 Tbit/s (200 Gbit/s × 7 channels) [90] and lately 3 Tbit/s (160 Gbit/s × 19 channels) capacity [91]. Figure 9.22 depicts the experimental set-up of the 100 Gbit/s transmission experiment over 200 km [73]. A wavelength-tunable mode-locked Er fibre laser

Table 9.4. Comparison of all-optical TDM and TDM/WDM transmission experiments

Signal Pulse Width (ps)	MUX	Fibre Length (km)	TW-SOA-PLL Principle/ Speed (GHz)	Jitter (ps)	DEMUX Control/ Pulse (ps)	Principle/ Speed (Gbit/s)	Total bit rate (Gbit/s)	Re
Tunable ML-EDFL 7.5	8-OTDM + 2-PDM	50	Gain mod. 6.3	2.2	Gain Sw. DFB-LD 7	NOLM 50	100	81
Tunable ML-EDFL 3.5	8-OTDM + 2-PDM	100 (1 Rep.)	Gain mod. 6.3	1.7	Gain Sw. DFB-LD 7	NOLM 50	100	82
Tunable ML-EDFL 3.5	16-OTDM	35	Gain mod. 6.3	0.8	Gain Sw. DFB-LD 5	NOLM 100	100	83
Tunable ML-EDFL 3.5	16-OTDM	50	Gain mod. 6.3	0.8	Tunable ML-EDFL 3.5	NOLM 100	100	84
Tunable ML-EDFL 3.5	16-OTDM	200 (4 Rep.)	PI-FWM 100	0.3	Tunable ML-EDFL 3.5	PI-FWM 100	100	73
Tunable ML-EDFL 3.5	16-OTDM	500 (11 Rep.)	PI-FWM 100	0.3	Tunable ML-EDFL 3.5	PI-FWM 100	100	85
SC pulse 2.1	32-OTDM	100 (2 Rep.)	FWM 200	0.24	Tunable ML-EDFL 3.5	FWM 200	200	86
SC pulse 1	40-OTDM	40	FWM 400	< 0.2	SC pulse 1.5	FWM 400	400	87
SC pulse 2.1	16-OTDM + 4-WDM	100 (Rep. less)	FWM 100	< 0.3	Tunable ML-EDFL 3.5	FWM 100	400	88
SC pulse 3.5	10-OTDM + 10-WDM	40	FWM 100	< 0.2	SC pulse 3.5	FWM 100	1000	89
SC pulse 2.1	20-OTDM + 7-WDM	50	FWM 200	0.14	SC pulse 3.5	FWM 200	1400	90
PM-SC pulses 2.1	16-OTDM + 19-WDM	40	FWM 160	0.14	SC pulse 3.5	FWM 160	3000	91

TW-SOA, travelling-wave semiconductor-optical amplifier;
ML-EDFL, mode-locked Er-doped fibre laser;
PDM, polarization division multiplexing; NOLM, nonlinear-optical loop mirror;
SC, supercontinuum; PI-FWM, polarization-independent four-wave mixing

(ML-EDFL) [15,16] provides stable 6.3 GHz, 3.5 ps TL pulses for external modulation ($2^{11} - 1$, pseudo-random binary sequence (PRBS)) by a LiNbO$_3$ modulator. A 16:1 planar lightwave circuit (PLC) multiplexer stably multiplexes the baseline 6.3 Gbit/s signal into 100 Gbit/s. The TDM 100 Gbit/s signal is then transmitted through five fibres connected via four in-line Er^{3+}-doped fibre amplifiers. The centre wavelength of the ML-EDFL is tuned to the zero-dispersion wavelength of the 200 km fibre. At the receiver side, a timing extraction PLL [72] using a TW-SOA, used as a phase detector,

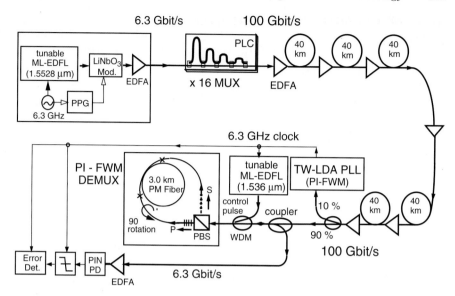

Fig. 9.22. 100 Gbit/s, 200 km OTDM transmission experiment

extracts the prescaled clock of 6.3 GHz from the received 100 Gbit/s signal. Using the FWM process instead of gain-modulation as adopted in the previous PLL [70,71], very stable polarization-independent timing extraction is achieved with jitter as low as 0.3 ps. The extracted 6.3 GHz clock is used to drive both the all-optical DEMUX and the optical receiver. Finally, a polarisation-insensitive FWM demultiplexer [31] (see Sect. 9.2.2) demultiplexes the 100 Gbit/s signal into the 6.3 Gbit/s original. With this configuration, a 100 Gbit/s optical signal, 16 × 6.3 Gbit/s, has been successfully transmitted through a 200 km-long fibre without any bit error. Figure 9.23 shows a 400 Gbit/s OTDM transmission experiment [87].

To generate sufficiently short TL optical pulses required for a baseline signal and a control (pump) pulse used for demultiplexing, the SC technique [18–22] (see Sect. 9.2.1) has been used. The initial, low-jitter 0.98 ps signal pulses at 1565 nm wavelength are generated by the SC generator pumped by the stabilized ML-EDF laser [16]. After random modulation by the LiNbO$_3$ modulator, the 10 Gbit/s signal pulses are time-division-multiplexed by a factor of 40 using the all-optical PLC-MUX (see Fig. 9.8a,b) to yield stable 400 Gbit/s test signals. Note that all the transmitter components including EDFAs are polarization-maintaining to assure long-term stability. The biggest problem associated with subpicosecond pulse transmission is the influence of the dispersion slope of the transmission fibre. This slope produces oscillations near the trailing edge of the transmitted pulses, even if their centre wavelength coincides with the zero-dispersion wavelength of the fibre. Thus the 40 km of dispersion-shifted fibre is followed by 8.2 km of dispersion-slope compensation

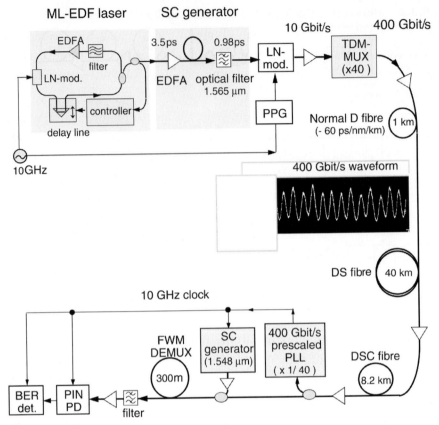

Fig. 9.23. 400 Gbit/s, 40 km OTDM transmission experiment

(DSC) fibre. Also, 1 km normal dispersion fibre is inserted to avoid nonlinearity. The fine dispersion adjustment is performed by measuring the 400 Gbit/s pulse waveform with the optical sampling method [78,79], as shown in the inset. After transiting the 40 km fibre, the 400 Gbit/s signal is split and introduced into the PLL timing extraction circuit and the all-optical DEMUX via EDFAs. The PLL successfully extracts the prescaled 10 GHz clock directly from the 400 Gbit/s signal with a timing jitter of < 0.2 ps, using the compressed local clock of 1.5 ps [74]. The all-optical DEMUX using FWM in 300 m fibre [33] demultiplexes the 400 Gbit/s signal to 10 Gbit/s. With this configuration, a 400 Gbit/s OTDM signal was successfully transmitted over 40 km for the first time. Recently, a 640 Gbit/s OTDM transmission experiment over 60 km has been conducted using a 640 fs pulse train from a mode-locked fibre laser and walk-off free, dispersion-flattened NOLM demultiplexer [92].

9.3.2   400 Gbit/s to 3 Tbit/s OTDM/WDM Transmission Experiments

The construction of large-capacity flexible optical networks, utilising both OTDM and WDM technologies, will be of vital importance. In these networks broadband low-noise optical sources, such as the SC pulse sources, are expected to play a major role. In 1996, the first OTDM/WDM transmission experiment with a 400 Gbit/s total capacity was demonstrated using 100 Gbit/s × 4 WDM channels generated from a single SC pulse source [88]. This experiment utilised the striking feature of the SC source: it could generate short pulses less than 0.3 ps over the continuous spectral range (200 nm), and multi-wavelength transform-limited short pulses could easily be selected by filtering with passive optical filters. Because the optical frequency characteristics, including frequency separation and stability, are determined by the filtering devices, it is easy to generate dense WDM pulses from a single SC pulse. Using an AWG filter as a WDM demultiplexer/multiplexer, 100 Gbit/s × 10 WDM channels (total capacity of 1 Tbit/s) OTDM/WDM transmission has successfully been demonstrated [89]. More recently, 1.4 Tbit/s transmission by 200 Gbit/s OTDM × 7 channel WDM [90] and 3 Tbit/s transmission by 160 Gbit/s OTDM × 19 channel WDM [91] have also demonstrated. Figure 9.24 illustrates the experimental set-up of 1 Tbit/s OTDM/WDM transmission. The SC pulse generator consists of a stabilized actively mode-locked EDF laser that outputs 10 GHz, 1572 nm, 3.5 ps pump pulses, EDFA to amplify the pump pulses to a peak power of 1.5 W, and a 3 km single-mode dispersion-shifted fibre (SC fibre) for SC generation. The WDM SC spectra filtered with the AWG DEMUX/MUX have more than 80 nm bandwidth, as shown in Fig. 9.24. The DEMUX/MUX consists of two AWGs directly connected, having 140 GHz channel bandwidth and 400 GHz (3.2 nm) channel spacing. The adjacent channel crosstalk was less than −25 dB for all 10 channels ranging from 1533.6 nm to 1562.0 nm. The 10 WDM channels are modulated by a common LiNbO$_3$ intensity modulator and time-division multiplexed to produce 100 Gbit/s × 10 channels OTDM/WDM signals by a 10 × PLC OTDM multiplexer. The OTDM/WDM signals are then amplified to 5 dB m per channel on average by a fluoride-based broadband EDFA. They are transmitted over 40 km dispersion-shifted fibre of zero-dispersion wavelength of 1561.3 nm. The output signals are filtered by three 3 nm optical bandpass filters in series to reduce the channel crosstalk. Dispersion-compensating fibres were used to compensate for the transmission fibre dispersion up to 74 ps/nm. The 3.5 ps, 100 Gbit/s pulse waveforms before and after transmission at channel 5 are also shown in Fig. 9.24. With dispersion compensation, the output pulse waveform recovered its original pulse shape and the 100 Gbit/s signals were fed into a prescaled PLL timing-extraction circuit and an all-optical FWM-based DEMUX. The demultiplexed 10 Gbit/s signal was directed to an optical receiver and the BER performance was measured. With this set-up, 1 Tbit/s error-free transmission

368   M. Saruwatari

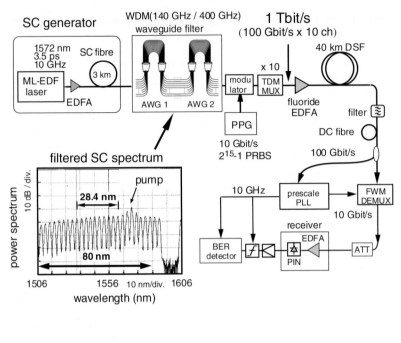

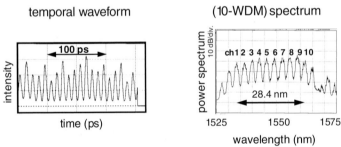

Fig. 9.24. 1 Tbit/s (100 Gbit/s × 10 channel) OTDM/WDM transmission experiment

over 40 km has successfully been demonstrated utilising the 100 Gbit/s × 10 channel OTDM/WDM technique. In order to increase the total transmission capacity, it is essential to increase the speed of the OTDM signal itself. By employing 200 Gbit/s OTDM signals together with 7 channel WDM, 1.4 Tbit/s signals have successfully been transmitted within the same optical bandwidth [90]. The main differences between the 1.4 Tbit/s and 1 Tbit/s experimental set-ups are as follows. The AWG filters have a 300 GHz bandwidth and 600 GHz channel spacing in order to increase the ratio of bit rate versus channel spacing. The shorter pulse width of 2.1 ps is used to apply twentyfold all-optical multiplexing. All the transmitter components such as AWG-WDM filters, PLC TDM-MUXs and booster EDFAs are polarization-maintaining in order to stabilize the 1.4 Tbit/s signal. All the channels are in the normal dispersion region of the transmission fibre to suppress spectral broadening caused by anomalous dispersion, and common dispersion compensation with a 1.3 μm zero-dispersion fibre and common WDM-DEMUX with an AWG filter are used. With this set-up, 1.4 Tbit/s OTDM/WDM signals have been transmitted over 50 km without any bit errors. More recently, 3 Tbit/s (160 Gbit/s × 19 channels) OTDM/WDM transmission through 40 km of dispersion-shifted fibre has been reported, employing two sets of low-noise, flatly broadened SC WDM pulse sources and ultra-broadband EDFAs [91]. The 3 Tbit/s signal generator consisted of two OTDM/WDM signal generators; one for the shorter wavelength region (1540−1566 nm) and one for the longer wavelength region (1570−1609 nm). Each generator was composed of a 3 ps, 10 GHz ML-EDF laser, optical modulator, optical amplifier, and SC fibre, all of which were polarization-maintaining so as to stabilize the SC output spectrum. The generated 10 Gbit/s SC signal pulses were time-division multiplexed by 16 times and were spectrally sliced and recombined by two sets of AWG filters with 450 GHz spacing to generate the 3 Tbit/s signal. The 3 Tbit/s OTDM/WDM signal was amplified by a 70 nm bandwidth tellurite-based EDFA consisting of two-stage EDFAs and an intermediate gain equalizer to yield flat-gain characteristics. The zero-dispersion wavelength of the transmission DSF was 1530 nm; thus all WDM channels were in the anomalous dispersion region. At the receiver, each channel was optically preamplified, demultiplexed with an AWG filter followed by a dispersion-compensating fibre, and then demultiplexed into a 10 Gbit/s signal by an all-optical DEMUX driven by a 10 GHz clock extracted from the 160 Gbit/s OTDM signal. With this set-up, 3 Tbit/s error-free transmission through a 40 km DS fibre has successfully been demonstrated utilising both sixteenfold OTDM and nineteenfold WDM techniques.

References

1. A. Takada, T. Sugie, and M. Saruwatari, "High-speed picosecond optical pulse compression from gain-switched 1.3-μm distributed feedback-laser diode (DFB-LD) through highly dispersive single-mode fiber," IEEE J. Lightwave Technol. **LT-5**, 1525 (1987)
2. M. Suzuki, H. Taga, H. Tanaka, N. Edagawa, K. Utaka, S. Yamamoto, Y. Matsushima, K. Sakai, and H. Wakabayashi, "Transform-limited optical pulse generation up to 20 GHz repetition rate by sinusoidally driven InGaAsP electroabsorption modulator," *Conf. Lasers and Electro-Optics (CLEO'92)*, postdeadline paper CPD 26 (1992)
3. K. Suzuki, K. Iwatsuki, S. Nishi, M. Saruwatari, K. Sato, and K. Wakita, "2.5 ps soliton pulse generation at 15 GHz with monolithically integrated MQW-DFB-LD/MQW-EA modulator and dispersion decreasing fiber," *Techn. Digest of Optical Amplifiers and their Applications*, OSA Techn. Digest Series 14 (Opt. Soc. America, Washington, DC 1993), p. 314

4. Y.K. Chen, M.C. Wu, T. Tanbun-Ek, R.A. Logan, and M.A. Chin, "Subpicosecond monolithic colliding-pulse mode-locked multiple quantum-well lasers," Appl. Phys. Lett. **58**, 1253 (1991)

5. P.B. Hansen, G. Raybon, U. Koren, B.I. Miller, M.G. Young, M. Chien, C.A. Burrus, and R.C. Alferness, "5.5-mm long InGaAsP monolithic extended-cavity laser with an integrated Bragg-reflector for active mode-locking," IEEE Photon. Techol. Lett. **4**, 215 (1992)

6. A. Takada, K. Sato, M. Saruwatari, and M. Yamamoto, "Pulse width tunable subpicosecond pulse generation from an actively modelocked monolithic MQW laser/MQW electroabsorption modulator," Electron. Lett. **30**, 898 (1994)

7. Y. Katagiri, A. Takada, S. Nishi, H. Abe, Y. Uenishi, and S. Nagaoka, "Passively mode-locked micromechanically-tunable semiconductor lasers," IEICE Trans. Electron. **E81-C**, 151 (1998)

8. R. Ludwig, S. Diez, A. Ehrhardt, L. Küller, W. Pieper, and H.G. Weber, "A tunable femtosecond modelocked semiconductor laser for applications in OTDM-systems," IEICE Trans. Electron. **E81-C**, 140 (1998)

9. S. Arahira, S. Kutsuzawa, Y. Matsui, and Y. Ogawa, "Higher order chirp compensation of femtosecond mode-locked semiconductor lasers using optical fibers with different group-velocity dispersions," IEEE J. Select. Topics Quantum Electron. **2**, 480 (1996)

10. K. Sato, I. Kotaka, A. Hirano, M. Asobe, Y. Miyamoto, N. Shimizu, and K. Hagimoto, "High-repetition frequency pulse generation at 102 GHz using modelocked lasers integrated with electroabsorption modulators," Electron. Lett. **34**, 790 (1998)

11. A. Takada and H. Miyazawa, "30 GHz picosecond pulse generation from actively mode-locked erbium-doped fibre laser," Electron. Lett. **26**, 216 (1990)

12. G.T. Harvey and L.F. Mollenauer, "Harmonically mode-locked fiber ring laser with an internal Fabry-Perot stabilizer for soliton transmission," Opt. Lett. **18**, 107 (1993)

13. X. Shan and D.M. Spirit, "Novel method to suppress noise in harmonically modelocked erbium fibre lasers," Electron. Lett. **29**, 979 (1993)

14. H. Takara, S. Kawanishi, M. Saruwatari, and K. Noguchi, "Generation of highly stable 20 GHz transform-limited optical pulses from actively mode-locked Er^{3+}-doped fibre lasers with an all-polarization maintaining ring cavity," Electron. Lett. **28**, 2095 (1992)

15. H. Takara, S. Kawanishi, and M. Saruwatari, "20 GHz transform-limited optical pulse generation and bit-error-free operation using a tunable, actively modelocked Er-doped fibre ring laser," Electron. Lett. **29**, 1149 (1993)

16. H. Takara, S. Kawanishi, and M. Saruwatari, "Stabilisation of a modelocked Er-doped fibre laser by suppressing the relaxation oscillation frequency component," Electron. Lett. **31**, 292 (1995)

17. H. Takara, S. Kawanishi, and M. Saruwatari, "Multiple-frequency modulation of an actively modelocked laser used to control optical pulse width and repetition frequency," Electron. Lett. **30**, 1143 (1994)

18. K. Mori, T. Morioka, H. Takara, and M. Saruwatari, "Continuously tunable optical pulse generation utilizing supercontinuum in an optical fiber pumped by an amplified gain-switched LD pulses," *Techn. Digest of Optical Amplifiers and their Applications*, OSA Techn. Digest Series 14 (Opt. Soc. America, Washington, DC 1993), p. 190

19. T. Morioka, S. Kawanishi, K. Mori, and M. Saruwatari, "Nearly penalty-free, < 4 ps supercontinuum Gbit/s pulse generation over 1535−1560 nm," Electron. Lett. **30**, 790 (1994)

20. T. Morioka, S. Kawanishi, K. Mori, and M. Saruwatari, "Transform-limited, femtosecond WDM pulse generation by spectral filtering of gigahertz super-continuum," Electron. Lett. **30**, 1166 (1994)

21. T. Morioka, K. Uchiyama, S. Kawanishi, S. Suzuki, and M. Saruwatari, "Multi-wavelength picosecond pulse source with low jitter and high optical frequency stability based on 200 nm supercontinuum filtering," Electron. Lett. **31**, 1064 (1995)

22. K. Mori, H. Takara, S. Kawanishi, M. Saruwatari, and T. Morioka, "Flatly broadened supercontinuum spectrum generated in a dispersion decreasing fibre with convex dispersion profile," Electron. Lett. **33**, 1806 (1997)

23. S. Kawanishi and O. Kamatani, "All-optical time division multiplexing using four-wave mixing," Electron. Lett. **30**, 1697 (1994)

24. S. Kawanishi, K. Okamoto, M. Ishii, O. Kamatani, H. Takara, and K. Uchiyama, "All-optical time-division-multiplexing of 100 Gbit/s signal based on four-wave mixing in a travelling-wave semiconductor laser amplifier," Electron. Lett. **33**, 976 (1997)

25. I. Shake, H. Takara, S. Kawanishi, and M. Saruwatari, "High-repetition-rate optical pulse generation by using chirped optical pulses," Electron. Lett. **34**, 792 (1998)

26. H. Takara, I. Shake, K. Uchiyama, O. Kamatani, S. Kawanishi, and K. Sato, "Ultrahigh-speed optical TDM signal generator utilizing all-optical modulation and optical clock multiplication," Conf. Opt. Fiber Commun. (OFC '98), OSA Techn. Digest Series (Opt. Soc. America, Washington, DC 1998), postdeadline paper PD16

27. T. Morioka, M. Saruwatari, and A. Takada, "Ultrafast optical multi/demulti-plexer utilising optical Kerr effect in polarization-maintaining single-mode fibres," Electron. Lett. **23**, 453 (1987)

28. T. Morioka, H. Takara, K. Mori, and M. Saruwatari, "Ultrafast reflective optical Kerr demultiplexer using polarization rotation mirror," Electron. Lett. **28**, 521 (1992)

29. D.M. Patrick and A.D. Ellis, "Demultiplexing using crossphase modulation-induced spectral shifts and Kerr polarization rotation in optical fibre," Electron. Lett. **29**, 227 (1993)

30. P.A. Andrekson, N.A. Olsson, J.R. Simpson, T. Tanbun-Ek, R.A. Logan, and M. Haner, "16 Gbit/s all-optical demultiplexing using four-wave mixing," Electron. Lett. **27**, 922 (1991)

31. T. Morioka, S. Kawanishi, K. Uchiyama, H. Takara, and M. Saruwatari, "Polar-isation-independent 100 Gbit/s all-optical demultiplexer using four-wave mixing in a polarization-maintaining fibre loop," Electron. Lett. **30**, 591 (1994)

32. T. Morioka, S. Kawanishi, H. Takara, and M. Saruwatari, "Multiple output, 100 Gbit/s all-optical demultiplexer based on multichannel four-wave mixing pumped by a linearly-chirped square pulse," Electron. Lett. **30**, 1959 (1994)

33. T. Morioka, H. Takara, S. Kawanishi, T. Kitoh, and M. Saruwatari, "Error-free 500 Gbit/s all-optical demultiplexing using low-noise, low-jitter supercon-tinuum short pulses," Electron. Lett. **32**, 833 (1996)

34. R. Ludwig and G. Raybon, "All-optical demultiplexing using ultrafast four-wave-mixing in a semiconductor laser amplifier at 20 Gbit/s," Proc. 19th Europ. Conf. Opt. Commun. (ECOC '93), Montreux, Switzerland, Vol. **3**, 57 (1993)

35. S. Kawanishi, T. Morioka, O. Kamatani, H. Takara, J.M. Jacob, and M. Saruwatari, "100 Gbit/s all-optical demultiplexing using four-wave mixing in a travelling wave laser diode amplifier," Electron. Lett. **30**, 981 (1994)

36. T. Morioka, H. Takara, S. Kawanishi, K. Uchiyama, and M. Saruwatari, "Polarisation-independent all-optical demultiplexing up to 200 Gbit/s using four-wave mixing in a semiconductor laser amplifier," Electron. Lett. **32**, 840 (1996)

37. T. Morioka, K. Mori, and M. Saruwatari, "Ultrafast polarization-independent optical demultiplexer using optical carrier frequency shift through crossphase modulation," Electron. Lett. **28**, 1070 (1992)

38. D.M. Patrick and A.D. Ellis, "Demultiplexing using crossphase modulation-induced spectral shifts and Kerr polarization rotation in optical fibre," Electron. Lett. **29**, 227 (1993)

39. K. Uchiyama, S. Kawanishi, and M. Saruwatari, "Multiple-channel output all-optical demultiplexer based on TDM-WDM conversion utilizing time-division chirping control of chirped clock pulse," *Proc. 23rd Europ. Conf. Opt. Commun. (ECOC '97)*, Edinburgh, UK, Vol. **3**, 71 (1997); K. Uchiyama, S. Kawanishi, M. Saruwatari, "Multiple-channel output all-optical OTDM demultiplexer utilizing XPM-induced chirp compensation (MOXIC)," Electron. Lett. **34**, 575 (1998)

40. N.J. Doran and D. Wood, "Nonlinear-optical loop mirror," Opt. Lett. **13**, 56 (1988)

41. K.J. Blow, N.J. Doran, and B.P. Nelson, "Demonstration of the nonlinear fibre loop mirror as an ultrafast all-optical demultiplexer," Electron. Lett. **26**, 962 (1990)

42. M. Jinno and T. Matsumoto, "Ultrafast low-power and highly stable fiber Sagnac interferometer," IEEE Photon. Technol. Lett. **2**, 349 (1990)

43. A. Takada, K. Aida, and M. Jinno, "Demultiplexing of a 40 Gb/s optical signal to 2.5 Gb/s using a nonlinear fiber loop mirror driven by amplified, gain-switched laser diode pulses," *Conf. Opt. Fiber Commun. (OFC '91)*, OSA Techn. Digest Series (Opt. Soc. America, Washington, DC 1991), p. 50

44. P.A. Andrekson, N.A. Olsson, J.R. Simpson, D.J. Digiovanni, P.A. Morton, T. Tanbun-Ek, R.A. Logan, and K.W. Wecht, "Ultra-high speed demultiplexing with the nonlinear optical loop mirror," *Conf. Opt. Fiber Commun (OFC '92)*, OSA Techn. Digest Series (Opt. Soc. America, Washington, DC 1992), postdeadline papers, p. 343

45. D.M. Patrick, A.D. Ellis, and D.M. Spirit, "Bit-rate flexible all-optical demultiplexing using a nonlinear optical loop mirror," Electron. Lett. **29**, 702 (1993)

46. H. Bülow and G. Veith, "Polarization-independent switching in a nonlinear optical loop mirror by a dual-wavelength switching pulse,". Electron. Lett. **29**, 588 (1993)

47. K. Uchiyama, H. Takara, S. Kawanishi, T. Morioka, and M. Saruwatari, "Ultrafast polarization-independent all-optical switching using a polarization diversity scheme in the nonlinear optical loop mirror," Electron. Lett. **28**, 1864 (1992)

48. K. Uchiyama, S. Kawanishi, H. Takara, T. Morioka, and M. Saruwatari, "100 Gbit/s to 6.3 Gbit/s demultiplexing experiment using polarization-independent nonlinear optical loop mirror," Electron. Lett. **30**, 873 (1994)

49. K. Uchiyama, H. Takara, T. Morioka, S. Kawanishi, and M. Saruwatari, "100 Gbit/s multiple-channel output all-optical demultiplexer based on TDM-WDM conversion in a nonlinear optical loop mirror," Electron. Lett. **32**, 1989 (1996)

50. M. Eiselt, W. Pieper, and H.G. Weber, "All-optical high speed demultiplexing with a semiconductor laser amplifier in a loop mirror configuration," Electron. Lett. **29**, 1167 (1993)
51. S. Diez, R. Ludwig, and H.G. Weber, "All-optical switch for TDM and WDM/TDM systems demonstrated in a 640 Gbit/s demultiplexing experiment," Electron. Lett. **34**, 803 (1998)
52. J.P. Sokoloff, P.R. Prucnal, I. Glesk, and M. Kane, "A terahertz optical asymmetric demultiplexer (TOAD)," IEEE Photon. Technol. Lett. **5**, 787 (1993)
53. A.D. Ellis and D.M. Spirit, "Compact 40 Gbit/s optical demultiplexer using a GaInAsP optical amplifier," Electron. Lett. **29**, 2115 (1993)
54. K. Suzuki, K. Iwatsuki, S. Nishi, and M. Saruwatari, "Error-free demultiplexing of 160 Gbit/s pulse signal using optical loop mirror including semiconductor laser amplifier," Electron. Lett. **30**, 1501 (1994)
55. W. Pieper, E. Jahn, M. Eiselt, R. Ludwig, R. Schnabel, A. Ehrhardt, H.J. Ehrke, and H.G. Weber, "System applications for all-optical semiconductor switching devices," in *Photonic Networks: Advances in Optical Communications* (G. Prati, ed.), (Springer, London 1997), p. 473
56. A.D. Ellis, T. Widdowson, X. Shan, G.E. Wickens, and D.M. Spirit, "Transmission of a true single polarization 40 Gbit/s soliton data signal over 205 km using a stabilised erbium fibre ring laser and 40 GHz electronic timing recovery," Electron. Lett. **29**, 990 (1993)
57. M. Jinno and T. Matsumoto, "Optical tank circuits used for all-optical timing recovery," IEEE J. Quantum Electron. **28**, 895 (1992)
58. Y. Miyamoto, H. Kawakami, T. Kataoka, and K. Hagimoto, "New optical frequency comb gain spectrum generated by stimulated Brillouin amplifier," *Techn. Digest of Optical Amplifiers and their Applications*, OSA Techn. Digest Series 14 (Opt. Soc. America, Washington, DC 1993), p. 194
59. M. Jinno and T. Matsumoto, "All-optical timing extraction using a 1.5 µm self pulsating multielectrode DFB LD," Electron. Lett. **24**, 1426 (1988)
60. P.E. Barnsley, G.E. Wickens, H.J. Wickes, and D.M. Spirit, "A 4 × 5 Gb/s transmission system with all-optical clock recovery," IEEE Photon. Technol. Lett. **4**, 83 (1992)
61. D.J. As, R. Eggemann, U. Feiste, M. Möhrle, E. Patzak, and K. Weich, "Clock recovery based on a new type of self-pulsation in a 1.5 µm two-section InGaAsP/InP DFB laser," Electron. Lett. **29**, 141 (1993)
62. B. Sartorius, C. Bornholdt, O. Brox, H.J. Ehrke, D. Hoffmann, R. Ludwig, and M. Möhrle, "All-optical clock recovery module based on self-pulsating DFB laser," Electron. Lett. **34**, 1664 (1998); C. Bornholdt, B. Sartorius, and M. Möhrle, "All-optical clock recovery at 40 Gbit/s," *Proc. 25th Europ. Conf. Opt. Commun. (ECOC '99)*, Nice, France (1999), postdeadline papers, p. 54
63. K. Takayama, K. Habara, and A. Himeno: "High-frequency operation of an all-optical synchronization circuit", *Springer Series in Electron. and Photon.*, Vol. 29, Photonic Switching II (Springer, Berlin 1990), p. 374
64. T. Ono, Y. Yamabayashi, and Y. Sato, "5 Gbit/s all optical clock recovery circuit using dual mode locking technique," *Conf. Opt. Fiber Commun. (OFC '94)*, Vol. 4, OSA Techn. Digest Series (Opt. Soc. America, Washington, DC 1994), p. 233
65. T. Ono, T. Shimizu, Y. Yana, and H. Yokoyama, "Optical clock extraction from 10 Gb/s data pulses using monolithic mode-locked laser diode," *Proc. IEICE Japan National Conference*, Mar. 1995, paper no. B-1157 (in Japanese)

66. K. Smith and J.K. Lucek, "All-optical clock recovery using a mode-locked laser," Electron. Lett. **28**, 1814 (1992)

67. A.D. Ellis, K. Smith, and D.M. Patrick, "All optical clock recovery at bit rates up to 40 Gbit/s," Electron. Lett. **29**, 1323 (1993)

68. D.M. Patrick, "Modelocked ring laser using nonlinearity in a semiconductor laser amplifier," Electron. Lett. **30**, 43 (1994)

69. K. Takayama, K. Habara, H. Miyazawa, and A. Himeno, "High-frequency operation of a phase-locked loop-type clock regenerator using a 1 × 2 optical switch as a phase comparator," IEEE Photon. Technol. Lett. **4**, 99 (1992)

70. S. Kawanishi and M. Saruwatari, "New-type phase-locked loop using travelling-wave laser-diode optical amplifier for very high-speed optical transmission," Electron. Lett. **24**, 1452 (1988); S. Kawanishi and M. Saruwatari, "10 GHz timing extraction from randomly modulated optical pulses using phase-locked loop with travelling-wave laser-diode optical amplifier using optical gain modulation," Electron. Lett. **28**, 510 (1992)

71. S. Kawanishi, H. Takara, M. Saruwatari, and T. Kitoh, "Ultrahigh-speed clock recovery circuit using a traveling-wave laser diode amplifier as a 50 GHz phase detector," *Topical Meeting on Optical Amplifiers and Their Applications '93*, Yokohama, Japan (1993), postdeadline papers PD5

72. O. Kamatani, S. Kawanishi, and M. Saruwatari, "Prescaled 6.3 GHz clock recovery from 50 Gbit/s TDM optical signal with 50 GHz PLL using four-wave mixing in a traveling-wave laser diode optical amplifier," Electron. Lett. **30**, 807 (1994)

73. S. Kawanishi, T. Morioka, O. Kamatani, H. Takara, and M. Saruwatari, "Time-division-multiplexed 100 Gbit/s, 200 km optical transmission experiment using PLL timing extraction and all-optical demultiplexing based on polarization insensitive four-wave-mixing," *Conf. Opt. Fiber Commun. (OFC '94)*, OSA Techn. Digest Series (Opt. Soc. America, Washington, DC 1994), postdeadline paper PD23, p. 108

74. O. Kamatani and S. Kawanishi, "Prescaled timing extraction from 400 Gb/s optical signal using a phase lock loop based on four-wave-mixing in a laser diode amplifier," IEEE Photon. Technol. Lett. **8**, 1094 (1996)

75. T. Saito, Y. Yano, and N. Henmi, "10-GHz timing extraction from a 20-Gbit/s optical data stream by using a newly proposed optical phase-locked loop," *Conf. Opt. Fiber Commun. (OFC '94)*, vol. **4**, OSA Techn. Digest Series (Opt. Soc. America, Washington, DC 1994), p. 61

76. M. Jinno and M. Abe, "All-optical regenerator based on nonlinear fibre Sagnac interferometer," Electron. Lett. **28**, 1350 (1992)

77. K. Smith and J.K. Lucek, "All-optical signal regenerator," *Conf. Lasers and Electro-Optics (CLEO'93)*, postdeadline paper CPD23

78. H. Takara, S. Kawanishi, T. Morioka, K. Mori, and M. Saruwatari, "100 Gbit/s optical waveform measurement with 0.6 ps resolution optical sampling using subpicosecond supercontinuum pulses," Electron. Lett. **30**, 1152 (1994)

79. H. Takara, S. Kawanishi, A. Yokoo, S. Tomaru, T. Kitoh, and M. Saruwatari, "100 Gbit/s optical signal eye-diagram measurement with optical sampling using organic nonlinear optical crystal," Electron. Lett. **32**, 2256 (1996)

80. S. Tomaru, S. Matsumoto, T. Kurihara, H. Suzuki, N. Ooba, and T. Kaino, "Nonlinear optical properties of 2-adamantylamino-5-nitropyridine crystal," Appl. Phys. Lett. **58**, 2583 (1991)

81. S. Kawanishi, H. Takara, K. Uchiyama, T. Kitoh, and M. Saruwatari, "100 Gb/s, 50 km optical transmission employing all-optical multi/demultiplexing and PLL timing extraction," *Conf. Opt. Fiber Commun. (OFC '93)*, OSA Techn. Digest Series (Opt. Soc. America, Washington, DC 1993), postdeadline paper PD2, p. 13

82. S. Kawanishi, H. Takara, K. Uchiyama, M. Saruwatari, and T. Kitoh, "100 Gbit/s, 100 km optical transmission with in-line amplification utilizing all-optical multi/ demultiplexing and improved PLL timing extraction," *Proc. 19th Europ. Conf. Opt. Commun. (ECOC '93)*, Montreux, Switzerland, Vol. **2**, 13 (1993)

83. S. Kawanishi, H. Takara, K. Uchiyama, M. Saruwatari, and T. Kitoh, "Single polarization completely time-division-multiplexed 100 Gbit/s optical transmission experiment," *Proc. 19th Europ. Conf. Opt. Commun. (ECOC '93)*, Montreux, Switzerland, Vol. **3**, 53 (1993)

84. H. Takara, S. Kawanishi, K. Uchiyama, M. Saruwatari, and T. Kitoh, "Nearly-penalty-free, fully TDM 100-Gbit/s optical transmission by using two tunable mode-locked Er-doped fiber lasers," *Conf. Opt. Fiber Commun. (OFC '94)*, Vol. **4**, OSA Techn. Digest Series (Opt. Soc. America, Washington, DC 1994), p. 15

85. S. Kawanishi, H. Takara, O. Kamatani, and T. Morioka, "100 Gb/s, 500 km Optical Transmission Experiment," *Conf. Opt. Fiber Commun. (OFC '95)*, Techn. Digest OFC '95, ThL2 (1995), p. 287; S. Kawanishi, H. Takara, O. Kamatani, T. Morioka, and M. Saruwatari, "100 Gbit/s, 560 km optical transmission experiment with 80 km amplifier spacing employing dispersion management," Electron. Lett. **32**, 470 (1996)

86. S. Kawanishi, H. Takara, T. Morioka, O. Kamatani, and M. Saruwatari, "200 Gbit/s, 100 km time-division-multiplexed optical transmission using supercontinuum pulses with prescaled PLL timing extraction and all-optical demultiplexing," Electron. Lett. **31**, 816 (1995)

87. S. Kawanishi, H. Takara, T. Morioka, O. Kamatani, K. Takiguchi, T. Kitoh, and M. Saruwatari, "Single channel 400 Gbit/s time-division multiplexed transmission of 0.98 ps pulses over 40 km employing dispersion slope compensation," Electron. Lett. **32**, 916 (1996)

88. T. Morioka, S. Kawanishi, H. Takara, O. Kamatani, M. Yamada, T. Kanamori, K. Uchiyama, and M. Saruwatari, "100 Gbit/s × 4ch, 100 km repeaterless TDM-WDM transmission using a single supercontinuum source," Electron. Lett. **32**, 468 (1996)

89. T. Morioka, H. Takara, S. Kawanishi, O. Kamatani, K. Takiguchi, K. Uchiyama, M. Saruwatari, H. Takahashi, M. Yamada, T. Kanamori, and H. Ono, "1 Tbit/s (100 Gbit/s × 10 channel) OTDM/WDM transmission using a single supercontinuum WDM source," Electron. Lett. **32**, 906 (1996)

90. S. Kawanishi, H. Takara, K. Uchiyama, I. Shake, O. Kamatani, and H. Takahashi, "1.4 Tb/s (200 Gb/s × 7 ch), 50 km OTDM-WDM transmission experiment," Techn. Digest OECC '97, PDP2-2 (1997), p. 14

91. S. Kawanishi, H. Takara, K. Uchiyama, I. Shake, K. Mori, "3 Tbit/s (160 Gbit/s × 19ch) OTDM/WDM transmission experiment," *Conf. Opt. Fiber Commun. (OFC '99)*, OSA Techn. Digest Series (Opt. Soc. America, Washington, DC 1999) postdeadline paper PD1

92. M. Nakazawa, E. Yoshida, T. Yamamoto, E. Yamada, and A. Sahara, "TDM single channel 640 Gbit/s transmission experiment over 60 km using 400 fs pulse train and walk-off free dispersion flattened nonlinear optical loop mirror," Electron. Lett. **34**, 907 (1998)

10 Optical Hybrid Integrated Circuits

Yasufumi Yamada, Yuji Akahori, and Hiroshi Terui

10.1 Introduction

There is an increasing demand for advanced optical communication systems to deal with the enormous data flow that will arise in the emerging multimedia era. These systems require low-cost and highly functional optical modules. However, conventional optical modules based on micro-optics consist of a large number of components including laser diodes (LD), lenses, mirrors, prisms and fibres. The use of these elements also results in the need for specially designed stems or subcarriers for constructing the modules. Moreover, these components are assembled by an active alignment procedure which requires monitoring via the activation of an LD and a photodiode (PD) or the excitation of guided light in the optical components. Although this procedure ensures a micron or submicron order alignment accuracy, it tends to become complicated and time-consuming. Therefore, there is a limit to the possibility of cost reduction with the conventional optical module. In addition, it is difficult to use micro-optics type modules as highly functional devices such as multi-channel wavelength or frequency selectors.

To overcome this problem, it appears essential to develop a new optical module technique which meets the following requirements. First, the new technique should reduce the optical-module assembly cost and be suitable for mass production. Second, it should be applicable to a wide range of applications. In other words, it should be possible to use the same technique to fabricate not only simple optical transmitters and receivers but also highly functional modules for optical signal processing. Third, the module configuration should be compatible with electronic assembly technology. This is because the optical module is typically used with electronic circuits after they have been assembled on an electronic board.

In the field of electronics, integrated-circuit technology has been successful in providing highly functional and low-cost large-scale integrated circuits (LSIs). By analogy with electronics, optical integrated circuits are expected to meet these requirements via the integration of passive and active optical elements on a substrate or on a chip. There are two approaches to optical integration: monolithic and hybrid. With monolithic optical integration all the optical elements are formed on a semiconductor chip (see Chap. 11). This is based on the same concept as electronic ICs or LSIs. However, there are some significant differences between electronic and optical integration, as shown in Table 10.1, and these differences seem to make monolithic integration rather difficult. With electronic integration, most functions are implemented on a silicon chip and their integration density is becoming higher and higher. By contrast, optical functions are currently implemented by using various kinds of materials such as semiconductors, silica-glass or LiNbO$_3$. Therefore, in order to fabricate a monolithic integrated circuit, the performance of the circuit elements formed on a semiconductor chip should first be as good as that of the optical circuits formed by silica waveguides or the LiNbO$_3$. In addition, the density that can be achieved in optical integration is limited because optical circuit elements require a certain size. For example, waveguides (WGs) must be large enough to maintain the propagation mode. Circuit elements such as Y-branches or directional couplers also require a certain length. The input and output WG spacing should be larger than the fibre diameter, which is 125 μm. As a result, a larger chip is required for an optical integrated circuit than for an electronic circuit. Therefore, the fabrication yield of semiconductor circuits should also be high enough to provide an integrated circuit with a larger chip size. However, optical semiconductor device technology is not yet sufficiently mature. Finally, optical components need a fibre connection so that signals can be input and output. This requires a submicron-order accuracy in fibre-to-semiconductor device alignment unless special means are taken to relax these requirements, e.g. by monolithically integrated spot size converters (see Sect. 10.2.3 and Chap. 11, Sect. 11.3).

These difficulties can be overcome in the case of hybrid integrated circuits, where optoelectronic (OE) devices such as LDs, PDs, or semiconductor optical amplifiers (SOAs) are directly assembled on a passive WG substrate such as a silica planar lightwave circuit (PLC) as shown in Fig. 10.1. This approach

Table 10.1. Comparison of optical and electronic integration

	Electronic	Optical
Materials	Single material	Multiple materials (III-V semiconductor, SiO$_2$, LiNbO$_3$, etc.)
Integration density	As high as possible	Limited by - Waveguide dimensions - Circuit elements (bends, directional couplers, etc.) - Fibre dimensions
Signal I/O	Electrical connection	Fibre connection - Difficulty in alignment and attachment

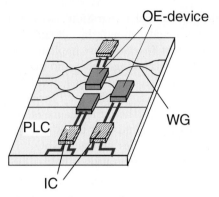

Fig. 10.1. Concept of optical hybrid integrated circuit

seems to have the following advantages over the monolithic approach. First, hybrid integration allows us to select the best combination of optical devices for achieving a desired optical function. Second, the substrate size can be increased if it is necessary to realize a desired optical function. Finally, when a silica WG is used as a passive component, fibre connection can be easily achieved even if arrayed fibres are used. However, hybrid-integration technology poses two serious difficulties. The first problem is to provide a substrate for optical hybrid integration. To achieve hybrid integrated circuits requires a substrate which has both a passive WG function and an OE-device mounting function. Here, we refer to such a substrate as a hybrid-integration platform. Another problem has been optical coupling between the active devices and the passive WGs on the substrate. Because of these difficulties, no hybrid integrated circuit has yet been realized.

Fortunately, considerable progress has been made in optical hybrid-integration technology in recent years and the above problems are now being overcome. This progress makes optical hybrid integration a realistic approach. This chapter first describes the key technologies used to overcome the difficulties outlined above. Then, recent achievements are introduced in order to show that the hybrid-integration technique has a wide range of applications. Finally, future prospects for hybrid integration will be discussed.

10.2 Key Technologies for Hybrid Integration

10.2.1 Platform for Hybrid Integration

To achieve hybrid integration as shown in Fig. 10.1, it is essential to develop a hybrid-integration platform which functions both as a passive WG and as an optical bench for OE-devices. The passive WG provides various kinds of PLCs including splitter/combiners, Mach–Zehnder interferometers and arrayed WG gratings. The optical bench acts as both a heat sink and an alignment plane for OE-devices. In addition, the platform should function as an electrical wiring board for OE-devices in order to assure their high-speed operation.

Silicon Optical Bench. Silicon (Si) is considered to be the best material for the optical bench. This is because its thermal conductivity is sufficient for it to be used as a heat sink for OE-devices. Furthermore, since the thermal expansion coefficient of Si is close to that of semiconductor devices, the thermal stress of a semiconductor device mounted on Si is less than if it were mounted on a conventional heat sink such as Cu. A precise V-groove in which to align the optical fibre can also easily be formed on a (100) Si surface by anisotropic etching. These characteristics have motivated attempts to employ Si as an optical bench.

In 1977, J.D. Crow et al. (IBM) fabricated a GaAs LD array package using V-grooved Si [1]. V-grooves for multimode optical fibres and a cylindrical lens were formed on a Si substrate. An LD was bonded to the Si substrate by using In solder. This study showed the potential of Si as an optical bench. In 1986, H. Kaufmann et al. (Swiss Federal Institute of Technology) succeeded in realizing passive alignment between a fibre and a GaAs WG assembled on a V-grooved Si substrate [2]. Since around 1990, there have been many attempts to use a V-grooved Si substrate to fabricate low-cost optical modules for optical interconnection.

Silica-waveguide on Silicon Substrate. In terms of passive optical WGs, silica WGs are superior to other WGs such as polymer or ion-diffused waveguides. The silica WG has good optical characteristics such as a low propagation loss, low coupling loss with optical fibre and good environmental stability. Therefore, a silica WG on a silicon substrate (SiO_2/Si) is considered to be a highly promising hybrid-integration platform.

Several types of SiO_2/Si substrate have been developed. Table 10.2 shows two conventional SiO_2/Si WG structures. A thin-cladding type silica-WG consists of an under-cladding layer deposited on a silicon substrate and a core ridge which is covered with a thin over-cladding layer typically $2-3\,\mu m$ thick. In terms of hybrid integration, an advantage of this WG is that its thin over-cladding layer facilitates the formation of an alignment plane for OE-devices. This is because the alignment plane is formed by etching the silica-glass layer and its height can be controlled more easily when the etching depth is smaller. Using this type of WG, H. Terui et al. (NTT) demonstrated the hybrid integration of an LD on a multimode SiO_2/Si substrate in 1985 [3]. In 1989, Y. Yamada et al. (NTT) achieved the hybrid integration of an LD and a PD on a single-mode WG with $5\,\mu m \times 5\,\mu m$ core and a $3\,\mu m$ thick over-cladding layer [4]. A single-mode 4×4 matrix switch was also fabricated by using a silica-WG and LD gate arrays in 1992 [5]. E.E.L. Friedrich et al. (LETI) [6] and C.A. Jones (BT) [7] also succeeded in LD integration using SiO_2/Si WGs with smaller cores of $2\,\mu m$ thickness in 1992 and 1994, respectively.

Table 10.2. Examples of conventional silica-waveguide structures

	bedded t pe	Thin cladding t pe
aveguide Structure	over — cladding under — cladding	over cladding
	etching depth 4	etching depth 1 2
aveguide ualit	oss 1 d c	oss 1 d c
Align ent plane	ifficult	ossible
Si heat sin	annot be used	ifficult

However, the WGs used in these early attempts of hybrid integration had
a serious problem in that their optical characteristics, such as a propagation
loss of $0.5-1\,\mathrm{dB/cm}$, made them inadequate for use as PLCs. Moreover, it
is difficult to form a directional coupler. This is an essential circuit element
for PLCs. Therefore, the thin cladding-type silica-WG can find only limited
applications.

An embedded-type silica WG whose core is surrounded by thick under-
cladding and over-cladding layers has excellent optical characteristics. It type
of WG has a low propagation loss of less than $0.1\,\mathrm{dB/cm}$, and can provide optical interferometer circuits including directional couplers. Therefore, conventional commercially available PLCs employ this embedded-type silica-WG.
However, the thick over-cladding layer makes it difficult to form an accurate
alignment plane for OE-devices. This also prevents the Si from being used as
heat sink. These aspects make the embedded-type silica-WG unsuitable as
a hybrid-integration platform.

Therefore, although the $\mathrm{SiO_2/Si}$ substrate has long been standard as
a hybrid-integration platform, there has been no WG substrate which has
both a high-quality passive WG function and a Si optical-bench function at
the same time.

PLC Platform. To solve the platform problem, a new $\mathrm{SiO_2/Si}$ substrate was
developed in 1993 as shown in Fig. 10.2 [8]. This is called a PLC platform
with a silica-on-terraced-silicon (STS) structure. The silicon with terraced
regions is used as the substrate instead of a conventional flat silicon substrate.
The STS-type PLC platform consists of three parts: a PLC region, a device-
assembly region and an electrical-wiring region.

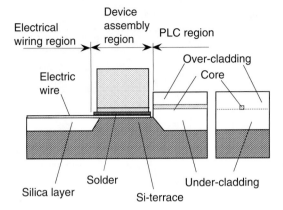

Fig. 10.2. Basic structure of the silica-on-terraced-silicon (STS) PLC platform

In the PLC region, the embedded-type silica WG is formed on the ground plane of the terraced silicon substrate. The quality of the WG is as good as that of the conventional silica-WG used for PLCs. The propagation loss of the WG on the PLC platform is less than 0.1 dB/cm and a fibre-to-WG coupling loss of 0.1 dB/facet is also achieved.

The device assembly region consists of a silicon terrace. The silicon terrace works as an alignment plane and heat sink for OE-devices. A SiO_2 layer about 0.5 μm thick is formed on the silicon terrace as an electrically insulating layer, and Au electrodes and thin AuSn solder patterns are formed on the thin SiO_2 layer on the silicon terrace. By designing the height of the WG core from the solder surface to be identical to that of the active layer of the OE-devices, the silica-WG and the OE-device can be vertically aligned simply by placing the OE-device upside-down on the solder.

In the electrical-wiring region, electrical wiring can be made on the silica layer instead of on the silicon substrate. This is an advantage, because an electrical wire on a silicon substrate has a large propagation loss for high-speed electrical signals and also has a large capacitance. This is mainly due to the slight conductivity of the silicon surface. This problem can be solved in the PLC platform by forming electrical wires on the silica layer. Therefore, the PLC platform can also function as an excellent wiring board for OE-devices.

Now that the PLC platform has been realized, one of the barriers to optical hybrid integration has been successfully removed.

10.2.2   Passive Alignment Technique

Active alignment is commonly used to achieve precise alignment between conventional optical components. For LD-to-fibre alignment, the optimum position is found by driving the LD and monitoring the output power from the fibre. Although submicron-order alignment accuracy can be achieved

Table 10.3. Passive alignment techniques

Method		Mechanical contact alignment	Index alignment	Solder-bump self-alignment
Configuration				
Alignment mechanism	Horizontal	Micromachined guiding structure	Mark alignment with help of manipulator	Self-alignment effect of molten solder
	Vertical	Controlled layer thickness of OE-device and waveguide		Solder-bump height
Reference		[6,7,9-11]	[12-14]	[15,16]

by this procedure, it is time consuming. The device alignment problem is more serious with the optical hybrid integrated circuit shown in Fig. 10.1. Although many optical chips must be assembled on the platform to realize a hybrid integrated circuit, this seems to be almost impossible as long as the conventional active alignment technique is used. Therefore, the development of a simple alignment technique is essential for optical hybrid integration.

This assembly difficulty is now being overcome by the development of a passive alignment technique. The basic approach with the passive alignment is to determine the position of an OE-device by using guiding mechanisms formed both on the platform and on the OE-device. This technique can be classified into three groups according to the guiding mechanism: mechanical contact alignment, index alignment and solder-bump self-alignment, as shown in Table 10.3.

Mechanical Contact Alignment. The concept of mechanical contact alignment is that the position of an OE-device is determined by using guiding structures such as notches or pedestals formed both on the platform and on the OE-device by micromachining. The OE-device position in the horizontal direction is determined by bringing guiding structures on the OE-device into contact with those on the Si bench. The vertical position is also determined by placing the device on a vertical alignment plane formed on the platform. Therefore, the major factor that determines the alignment accuracy in the horizontal direction is the accuracy with which guiding structures are fabricated both on the platform and on the OE-device. The vertical alignment accuracy depends mainly on how accurately the semiconductor layer is deposited as regards its thickness and the accuracy of core height of the fibre or the WG. Therefore, a careful fabrication process is needed to achieve a high alignment accuracy.

Figure 10.3 shows the technique for the Si optical bench which was developed by C.A. Armiento et al. (GTE) in 1991 [9]. Alignment pedestals for the horizontal alignment and stand-offs for the vertical alignment are formed on the Si optical bench in order to assemble a 4-channel LD array and single-mode fibres. These pedestals, which are fabricated by reactive ion etching (RIE) of the silicon surface, are about 9 μm high and have vertical walls. The stand-offs are made of polyimide. V-grooves for fibre alignment are also formed on the silicon surface with an accuracy of ±0.5 μm. An LD array has a notched edge which is at a controlled distance from the active region of the LD. Horizontal alignment is achieved simply by bringing the pedestals on the optical bench into contact with the etched notch on the LD. Since the height distance from the stand-offs and the fibre is designed to be identical to the height of the active layer of the LD, vertical alignment is also accomplished by bringing the LD surface into contact with the stand-offs. The LD is then bonded to the Si bench by In solder. Fibres are also placed in the V-grooves and attached with epoxy adhesives. A fibre-to-LD coupling loss of 12 dB was achieved.

The same technique was employed to assemble an LD on a SiO$_2$/Si WG by E.E.L. Friedrich et al. (LETI) in 1992 [6]. The achieved LD-to-WG coupling loss was 15.5 dB, which included an excess loss of 6 dB due to misalignment. This loss is higher than that in the Si optical bench. The larger excess loss indicates that it is more difficult to fabricate an accurate guiding structure in a WG substrate than in a Si optical bench. This is because the silica-glass etching must be deeper to form the guiding structure on the SiO$_2$/Si substrate. This tends to degrade the fabrication accuracy of the guiding structures both for vertical and horizontal alignment. Therefore, a significant

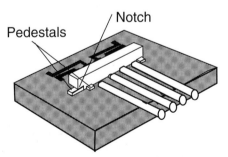

Fig. 10.3. Example of mechanical contact aligment on a silicon optical bench

improvement in the etching technique seems to be necessary before applying mechanical contact alignment to a WG substrate.

Index Alignment. The index alignment technique uses alignment marks formed both on the platform and on the LD. The horizontal position of the LD is determined by aligning these marks with the help of a high-precision manipulator. This procedure is very similar to mask alignment in the photolithographic process. Vertical alignment is accomplished simply by placing the LD on the platform. The horizontal alignment accuracy is determined by the accuracy with which the marks are formed both on the LD and on the platform, and by the placement accuracy of the manipulator. The factor that dominates the vertical alignment accuracy with the index-alignment technique is identical to that relevant for mechanical contact alignment. This technique was first demonstrated by M.S. Cohen et al. (IBM) in 1991 [12]. In 1995, K. Kurata et al. (NEC) succeeded in the index alignment of an LD on a Si bench as shown in Fig. 10.4 [13]. The key feature of this work is the use of an infrared light to observe alignment marks formed both on the LD and on the Si bench. This procedure achieves an average alignment accuracy to the marks of better than ±1 μm.

Although index alignment requires a high-precision assembly manipulator, this technique has a significant advantage over mechanical contact alignment in that it does not require a micromachining process to fabricate the guiding structure. The only element needed for the LD and the platform is alignment marks. Since these marks can be formed lithographically together with the electrodes, no additional process is required to form the marks on the LD and the platform. Therefore, this technique can be easily applied to LD assembly on a WG substrate. T. Hashimoto et al. (NTT) applied the index alignment technique to a PLC platform in 1996 [14]. Figure 10.5

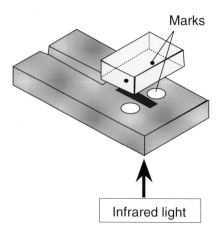

Fig. 10.4. Example of index alignment on a silicon optical bench

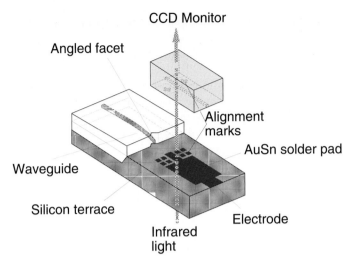

CCD Monitor

Angled facet

Alignment
marks

AuSn solder pad

Waveguide

Silicon terrace

Electrode

Infrared
light

Fig. 10.5. Index alignment on a PLC platform

shows the PLC platform for the index alignment. The alignment marks are
formed on the silicon terrace near the WG facet with a lithographic procedure
employing a Au film together with an electrode and a AuSn solder pattern.
The alignment marks are also formed on the LD surface. Therefore, horizontal alignment is achieved by observing and aligning these marks. Vertical
alignment is accomplished by using the solder surface as an alignment plane
for the vertical direction. The OE-device (e.g. LD) is first placed on the solder
formed on the Si terrace and is then fixed in position by reflowing the solder.
This procedure achieves an average alignment accuracy of better than $\pm 2\,\mu$m.
This value comprises a fabrication accuracy of the alignment marks on the
PLC platform of $\pm 1\,\mu$m and on the LDs of better than $\pm 0.5\,\mu$m, and an LD
bonding accuracy to the marks of $\pm 1\,\mu$m.

Solder-bump Self-alignment. The solder-bump self-alignment technique
utilizes the surface tension of molten solder bumps to align the LD. When
there is mis-alignment after the LD has been positioned, the surface tension
of the molten solder moves the LD toward the desired position during the
solder reflow operation, and thus the chip is bonded accurately without the
need for positioning adjustment. For surface tension to be used, the ratio of
bump height to pad diameter (aspect ratio) should be carefully controlled.
The bumps require an aspect ratio of about 1 to obtain a surface tension
above that needed for horizontal self-alignment. This means that spherical
bumps are necessary to achieve the self-alignment effect. By using a spherical
solder-bump with a diameter of about 50 μm, a horizontal accuracy of about
$\pm 1\,\mu$m can be achieved [16]. However, the vertical alignment accuracy tends
to be affected by unavoidable deviations in solder volume. As a result, the

achievable vertical alignment accuracy with spherical solder-bumps is about $\pm 2\,\mu m$.

To achieve precise alignment in the vertical direction, a stripe-type bump-bonding technique has also been developed. By using the stripe-type solder-bumps $20\,\mu m$ in height, an LD-to-fibre coupling loss of $9.5\,dB$ was achieved on a Si optical bench. This corresponds to an alignment accuracy of $\pm 1\,\mu m$ [16]. Attempts are also being made to apply this technique to SiO_2/Si WGs.

Since the development of the passive alignment techniques, Si optical benches have been used to provide low-cost LD and PD modules for optical interconnections. Although passive alignment on a SiO_2/Si WG seems to be more difficult than on a Si optical bench, the index-alignment technique also makes it easier to integrate OE-devices on the PLC platform.

10.2.3   OE-device for Hybrid Integration

The PLC platform and the passive alignment technique are of great importance. Of equal significance has been the development of OE-devices suitable for hybrid integration, namely the spot-size converter integrated LD and the waveguide PD.

Spot-size Converter Integrated LD. As mentioned above, an LD module fabricated by the passive alignment technique has a coupling loss of more than $10\,dB$. This large coupling loss is mainly due to spot-size mismatch between the LD and the WG. A conventional way to reduce the LD-to-fibre coupling loss is to use a lens. The lens effectively makes the fibre spot-size smaller thus reducing the spot-size mismatch between the LD and the fibre. A lensed fibre can reduce the coupling loss to less than $1\,dB$. However, this requires a submicron alignment accuracy. Since the alignment accuracy guaranteed by the passive alignment technique is considered to be $1-2\,\mu m$, lens coupling is not suitable for hybrid integration.

The only way to improve the coupling efficiency without reducing the alignment tolerance is to enlarge the spot-size of the LD rather than decrease the WG spot-size. Various types of spot-size converter integrated LD (SSC-LD) have been developed in recent years with a view to achieving this. Experiments on the hybrid integration of an SSC-LD on a platform have shown that the SSC-LD greatly improves the coupling efficiency.

Table 10.4 shows examples of the SSC-LDs which have already been applied to optical hybrid integration. The hybrid integration of an SSC-LD on a Si optical bench was undertaken by J.V. Collins et al. (BT) in 1995 [17]. The SSC-LD used in the experiment was a $1.55\,\mu m$ Fabry-Perot LD with a tapered active layer [18]. As the optical mode encounters the tapered region, it expands adiabatically into the underlying passive guide. The far-field angle of the LD is $\sim 11°$ both horizontally and vertically. The intrinsic coupling loss with a flat-ended single-mode fibre is $2.1\,dB$. The Si optical bench structure

Table 10.4. Examples of spot-size converter integrated LDs used for hybrid integration

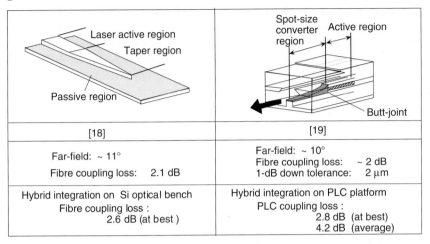

[18]	[19]
Far-field: ~ 11° Fibre coupling loss: 2.1 dB	Far-field: ~ 10° Fibre coupling loss: ~ 2 dB 1-dB down tolerance: 2 µm
Hybrid integration on Si optical bench Fibre coupling loss : 2.6 dB (at best)	Hybrid integration on PLC platform PLC coupling loss : 2.8 dB (at best) 4.2 dB (average)

used in this experiment is almost the same as that shown in Fig. 10.3. The LD was assembled on the Si optical bench by using mechanical contact alignment. The best coupling loss achieved in this experiment was 2.6 dB.

SSC-LD integration on a PLC platform was undertaken by T. Hashimoto et al. (NTT) in 1996 [14] in a way similar to that shown in Fig. 10.5. A 1.3 µm SSC-LD was used in the experiment. It had a vertically tapered passive WG which was butt-jointed to an active region [19]. The FWHM of its far field was ~ 10°, and the intrinsic coupling loss to a flat-ended fibre was ~ 2 dB. The 1 dB down-alignment tolerance was ±2 µm. This SSC-LD was assembled on the PLC platform by using index alignment. Figure 10.6 shows a histogram of the coupling loss between the SSC-LD and the silica-WG on the platform. The average and the best coupling loss was 4.2 dB and 2.8 dB, respectively.

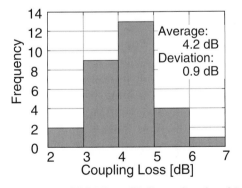

Fig. 10.6. SSC-LD-to-PLC coupling-loss histogram

A coupling loss of less than 5 dB was achieved with a probability of better than 85%.

These experiments make it clear that the combination of an SSC-LD and the passive alignment technique provides a comprehensive solution to the device assembly problem as regards optical hybrid-integration technology. The spot-size conversion technique is not only applicable to the LD but to various kinds of OE-device such as SOAs and modulators.

Photodiode for Hybrid Integration. So far, planar PDs have been most commonly used in hybrid integration. However, this type of PD requires additional optical elements such as a reflection mirror to enable it to be coupled to an optical fibre or a WG. To simplify the device-assembly procedure, it is desirable that PDs are integrated on the platform in the same way as in LD integration.

One possible way to achieve this is to use a side-illuminated waveguide PD (WGPD) [20]. Figure 10.7 shows an example of a WGPD. The WGPD has a responsivity of 0.85 A/W with a flat-ended fibre. The 1 dB down-tolerance is ±17 μm horizontally and ±2 μm vertically. The hybrid integration of the WGPD on the PLC platform is also achieved by using the index-alignment technique, and an average PD-to-WG coupling loss of 0.5 dB is achieved.

Figure 10.8 shows another type of PD designed for hybrid integration, the corner-illuminated PD. Here, an angled surface is formed on the reverse side of the substrate to guide light from a corner of the PD [21]. The light beam is reflected at the angled surface and directed to the absorption region, resulting in a high coupling efficiency leading to more than 0.85 A/W responsivity and good uniformity. This PD also has wide tolerance widths of 25 μm and 65 μm in the vertical and horizontal directions, respectively. Moreover, it has been assembled on a PLC platform with an average coupling loss of less than 0.5 dB.

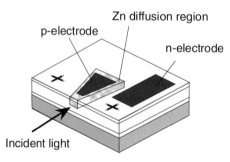

Fig. 10.7. Schematic structure of a waveguide PD

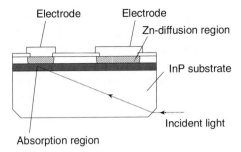

Electrode   Electrode

Zn-diffusion region

InP substrate

Incident light

Absorption region

Fig. 10.8. Schematic structure of a corner-illuminated PD

10.3   Contributions of Hybrid Integration to Optical Communication Technology

10.3.1   Application of Hybrid-Integration Technology

Since the device-assembly problem has almost been eliminated, optical hybrid-integration technology is now ready to provide a variety of optical modules. Nowadays, optical communication systems require various kinds of optical modules, which are not only low cost but highly functional as well.

Several kinds of optical-access network have been proposed and developed with the goal of realizing optical telecommunication networks called fibre-to-the-floor (FTTF) / fibre-to-the-office (FTTO) or fibre-to-the-home (FTTH). The key to the successful widespread deployment of these networks is finding a way to reduce construction costs. A barrier to this has been the high cost of conventional optical modules and this must be dealt with as a matter of urgency.

In addition, the growth of communication traffic as the multimedia society approaches requires high-capacity optical transmission, interconnection and switching systems. These systems use both wavelength-division multiplexing (WDM) and time-division multiplexing (TDM) to expand their operation throughput into the terabit-per-second range. Therefore, they require highly functional optoelectronic modules that can handle optical signals using WDM and TDM. These modules have to offer versatile functions such as providing an interface between optical and electrical circuits, optical path switching, and selection and conversion among multiple-wavelength optical signals.

Hybrid-integration technology is expected to provide optical modules for these optical systems. Not only a simpler transmitter module but also a highly functional module for optical signal processing can be realized by flip-chip bonding OE devices on the platform utilizing the same fabrication facilities. Recently, several modules that employ the PLC platform have been demonstrated successfully, thus proving that this technology has a wide range of applications. This section describes modules that have been fabricated for three different types of application. The first is an optical transceiver module for the FTTH network designed with the aim of greatly reducing costs. The second type consists of modules for WDM systems. It includes the hybrid external cavity laser diode (LD) module and the wavelength converter. These modules use the interaction between OE-devices and a PLC to provide high functionality, and they are applicable to systems using WDM technology. The third type is for TDM applications. It includes optical transmitter and receiver modules for high-speed optical interconnections. These modules showed that the PLC platform can be used for wide-bandwidth applications using TDM technology.

10.3.2   Optical Module for Fibre-optic Subscriber System

Key Issues for Cost Reduction. The passive double-star (PDS) architecture, which employs star couplers in access fibres, is currently being developed as the most cost-effective optical-access network, in which transmission facilities can be shared with many subscribers [22]. In the PDS system, it is planned that $1.3\,\mu m$ bidirectional time compression multiplexing for bidirectional transmission, time-division multiple access for point-to-multipoint transmission (TCM-TDMA) signals and $1.55\,\mu m$ one-way video signals will be transmitted simultaneously in one single-mode fibre. Therefore, a low-cost WDM transceiver module, which transmits and receives $1.3\,\mu m$ signals and multi/demultiplexes $1.3\,\mu m/1.55\,\mu m$ signals, is urgently required for the optical network unit (ONU) located at each home and office [23].

The key problems which must be solved in order to realize low-cost optical modules are summarized in Fig. 10.9. The major factors affecting the cost of the module relate to the assembly process. The first requirement is to reduce the number of module components. It is also essential to simplify the alignment technique. The most likely way to achieve these goals is via the introduction of optical hybrid integration.

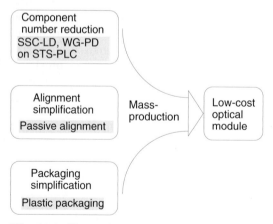

Fig. 10.9. Key problems regarding low-cost optical modules

Package simplification is also an important issue. Most of the commercially available devices for optical transmission in the 1.3 µm to 1.55 µm wavelength band are contained in hermetically-sealed packages. These are limited in terms of cost reduction due to their complicated airtight structure. By contrast, in the field of consumer electronics, package simplification has already been achieved by the adoption of plastic packaging. Plastic packaging has provided the low-cost optical devices for printers, auto-focusing cameras, and optical disk systems. Therefore, the non-hermetic plastic packaging of the optical transmission devices needs to be developed to achieve cost reduction.

Recently, a hybrid integrated WDM transceiver module has been developed in which an attempt has been made to overcome these key problems.

Module Fabrication. Figure 10.10 shows the configuration of a hybrid integrated WDM transceiver module with a PLC platform [24]. The parameters of the components are listed in Table 10.5. A 1.3 µm SSC-LD and two WG-PDs, one of which was used as a receiver PD (R-PD) and the other as a monitor PD (M-PD), were flip-chip bonded onto the Si terraces formed in the PLC platform using a passive alignment technique. The PLC consists of a 1.3 µm/1.55 µm WDM circuit and a Y-branch circuit for 1.3 µm light. A fibre block containing two fibre arrays was attached to one facet of the PLC. The LD- and PD-mounted part of the PLC was molded from epoxy resins to provide a non-hermetic seal. The resin moulding will be described in a later section.

The WDM circuit consists of a Y-shape WG with a groove set at the cross-point and a thin-film WDM filter inserted in the groove. The filter is composed of a SiO_2/TiO_2 multilayer evaporated on a polyimide film and its total thickness is 20 µm. The 1.55 µm light incident from the 1.3 µm/1.55 µm common port is reflected at the filter and exits the 1.55 µm port. This configuration has the following advantages. Two fibre connection operations can be simplified into one dual-fibre connection operation. Moreover, this fibre-one-side-connected structure makes module handling easy and provides flexibility as regards the layout design of ONU devices, because the available space for

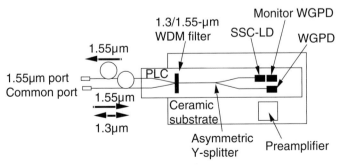

Fig. 10.10. Configuration of a WDM transceiver module

Table 10.5. Parameters of devices used in the WDM transceiver module

Device	Parameters
PLC	$\Delta = 0.45\%$, Spot size $= 4.5\,\mu\mathrm{m}$, Size$= 3\,\mathrm{mm} \times 15\,\mathrm{mm}$
SSC-LD	Total length $= 600\,\mu\mathrm{m}$, SSC $= 300\,\mu\mathrm{m}$, Horizontal spot size $= 4.0\,\mu\mathrm{m}$, Vertical spot size $= 2.6\,\mu\mathrm{m}$, $I_{\mathrm{th}} = 6\,\mathrm{mA}$
WGPD	Length $= 100\,\mu\mathrm{m}$, Width $= 30\,\mu\mathrm{m}$, Responsivity $= 0.8\,\mathrm{A/W}$
Preamplifier	Transimpedance $= 86\,\mathrm{dB}\,\Omega$, $I_{\mathrm{ep}} = 1.5\,\mathrm{pA}/\sqrt{\mathrm{Hz}}$

the optical module becomes larger than that for the fibre-both-side-connected structure. These factors lead to cost reduction.

A newly developed asymmetric Y-branch structure was adopted for the Y-branch circuit. This made it possible to design a splitting ratio of 50:50 to 25:75 with low excess loss. In order to achieve high responsivity for the receiver port, a 75:25 splitting ratio was selected for the PD:LD port. In this case, the total loss from the common port to the R-PD was estimated at 2.7 dB, which is the sum of the 75% Y-branch loss of 1.2 dB, the coupling loss of 0.5 dB between the R-PD and the PLC, and the excess loss of 1.0 dB including the radiation loss and the fibre coupling loss. A responsivity of more than 0.4 A/W was obtained when the responsivity of the R-PD was 0.8 A/W. On the other hand, the total loss from the LD to the common port was estimated to be around 11 dB, which is the sum of the 25% Y-branch loss of 6.0 dB, the average coupling loss of 4 dB between the LD and the PLC and the excess loss of 1.0 dB. This loss increase is acceptable because an LD is capable of emitting over 11 dB m at a driving current of 60 mA.

Capacitance increases associated with electric wiring between the R-PD and the preamplifier degrade the receiver performance. This parasitic capacitance increases the noise current of the preamplifier and decreases the bandwidth of the receiver. To minimize this, a bare preamplifier chip was directly mounted on the ceramic carrier as close as possible to the PLC, as shown in Fig. 10.11.

Fig. 10.11. Photograph of fabricated module

Module Characteristics. The obtained module characteristics are summarized in Table 10.6.

In terms of the transmitter characteristics, the optical output versus the driving current in the 0−75 °C temperature range is shown in Fig. 10.12. The output power at a driving current of 60 mA and a temperature of 25 C was about 1 mW, and more than 0.5 mW was achieved over the entire temperature range. No kink was observed in the I–L curve, which means that backward reflection into the LD itself was sufficiently suppressed.

Table 10.6. Module performance

Item	Performance
Responsivity of 1.3 μm signal	0.41 A/W
Minimum and maximum received power (50 Mb/s burst signal)	Min: −37 dB m Max: − 12.2 dB m
Optical output power of 1.3 μm signal (60 mA of drive current)	0 dB m
Deviation of output power in 0−70 °C range under automatic power control	±0.5 dB
Insertion loss of 1.55 μm signal (common port → 1.55 μm port)	1.0 dB
Isolation of 1.3 μm signal (LD → 1.55 μm port)	40 dB

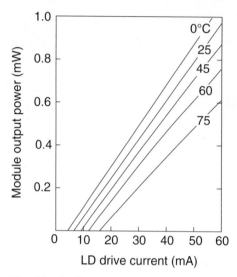

Fig. 10.12. Output power of module versus LD drive-current

As regards the receiver characteristics, a responsivity of 0.41 A/W was obtained.

The WDM characteristics of the module are shown in Fig. 10.13. Insertion loss as low as 1.0 dB was obtained. The rectangular characteristics of this filter-type WDM circuit are ideal for practical applications, because it allows extensive use of the 1.3 µm and 1.5 µm bands.

In a bidirectional TCM-TDMA transmission, the module has to respond quickly to burst signals. To evaluate the 1.3 µm transceiver performance for a burst signal, the module was combined with an instantaneous AGC CMOS

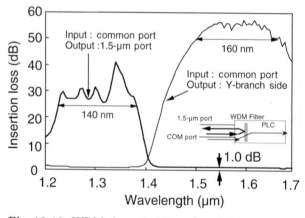

Fig. 10.13. WDM characteristics of module

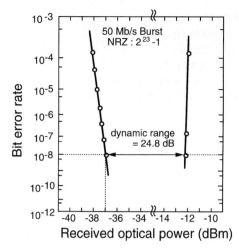

Fig. 10.14. Bit error rate characteristics of module for 50 Mb/s signal

amplifier IC and a driver IC. The bit error rate (BER) performance of the receiver for a 50 Mb/s burst signal is shown in Fig. 10.14. The minimum received optical power at a BER of 10^{-8} was -37 dB m, and the dynamic range was 24.8 dB. These characteristics are good enough for practical applications.

Plastic Packaging. The development of plastic packaging is not at such an advanced stage as the other two concepts shown in Fig. 10.9. The main reason is that it takes a comparatively long time to ensure its reliability. As for the transceiver module, the results of preliminary reliability tests have just been obtained.

Figure 10.15 is a schematic diagram of the plastic packaging structure adopted for the module [25]. An LD and PDs were mounted on the PLC platform using a passive alignment technique and then covered with a silicone that was transparent to the lasing light. They were then covered again with

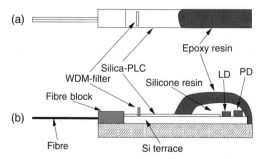

Fig. 10.15. Schematic diagram of plastic module;
(**a**) : top view, (**b**) : cross-sectional view

an epoxy conventionally used in silicon ICs. This epoxy resin protects the device to some extent from humidity and mechanical stress from the outside. The mechanical stress originating from the epoxy itself is released by the inside silicone. Before undertaking the reliability tests on this doubly molded module, some long-term reliability tests were carried out under constant output powers of 10 mW at 25 °C and 5 mW at 70 °C for over 10 000 h to examine the stability of the silicone. The test showed that the LD lifetime can be estimated as more than 10^5 h even at 70 °C at 5 mW, when the failure criterion was set at a 50% increase in the initial operating current. So, the silicone used was confirmed to be quite stable against both the light output power and heating at the laser facet.

The following describes the results of module storage tests. In a test at 85 °C and 85% relative humidity, no noticeable degradation was observed for over 4000 h. The threshold current of the SSC-LD scarcely changed, and the dark current of the receiver WG-PD was kept below 1 nA at 25 °C during the test. The modules also showed stable operating characteristics during temperature cycling tests. No failure was observed over 1500 temperature cycles ranging from −40 to 85 °C without humidity control.

The plastic moulding structure eliminates the need for a cap or box for hermetic sealing and no equipment is required for soldering and welding. This greatly simplifies the packaging fabrication process. In addition, the reliability test results indicate that the plastic packaging is potentially applicable to optical transmission devices.

These results indicate that it will be possible to use this moulding technology for producing low-cost laser modules. This will raise the curtain on a new stage of laser-module fabrication technology.

10.3.3   Optical Modules for WDM Applications

High Optical Functionality Using a PLC and Optoelectronic Devices. Versatile optical functionality can be realized by combining optical WG circuits and semiconductor optoelectronic devices. For example, the combination of a fibre grating and an LD creates an external-cavity LD module whose oscillation wavelength is stable against temperature. The combination of an optical interferometer circuit and two semiconductor optical amplifiers (SOAs) can form a Michelson-interferometric wavelength converter (MI-WC) which employs cross-phase modulation (XPM). However, when these functional modules are composed of discrete components such as LD modules and fibre couplers, the total module size increases and performance stability deteriorates. The hybrid integration of a PLC and OE-devices on the same substrate can provide a more compact and stable module. This section describes two types of module that offer optical functionality via the hybrid integration of optical devices on a PLC platform. The first module is a 2.5 Gb/s 4-channel multi-wavelength light source composed of UV-written WG gratings and LDs. It functions as an arrayed external cavity LD by employing a compact substrate that can transmit light signals at four different wavelengths. The second is a hybrid-integrated wavelength-converter module using a spot-size converter integrated-semiconductor optical-amplifier (SSC-SOA) array on a PLC platform. This module functions as a wavelength converter at a low insertion loss of 4.2 dB. These high-performance modules show the benefit of hybrid integration using a PLC platform which is a promising technology for combining optical signal processing functions and semiconductor optoelectronic devices.

External Cavity LD Module Using Waveguide Grating and LDs.
An external cavity LD array which employs a WG grating has been demonstrated [26]. Figure 10.16 shows the configuration of a hybrid-integrated external-cavity LD which uses UV-written WG gratings. In this module four Fabry-Perot LDs are flip-chip bonded using a AuSn solder film. The gratings were written into the WGs by excimer laser irradiation through phase masks. Each diode chip was monolithically integrated with a spot-size converter to improve the coupling efficiency with the silica WG. The rear and front facets of the LD were high-reflection (HR) and anti-reflection (AR) coated, respectively. These coatings provide a low-loss laser cavity between the LD rear facet and the grating. The oscillation wavelength of each channel is determined by the phase-mask design. The fabricated submodule had a channel-to-channel distance of 3 mm and the substrate was 12 mm × 15 mm. This submodule was assembled on a ceramic carrier for experimental convenience. 45 Ω chip resistors were also inserted in the signal lines on the carrier to provide an input impedance of 50 Ω.

Figure 10.17 shows the output spectra of the LD module. The measured oscillation wavelengths were 1249.7 nm, 1296.8 nm, 1298.8 nm, and 1301.0 nm, exhibiting a spacing of about 2 nm. Each LD exhibits single-mode operation. Figure 10.18 shows the oscillation wavelengths at substrate temperatures between 14 °C and 36 °C. The average temperature dependence was as low as 0.01 nm/°C. As this value coincides with the temperature dependence of the silica WG, the wavelength was successfully locked to the reflection

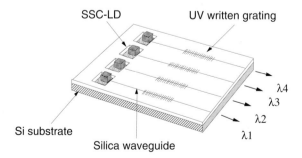

Fig. 10.16. Schematic structure of external-cavity LD module

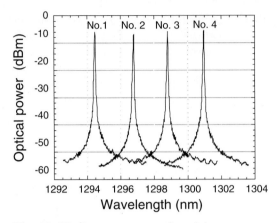

Fig. 10.17. Output spectra of module

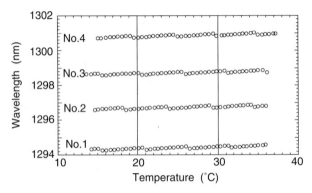

Fig. 10.18. Temperature dependence of oscillation wavelength for four lasers

peak of the grating. The small steps every 5 °C are due to mode jumps. At these temperatures, the oscillation mode jumps to the neighbouring mode due to the refractive index change in the laser diode caused by the injection-current modulation. The mode jump degrades the receiver sensitivity at the photoreceiver, but the imposed power penalty due to the step was as low as 1.5 dB at 1 Gb/s non-return-to-zero (NRZ) signal. Figure 10.19 shows the bit error rate characteristics at 2.488 Gb/s, pseudo-random binary sequence (PRBS) $2^7 - 1$. The substrate temperature was 22 °C, at which none of the four LDs suffered from mode jump. The amplitude of the injection current was 20 mA. The figure shows that a bit error rate of 10^{-9} was obtained for all the LDs. The operating speed of the LD array module is limited by the resonance due to the parasitic inductance at the bonding wires and the electrical circuit outside the module. The bandwidth will be improved by using wiring technology for high-speed operation and an appropriate chip carrier. The channel-to-channel distance will also be reduced without increasing the

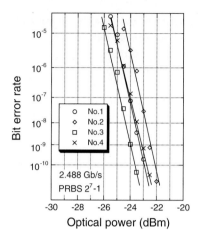

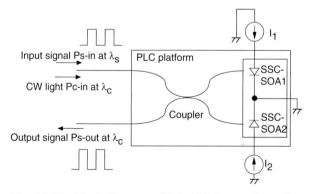

Fig. 10.19. Bit error rate for four wavelengths at 2.488 Gb/s

crosstalk among the channels by using coplanar WGs or microstrip lines on the PLC platform.

Michelson-Interferometric Wavelength Converter Using PLC Platform and SOA Array. The Michelson-interferometric wavelength converter (MI-WC) was realized using the hybrid integration of an SOA array on a PLC platform [27]. Figure 10.20 shows the configuration of the MI-WC module. The module converts an input signal Ps-in at wavelength λ_s into an output signal Ps-out at a wavelength λ_c. Operation of the module is as follows. A CW-light Pc-in at wavelength of λ_c is fed into the upper input port of the coupler. This CW-light is split to the two arms and reflected at the two SOAs. The reflected light is fed into the coupler backward and appears at the lower output port. The magnitude of the output signal Ps-out at wavelength λ_c depends on the relative phase difference of the reflected lightwave. When the modulated light signal with a wavelength of λ_s is fed into the input port,

Fig. 10.20. Block diagram of hybrid-integrated wavelength-converter module

this lightwave changes the refractive index of the SOAs. As the magnitude of the refractive index change that is induced by the signal at wavelength λ_s is different for the two SOAs, the phase shift of the reflected CW light at the interferometric arms changes accordingly. This results in an output signal modulation at wavelength λ_c.

The hybrid MI-WC module employs a spot-size converter integrated semi-conductor optical amplifier (SSC-SOA) to reduce the coupling loss between the silica WG and the SOA. The SSC-SOA consists of two identical SOAs to shorten the flip-chip bonding step and realize uniform coupling efficiency with the silica WG. The structure of the assembly region is identical to that of the lower-cost WDM transceiver module described in Sect. 10.3.2 because only a DC power supply is required. The SSC-SOA is 600 μm long including the active region and the butt-jointed spot-size converter. The front and rear facets were AR- and HR-coated, respectively. The chip was flip-chip bonded using index-alignment technology. A coupling loss of 2.5 dB was obtained between the SSC-SOA and the silica WG.

Figure 10.21 shows the interferometric operation of the fabricated MI-WC. CW-light at $\lambda_c = 1316$ nm and a signal lightwave at $\lambda_s = 1280$ nm was generated by DFB-LDs. The output power at λ_c was measured while changing the SSC-SOA2 injection current I_2 from 3 mA to 11 mA. In this measurement, the light at wavelength λ_s was removed by an optical filter. The injection current of the SSC-SOA1, I_1, was fixed at 11.8 mA. Figure 10.21 shows the interferometric output power while feeding signal lightwaves of 5.5 dB m, -1 dB m and no input light. No temperature controller was used for the MI-WC module. It is clearly seen that the output power at λ_c varies as the injection power of the signal lightwave increases. A maximum static extinction ratio of 14 dB was obtained as well as an out-of-phase conversion at $I_2 = 5.8$ mA.

Figure 10.22 shows the relationship between the wavelength-converted output power and the input-signal power. The injection currents into SSC-

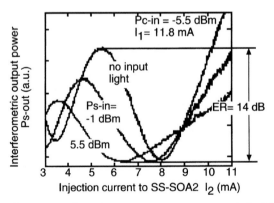

Fig. 10.21. Interferometric output power of the MI-MC module

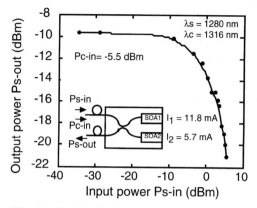

Fig. 10.22. Static characteristics of MI-WC module

SOA1, I_1, and SSC-SOA2, I_2, were 11.8 mA and 5.7 mA, respectively. The wavelengths of the CW light and the input signal were 1316 nm and 1280 nm, respectively. The CW light input power Pc-in was set to -5.5 dB m. This figure reveals that the wavelength-conversion efficiency, that is the ratio between the input signal power and the output signal power, was -15 dB when the input signal power Ps-in was 5.5 dB m. The extinction ratio of the output signal power Ps-out was 11 dB. This graph also shows that the insertion loss calculated by using the input CW light power Pc-in and the output signal power Ps-out was as low as 4.2 dB when there was no input signal Ps-in.

10.3.4   Optoelectronic Hybrid Modules for High-speed Applications

High Speed Operation: Requirements and Solutions. In order to apply hybrid-integration technology to TDM systems, the platform should have excellent electrical characteristics. Although the electrical characteristics of the PLC platform shown in Fig. 10.3 have proved to be good enough for up to ~ 2.5 Gb/s operation, they are still insufficient for TDM applications, where an operation bandwidth of more than 10 Gb/s is required. To overcome this limitation, the PLC platform structure must be modified for higher speed applications.

The operation bandwidth of the platform is limited by the RC time constant in the assembly region and the propagation loss in the signal lines in the electrical wiring region. The RC time constant in the assembly region is due to the parasitic capacitance between the electrodes and the silicon terrace. The electrodes on the silicon terrace have a larger parasitic capacitance, because the insulation layer between the electrodes and the terrace is as thin as 0.5 μm in order to achieve high-precision alignment in the vertical direction. Therefore, the passivation layer thickness has to be thicker to reduce the capacitance without degrading the alignment accuracy.

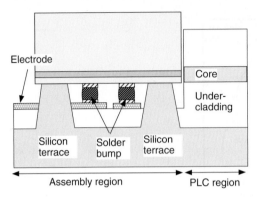

Fig. 10.23. Schematic cross-section of PLC platform for high-speed operation

In order to satisfy this demand, the assembly-region structure is modified as shown in Fig. 10.23 [28]. All the electrodes are formed on the under-cladding layer and solder bumps are introduced. The thick silica layer under the electrodes can reduce the parasitic capacitance. The solder-bumps enable an electrical connection to be established between the electrodes on the under-cladding layer and that on the OE-device, while the silicon terrace is used as the alignment plane and heat sink. Therefore, the OE- devices are brought into mechanical contact with the silicon terrace to achieve alignment with the silica WG in the vertical direction.

The characteristic impedance in the signal lines must also match that of the circuit outside the platform for higher-speed applications. Therefore, a coplanar WG (CPW) and a microstrip line (MSL) are formed on the PLC platform. A cross-section of the CPW is shown in Fig. 10.24. The propagation loss of the CPW increases when the electrical field reaches to the silicon substrate. For example, the propagation loss of the CPW on a 1.5 μm thick silica layer is 17 dB/cm at 10 GHz. This is crucial when the wire length is near

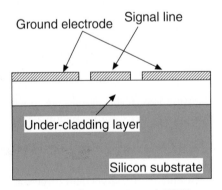

Fig. 10.24. Cross-section of CPW

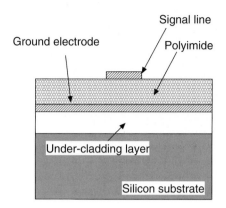

Fig. 10.25. Cross-section of MSL

1 cm or an operating bandwidth over 10 GHz is needed. This high propagation loss is due to the larger loss tangent ($\tan \delta$) of the silicon substrate and the thermal donors generated by the high-temperature treatment during the silica-consolidation process at more than 1000 °C. In order to reduce propagation losses, the CPW is formed on the silica under-cladding. The adverse effect of the thermal donors is also reduced by quenching. These technologies were used to achieve a low propagation loss of 1.1 dB/cm at 10 GHz employing a coplanar line on a 12 μm thick under-cladding layer. A cross-section of the MSL is shown in the Fig. 10.25. The MSL on a PLC platform consists of a polyimide insulation layer sandwiched between two electrodes. A lower electrode on the platform acts as the ground and the upper electrode acts as the signal line. Low propagation loss is obtained, because the lower electrode can terminate the electric field generated by the signal line before it reaches the silicon substrate. For example, an MSL with a 29 μm-thick polyimide layer had a propagation loss of less than 1 dB/cm. The MSL can be used for a long signal line on a PLC platform. The MSL structure is also used for two-layer wiring that can reduce the parasitic inductance in the ground electrode and the power supply lines.

Several PLC platforms designed for high-speed operation have been demonstrated. They employ solder bumps for bonding OE-devices, and a CPW and an MSL for electrical interconnection. This section describes a photoreceiver module and a transmitter module. The photoreceiver module consists of a monolithically integrated photoreceiver array and is operated at 10 Gb/s. The transmitter module consists of an LD array and a GaAs metal-semiconductor field-effect transistor (MESFET) driver IC and is operated at 9 Gb/s.

Photoreceiver Array Module with Monolithic OEIC. Figure 10.26 is a block diagram of a photoreceiver array submodule. This submodule consists of a PLC platform and a monolithically integrated optoelectronic integrated

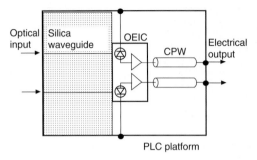

Fig. 10.26. Block diagram of photoreceiver array module

circuit (OEIC) [28]. The OEIC consists of two-channel photoreceiver circuits. Each channel uses a waveguide p-i-n photodiode (WGPD) that has a multimode WG structure consisting of an i-InGaAs photo-absorption layer, n^+- and p^+-InGaAsP intermediate layers, n^+- and p^+-InP cladding layers, and an InGaAs contact layer. This structure has the advantage of being able to achieve high coupling efficiency without bandwidth degradation for operation above 10 GHz. The OEIC is also integrated with transimpedance preamplifiers that consist of InGaAs/InAlAs High-Electron-Mobility Transistors (HEMTs). The preamplifiers have on-chip metal-insulator-metal capacitors to suppress the resonance at the frequency response caused by the parasitic inductance in the power lines. The monolithic photoreceiver chip was flip-chip bonded using 16 In solder bumps: two for the WGPD, six for the power supply and 8 for the ground. Vertical alignment was performed using the mechanical contact of the chip and the silicon terrace. The PLC waveguide had a refractive-index difference of 0.75% and employed an angled facet at the output port to reduce the optical reflection from the WGPD. The wiring region used CPW to transmit the electrical signals from the OEIC. The CPW was as short as 1.5 mm to reduce the insertion loss. In addition, the power lines employed a two-layer structure consisting of a ground electrode on the under-cladding layer, a polyimide insulation layer and an upper electrode that acted as the power lines. This structure offers an electrically stable ground electrode when many power lines are required. Figure 10.27 shows a photograph of the fabricated photoreceiver submodule which was $14 \times 12 \, \text{mm}^2$ in size.

The DC responsivity of the module was measured at 1.55 μm. The module using the WGPD without an AR coating showed responsivities of 0.35 A/W and 0.21 A/W for the two channels. The difference between the DC responsivities of the two channels was 4.2 dB. The responsivity includes the coupling loss between the fibre and the silica WG, the WG propagation loss and the coupling loss between the WG and the WGPD. The calculated responsivity of the discrete WGPD to the light from the WG was 0.5 A/W. Therefore, the excess loss at the submodule was between 1.5 and 3.8 dB.

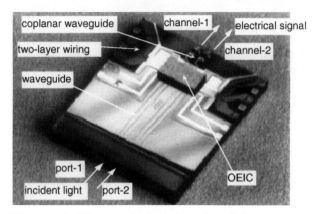

Fig. 10.27. Photograph of fabricated module

Figure 10.28 shows the frequency response of the receiver module measured with a 1.3 μm wavelength-modulated signal. The characteristics were measured using an on-wafer probe and a lightwave component analyzer. The 3 dB bandwidth was 8 GHz. This is almost the same as the value for a discrete receiver OEIC chip. The crosstalk was well suppressed to below −25 dB at frequencies below 6 GHz. This spectrum is similar to that of the discrete device measured with the on-wafer probe. The crosstalk measured at the hybrid integrated submodule was about 5 dB lower than that of the discrete device at 6 GHz. This shows that the electrically stable ground electrode on the PLC platform successfully suppressed the crosstalk due to the electrical potential instability at the ground electrode during operation.

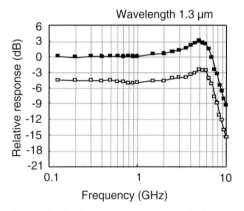

Fig. 10.28. Frequency response of photoreceiver array module

Transmitter Module with Integrated LD Array and Driver IC. In order to expand the functionality of the electronic circuits on the platform, it is necessary to develop technologies for the hybrid integration of opto-electronic devices and integrated circuits (ICs) on a PLC platform [29]. Figure 10.29 is a block diagram of a transmitter array submodule. This submodule consists of an LD array and an LD-driver IC flip-chip bonded on a PLC platform. The two-channel LD array was made of 1.55 µm InGaAsP/InP buried-heterostructure (BH) lasers embedded in a high-resistance InP epitaxial layer on a semi-insulating substrate. The LD-driver IC used 0.2 µm-gate-length GaAs MESFETs. This IC consisted of a two-channel LD-driver circuit which had a 3 dB bandwidth of 5.3 GHz, from on-wafer probe measurements. The driver IC had two high-frequency input ports and two output ports whose input impedance was 50 Ω and 25 Ω, respectively. It also needed 24 interconnections for the power supply and operation monitors.

Figure 10.30 shows a cross-section of the PLC platform for the transmitter module. The LD array was flip-chip bonded on the platform using AuSn solder bumps. The assembly region had the same structure as the photoreceiver submodule described in the previous section. The heat from the LD array is mainly dispersed through the solder pads formed near the cavity. The LD

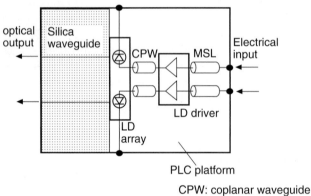

CPW: coplanar waveguide

MSL: microstrip line

Fig. 10.29. Block diagram of transmitter module

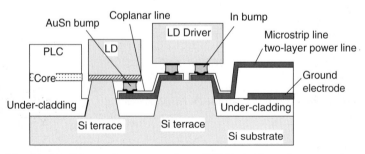

Fig. 10.30. Schematic cross-section of PLC platform for transmitter module

driver chip was flip-chip bonded on the silicon terrace using In solder bumps. The solder pads for the LD driver had the same configuration as those for the LD array. The silicon terrace acted as the heat sink of the driver chip. The use of the different solder materials made it possible to achieve high optical-coupling accuracy with the LD array after the IC assembly steps. The input electrical signals were fed through the MSL to suppress the insertion loss of the signal lines. The electrical interconnection between the driver and the LD array used CPW, thus providing a simple structure.

Figure 10.31 is a photograph of a fabricated submodule 12 mm × 26 mm in size. The many power-supply lines employed two-layer wiring to offer an electrically stable ground. The CPW between the LD driver and the LD array was as short as 1 mm to lower the propagation loss. MSLs were used as input signal lines because their length was about 1 cm. The LD array used a subcarrier to ease the handling during the assembly processes. Two chip resistors were also inserted at the CPW to match the impedance of the LD array to the characteristic impedance of the LD driver.

The submodule was assembled in a package. Figure 10.32 shows the optical output power versus current characteristics of the fabricated module. These data were measured by feeding the injection current from the LD-driver IC. The output optical power levels of 0.83 mW and 0.61 mW at 40 mA injection current are close to the results obtained from a transmitter module without an IC assembly. Therefore, the IC bonding process did not increase the coupling loss between the laser diode and the silica WG. The frequency response of the transmitter module is shown in Fig. 10.33. The circles and squares show the response of channel 1 and channel 2, respectively. The graph shows that the 3 dB bandwidths were 6.5 GHz and 5.2 GHz in the two channels. These bandwidths almost coincide with the bandwidth of the LD driver measured by the on-wafer probe.

This section has made clear that optical hybrid-integration technology based on the PLC platform can provide low-cost optical modules, because it can reduce the number of components in the module and simplifies their

Fig. 10.31. Photograph of transmitter module

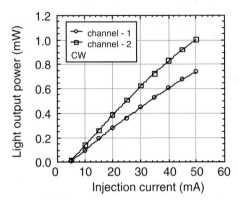

Fig. 10.32. I–L characteristics of transmitter module

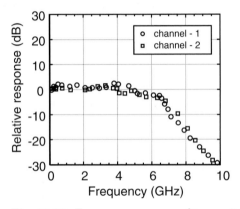

Fig. 10.33. Frequency response of transmitter module

assembly. The same technology enables also highly functional optical modules which can be used in WDM or TDM optical networks.

10.4   Future Prospects

Hybrid integration is now solving many of the problems related to optical component technology. However, there remains one important issue for optical-module technology to overcome. That issue is how to simplify the procedure for assembling optical modules on an electronic board. Optical modules are usually used together with ICs or LSIs after being assembled on an electrical board. However, it requires rather intricate work to assemble optical components on a board due to the difficulty in dealing with optical fibres. Progress on optical networks will lead to an increase in the number of optical components on a board, thus aggravating the above difficulty.

To overcome this, first the fibre pigtails attached to conventional optical modules need to be eliminated. These fibre pigtails make it difficult to assemble optical modules on an electronics board automatically. Today, many studies are being undertaken to develop a receptacle-type optical module. When this is achieved, optical modules will be handled as if they were electrical modules and the assembly procedure will be greatly simplified.

Moreover, it is important to minimize the number of optical fibres on the board. If the number of optical modules assembled on the board is increased, more and more fibres will be required to interconnect them. Therefore, we will need to reduce the number of optical modules by introducing an optoelectronic multi-chip module (OE-MCM) in the device assembly hierarchy, as shown in Fig. 10.34. Here, the OE-MCM is a large-scale hybrid integrated circuit on which many OE-devices are assembled. To establish the OE-MCM, it is necessary to develop a technique for assembling optical multi-chips on a platform. A trial aiming at optical multi-chip integration is currently being undertaken. Therefore, the OE-MCM will be realized before long.

However, this approach will be limited in terms of integration scale, because any increase in the number of OE-chips on the platform will lead to a larger OE-MCM size. So, as a second step, the module size will need to be made compact. To achieve this, semiconductor device fabrication technology should also be upgraded to make it capable of realizing monolithic integrated circuits. Figure 10.34 explains this by using an 8×8 matrix switch as an example. If we fabricate a matrix switch using 4-channel SOA array chips, only 4 SOA chips need to be integrated on the platform for a 4×4 switch. This is considered to be a realistic way to provide a 4×4 matrix switch. However, the number of SOA chips integrated on the platform increases to as many as 16 to realize an 8×8 matrix switch because 64 SOA gates are needed. Moreover, 8 arrays of 1×8 splitters (or combiners) are necessary as input and output passive optical circuits. To achieve this will require an over-large platform. When a monolithically integrated 4×4 optical switch is realized, the 8×8 switch will be easily fabricated by using hybrid-integration technology as shown in Fig. 10.35. Therefore, it must be stressed here finally that hybrid and monolithic integration techniques should be combined to

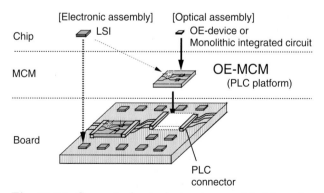

Fig. 10.34. Concept of optoelectronic assembly hierarchy

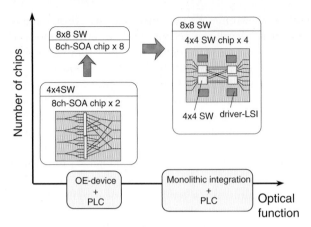

Fig. 10.35. Approach to large-scale OE-MCM

achieve a compact and large-scale integrated OE-MCM for application to future optical-component technology.

10.5   Summary

This chapter has described key technologies for achieving optical hybrid integrated circuits and the role of optical hybrid integration in optical communication systems. Today, new optical-module technology must be developed to support future optical networks. This technology should enable the module-assembly cost to be reduced and also have a wide range of applications. The optical module should also be compatible with conventional electronic device assembly processes. Although the hybrid optical integration technique is considered to be the most promising approach, the difficulty in assembling optical active devices on a passive WG substrate has been an obstacle to its realization. This difficulty is being overcome today by the development of the PLC platform, a passive alignment technique and semiconductor devices suitable for hybrid integration.

Hybrid integration based on the above techniques has significant advantages in that it can greatly simplify the module assembly procedure and can also provide optoelectronic modules with high functionality. Therefore, hybrid integration can make a great contribution to future optical networks. It will be able to greatly reduce the cost of the WDM transceiver module, which is the key component for the construction of FTTH networks. This technology also provides highly functional optical modules for WDM networks. A module employing a combination of a UV-written Bragg grating in a silica WG and an SSC-LD has been used to produce a four-channel multi-wavelength light source. A module employing a PLC platform with a Michelson interferometer and an SSC-SOA array has been used to produce a wavelength converter.

High-speed optical modules for optical TDM applications have also been achieved by employing assembly and wiring technologies for wide-bandwidth operation at more than 10 Gb/s.

The major remaining issue as regards the optical-module technique is to simplify the procedure for assembling a module on an electronic board. This problem will be overcome by the development of receptacle-type optical modules and large-scale OE-MCMs. In the future, the hybrid and monolithic approaches should be combined to provide compact and highly functional OE-MCMs.

References

1. J.D. Crow, L.D. Comerford, R.A. Laff, M.J. Brady, and J.S. Harper, "GaAs laser array source package," Opt. Lett. **1**, 40 (1977)
2. H. Kaufmann, P. Buchmann, R. Hirter, H. Melchior, and G. Guekos, "Self-adjusted permanent attachment of fibers to GaAs waveguide components," Electron. Lett. **22**, 642 (1986)
3. H. Terui, Y. Yamada, M. Kawachi, and M. Kobayashi, "Hybrid integration of a laser diode and a high silica multimode optical channel waveguide on Si," Electron. Lett. **21**, 646 (1985)
4. Y. Yamada, M. Yamada, H. Terui, and M. Kobayashi, "Optical interconnections using a silica-based waveguide on a silicon substrate." Opt. Eng. **28**, 1281 (1989)
5. Y. Yamada, H. Terui, Y. Ohmori, M. Yamada, A. Himeno, and M. Kobayashi, "Hybrid integrated 4 × 4 optical gate matrix switch using silica-based optical waveguides and LD array chips," IEEE J. Lightwave Technol. **10**, 383 (1992)
6. E.E.L. Friedrich, M.G. Oberg, B. Broberg, S. Nilsson, and S. Valette, "Hybrid integration of semiconductor lasers with Si-based single-mode ridge waveguide," IEEE J. Lightwave Technol. **10**, 336 (1992)
7. C.A. Jones, K. Cooper, M.W. Nield, J.D. Rush, R.G. Waller, J.V. Collins, and P.J. Fiddyment, "Hybrid integration of a laser diode with a planar silica waveguide," Electron. Lett. **30**, 215 (1994)
8. Y. Yamada, A. Takagi, I. Ogawa, M. Kawachi, and M. Kobayashi, "Silica-based optical waveguide on terraced silicon substrate as hybrid integration platform," Electron. Lett. **29**, 444 (1993)
9. C.A. Armiento, M. Tabasky, C. Jagannath, T.W. Fitzgerald, C.L. Shieh, V. Barry, M. Rothman, A. Nergi, P.O. Hausjaa, and H.F. Lockwood, "Passive coupling of InGaAsP/InP laser array and single-mode fibers using silicon waferboard," Electron. Lett. **27**, 1109 (1991)
10. W. Hunziker, W. Vogt, R. Hess, and H. Melchior, "Low-loss, self-aligned flip-chip technique for interchip and fiber array to waveguide OEIC packaging," *Proc. IEEE Lasers Electro-Optics Soc. Conf. (LEOS '94)*, Vol. **2** (Boston, MA 1994), p. 269
11. K.P. Jackson, E.B. Flint, M.F. Cina, D. Lacey J.M. Trewhella, T. Caulfield, and S. Sibley, "A compact multichannel transceiver module using planar-processed optical waveguides and flip-chip optoelectronic components," *Proc. 42nd Electronic Comp. Technol. Conf. (ECTC '92)*, (San Diego, CA 1992), p. 93

12. M.S. Cohen, M.F. Cina, E. Bassous, M.M. Oprysko, and J.L. Speidell, "Passive laser-fiber alignment by index method," IEEE Photon. Technol. Lett. **3**, 985 (1991)
13. K. Kurata, K. Yamauchi, A. Kawatani, H. Tanaka, H. Honmou, and S. Ishikawa, "A surface mount type single-mode laser module using passive alignment," *Proc. 45th Electronic Comp. Technol. Conf. (ECTC '95)*, (Las Vegas, NV 1995), p. 759
14. T. Hashimoto, Y. Nakasuga, Y. Yamada, H. Terui, M. Yanagisawa, K. Moriwaki, Y. Suzaki, Y. Tohmori, Y. Sakai, and H. Okamoto, "Hybrid integration of spot-size converted laser diode on planar lightwave circuit platform by passive alignment technique," IEEE Photon. Technol. Lett. **8**, 1504 (1996)
15. T. Hayashi, "An innovative bonding technique for optical chips using solder bumps that eliminate chip-positioning adjustments," IEEE Trans. Comp. Hybrids Manuf. Technol. **15**, 225 (1992)
16. M. Itoh, J. Sasaki, A. Uda, I. Yoneda, H. Honmou, and K. Fukushima, "Use of AuSn solder bumps in three-dimensional passive aligned packaging of LD/PD arrays on Si optical benches," *Proc. 46th Electronic Comp. Technol. Conf. (ECTC '96)*, (Orlando, FL 1996), p. 1
17. J.V. Collins, I.F. Lealman, P.F. Fiddyment, C.A. Jones, R.G. Waller, L.J. Rivers, K. Cooper, S.D. Perrin, M.W. Nield, and M.J. Harlow, "Passive alignment of a tapered laser with more than 50% coupling efficiency," Electron. Lett. **31**, 730 (1995)
18. I.F. Lealman, C.P. Seitzer, L.J. Rivers, M.J. Harlow, and S.D. Perrin, "Low threshold current 1.6 μm InGaAsP/InP tapered active layer multiquantum well laser with improved coupling to cleaved singlemode fiber," Electron. Lett. **30**, 973 (1994)
19. Y. Tohmori, Y. Suzaki, H. Fukano, M. Okamoto, Y. Sakai, O. Mitomi, S. Maysumoto, M. Yamamoto, M. Fukuda, M. Wada, Y. Itaya, and T. Sugie, "Spot-size converted 1.3 μm laser with a butt-jointed selectively grown vertically tapered waveguide," Electron. Lett. **31**, 1069 (1995)
20. K. Kato, M. Tuda, A. Kozen, Y. Muramoto, K. Noguchi, and O. Nakajima, "Selective-area impurity doped planar edge-coupled waveguide phtotodiode (SIMPLE-WGPD) for low-cost, low-power-consumption optical hybrid modules," Electron. Lett. **32**, 2078 (1996)
21. K. Tanaka, M. Makiuchi, M. Norimatsu, C. Sakurai, N. Yamamoto, K. Miura, and M. Yano, "A corner-illuminated structure PIN photodiode suitable for planar lightwave circuit," *Proc. IEEE Lasers Electro-Optics Soc. Conf. (LEOS '96)*, Vol. 1 (Boston, MA 1996), p. 14
22. N. Shibata and I. Yamashita, "System and component technologies toward full access network opticalization,". IEICE Trans. Electron. **E80-C**, 3 (1997)
23. J. Yoshida, M. Kawachi, T. Sugie, M. Horiguchi, Y. Itaya, and M. Fukuda, "Strategy for developing low-cost optical modules for the ONU of optical subscriber systems," *VII Int. Workshop on Optical Access Networks* (1996), p. 1
24. N. Uchida, Y. Yamada, Y. Hibino, Y. Suzuki, and I. Ishihara, "Low-cost hybrid WDM module consisting of a spot-size converted integrated laser diode and a wavelength photodiode on a PLC platform for access network systems," IEICE Trans. Electron. **E80-C**, 88 (1997)
25. M. Fukuda, F. Ichikawa, Y. Yamada, Y. Inoue, K. Kato, H. Sato, T. Sugie, H. Toba, and J. Yoshida, "Highly reliable plastic packaging for laser diode and photodiode modules used for access network," Electron. Lett. **33**, 2158 (1997)

26. H. Takahashi, T. Tanaka, Y. Akahori, T. Hashimoto, Y. Yamada, and Y. Itaya, "A 2.5 Gb/s, 4-channel multiwavelength light source composed of UV written waveguide gratings and laser diodes integrated on Si," *Proc. 11th Int. Conf. Integr. Optics Opt. Fibre Commun. / 23rd Europ. Conf. Opt. Commun. (IOOC-ECOC '97)*, Vol. **3** (Edinburgh, UK 1997), p. 355

27. R. Satoh, Y. Sakai, S. Sekine, Y. Tohmori, Y. Inoue, K. Shuto, M. Yamada, and T. Kanamori, "Hybrid integrated-wavelength-converter module using a spot-size converter integrated semiconductor optical amplifier array on a PLC platform," *Techn. Digest 2nd Optoelectron. Commun. Conf. (OECC '97)*, (Seoul, South Korea 1997), p. 182

28. T. Ohyama, S. Mino, Y. Akahori, M. Yanagisawa, T. Hashimoto, Y. Yamada, Y. Muramoto, and H. Tsunetusgu, "10 Gbit/s hybrid integrated photoreceiver array module using a planar lightwave circuit platform," Electron. Lett. **32**, 845 (1996)

29. Y. Akahori, T. Ohyama, M. Yanagisawa, Y. Yamada, H. Tsunetsugu, Y. Akatsu, M. Togashi, S. Mino, and Y. Shibata, "A hybrid high-speed silica-based planar lightwave circuit platform integrating a laser diode and a driver IC," *Proc. 11th Int. Conf. Integr. Optics Opt. Fibre Commun. / 23rd Europ. Conf. Opt. Commun. (IOOC-ECOC '97)*, Vol. **3** (Edinburgh, UK 1997), p. 359

11   Monolithic Integration

Herbert Venghaus, Heinz-Gunter Bach, Frank Fidorra,
Helmut Heidrich, Ronald Kaiser, and Carl Michael Weinert

11.1   Introductory Remarks

Generic benefits of monolithic integration are commonly known from silicon
ICs, which have experienced a continuous, intense technological and economic
development over several decades. As a consequence, the development of
monolithic optoelectronic ICs (OEICs) has been considered equally rewarding
for quite a while[1]. However, there is a fundamental difference between purely
electronic and optoelectonic (OE) integration: Si-integration is essentially
the engineering of electrical functionalities by incorporation of dopants into
the host material, the key element is the transistor, and a single chip may
contain millions of transistors. On the other hand, OE integration requires
the epitaxial growth of materials, which have different compositions and thus
exhibit different energy gaps and different refractive indices. OE integration
further includes vertical and lateral structuring, and one can achieve optical,
optoelectronic and electronic functionalities as well. The key component is
the semiconductor laser, and in contrast to Si-integration, the components to
be integrated on a single OE chip, are rather different. In comparison to Si-
integration, OE integration is considerably more complex by its very nature
and OE subcomponents have significantly larger geometrical dimensions than
integrated (Si-) transistors. Thus the number of different subcomponents to
be integrated on a single OE chip is fairly limited (several tens at most, and
frequently only a few).

OE modules can be implemented in many different fashions ranging from
hybrid over partly integrated to monolithically integrated (see also Chap. 10).
According to their complex functionality most modules in optical communication systems are very likely to be, even in the long run, hybrid ones, in
the sense that monolithic OEICs are combined with other elements based
on SiO_2/Si, $LiNbO_3$, glass, polymers or Si, for example, as outlined in the
preceding chapter. In agreement with such expectations the goal of monolithic

---

[1] The term OEIC refers to the integration of optoelectronic devices such as lasers and
detectors with electronic ones, transistors for example. The frequently used term Photonic
Integrated Circuits (PICs) designates a subset of OEICs, namely optically interconnected
guided-wave optoelectronic devices.

OE integration is to (1) integrate an optimum number of different components on a single chip and (2) take care of enabling an efficient hybrid coupling with complementary optoelectronic and electronic chips.

A significant part of the optoelectronic module cost is due to mounting and packaging. Monolithic integration lowers the mounting and packaging cost by a reduction of cost-driving alignment and fixing procedures. Even the mounting and packaging costs of discrete lasers can be lowered by monolithic integration, i.e. by combining the laser with a spot-size converter, which relaxes otherwise very tight alignment tolerances.

Compared to hybrid modules monolithic chips are inherently more stable, which is equivalent to higher reliability and lower cost. Higher functionality of OEICs is another asset: very high-speed or interferometric devices can only be made to operate satisfactorily when integrated monolithically.

Finally, the channel separation of monolithic integrated multiwavelength sources can easily be assured with high precision, which is another example of benefits offered by integration.

However, OEICs will be (or become) competitive with hybrid solutions only, if overall processing cost and chip size can be kept sufficiently low, so that the savings in mounting and assembly are not annihilated by excessive OEIC costs.

Due to the transmission characteristics of optical fibres the operation wavelengths of today's long- or medium-distance optical communication systems are in the wavelength range around $1.3\,\mu\mathrm{m}$ and $1.55\,\mu\mathrm{m}$. The appropriate material system for corresponding lasers (or semiconductor optical amplifiers) is InP, including its ternary and quaternary lattice-matched alloys; InP-based integration will be treated in more detail in the following. Most of the concepts and technological approaches apply also to the GaAs/AlGaAs material system, which enables lasers with emission wavelengths $< 1\,\mu\mathrm{m}$. These are suited for short-distance optical communication systems, which, however, will not be covered in greater detail here. In the long run additional material systems may emerge. Unfavourable temperature characteristics found in GaInAsP/InP can for example be overcome by using GaInAsN/GaAs, but a more detailed treatment will not be attempted here.

In the following paragraphs we will present the essentials of monolithic integration rather than compile all OE integrations realized so far. We will treat waveguides, which connect various subcomponents on a chip or represent the basic element of several components themselves. Integrated spot-size converters enable relaxed alignment tolerances and are a must for future OEICs. Another important group of devices are integrated transmitters. Here we will focus particularly on the link between passive and active waveguides (where the latter refer to lasers or semiconductor amplifiers). Integrated receivers are an example of combining rather different subcomponents into one OEIC. Finally, we will discuss crosstalk (electrical, optical, thermal), which needs particular attention if different components are implemented close to each other on a single chip. The chapter on monolithic integration ends by highlighting different integrations realized up to now and sketches possible lines of future development.

11.2   Waveguides

Dielectric waveguides (WG) as part of PICs or OEICs serve essentially two purposes: they are part of subcomponents themselves, and they link different subcomponents on a chip. The former topic, i.e. WG-based devices such as couplers, splitters, and wavelength multiplexers/demultiplexers, are covered in greater detail in Chap. 7, and only the latter aspect will be treated here.

WGs for OEIC applications are primarily selected by the requirement of monomode behaviour, low intrinsic loss, low loss in WG bends, sufficiently low polarization dependence, and high coupling efficiency to the components on the chip. A great variety of dielectric WGs exist as outlined in standard textbooks, e.g. [1], falling into two main categories of WGs, buried WGs and WGs at the chip surface.

Monomode operation of the waveguide is a prerequisite for the precise operation of WG components and for effective coupling to the monomode fibre. Due to technological reasons WG widths w are typically chosen between $2\,\mu m$ and $5\,\mu m$. The parameters to be determined consecutively by the monomode requirement are WG thickness t and composition λ_Q of the core material which determines the core refractive index. Typical thickness and composition control correspond to $\Delta t/t = \pm 2\%$ and $\Delta\lambda_Q = \pm 4\,nm$, leading to uncertainties in the propagation constants of the waveguide of about 7%.

The intrinsic waveguide losses depend crucially on the semiconductor growth and processing techniques. With today's standard Metal-Organic Vapour Phase Epitaxy (MOVPE) and optimized etching techniques intrinsic WG losses can be kept below $1\,dB/cm$.

The parameter that determines the radiation loss in WG bends is the lateral confinement of the lightwave. The lateral confinement describes how strongly the fundamental WG mode is confined laterally to the WG core. It can be measured by $\Delta n_{eff} = n_{eff}(2) - n_{eff}(1)$, where $n_{eff}(2)$, $n_{eff}(1)$ are the slab effective indices calculated in the core and cladding region, respectively (see Fig. 11.1). For all types of WGs radiation losses in waveguide bends increase strongly with decreasing bend radius. In Fig. 11.1 we depict the radius for which the radiation loss in a quarter-bend has decreased down to $-20\,dB$ and can be neglected. This radius can therefore be considered as the minimum radius, for which the WG is lossless. The curves in Fig. 11.1 show that for strong lateral confinement (i.e. large Δn_{eff},) the minimum bend radius can be made rather small and thus compact integration on the chip is possible. The calculations show only minor differences between TE and TM polarized modes (TE, TM: Transverse Electric, Transverse Magnetic, see Chap. 6, Sect. 6.2.2). Also, the minimum radius for $1.3\,\mu m$ operation is only slightly lower than that for $1.55\,\mu m$ operation. An alternative to bent

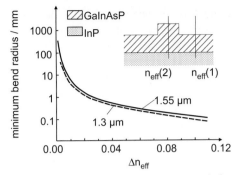

Fig. 11.1. Minimum allowed waveguide bend radii as a function of Δn_{eff} for 1.3 μm and 1.55 μm operation. Inset illustrates determination of Δn_{eff} evaluated by difference of slab calculations along vertical lines (2) and (1)

WGs are dielectric mirrors. Typical (optimum) losses per mirror are around -1 dB (-0.5 dB), and these results hold for the InP- and the GaAs-material system as well [2].

WG birefringence depends on the WG geometry and on whether the WG is buried or at the chip surface. Buried WGs can in principle be made polarization-independent by making the core square-shaped. However, square-shaped monomode WGs in GaInAsP/InP require core heights and widths < 1.5 μm, and lateral dimensioning of such geometries is extremely difficult with currently available technologies. As a consequence, WGs are normally made with a rectangular core leading to a residual polarization dependence. Typical values are shown in Fig. 11.2.

WGs at the chip surface like ridge or rib WGs are always polarization dependent. This may be advantageous for the fabrication of integrated polarization splitters or rotators, but normally it is unwanted. On the other hand, in some cases integration technologies are simpler for uncladded WGs than for buried WGs.

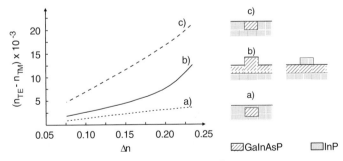

Fig. 11.2. Calculated effective index difference $n_{\text{TE}} - n_{\text{TM}}$ versus refractive index difference Δn between WG core and cladding. All waveguides are monomode with $w = 3$ μm and have equal Δn_{eff}

In Fig. 11.2 the calculated effective index difference $n_{TE} - n_{TM}$ is depicted as a measure of the birefringence for three different types of WGs. Polarization dependence increases as the refractive index difference Δn between WG core and cladding gets larger.

For integration purposes there are, in general, conflicting demands on WG parameters: As InP-based OEICs normally contain Laser Diodes (LD) or Semiconductor Optical Amplifiers (SOA), WGs with efficient coupling to these devices require small spot size and large Δn, which may lead to unwanted WG birefringence. Additionally, although strong lateral confinement is necessary for small bend radii and compact structures, it is extremely unfavourable for efficient coupling to the optical fibre. Whereas in most cases the WG requirements of different components on the chip can be solved by a compromise, the spot-size difference between the chip WG (typically $3\,\mu m \times 1\,\mu m$) and the fibre (typically $10\,\mu m \times 10\,\mu m$) remains a serious problem. It leads to chip-fibre coupling losses of the order of $-10\,dB$ and, even worse, decreases the fibre chip alignment tolerances to below $1\,\mu m$. In order to overcome this problem, spot-size converters are necessary. This topic is discussed in more detail in the following paragraph, but also in Chap. 10.

11.3   Integrated Spot-Size Converters

As already mentioned the spot size of the optical field of monomode InP-WGs neither matches the spot size of standard single-mode fibres nor that of SiO_2/Si WGs on optical boards. This leads to high coupling losses and to very sensitive alignment tolerances, which are detrimental for cost-efficient packaging of OEICs. The required field matching can be provided by adiabatic (i.e. lossless) spot-size converters (SSC), frequently designated as 'tapers' also. SSCs can be realized on-chip or alternatively as a tapered fibre to assure a high coupling efficiency. If, in addition to efficient coupling, alignment tolerances larger than $1\,\mu m$ are required to enable passive, reliable and thus economic single-mode fibre-chip coupling or optical board-chip mounting, SSCs have to be implemented on-chip.

Adiabatic spot-size expansion is achieved by continuously reducing the size of the core WG. In the horizontal direction, this reduction can be accomplished with standard lithographic techniques. WG widths as narrow as $0.5\,\mu m$ have been fabricated in this way. In the case of buried WGs, proper WG narrowing provides horizontal and vertical beam expansion simultaneously. An example of an SSC fabricated by this cost efficient technique [3] is shown in Chap. 10 in Table 10.4, upper left.

Alternatively, WGs can be vertically shaped by a smooth decrease in the WG thickness. Such WG ramps have been fabricated by different methods. The most common technologies include selective-area MOVPE. This technique exploits the enhanced MOVPE growth in the vicinity of masked regions (see also Sect. 11.4, Figs. 11.8 and 11.9). Altogether, selective-area MOVPE is a simple and versatile epitaxial process; however, it is reactor specific, and the growth rates are mask dependent.

Another approach is based on etching a ramp into the chip waveguide. This technique uses a subsequent growth of material with different composition or quantum-wells followed by a special dry etching process for ramp formation. The WG ramp fabrication can be performed by so-called shadow masks [4] or semi-transparent masks [5]. The former technique utilizes the gradual dry etching in the shadow region of a mask. It relies on a critical wafer adjustment process not suited for mass production. In the latter case a transmission gradient of the mask is achieved by a varying density of submicron holes (e.g. 10 nm in diameter). Upon e-beam exposure, the resist is developed according to the hole density. The WG ramp is fabricated by subsequent dry etching. Both techniques require an increased technological effort.

An example of a buried SSC with a vertical ramp integrated to an edge-emitting laser is illustrated in Chap. 10 in Table 10.4, upper right. As the laser is fabricated independently from the SSC, the material composition of the tapered WG can be freely chosen (e.g. with a larger bandgap than that of the laser-active WG).

SSCs on WGs at the chip surface are particularly important for more complex PICs or OEICs, where device degradation by hot epitaxial overgrowth processes should be avoided. For application of self-aligning techniques vertical and lateral tolerances of $\pm 2\,\mu$m are required. For uncladded SSCs (in contrast to buried WGs + SSCs) a vertical ramp of the WG core is in general necessary to achieve a vertical beam expansion. For the uncladded rib the mode expands with decreasing core thickness and approaches vertical cut-off. Therefore, additional means to assure well-defined weak guiding are needed. Weak guiding in the SSC region can be provided by properly designed so-called guiding layers incorporated into a buffer and lower cladding-layer zone of the uncladded rib WG. A corresponding SSC is illustrated in Fig. 11.3, (see also Chap. 10, Table 10.4).

Coupling efficiency and alignment tolerances of this SSC are shown in Fig. 11.4. The residual insertion loss can be of the order of -1 dB according to theoretical calculations [7]. The current status of InP technology enables optimum insertion loss values for SSCs of 2 dB, with typical values

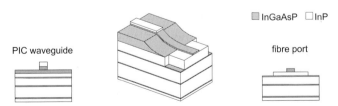

Fig. 11.3. Spot-size converter for uncladded rib waveguide [6]

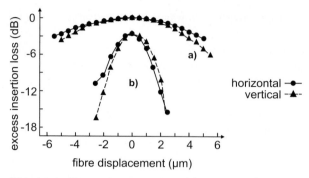

Fig. 11.4. Excess insertion loss (**a**) with and (**b**) without SSC for TE and TM polarized radiation [6]

of -3 to $-4\,\mathrm{dB}$. In general, SSCs are possible with almost no loss due to mode-mismatch, but at the expense of increasing technological effort.

For specially designed WG structures for which the guided wave is already close to cut-off, vertical mode expansion into a guiding layer can be achieved by lateral tapering already. A corresponding example is a ridge laser structure with integrated taper [8], which enabled -3-dB coupling loss with a cleaved optical fibre. Note, however, that such close-to-cut-off WGs are strongly polarization dependent and rather lossy and are therefore only well-suited for a simple laser taper integration, where only one polarization is present and loss is compensated by laser gain.

In general SSCs should operate for optical waves with arbitrary polarization and for a certain wavelength range. Using a specific taper design [7] the coupling loss can be made equal for TE and TM polarization. Wavelength independent coupling is usually guaranteed for closely spaced channels (typically a few nanometers) within the same spectral window (second or third optical window). For wavelengths lying in different spectral windows SSCs can be designed to achieve similar coupling losses for both wavelengths. Tapers of this type are needed for example for bidirectional optical WDM transmit/receive ICs (transceivers) in optical-access network links using the $1.3\,\mu\mathrm{m}$ and $1.5\,\mu\mathrm{m}$ wavelength range for the upstream and downstream transmission, respectively, (see also Sect. 11.7). Experimental investigations of such SSCs exhibit about $1\,\mathrm{dB}$ difference in the coupling efficiency for both wavelengths.

11.4   Monolithic Laser Integration

Monolithic integration of semiconductor LDs or SOAs with simple dielectric WG networks or other more complex WG devices is of great importance for a successful and economic PIC or OEIC fabrication. Examples for such photonic ICs are LDs/SOAs with integrated spot-size converters, monolithic modulator-laser chips, multi-wavelength light sources, bidirectional transceivers, heterodyne receiver ICs, wavelength converters and high speed optical multiplexers/demultiplexers (see Sect. 11.7).[2]

Another monolithic integration approach includes the parallel formation and interconnection of LDs with electronic driver circuits, which incorporate different electronic devices (e.g. heterojunction bipolar transistors (HBTs), resistors and/or capacitors). This kind of integration is less important for today's commercial applications, and as a consequence this chapter will only include a brief description of a monolithic LD-HBT integration.

A rather large number of different technological approaches has been developed in the past to accomplish low-loss and low-reflective integration of LDs with dielectric WGs [9], and the most important approaches are described in the following. Each of these different techniques is more or less developed to date and has its own advantages and shortcomings. A single technique representing a best solution for the fabrication of any kind of LD-integrated OEIC is not available. Instead, the best choice in a given situation depends upon the desired IC architecture, functionality, performance specifications and economic requirements.

11.4.1   Vertical Laser–Waveguide Coupling

All LD-WG coupling schemes are essentially based on two principles: (a) vertical and (b) butt coupling as illustrated in Fig. 11.5.

In the case of vertical coupling, light generated in the active layer stack of an LD is fed vertically into a passive optical WG layer, which is part of both subelements to be integrated (Fig. 11.5a). In the simplest case the active layers within the LD section are located directly on top of this optical WG. The concept has been known for a long time and is used for example to couple active and passive sections within DBR lasers [10]. LD-integrated components based upon this kind of vertical coupling are relatively simple to fabricate. One can use standard full wafer (re)growth techniques and there

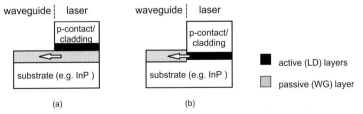

Fig. 11.5. Schematic cross-sectional views of (**a**) vertical and (**b**) butt coupling for LD-WG integration

---

[2] In the following we shall talk about LD integration only, but all concepts apply in an identical fashion to the integration of SOAs.

is no need for selective-area regrowth. Furthermore, vertical and lateral self-alignment of the light-guiding structures of both elements is provided. On the other hand, there are conceptual shortcomings. First, the passive WG layers in the LD section have to be n-doped for optimum laser characteristics, because these layers serve also as contact layers. This induces additional losses, which is a potential drawback for integrated, extended WG-based devices (e.g. multiplexers, filters, couplers or combiners). Second, this integration approach has limited flexibility in the choice of architectures to be integrated. An integration of two individually optimized subelements is often not possible, because their layer stacks are frequently completely different in terms of semiconductor material and geometrical dimensions.

Other vertical coupling techniques are based upon specially designed vertical WG coupler architectures, in which an additional layer is usually located between the active and the passive WG layers. Those coupling schemes have been demonstrated experimentally to date for the fabrication of various widely tunable laser devices. These include the tunable-twin-guide (TTG) laser, the amplifier-coupler-absorber (ACA) laser, the vertical-coupler-filter (VCF) laser, the distributed-forward-coupled (DFC) laser, the grating-assisted codirectional-coupler laser with rear-sampled grating reflector (GCSR) and the vertical Mach-Zehnder (VMZ) laser. A comprehensive treatment of this topic is given in [11] for example.

11.4.2   Laser–Waveguide Butt Coupling

In the case of butt-coupling, light generated in the active LD layer(s) is end-fire launched into the second WG (device), which can be completely different in terms of geometrical dimensions and material composition (Fig. 11.5b). The butt-coupling approach is rather straightforward from a conceptual point of view and offers, at least theoretically, maximum flexibility in the choice of the two integrable devices. In reality, however, processing capabilities may introduce constraints, and technological demands and efforts are generally higher compared to vertical coupling.

Butt coupling can be accomplished in many different ways, as illustrated in Fig. 11.6 and outlined in the following paragraphs.

Non-Selective Full-Wafer Regrowth over Mesa-Shaped LD Sections.
The approach illustrated in Fig. 11.6a starts with the non-selective growth of a complete LD layer stack including the top contact and cladding layers. In a next step these layers are removed in most parts of the wafer, leaving only isolated, but unmasked (!) mesa structures at the LD sites. Subsequently, the layer stack of the WG (or WG device) is grown, again non-selectively, over the full wafer [12], and finally the overgrown layers are removed from the LD mesa.The latter process implies a critical alignment of the required etching mask, and as a consequence either a local removal of the regrown layers at the

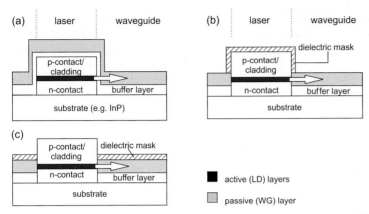

(a) laser   waveguide

p-contact/
cladding

n-contact | buffer layer

substrate (e.g. InP)

(b) laser   waveguide

dielectric mask

p-contact/
cladding

n-contact | buffer layer

substrate

(c)

p-contact/ | dielectric mask
cladding

n-contact | buffer layer

substrate

■ active (LD) layers

▨ passive (WG) layer

Fig. 11.6. Schematic cross-sections of butt joints fabricated by (**a**) non-selective full-wafer regrowth, (**b**) selective-area regrowth around mesa structures, and (**c**) selective-area infill regrowth

butt-joint interface or an unwanted strong topography on the mesa surface are frequently observed.

Selective Area Regrowth around Mesa-Shaped LD Sections. Select-ive-area regrowth around LD mesas starts again with a non-selective growth of a complete LD layer stack and subsequent removal of these layers except at the laser sites. However, the mesa-shaped LD sections are formed by par-ticular dry and wet chemical etching in combination with dielectric masks (e.g. SiO_2, SiN_x). Afterwards the WG layer stack is grown around these mesas (Fig. 11.6b). The cross section of a real butt joint fabricated by this technique and utilizing selective-area MOVPE is shown in Fig. 11.7.

In the case of MOVPE an enhanced material regrowth in the vicinity of the masked mesas, or more generally, masked areas, is obtained, which

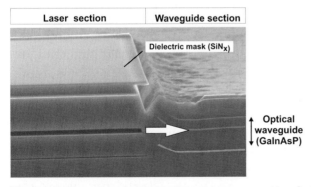

| Laser section | Waveguide section |

Dielectric mask (SiN_x)

Optical
waveguide
(GaInAsP)

Fig. 11.7. Scanning electron microscope cross-sectional view of a laser–waveguide butt joint fabricated by selective-area MOVPE regrowth on InP

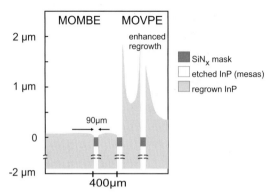

Fig. 11.8. Comparison of MOVPE and MOMBE selective-area regrowth around masked, mesa-shaped areas (800 μm length, 90 μm width, 2100 nm height). Regrown layer sequence: 1000 nm InP, 1000 nm GaInAsP, ($\lambda_Q = 1.06$ μm), 400 nm GaInAsP; growth temperatures: 650 °C (MOVPE), 465 °C (MOMBE)

depends strongly on the geometrical and technological parameters. This effect can either be exploited (see below) or suppressed. In particular, it is generally not observed, if selective-area regrowth is accomplished by Metal Organic Molecular Beam Epitaxy (MOMBE). These different MOVPE and MOMBE characteristics are illustrated in Fig. 11.8.

The optical efficiency of fabricated butt joints is mainly determined by the accuracy of the alignment of both WGs (or, more precisely, both optical fields) and the crystal quality at the butt-joint interface (e.g. (no) voids in this area). An almost perfect alignment can be guaranteed, if special etch-stop or marker layers are used during LD mesa shaping, and under such conditions the required layer thicknesses can be grown exactly and uniformly from wafer run to wafer run.

Selective-Area Regrowth of LD Layers into Etched Grooves. In contrast to the selective-area regrowth technique described above, LD layers can also be grown into etched grooves [13]. In this case the layer stack of the WG device has to be grown first, and properly designed grooves with geometrical dimensions of the LD sections are etched into these layers subsequently. The final step is a selective-infill regrowth (Fig. 11.6c). The geometrical dimensions of the recessed areas are generally very small and the ratio of masked/unmasked areas on the wafer surface is very high. Therefore, selective-area MOMBE is the technique of choice for this kind of butt-coupling technology, because an extremely strong enhanced regrowth would be obtained in case of selective-area MOVPE.

Bandgap Energy Control by Selective-Area MOVPE. The already mentioned enhanced deposition rates and material composition changes close

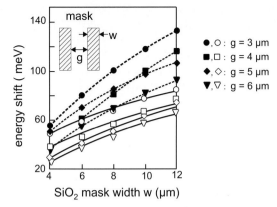

Fig. 11.9. Gap wavelength shift vs. mask width w for GaInAs/InP MQWs for various gap widths g. Full (open) symbols: atmospheric (100 Torr) pressure, after [14]

to masked areas during MOVPE selective-area regrowth can be used to locally change the bandgap energy of a multi-quantum well (MQW) layer stack. Bandgap variations depend upon (1) the geometrical dimensions of the masked regions, (2) the ratio of the masked to unmasked areas, (3) epitaxial parameters (growth temperature, pressure, III/V-ratio, etc.), and (4) equipment technical data. The bandgap energy of a selectively grown stripe of GaInAsP MQW layers, for example, decreases with increasing width of the dielectric masks on both sides of this stripe. Consequently, a butt coupling between an active MQW LD and another MQW waveguide device with different bandgap energy can be obtained simply by a proper design of masked areas and utilizing only one single regrowth step. Due to this technique controlled bandgap equivalent wavelength shifts of up to $\Delta\lambda_{\text{gap}} \leq 130\,\text{nm}$ have been achieved with properly optimized MOVPE process parameters. Figure 11.9 illustrates characteristic dependencies [14], but it should be kept in mind that these are process- and plant-specific.

Laser-modulator integration is one of the best-known examples for this kind of butt-coupling technique [15]. Other examples are laser arrays, multi-wavelength sources, SOAs and LDs with integrated SSCs, and transceiver PICs. The simplicity of the fabrication process is very attractive for many kinds of PIC components, but implies also several performance compromises, because it is in general impossible to optimize all important parameters for each subelement independently.

Quantum-Well Intermixing. Butt joints can also be fabricated by utilizing so-called quantum-well (QW) intermixing techniques, if both subelements consist of a MQW waveguide of the same kind [16]. At first, the active MQW LD layer stack with bandgap energy $E_{g,\text{act}}$ is grown non-selectively on the wafer surface. Then, the quantum wells and barriers of these layers are intermixed to form an alloy semiconductor with a bandgap energy $E_{g,\text{WG}} > E_{g,\text{act}}$ in those regions, where the passive WG device should be placed. Quantum-well intermixing generally results in an increase of the bandgap energy and is accomplished experimentally by selective-area impurity diffusion or ion implantation (Impurity Induced Disordering, (IID)), dielectric capping and laser annealing (Vacancy Induced Disordering, (VID) [17,18]), or photoabsorption disordering [19].

Identical Layer Concept. The 'identical layer concept' [20] can be considered the most straightforward laser-WG coupling, as two different subelements share the same WG. An example is an electro-absorption modulator based on the quantum-confined Stark effect monolithically integrated with a laser. A reverse bias applied to the modulator induces a redshift, while the laser wavelength is shifted into the region of low modulator absorption by a Bragg grating (typically 30−50 nm detuning). Advantages of this approach include very simple technological processing, efficient modulation by relatively small wavelength detuning and the possibility to achieve negative chirp. On the other hand, the laser and the modulator cannot be individually optimized.

11.4.3   Laser-HBT Integration

The laser integration treated in the foregoing section focused on the optical functionality of the integrated chip, and this integration is of prime importance for PIC or OEIC development. However, there are additional varieties of laser integration, where the laser is to be integrated with elements, which are significantly different from the laser. Examples are the integration of a laser with transistors (FETs or HBTs) or an integrated heterodyne receiver (see Fig. 11.16, below). Under such circumstances there are essentially two complementary approaches: one can either use different epitaxial structures for the laser and the other component(s) to be integrated (separate structure approach), or a common layer sequence is used for both (all) devices.

The key problem for the *separate structure* approach is processing crosstalk, which means that the growth of the subsequent structure(s) potentially degrades the one(s) grown before. The main problem is unwanted diffusion of dopants during later high-temperature process steps. It can be avoided (or reduced to an acceptable extent) by using less-mobile dopants or alternatively by changing the overall processing sequence. In general, the subsequent processing of different structures is significantly more complex than just adding the individual processing steps, and this is the price to be paid for the advantages offered by the integrated device in terms of compactness, reliability, and easier module assembly, for example.

The *common structure* approach avoids the process crosstalk just described, while its main challenge is the development of a layer sequence which

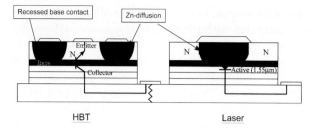

Fig. 11.10. Schematic cross-section of HBT and laser

is equally suited for the fabrication of the laser and the other subelement(s) and with negligible (or acceptable) compromises for individual device performance. Figure 11.10 shows an example of such an integration, namely that of a laser and an HBT [21]. The same basic structure is processed either into a multiquantum-well laser or an HBT. By utilising zinc diffusion to convert n-type into p-type layers, it is possible to maintain close to optimum design and performance for both components. This has already been proven for the DC characteristics and is expected for the high-frequency behaviour as well.

11.5   Integrated Receiver

Integrated receivers offer several benefits compared to hybrid solutions: they need smaller chip area and exhibit reduced parasitics, which leads to lower noise and thus higher sensitivity. Furthermore, integration enables higher reliability and lower assembly cost, and, finally, in the last few years monolithics have even surpassed commercially available hybrids in terms of bit rate up to the 50 Gbit/s regime [22–25].

Constituting elements of integrated receivers are dielectric waveguides, photodiodes (PD) and transistors/amplifiers. To qualify as an integrated device a chip should comprise at least two of these elements; in many cases all three are combined to form a monolithically integrated photoreceiver.

The most common designs for photodetectors are the pin structure or the Metal-Semiconductor-Metal (MSM) approach (see also Chap. 4). If these devices are top-illuminated, they are subject to a bandwidth-efficiency trade-off: A small absorption layer thickness ($< 0.6\,\mu$m) leads to a short transit time, but at the same time gives rise to reduced responsivity. Nevertheless, up to 20 Gbit/s the concept is normally considered acceptable. For higher speed devices one means to evade the bandwidth-efficiency limitation is edge illumination (i.e. absorption path perpendicular to electric field); another alternative is the evanescently coupled WG integrated PD. The latter approach lends itself to monolithic integration, and it is particularly suited for ultra-high-speed devices, if it is combined with the travelling-wave concept.

Edge-illuminated photodiodes realize a high bandwidth-efficiency product ($\approx 32\,\mathrm{GHz}$) and have been integrated with HEMT-based transimpedance [26] and travelling-wave amplifiers [23]. In both cases the waveguide photodiode (WG-PD) concept, based on single-step MOVPE-growth of the optical layers after the HEMT-growth comprises multimode optical waveguiding in the absorption layer (GaInAs) and adjacent higher bandgap cladding layers (GaInAsP, $\lambda_Q = 1.3\,\mu\mathrm{m}$), to adapt the optical field size of the WG-PD to that of the fibre. This leads to good responsivity values up to $0.8\,\mathrm{A/W}$ for $1.55\,\mu\mathrm{m}$ wavelength and simultaneously to high bandwidths of more than $50\,\mathrm{GHz}$. With respect to extended optical integration this PD type is only applicable as a single device combined with electronics, because the elements have to be placed at precisely cleaved wafer edges.

An optical, monolithic WG-PD integration may be an alternative to surpass the bandwidth-efficiency limitation of top-illuminated pin- or MSM-PDs. The integrated WG-PD may be a component in its own right, or a substructure of a more complex PIC or OEIC (transceiver, optical mm-wave source, monolithic heterodyne receiver). Details of WG-PD coupling are treated in [27] for example.

The WG-PD coupling can be made in essentially two fashions: by (1) butt coupling or (2) vertical or evanescent field coupling (see Fig. 11.5 and Sects. 11.4.1, 11.4.2, as similar arguments apply for LD-WG and WG-PD coupling). Butt coupling generally provides higher design flexibility and small detector capacitances (several 10 fF). However, these benefits are achieved at the expense of more complex fabrication including an additional (selective or non-selective) vertically aligned epitaxial-regrowth step. The technological requirements are tight, and there is potential for problems at the regrown WG-PD interface. On the other hand, evanescent-field coupling offers simple processing, while the PD area (and capacitance) is larger (35 fF), and the total structure gets increasingly non-planar.

The moderate coupling efficiency between the WG and the absorbing layer in the latter case can be improved by inserting a transparent layer. Its refractive index and geometrical dimensions as well as those of the PD section have to be carefully chosen. ('vertical impedance matching', [27]). Thus internal coupling efficiencies of more than 80% can be achieved. Consequently the efficiency of waveguide-integrated photodiodes is determined mainly by the fibre-chip coupling, which may be improved by applying integrated spot-size converters.

Effective absorption coefficients of WG-integrated PDs of $0.6\,\mathrm{dB/\mu m}$ have been observed in GaInAsP/InP for $1.55\,\mu\mathrm{m}$ operation wavelength. If the PD length is $20\,\mu\mathrm{m}$, then 50 or 100 GHz bandwidth can be achieved for 5 or $3\,\mu\mathrm{m}$ diode width, respectively.

If more than one wavelength is propagating in an OEIC, efficient light absorption may not be the only requirement, but wavelength selectivity may be equally important. An example is $1.3\,\mu\mathrm{m}/1.55\,\mu\mathrm{m}$ transceivers (see Fig. 11.17),

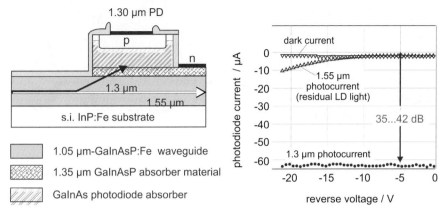

Fig. 11.11. Waveguide-integrated photodiode (*left*) and photocurrent as function of reverse voltage (*right*); for further details see text

where the absorption of the unwanted wavelength gives rise to crosstalk, which limits the overall system performance. 1.55 μm signal absorption and 1.3 μm suppression can be assured, if the 1.55 μm PD section is preceded by a region with evanescent coupling to a 1.3 μm absorber. For 150 μm length of the latter > 40 dB suppression of the 1.3 μm signal has already been achieved and > 60 dB is expected. A solution to the complementary requirement, PD absorption of a 1.3 μm signal and suppression of 1.55 μm radiation, is shown in Fig. 11.11. By making the PD from a material with $\lambda_Q = 1.35$ μm, the unwanted absorption at 1.55 μm can be kept sufficiently small (crosstalk below -40 dB).

Integrations including amplifying elements realized so far comprise pin PD-HEMT-, pin PD-HBT-, MSM PD-HEMT-, and WG-pin PD-HEMT-receivers. Devices can be fabricated either by a single growth run or alternatively by multiple growth runs.

For moderate bit-rate applications pin PD-amplifier integrations can be accomplished with a single epitaxial layer stack and less than 10 lithography processing steps. In the case of integrated pin PD-HBTs, integration benefits stem from using the base-collector junction as the photodiode also. The collector-layer-thickness is a trade-off between sensitivity and bandwidth, but with the availability of erbium doped fibre amplifiers (EDFA) sensitivity becomes less important and the design can be focused on speed. 50 GHz bandwidth for an integrated pin/SHBT (Single-HBT) receiver [28,29] and the successful detection of 40 Gbit/s return-to-zero signals with a monolithic pin-DHBT (Double-HBT) receiver [30] have already been reported.

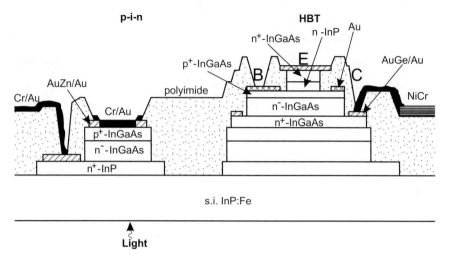

Fig. 11.12. Integrated pin PD-HBT (after [31])

For pin PD-FETs multiple use of epitaxial layers has not been reported so far, but the FET contact and gate metals can be reused for PD n- and p-contacts, and the FET and pin passivation insulator (silicon nitride) may also function as a capacitance dielectric and an antireflection coating. In general, the 'single epitaxy' approach is particularly cheap, but the associated non-planar chip surface may be critical for high-resolution optical lithography. An example of such a structure is shown in Fig. 11.12 [31].

Integrated 1.55 μm pin-HEMT photoreceivers have also been successfully fabricated on s.i. GaAs substrate, although with a topography different from that shown in Fig. 11.12. The single epitaxial layer stack starts with Al-GaAs/GaInAs/GaAs HEMT layers, followed by a buffer layer of composition-ally graded AlGaInAs and GaInAs photodiode layers on top of that [32]. The integrated receivers exhibited 36.5 GHz bandwidth and enabled the successful detection of 40 Gbit/s non-return-to-zero data.

If the detector-transistor integration is accomplished by two epitaxial steps, both devices can be optimized individually. A planarized chip surface is obtained by growing the PD layers into a properly recessed area, or alterna-tively, by growing the transistor layers on top of an appropriate buffer. Both approaches have successfully been demonstrated, and examples are shown in Figs. 11.13 [33] and 11.14 [34,35] (17 lithography processing steps).

Even if the double-epitaxial growth approach has been chosen, monolithic integration offers the additional benefit of multiple use of several material layers without noticeable performance compromise.

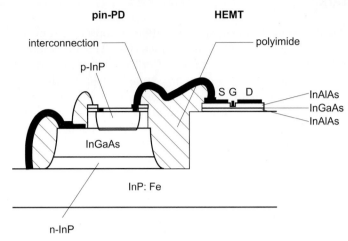

Fig. 11.13. Integrated pin PD-HFET (after [33])

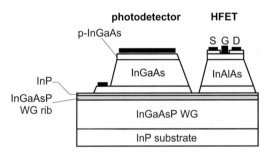

Fig. 11.14. Integrated WG-pin PD-HFET (after [34,35])

The capabilities of monolithic integration can be fully exploited if multichannel receivers are fabricated with essentially no extra effort, and four- and eight-channel receivers have been fabricated successfully in various laboratories with bit rates ranging from 0.62 to 10 Gbit/s [36,37]. The key point of such receivers is power consumption and crosstalk. Excellent results have been reported for HBT preamplifiers integrated with pin PDs, and to a lower extent for integrations including HEMTs. However, strong competition for monolithic integration arises from hybrid solutions based on GaInAs/InP detector arrays combined with GaAs-based multichannel preamplifiers. At present it is an open question to what extent OEICs or hybrid devices will dominate the field of next-generation 40 Gbit/s multichannel receivers.

11.6   Crosstalk

The potential of OEICs stems from the fact that different optoelectronic devices can be fabricated close to each other on a single chip. This is favourable for chip reliability, and it reduces the number of critical interfaces, and as a consequence mounting and packaging costs are strongly reduced. However, as an unwanted side effect, crosstalk between the various subcomponents may become an important issue [38]. Most relevant are electrical and optical crosstalk, while thermal crosstalk is normally less of a problem, but may become relevant under special circumstances.

The effect of crosstalk on the chip performance depends strongly on its architecture, functionality, and the operation frequency. In the case of integrated multi-laser chips electrical and thermal crosstalk tend to be relevant. Chips including a transmitter and a photodiode rely on low electrical and optical crosstalk. A corresponding example is the optical full-duplex WDM transceiver, where the received signal current is orders of magnitude below the transmitter driving current.

11.6.1   Electrical Crosstalk

Electrical crosstalk comprises essentially three contributions: (a) conductive, via conductive and capacitive paths, leading to additional currents in the distorted circuit part; (b) inductive coupling, leading to additional voltages; and (c) radiative coupling, especially at the highest frequencies, adding unwanted electrical signal power to remote locations of the circuit. This is illustrated in Fig. 11.15, which schematically addresses sources of crosstalk in a fictitious transmitter-receiver (Tx-Rx) circuit. The Rx normally uses a different optical wavelength to receive signals and has to be well isolated from the normally simultaneously operating Tx over a broader frequency range up to the bit rate or even higher.

Conductive Crosstalk. Conductive crosstalk can be reduced to acceptable values by the following general approaches:

1. One can provide a very high degree of isolation between active components on the chip in cases where high distorting voltages have to be isolated. This strategy necessitates a semi-insulating substrate and, in addition, all other layers have either to be interrupted physically or must be grown non-conducting. An all-insulating OEIC approach usually adds complexity to the growth and it also renders active component design increasingly complicated, since extra connections are needed, which could otherwise be made simply via the substrate.
2. All signal routeing may be based on a true differential layout in contrast to single-ended designs, so that capacitively or radiatively induced currents cancel each other at sensitive amplifier inputs to first order.

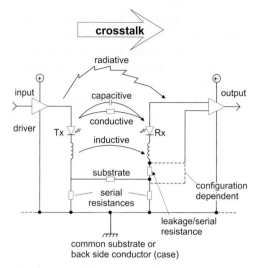

Fig. 11.15. Different contributions to electrical crosstalk (schematic), adding spurious currents, voltages or unwanted signal power to remote circuit parts in a fictitious transmitter-receiver OEIC

3. A common earth to all components may suppress or minimize the otherwise uncontrolled floating of the various common connections. In this case the chip is fabricated on a conductive substrate and the back side of the chip serves as the common earth connection. Such a design requires that the common semiconductor provides a very efficient route to earth, and this is only simple at low frequencies, but becomes more of a problem as the operating frequency increases. For example, an extra impedance of less than $1\,\Omega$ between the true (generator) earth and the common point, where 'transmit' and 'receive' earth meet, may impair the crosstalk coupling by several $10\,\mathrm{dB_{elec}}$ compared to the case of negligible resistance. However, a common conductive-substrate approach leads to additional capacitive loading of interconnection lines between integrated components, because the isolating dielectrics can be deposited only with limited thicknesses; thus, the dielectric constants should be chosen to be as low as possible, e.g. by applying polyimides.

Inductive Coupling. Inductive coupling, especially from higher signal currents into transmitter stages, may be reduced by chosing an orthogonal line design between circuit parts for good isolation. Also, lines should be kept as short as possible to reduce the induced voltage. A differential input-signal layout is much better suited to reduce induced-voltage contributions compared to single-ended designs.

Radiative Coupling. Transmitter lasers are typically driven with a modulated signal current of the order of tens of mA. The feed tracks to the OEIC, the bond wires and the laser all radiate electromagnetically, and this gives rise to spurious receiver signals. In order to keep such crosstalk sufficiently low, different measures can be taken, and these include: placing the laser and the photodiode as far as possible from each other, implementing on-chip screening and placing guard rings around the photodiode, or replacing bond wires by flip-chip mounting.

For efficient, although technologically demanding, on-chip screening an additional spaced metallization layer is applied across most parts of the circuit to be protected [39]. This measure avoids at the same time capacitively and radiatively induced crosstalk between neighbouring circuits, e.g. arrayed receivers, at the expense of somewhat increased stray capacitance.

In summary, fully insulated devices are the most favourable choice to minimize electrical crosstalk, at the cost, however, of more complex device design and processing. On the other hand, a common substrate connection is simpler; however, small imperfections in earthing may cause dramatic increases in crosstalk.

11.6.2   Optical Crosstalk

Optical crosstalk may be due to (a) reflected (guided) or (b) scattered (unguided) light. Reflections can occur at various sites on the chip, but usually only the facet reflection is important. With single-layer antireflection (AR) coatings the edge reflectance can be reduced to 1% typically (-20 dB), and two-layer coatings enable values as low as 10^{-4} (-40 dB). However, such low values are normally restricted to a rather narrow wavelength range. Alternatively, one can use an angled facet, where the reflection suppression gets better with increasing facet angle. At the same time, however, the fibre-chip coupling gets increasingly difficult, as the fibre has to be tilted with respect to the angled chip facet. A good compromise is the combination of a moderate-effort AR coating plus a small facet angle, which enables a reduction of the edge reflectance to values meeting the systems demands ($2°$–$3°$ facet angle provides 10^{-2} to 10^{-3} extra isolation in the 1.3–$1.55\,\mu$m wavelength range [38,40]).

Optical crosstalk due to unguided light is primarily caused by scattered coherent light, but laser spontaneous emission acts as another source of unwanted stray light [41]. Scattered coherent light may escape from waveguides due to material imperfections, roughness of WG walls, insufficient mode confinement in bends, or mode mismatch between components, including the laser–waveguide joint. The stray light can be prevented from causing unwanted detector signals by (a) placing the photodiode as far as possible from the laser and off the cavity, (b) providing absorbing regions on the chip where ever possible, (c) adding reflection features, e.g. etched slots, which deflect light away from the photodiode, or (d) using a detector absorbing material which is transparent for the unwanted light (if possible).

11.6.3   Thermal Crosstalk

Thermal crosstalk gets important in multi-wavelength emitters (multi-laser arrays), if several lasers are operated simultaneously, as a laser is a very efficient heat emitter and its performance depends on the local temperature. Thermal crosstalk depends on the distance between the integrated lasers, the total heat load and the quality of the mounting (heat transfer to a heat sink), and it is particularly relevant for dense WDM sources, as a temperature variation of $1\,K$ varies the emission wavelength of a DFB laser by roughly $0.1\,nm$ ($\cong 12\,GHz$) in the $1.5\,\mu m$ wavelength range. The typical distance-dependence is a pronounced temperature increase close to a laser acting as a heat source and a floor of elevated temperature all over the chip. The magnitude of this floor depends essentially upon the heat transfer between the laser submount and the heatsink, and to a minor extent upon the heat transfer between the laser and its submount, and upon the thermal resistivity of the semiconductor material itself [42,43].

Several approaches exist for overcoming thermal crosstalk in laser arrays: to operate only one laser at a time (wavelength-selectable transmitter) or to assure constant average dissipated power for each laser (by chosing a proper encoding scheme, and this corresponds to frequently chosen operation conditions). Finally, an efficient means to eliminate (or to significantly reduce) thermal effects of OEICs under varying operating conditions is a monolithically integrated on-chip temperature sensor (in contrast to the standard temperature sensor attached to the heat sink) in combination with a thermoelectric cooler.

11.7   Current Status of Optoelectronic Integration

About one decade ago one widely followed guideline for monolithic OE integration was a steady raise of complexity of OEICs. This has been changed in the last few years towards lower OEIC complexity and a simultaneous strong focus on module cost reduction. An alternative goal are functionalities, which rely on monolithic integration. As a consequence of such tendencies one of the most complex InP-based OEICs realized so far is the polarization-diversity heterodyne-receiver chip shown in Fig. 11.16 and reported already some time ago [44]. In addition to a laser, balanced receivers and a WG network, the heterodyne receiver comprises polarization splitters [45] and a polarization rotator [46] as subcomponents, whereas an optical isolator is not needed due to the small chip dimensions, in accordance with theory [47].

An integrated transceiver represents an OEIC, which is currently optimized with respect to eventual mass production. Two complementary varieties are needed, one for the exchange office (Optical Line Termination, OLT), one for the residential customer (Optical Network Unit, ONU). For an economic, commercial fabrication the minimization of the chip size is

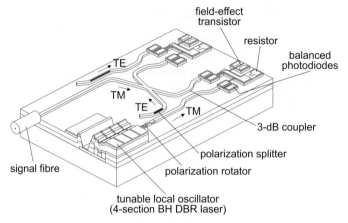

Fig. 11.16. Monolithic polarization-diversity heterodyne-receiver chip (schematic) [44]

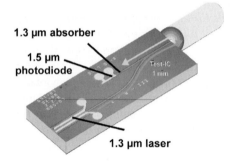

Fig. 11.17. Photograph of an integrated bidirectional 1.3 μm/1.55 μm transceiver [48]

a crucial demand. At the same time, this increases the challenge to achieve the required electrical and optical crosstalk suppression (30−50 dB optical crosstalk suppression is typically required, a specific value depends on the particular application under consideration, and these values get even larger by the required systems margin). Figure 11.17 shows a photograph of a Y-junction 1.3/1.5 μm transceiver [48]. With a 1.3 μm absorber section located in front of the photodiode a crosstalk-suppression of 33 dB has recently been achieved (36 dB best value), the 1.3 μm (1.5 μm) coupling losses are −5 dB (−2 dB) with essentially no polarization dependence.

An alternative variant is the in-line-type transceiver [49]. The in-line geometry yields a particularly compact chip, and a −27 dBm full-duplex sensitivity at 155 Mbit/s data rate has been recently reported for a transceiver comprising a 1.3 μm laser, a 1.5 μm detector, and a light absorptive section in between, which prevents laser light from being detected in the PD section.

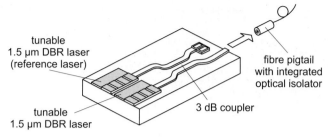

Fig. 11.18. Integrated optical mm-wave source [50]

An OEIC, which takes advantage of the simple control of relative emission wavelengths/frequencies on a single chip is the integrated optical mm-wave source shown in Fig. 11.18 [50]. Frequency differences in the GHz regime (e.g. 38 or 60 GHz) can be generated easily and with high stability (temperature sensitivity: 5 MHz/K), and the device can be used to optically feed base stations in mobile (broadband) communication systems.

Integrated multiwavelength transmitters are particularly appealing, since integration enables a very easy but accurate definition of the relative difference of emission wavelengths, and this difference is essentially independent of chip temperature. A recently reported example is a multi-wavelength emitter comprising eight lasers, monolithically integrated with a combiner and a semiconductor optical amplifier [51]. The chip is very compact ($0.6 \times 2\,\text{mm}^2$) and performance of the integrated transmitter is close to that of discrete DFB lasers (7.9 mA laser threshold current, +10 dBm fibre launched power at 50 mA laser and 100 mA SOA current, and 50 dB sidemode suppression ratio).

A still higher degree of chip functionality is obtained if an electro-absorption (EA) modulator is added to a multi-wavelength source. Corresponding transmitter chips with six [52] or eight [53] channels have already been reported. Characteristics of the latter OEIC are: emission wavelengths between

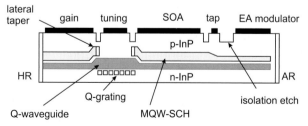

Fig. 11.19. Schematic drawing of an integrated 20 channel DBR laser with integrated semiconductor optical amplifier, photodetector and electroabsorption modulator [54]

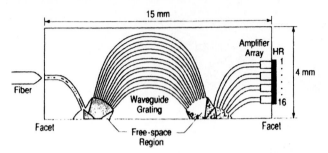

Fig. 11.20. Schematic drawing of 16-channel multi-frequency laser [55]

1552.5 nm and 1567.8 nm (by proper channel selection and changing the operation temperature between 5 °C and 25 °C), 7 mW output power at 100 mA laser and 100 mA SOA current (with the modulator unbiased) and 2.5 Gbit/s modulation capability.

Comparable functionality of a single chip, however with a tunable DBR-laser instead of several individual lasers, has also been demonstrated recently [54]. The chip is illustrated in Fig. 11.19, its dimensions are 1.700 × 500 μm². The chip was packaged into a single industry standard module and operated at 2.5 Gbit/s. Performance characteristics are 20 channels with 50 GHz separation, > 0 dBm average output power, and < 9 MHz linewidth for all channels.

An alternative multi-wavelength emitter concept relies on the combination of a PHASAR or AWG (see Chap. 7, Sect. 7.7) with active sections. This approach circumvents the problem of ever-increasing combiner losses with increasing numbers of channels, and in addition, such devices enable the simultaneous, individual modulation of all channels. Several different varieties have already been reported (see [55,56] and references given herein), and a corresponding 16-channel emitter is shown in Fig. 11.20 [55]. Typical laser cavity lengths of such multichannel emitters are in the range of 1.5 cm, which limits the maximum modulation speed to about 1 Gbit/s.

Another OEIC comprising AWGs and active sections is a multi-channel wavelength selector fabricated with four [57] and sixteen channels [58] recently. The layout of the wavelength selector is shown in Fig. 11.21, for the sake of clarity the four-channel variety has been chosen. The two AWGs provide wavelength demultiplexing and multiplexing, respectively, the SOAs serve as optical gates providing both space and wavelength switching, with time responses in the ns range. In addition to representing gates the SOAs also enable zero-loss operation or may even provide overall gain. Key issues in device fabrication are a low-loss, low-reflectivity coupling between the active and passive sections, low propagation losses in the passive waveguides, and compact device dimensions (4.2 × 4.6 mm² for the sixteen-channel device).

Lasers integrated monolithically with EA-modulators are already commercially available for 2.5 and 10 Gbit/s operation, and they are under devel-

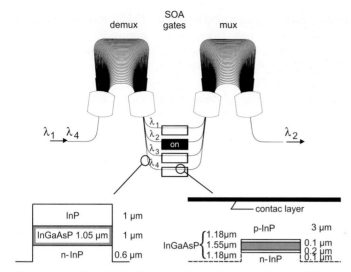

Fig. 11.21. Monolithic integrated-wavelength selector [57]

opment for 40 Gbit/s [59]. Integrated laser-modulator chips have been realized in many different fashions (see Sect. 11.4).

One example is shown in Fig. 11.22 [60]. It is important to note that 40 different laser-modulators, where each laser has its own wavelength evenly spaced from the neighbouring ones, have been fabricated on a single wafer. The different bandgaps in the laser and the modulator sections have been realized by selective-area MOVPE on InP with a patterned SiO$_2$ mask.

A single EA-modulator has a useful wavelength operation range of about 5 to 10 nm, which is sufficient for many applications. Several 10 nm wavelength

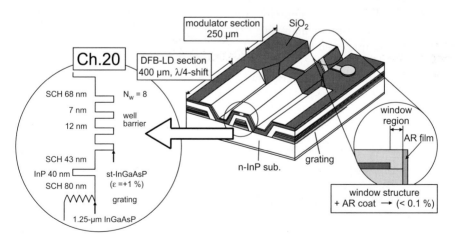

Fig. 11.22. Monolithic integrated laser-modulator [60]

coverage are offered by Mach-Zehnder interferometer (MZI) modulators, and corresponding GaAs- or LiNbO$_3$-based devices are widely used at present. However, InP-based MZI-modulators have also attracted recent interest, as they offer the potential for monolithic integration. One corresponding example is the integration of an MZI modulator with a FP laser and an on-chip mirror [61]. The laser and the modulator section are butt-coupled. In addition to the laser, the mirror, and the modulator the chip also comprises an integrated spot size converter for efficient coupling of the active section to a fibre which contains a Bragg grating. Data transmission at 2.5 and 10 Gbit/s in the wavelength range from 1540 nm to 1560 nm has already been demonstrated [62].

The relatively large number of WDM-related OEICs, which have been developed and reported recently reflects the fact that WDM is the key enabling technology for current system capacity upgrades. Nevertheless there is also work on OTDM-related components, which are mandatory if the individual channel bit rates are to be raised further (up to 160 Gbit/s or beyond in the years to come).

The following requirements for high-speed transmitters are of prime importance: The width of the pulses should be sufficiently small (several ps), they should be (close to) transform-limited, the repetition frequency must be synchronizable to the systems clock, and the emission wavelength has to be tunable within a useful range. The monolithic integration of such transmitters is particularly attractive for stability and reliability reasons, and a corresponding wavelength tunable picosecond pulse source comprising a DBR-laser and an electroabsorption modulator section for active mode-locking at 10 Gbit/s has already been reported [63]. Proper SiO$_2$-masking provides different bandgaps for the passive waveguide-, the intracavity EA-modulator-, and the amplifier section in a single epitaxial run. Nearly transform-limited 8 ps pulses tunable over 4.3 nm (either by current or reverse biasing of the DBR-section) have been demonstrated.

The complement to high-speed transmitters are OTDM demultiplexers, and one example is shown in Fig. 11.23 [64]. The operation of the demultiplexer is as follows. In the absence of control pulses the Mach-Zehnder-type structure with two asymmetrically located SOAs is balanced for operating in the bar-state. Counter-propagating control pulses (fed into port 3) saturate

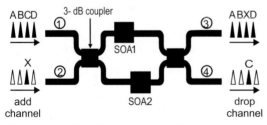

Fig. 11.23. Monolithic integrated optical time-domain demultiplexer [64]

the SOAs at different times as a consequence of the asymmetric geometry. The corresponding time interval represents a gate, in which original input data are routed to port 4 (instead of port 3), while add data are directed to port 3 (instead of port 4).

A number of different OTDM demultiplexers has been realized in the last few years including so-called band-gap shifted SOAs [65], but their operation relies always on the same principle, and as a consequence this topic will not be discussed in more detail here.

A number of other InP OEICs have also been developed and reported in the last few years, but they are comparable to those shown here as far as integration complexity, functionality, and maturity are concerned. Most of these OEICs are still in the laboratory stage, although others, such as integrated laser-modulators for example, have already become commercial products.

11.8   Outlook

In the long run, the relevance of OEICs or PICs will be valuated by their cost effectiveness. The comparison will not focus on an OEIC/PIC and a competing hybrid device, but will cover the complete solution to a systems demand, including supervision, operation and maintenance (OAM), etc. At present, OEICs of moderate complexity seem to be most promising. A minimum degree of complexity is needed as a prerequisite to have a gain in packaging and overall processing effort. OEICs with good prospects for the future include the following. Devices with integrated spot-size converters will be ubiquitous shortly, as an integrated SSC is a must for economic hybrid chip mounting or fibre-chip coupling as well. Integrated lasers + modulators or lasers + drivers could also become a standard commercial product soon. An example of a slightly more complex OEIC with good potential for the future is the bidirectional transceiver. Its complexity is manageable with currently available technologies, and a demand for large numbers can be expected in accordance with the increasing number of residential customers that will be connected to optical communication systems. OEICs comprising a multi-wavelength source monolithically integrated with a modulator [52–54] might also become commercially available soon.

The development and subsequent commercialisation of highly complex OEICs faces the problem of getting the chip cost sufficiently low. On the one hand, the development cost cannot be shared among a large enough number of OEICs as long as the total number of very complex OEICs needed in communication systems remains limited. In addition, the overall yield tends to decrease as the chip complexity increases. On the other hand, the tremendous growth of Internet traffic has stimulated a strong interest in advanced optical techniques such as optical packet switching or all-optional 3R signal regeneration, and this may evolve into a strong stimulus to develop high-end OEICs, where performance is of prime significance and cost is less an issue.

In general, the long-term economic success of InP-based OEICs can be assured if several general guidelines for device development are followed:

1. One should develop different integrable fundamental building blocks, which can serve as a basis for various OEICs of different functionality. This design concept reduces development effort and time and, as a consequence, development cost as well.
2. Subcomponents of an OEIC should be replaceable by more advanced ones or ones fabricated differently, without any need to modify the whole OEIC fabrication process.
3. The overall fabrication process should comprise the minimum possible number of processing steps; in particular, the number of epitaxial growth runs should be as low as possible.
4. The whole fabrication process must be based on well established technologies and use commercially available and proven equipment.
5. A compact chip design is extremely important, in particular for high-volume applications, since chip cost is essentially proportional to chip size. Integrated solutions have to compete with hybrid ones, and the latter have aquired a very strong postition for many applications, as the cost for mounting and assembly has been steadily and significantly reduced in the last few years.
6. Devices with more complex functionality should be developed without increasing the processing complexity. This requires a strong focus on device design relying on multiple use of individual layers without sacrificing performance of the various devices involved.

Additional technological guidelines are more specific and are consequently the topic of more technology-oriented and stronger focused presentations.

Acknowledgments

The authors are deeply indebted to all their colleagues of the Photonics division of the Heinrich Hertz Institute for the communication of a large number of specific results, many helpful comments and suggestions, and numerous valuable discussions.

References

1. K.J. Ebeling, *Integrated Optoelectronics* (Springer, Heidelberg 1993)
2. U. Niggebrügge, P. Albrecht, W. Döldissen, H.-P. Nolting, and H. Schmid, "Self-aligned low-loss totally reflecting waveguide mirrors in InGaAsP/InP," *Proc. 4th Europ. Conf. Integr. Optics (ECIO '87)*, Glasgow, Scotland, 90 (1987)

3. I.F. Lealman, C.P. Seltzer, L.J. Rivers, M.J. Harlow, and S.D. Perrin, "Low threshold current 1.6 μm InGaAsP/InP tapered active region layer multiquantum well laser with improved coupling to cleaved single-mode fibre," Electron. Lett. **30**, 973 (1994)

4. G. Wenger, L. Stoll, B. Weiss, M. Schienle, R. Müller-Nawrath, S. Eichinger, J. Müller, B. Acklin, and G. Müller, "Design and fabrication of monolithic optical spot size transformers (MOST's) for highly efficient fibre-chip coupling," IEEE J. Lightwave Technol. **12**, 1782 (1994)

5. J. Wengelink, H. Engel, and W. Döldissen, "Semitransparent mask technique for relief type surface topographies," Microelectron. Eng. **27**, 247 (1995); J. Wengelink and H. Engel, "Fabrication of waveguide tapers by semitransparent mask photolithography," Microelectron. Eng. **30**, 137 (1996)

6. P. Albrecht, H. Heidrich, R. Löffler, L. Mörl, F. Reier, and C.M. Weinert, "Integration of polarization independent mode transformers with uncladded InGaAsP/InP rib waveguides," Electron. Lett. **32**, 1196 (1996)

7. C.M. Weinert, "Design of fibre-matched uncladded rib waveguides on InP with polarization-independent mode matching loss of 1 dB," IEEE Photon. Technol. Lett. **8**, 1049 (1996)

8. H. Bissessur, C. Graver, O. Le Gouezigou, G. Michaud, and F. Gaborit, "Ridge laser with spot-size converter in a single epitaxial step for high coupling efficiency to single-mode fibres," IEEE Photon. Technol. Lett. **10**, 1235 (1998)

9. K. Kishino and S. Arai, "Integrated lasers", in *Handbook of Semiconductor Lasers and Photonic Integrated Circuits* (Y. Suematsu and A.R. Adams, eds.), Chap. 11 (Chapman & Hall, London 1994)

10. T.L. Koch and U. Koren, "Semiconductor photonic integrated circuits," IEEE J. Quantum Electron. **27**, 641 (1991)

11. M.-C. Amann and J. Buus, *Tunable Laser Diodes* (Artech House, Norwood, MA 1998)

12. D. Remiens, V. Hornung, B. Rose, and D. Robein, "Buried ridge stripe 1.5 μm GaInAsP/InP laser-waveguide integration by a simplified process," IEEE Proc. J. Optoelectron. **138**, 57 (1991)

13. B. Torabi, "Integrierbare langwellige Laserdioden," PhD thesis, Fakultät für Elektrotechnik und Informationstechnik der Technischen Universität München, München (1997)

14. C.H. Joyner, S. Chandrasekhar, J.W. Sulhoff, and A.G. Dentai, "Extremely large band gap shifts for MQW structures by selective epitaxy on SiO$_2$ masked substrates," IEEE Photon. Technol. Lett. **4**, 1006 (1992)

15. K. Komatsu, "Photonic integrated circuits using bandgap energy controlled selective MOVPE technique," *Proc. 22nd Europ. Conf. Opt. Commun. (ECOC '96)*, Oslo, Norway, vol. 4, 77 (1996)

16. J.H. Marsh, "Quantum well intermixing," Semicond. Sci. Technol. **8**, 1136 (1993)

17. J.H. Lee, S.K. Si, Y.B. Moon, E.J. Yoon, and S.J. Kim, "Bandgap tuning of In$_{0.53}$Ga$_{0.47}$As/InP multiquantum well structure by impurity free vacancy diffusion using In$_{0.53}$Ga$_{0.47}$As cap layer and SiO$_2$ dielectric capping," Electron. Lett. **33**, 1179 (1997)

18. D. Hofstetter, H.P. Zappe, and R. Dändliker, "Optical displacement measurement with GaAs/AlGaAs-based monolithically integrated Michelson interferometers," IEEE J. Lightwave Technol. **15**, 663 (1997)

19. A. McKee, C.J. McLean, G. Lullo, A.C. Bryce, R.M. De La Rue, and J.H. Marsh, "Monolithic integration in InGaAs-InGaAsP multiple-quantum-well structures using laser intermixing," IEEE J. Quantum Electron. **QE-33**, 45 (1997)
20. A. Ramdane, F. Devaux, N. Souli, D. Delprat, and A. Ougazzaden, "Monolithic integration of multiple-quantum-well lasers and modulators for high-speed transmission," IEEE J. Select. Topics Quantum Electron. **2**, 326 (1996)
21. U. Eriksson, "Technologies for monolithic indium phosphide opoelectronic circuits," PhD thesis, Royal Institute of Technology, Stockholm, Sweden (1999)
22. H.-G. Bach, W. Schlaak, G.G. Mekonnen, R. Steingrüber, A. Seeger, Th. Engel, W. Passenberg, A. Umbach, C. Schramm, and G. Unterbörsch, "50 Gbit/s InP-based photoreceiver OEIC with gain flattened transfer characteristics," *Proc. 24th Europ. Conf. Opt. Commun. (ECOC '98)*, Madrid, Spain, vol. **1**, 55 (1998)
23. K. Takahata, Y. Miyamoto, Y. Muramoto, H. Fukano, and Y. Matsuoka, "50-Gbit/s operation of monolithic WGPD/HEMT receiver OEIC module," *Proc. 24th Europ. Conf. Opt. Commun. (ECOC '98)*, Madrid, Spain, vol. **3**, 67 (1998)
24. Y. Muramoto, K. Kato, Y. Akahori, M. Ikeda, A. Kozen, and Y. Itaya, "High-speed monolithic receiver OEIC consisting of a waveguide p-i-n photodiode and HEMT's," IEEE Photon. Technol. Lett. **7**, 685 (1995)
25. K. Takahata, Y. Muramoto, H. Fukano, K. Kato, A. Kozen, O. Nakajima, and Y. Matsuoka, "46.5-GHz − bandwidth monolithic receiver OEIC consisting of a waveguide p-i-n photodiode and a HEMT distributed amplifier," IEEE Photon. Technol. Lett. **10**, 1150 (1998)
26. Y. Muramoto, K. Kato, Y. Akahori, M. Ikeda, A. Kozen, and Y. Itaya, "High-speed monolithic receiver OEIC consisting of a waveguide p-i-n photodiode and HEMTs," IEEE Photon. Technol. Lett. **7**, 685 (1995)
27. R.J. Deri, "Monolithic integration of optical waveguide circuitry with III-V photodetectors for advanced lightwave receivers," IEEE J. Lightwave Technol. **11**, 1296 (1993)
28. D. Huber, M. Bitter, E. Gini, A. Neiger, T. Morf, C. Bergamaschi, and H. Jäckel, "50 GHz monolithically integrated InP/InGaAs PIN/HBT-receiver," *Proc. 11th Internat. Conf. Indium Phosphide and Related Materials (IPRM '99)*, Davos, Switzerland, PDA-1 (1999)
29. D. Huber, M. Bitter, T. Morf, C. Bergamaschi, H. Melchior, and H. Jäckel, "46 GHz bandwidth monolithic InP/InGaAs pin/SHBT photoreceiver," Electron. Lett. **35**, 40 (1999)
30. E. Sano, K. Sano, T. Otsuji, K. Kurishima, and S. Yamahata, "Ultra-high speed, low power monolithic photoreceiver using InP/InGaAs double heterojunction bipolar transistors," Electron. Lett. **33**, 1047 (1997)
31. S. Chandrasekhar, L.D. Garrett, L.M. Lunardi, A.G. Dentai, C.A. Burrus, and E.C. Burrows, "Investigation of crosstalk performance of eight-channel p-i-n/HBT OEIC photoreceiver array modules," IEEE Photon. Technol. Lett. **8**, 682 (1996)
32. V. Hurm, W. Benz, W. Bronner, A. Hülsmann, T. Jakobus, K. Köhler, A. Leven, M. Ludwig, B. Raynor, J. Rosenzweig, M. Schlechtweg, and A. Thiede, "40 Gbit/s 1.55 μm pin-HEMT photoreceiver monolithically integrated on 3in GaAs substrate," Electron. Lett. **34**, 2060 (1998)
33. W. Kuebart, J.-H. Reemtsma, D. Kaiser, H. Grosskopf, F. Besca, G. Luz, W. Körber, and I. Gyuro, "High sensitivity InP-based monolithically integrated pin-HEMT receiver-OEIC's for 10 Gb/s," IEEE Trans. Microwave Theory Tech. **43**, 2334 (1995)

34. S. v. Waasen, A. Umbach, U. Auer, H.-G. Bach, R.M. Bertenburg, G. Janssen, G.G. Mekonnen, W. Passenberg, R. Reuter, W. Schlaak, C. Schramm, G. Unterbörsch, P. Wolfram, and F.-J. Tegude, "27 GHz bandwidth high speed monolithic integrated optoelectronic photoreceiver consisting of a waveguide fed photodiode and an InAlAs/InGaAs-HFET-traveling wave amplifier," IEEE J. Solid-State Circuits **32**, 1394 (1997)
35. A. Umbach, T. Engel, H.-G. Bach, S. van Waasen, E. Dröge, A. Strittmatter, W. Ebert, W. Passenberg, R. Steingrüber, W. Schlaak, G.G. Mekonnen, G. Unterbörsch, and D. Bimberg, "Technology of In-based 1.55-µm ultrafast OEM-MIC's: 40-Gbit/s broad-band and 38/60-GHz narrow-band photoreceivers," IEEE J. Quantum Electron. **QE-35**, 1024 (1999)
36. S. Chandrasekhar, L.M. Lunardi, R.A. Hamm, and G.J. Qua, "Eight-channel p-i-n/HBT monolithic receiver array at 2.5 Gb/s per channel for WDM applications," IEEE Photon. Technol. Lett. **6**, 1216 (1994)
37. H. Yano, M. Murata, G. Sasaki, and H. Hayashi, "A high-speed eight-channel optoelectronic integrated receiver array comprising GaInAs p-i-n PD's and AlInAs/GaInAs HEMTs," IEEE J. Lightwave Technol. **10**, 933 (1992)
38. G. Foster, "Near-end Crosstalk Issues in Indium Phosphide Opto-Electronic Integrated Circuits," PhD thesis, University of London, London (1994)
39. A.L. Gutierrez-Aitken, P. Bhattacharya, K.-C. Syao, K. Yang, G.I. Haddad, and X. Zhang, "Low crosstalk (< −40 dB) in 1.55 µm high-speed OEIC photoreceiver arrays with novel on-chip shielding," Electron. Lett. **32**, 1706 (1996)
40. J. Buus, M.C. Farries, and D.J. Robbins, "Reflectivity of coated and tilted semiconductor facets," IEEE J. Quantum Electron. **27**, 1837 (1991)
41. R. Kaiser, M. Hamacher, H. Heidrich, P. Albrecht, K. Janiak, S. Malchow, W. Rehbein, and H. Schroeter-Janßen, "Optical crosstalk within monolithic transceiver ICs on GaInAsP/InP," in *Integrated Photonics Research*, OSA Techn. Digest. (Opt. Soc. America, Washington, DC 2000), p. 257
42. B. Klepser and H. Hillmer, "Investigations of thermal crosstalk in laser arrays for WDM applications," IEEE J. Lightwave Technol. **16**, 1888 (1998)
43. F. Fidorra, D. Franke, M. Möhrle, W. Rehbein, A. Sigmund, R. Stenzel, H. Venghaus, M. Weickhmann, M. Töpfer, M. Schmidt, and H. Reichl, "Thermal crosstalk of integrated multiwavelength transmitters," *Proc. 16th IEEE Internat. Semicond. Laser Conf.*, Nara, Japan, 149 (1998)
44. R. Kaiser, D. Trommer, H. Heidrich, F. Fidorra, and M. Hamacher, "Heterodyne receiver PICs as the first monolithically integrated tunable receivers for OFDM system applications," Opt. Quantum Electron. **28**, 565 (1996)
45. P. Albrecht, M. Hamacher, H. Heidrich, D. Hoffmann, H.-P. Nolting, and C.M. Weinert, "TE/TM Mode Splitters on InGaAsP/InP," IEEE Photon. Technol. Lett. **2**, 114 (1990)
46. H. Heidrich, P. Albrecht, M. Hamacher, H.-P. Nolting, H. Schroeter-Janssen, and C.M. Weinert, "Passive Mode Converter with a Periodically Tilted InP/GaInAsP Rib Waveguide," IEEE Photon. Technol. Lett. **4**, 34 (1992)
47. K. Petermann, *Laser Diode Modulation and Noise* (Kluwer Academic Publishers, Dordrecht 1988)
48. M. Hamacher, R. Kaiser, H. Heidrich, P. Albrecht, B. Borchert, K. Janiak, R. Löffler, S. Malchow, W. Rehbein, and H. Schroeter-Janssen, "Monolithic integration of lasers, photodiodes, waveguides and spot size converters on GaInAsP/InP for Photonic IC applications," *Proc. 12th Internat. Conf. Indium Phosphide and Related Materials (IPRM '00)*, Williamsburg, VA, USA, 21 (2000)

49. H. Nakajima, A. Leroy, B. Pierre, P. Boulet, J. Charil, S. Grosmaire, A. Gloukhian, Y. Rafflé, S. Slempkes, F. Mallécot, and F. Doukhan, "1.3 µm/1.55 µm in-line transceiver assembly with −27 dBm full-duplex sensitivity," Electron. Lett. **34**, 303 (1998)

50. D. Trommer, R. Kaiser, R. Stenzel, and H. Heidrich, "Multi-purpose photonic integrated circuit based on InP: dual tunable laser/combiner to be used as dual wavelength transmitter & optical microwave generator," *Proc. 21st Europ. Conf. Opt. Commun. (ECOC '95)*, Brüssel, Belgium, vol. **1**, 83 (1995)

51. M. Bouda, M. Matsuda, K. Morito, S. Hara, T. Watanabe, T. Fujii, and Y. Kotaki, "Compact high-power wavelength selectable lasers for WDM applications," *Conf. Opt. Fibre Commun. (OFC '00)*, Baltimore, MA, USA, TuL1, 178 (2000)

52. M.G. Young, U. Koren, B.I. Miller, M. Chien, T.L. Koch, D.M. Tennant, K. Feder, K. Dreyer, and G. Raybon, "Six wavelength laser array with integrated amplifier and modulator," Electron. Lett. **31**, 1835 (1995)

53. K. Kudo, K. Yashiki. T. Sasaki, Y. Yokoyama, K. Hamamoto, T. Morimoto, and M. Yamaguchi, "1.55 µm wavelength-selectable microarray DFB-LD's with integrated MMI combiner, SOA, and EA-modulator," *Conf. Opt. Fibre Commun. (OFC '00)*, Baltimore, MA, USA, TuL5, 190 (2000)

54. L.J.P. Ketelsen, J.E. Johnson, D.A. Ackerman, L. Zhang, K.K. Kamath, M.S. Hybertsen, K.G. Glogovsky, M.W. Focht, W.A. Asous, C.L. Reynolds, C.W. Ebert, M. Park, C.W. Lentz, R.L. Hartman, and T.L. Koch, "2.5 Gb/s transmission over 680 km using a fully stabilized 20 channel DBR laser with monolithically integrated semiconductor optical amplifier, photodetector and electroabsorption modulator," *Conf. Opt. Fibre Commun. (OFC '00)*, Baltimore, MA, USA, Post-deadline papers, PD14 (2000)

55. R. Monnard, C.R. Doerr, C.H. Joyner, M. Zirngibl, and L.W. Stulz, "Direct modulation of a multifrequency laser up to 16 × 622 Mb/s," IEEE Photon. Technol. Lett. **9**, 815 (1997)

56. C.H. Joyner, C.R. Doerr, L.W. Stulz, M. Zirngibl, and J.C. Centanni, "Low-threshold nine-channel waveguide grating router-based continuous wave transmitter," IEEE J. Lightwave Technol. **17**, 647 (1999)

57. R. Mestric, M. Renaud, F. Pommereau, B. Martin, F. Gaborit, G. Lacoste, C. Janz, D. Leclerc, and D. Ottenwalder, "Four-channel wavelength selector monolithically integrated on InP," Electron. Lett. **34**, 1841 (1998)

58. R. Mestric, C. Porcheron. B. Martin, F. Pommereau, I. Guillemot, F. Gaborit, C. Fortin, J. Rotte, and M. Renaud, "Sixteen-channel wavelength selector monolithically integrated on InP," *Conf. Opt. Fibre Commun. (OFC '00)*, Baltimore, MA, USA, TuF6, 81 (2000)

59. H. Takeuchi, K. Tsuzuki, K. Sato, M. Yamamoto, Y. Itaya, A. Sano, M. Yoneyama, and T. Otsuji, "Very high-speed light-source module up to 40 Gb/s containing an MQW electroabsorption modulator integrated with a DFB laser," IEEE J. Select. Topics Quantum Electron. **3**, 336 (1997)

60. K. Kudo, M. Ishizaka, T. Sasaki, H. Yamazaki, and M. Yamaguchi, "1.52–1.59-µm range different-wavelength modulator-integrated DFB-LD's fabricated on a single wafer," IEEE Photon. Technol. Lett. **10**, 929 (1998)

61. H. Bissessur, C. Graver, O. Le Gouezigou, A. Vuong, A. Bodéré, A. Pinquier, and F. Brillouet, "WDM operation of a hybrid emitter integrating a wide-bandwidth on-chip mirror," IEEE J. Select. Topics Quantum Electron. **5**, 476 (1999)

62. H. Bissessur, A. Vuong, C. Graver, O. Le Gouezigou, G. Michaud, and A. Pinquier, "External grating laser-Mach-Zehnder emitter for 2.5 and 10 Gbit/s WDM systems," Electron. Lett. **36**, 139 (2000)

63. E. Zielinski, E. Lach, J. Bouayad-Amine, H. Haisch, E. Kühn, M. Schilling, and J. Weber, "Monolithic multisegment mode-locked DBR laser for wavelength tunable picosecond pulse generation," IEEE J. Select. Topics Quantum Electron. **3**, 230 (1997)

64. E. Jahn, N. Agrawal, H.-J. Ehrke, R. Ludwig, W. Pieper, and H.G. Weber, "Monolithically integrated asymmetric Mach-Zehnder interferometer as a 20 Gbit/s all-optical add/drop multiplexer for OTDM systems," Electron. Lett. **32**, 216 (1996)

65. T. Tekin, M. Schlak, W. Brinker, B. Maul, and R. Molt, "Monolithically integrated MZI comprising band gap shifted SOAs: a new switching scheme for generic all-optical signal processing," *Proc. 26th Europ. Conf. Opt. Commun. (ECOC 2000)*, Munich, Germany, Vol. **3**, 123 (2000)

Biographical Notes

Yuji Akahori was born in Tokyo, Japan in 1964. He received his B.E. and M.E. degrees from Waseda University, Tokyo, and his Ph.D. from Tokyo University, Tokyo, all in electronics, in 1978, 1980 and 1991, respectively. In 1980, he joined the NTT Electrical Communication Laboratories, where he was engaged in the research and development of superconductive logic circuits using Josephson junctions. During 1985–1993, he worked on the development of InP-based monolithically integrated photoreceivers. He is currently working on hybrid integrated PLCs.

Erland Almström received M.Sc. and Ph.D. degrees in electrical engineering from the Royal Institute of Technology in Stockholm in 1993 and 1999, respectively. In 1992 he joined Ellemtel Telecommunication System Laboratories working with reconfigurable WDM networks within the award-winning European RACE project MWTN (Multi-Wavelength Transport Network). In 1995 he joined Ericsson Telecom, Switchlab, where he coordinated the demonstration activities in the European ACTS project METON (Metropolitan Optical Network). Since 1998 he has been engaged in Ericsson optical cross-connect product development working with questions related to optical networking for datacom networks.

Heinz-Gunter Bach was born in Dresden, Germany, in 1953. He received his M.Sc. degree in electrical engineering from the Technical University, Berlin (TUB), in 1978. From 1978 to 1981 he worked at the Hahn-Meitner-Institute, Department of Data Processing and Electronics, working on the electrical characterization of BCCDs. He received his Ph.D. degree from the TUB in 1982. In 1981 he joined the Heinrich-Hertz-Institut für Nachrichtentechnik Berlin GmbH as the head of the electrical characterization group within the Photonics division. He has been engaged in III–V material characterization, heterostructure analysis, and device optimization (invertible bipolar transistors). He has given annual lectures at the Technical University on "measurement techniques for semiconductor devices" and supervises students and diploma work. Further he has been engaged in the project management of topics such as MBE-Al(Ga)InAs layer optimization for heterostructure FETs, opto-electronic packaging and, within the last six years, photoreceiver integration.

Herbert Burkhard received his M.Sc. and Ph.D. degrees in physics at the University of Köln in 1966 and 1969, respectively. From 1970 to 1977 he was an assistant professor at the Department of Physics at the Technical University, Aachen, where he worked on electron–phonon interaction in semiconductors. From 1974 to 1976 he was on leave at the High Magnetic Field Facilities of the Max Planck Institut für Festkörperforschung, Grenoble, France. In 1977 he joined the Deutsche Telekom Research Center. In 1988 he became the head of the research group Optoelectronic Devices, and since 1996 he has headed the Optoelectronics Department. From 1991 to 1999 he was active as a project manager of the EC programs RACE II, ACTS, ESPRIT and TMR, and other nationally funded programs. After leaving Deutsche Telekom in 1999 he became a consultant to Coherent, Semiconductor Group, Santa Clara, CA.

Frank Fidorra studied physics in Regensburg and Frankfurt a.M. and received his diploma degree (Dipl. Phys.) in 1986. For three years he worked for Philips-Kommunikations-Industrie (PKI) in Nürnberg. In 1989 he joined the Heinrich-Hertz-Institut für Nachrichtentechnik in Berlin. Since 1990 he has headed various projects to develop lasers on InP and to investigate integration concepts, including lasers. In 1995 he spent one year at AT&T/Lucent in Holmdel, NJ (USA), joining the group of Tom Koch. In 1996 he became head of a pilot project at HHI working in very close cooperation with an industrial customer. Since then he has worked on InP laser components which are near applications level.

Philip Garel-Jones co-founded JDS Optics, which is now JDS Uniphase Inc., in 1981, and held various positions within the company, including Vice-President of Switches. Prior to 1984 he was with Bell-Northern Research and Northern Telecom and worked in a number of fields, including solid-state emitters, optical fibre development and fibre-optic components. He has a number of patents and papers to his name.

Ken Garrett has over 20 years of experience in telecommunications and fiber optics. During this time, he has held senior-level positions at AT&T, US West, Astarte Fiber Networks (founder), LIGHTech Fiberoptics, and JDS Uniphase. Over the past twelve years he has focused on optical switching and development of optical crossconnect applications. He has developed and implemented optical switch solutions for carriers worldwide and major end-users, with systems being installed at locations that include NASA, The White House, the United Nations and other mission-critical networks. He holds an MS CE degree as well as an MBA degree.

Louis Giraudet was born in Paris in 1962. His Ph.D., obtained in 1988 from the University of Paris at Orsay, was on InP-based HFETs. He joined France Télécom-CNET in 1990 where he was involved in research on InP-based optoelectronics for optical fibre telecom applications. In particular, his work focused on integrated photoreceivers and OEICs. In 1995, he was in charge of a research group working on high-speed APD and PIN photodiodes. In 1998, he joined OPTO+, a joint laboratory funded by France Télécom and Alcatel, where he is now in charge of research into waveguided photodetectors.

Norbert Grote studied physics at the Technical University of Aachen, Germany, from which he received the M.Sc. degree in physics (Dipl. Phys.) and the Ph.D. degree (Dr. Ing.) in 1974 and 1977, respectively. His thesis focused on the LPE growth and LED applications of GaInP and AlInP. Subsequently, he worked on GaAs-based lasers and bipolar transistors. In 1980 he joined the Integrated Optics Division of the Heinrich-Hertz-Institut of Berlin, which had been founded at that time to pursue R&D in the field of photonic devices for optical communication applications. Since then he has been engaged in several national and European R&D projects, with his major research interests being InP-based III–V materials, epitaxy, and electronic and laser devices. Currently he is manager of the Materials Technology Department, supervising R&D projects on optoelectronic III–V materials, single lasers, and silica- and polymer-based photonic devices for optical communications.

Stefan Hansmann was born in Paderborn, Germany, 1961. He received his diploma degree in physics from Stuttgart University in 1988 and his Ph.D. degree from Darmstadt University in 1994. His dissertation was on the spectral properties of semiconductor lasers with distributed feedback. During 1989–1999 he was with the Deutsche Telekom Forschungszentrum, Darmstadt, where he was engaged in the simulation and technological realization of short- and long-wavelength semiconductor lasers and laser arrays. In 1999 he became the head of a research group of Deutsche Telekom, working on the application of photonic devices. Since 2000 he has headed the semiconductor laser group of Opto Speed Deutschland GmbH.

Helmut Heidrich (53) received his Dipl. Phys. and Dr. Ing. degree in physics from the Technische Universität Berlin, in 1973 and 1979, respectively. After three years work at Standard Electric Lorenz AG, Berlin, he joined the Heinrich-Hertz-Institut für Nachrichtentechnik Berlin GmbH, in 1982, where for the first five years he headed a group working on integrated optics LiNbO$_3$ devices. For over a decade, he has been engaged in research on photonic ICs based on InP, and he has headed several national and European projects (RACE, ACTS) within HHI on the realization of complex photonic ICs, e.g. heterodyne receivers, optical sweepers, and bidirectional WDM transceivers. He has (co)authored of more than 100 publications.

Harald Herrmann received his diploma degree in physics (Dipl. Phys.) from the University of Hannover, Germany, in 1984. The same year he joined the Applied Physics/Integrated Optics group of the Physics Department of the University of Paderborn. There he is engaged in the development of integrated optical devices in LiNbO$_3$. In 1991 he received his Ph.D. degree in physics (Dr. rer. nat.), with a thesis on nonlinear frequency generation in optical waveguides. In recent years his main interests have been concentrated on the development of integrated acousto-optical devices in lithium niobate, and their applications in optical communication systems and optical instrumentation.

Winfried H.G. Horsthuis, born in 1956, studied Electrical Engineering and Philosophy of Sciences at Twente University and Groningen University, respectively, both in The Netherlands. He received a Ph.D. at Twente University in April 1987, Following his research on periodic perturbations in optical waveguides. Winfried joined Akzo Nobel Research to continue R&D in optical waveguides. In 1991, he was appointed Program Manager Photonics, responsible for projects in the fields of waveguides, LCDs and optical recording media. He then was named General Manager of Akzo Nobel Photonics, a new business unit specifically created for the commercialization of optical switches based on polymer waveguides. In 1996, he joined JDS Uniphase as Vice President, Strategy and Business Development, with his main focus on Merger & Acquisition activities, as well as intellectual property and technology portfolio management.

Gerlas van den Hoven received his M.Sc. degree in experimental physics from the University of Groningen, The Netherlands, in 1991, where he studied the structure of amorphous silicon. He then moved to the FOM Institute for Atomic and Molecular Physics in Amsterdam, where he worked on planar Er-doped optical amplifiers and Er doping of silicon. In 1996 he obtained his Ph.D. degree in physics and was awarded the Else Kooi prize for microelectronics. In that year he joined Philips Research Laboratories, Eindhoven, engaging in the research and development of semiconductor optical amplifiers. Since 1999 he has been responsible for the product line of active components for optical amplification at JDS Uniphase in Eindhoven.

Sonny Johansson was born in 1961 in Stockholm, Sweden. He received his M.Sc. degree in electrical engineering in 1985 from the Royal Institute of Technology in Stockholm. During 1986–1989, he was a researcher at the Institute of Optical Research. Earlier, in 1981–1982, he participated in designing exchange terminals for the AXE-10 system at Ericsson. In 1989 he joined Ellemtel, Telecommunication System Laboratories. He led the company activities on optical switching and their applications in telecommunication networks. He represented Ellemtel in RACE II-MWTN (Multiwavelength Transport Network), where he coordinated the development activities of the consortium demonstrator, continuing the earlier work in RACE I-OSCAR (Optical Switching, Components and Application Research). By the end of 1995, Sonny moved to Ericsson Telecom, where he represented Ericsson as the prime contractor in the ACTS-METON (Metropolitan Optical Research) consortium. In 1998 he got married and took his wife's name (not an odd habit in Sweden), and is now called Sonny Thorelli. The same year he moved to Ericsson in Budapest, where he is the director of a software development unit.

Ronald Kaiser received his Dipl. Phys. degree and his Dr. rer. nat. from the Technische Universität Berlin (Germany) in 1984 and 1991, respectively. He joined the Heinrich-Hertz-Institut für Nachrichtentechnik Berlin GmbH (HHI) in 1984. At the beginning he was involved in the development, fabrication, and characterization of waveguide integrated photodetectors based on InP. Since 1991 he has been engaged in R&D on integrated lasers and optoelectronic/photonic integrated circuits (OEICs, PICs) on InP with different functionalities, suitable for applications in future optical communication systems.

Antoine Kévorkian received his M.Sc. degree from the University of California San Diego (USA) in 1982 and his Ph.D. degree from the National Polytechnic Institute of Grenoble (France) in 1987. His technical experience in the field includes both silica/silicon and ion-exchange waveguide technologies and the modeling of guided-wave structures. He structured and ran GeeO, an R&D consortium focusing on optical sensors and communications components, for 8 years. He has been President (and was a cofounder) of Teem Photonics, a spin-off from GeeO since Nov. 98. He has been actively involved in the initial research and development of integrated optics erbium amplifiers, now a flagship product of the company. He was also granted patents for other optical functions and power electronics applications. When not involved in operational activities, he has authored or coauthored publications and given short courses at international conferences.

Ton (Antonius Marcellus Jozef) Koonen obtained an M.Sc. degree (cum laude) in Electrical Engineering in 1979 from the Technical University Eindhoven, The Netherlands. In the same year, he joined Philips' Telecommunicatie Industry, part of which became Bell Laboratories, Lucent Technologies Nederland, in Huizen. He is presently a Technical Manager in the Forward Looking Work department. He has also been a part-time professor in Photonic Networks at the Centre for Telematics and Information Technology, University of Twente, in Enschede, since 1991. In June 1999, he received the Bell Labs Fellow degree, the most prestigious recognition within Bell Labs, as one of the two first BLFs outside the US. He has worked in various areas of optical-fibre communication, such as wavelength-multiplexed networks and related passive optical devices. He has been managing two ACTS projects on WDM access networks. His present interests are in broadband optical networking.

R. Ian MacDonald is a Senior Scientist at JDS Uniphase, and a Professor Emeritus of Electrical Engineering at the University of Alberta, Canada. While at the Communications Research Centre in Ottawa he made the first demonstrations of frequency-domain optical reflectometry and optoelectronic downconversion detection, and co-invented and developed the optoelectronic principle for broadband switching and signal processing. In 1985/86 he was guest scientist at the Heinrich Hertz Institut in Berlin. In 1986 he joined the faculty of the University of Alberta in Edmonton and was Director of Photonics at the TRLabs Institute from its inception until 1996. Dr. MacDonald is a Fellow of the Optical Society of America, a Senior Member of IEEE, a past Chairman of the Division of Optics and Photonics of the Canadian Association of Physicists, and a member of the Association of Professional Engineers of Ontario.

Edmond J. Murphy is currently Director of Technology with JDS Uniphase in Bloomfield, CT. He received his B.Sc. degree (1976) in chemistry from Boston College, Chestnut Hill, MA and his Ph.D. degree (1980) in chemical physics from the Massachusetts Institute of Technology, Cambridge, MA. He has worked for AT&T and Lucent Technologies Bell Labs where he worked on optical components, including lithium niobate and glass waveguide devices. He has also worked on optical transmitters and receivers and optical amplifiers. He holds 12 patents and is the author of more than 50 technical papers. He is a Fellow of the Optical Society of America and a senior member of the International Electrical and Electronic Engineers Society.

Masatoshi Saruwatari (born in 1945) received his B.Sc., M.Sc. and Ph.D. degrees in 1968, 1970 and 1980 from the University of Tokyo, Japan. In 1970, he joined the Musashino Electrical Communication Laboratory (ECL), NTT, and studied solid-state lasers and optical components. In 1978, he moved to the Yokosuka ECL and was engaged in the development of large-capacity single-mode fiber transmission systems including the F-400 M, F-1.6 G and FA-10 G systems. From 1990 to 1997, he led a research group at NTT Laboratories, and demonstrated the first successful 100 Gbit/s–1 Tbit/s transmission experiments using newly developed OTDM technologies. Presently, he is a Professor at the National Defense Academy, Yokosuka, Japan. Prof. Saruwatari has published more than 220 technical journal and international conference papers and holds more than 40 patents in the field of optical communication. He received three Excellent Paper Awards, the Distinguished Achievement Award, and the Kobayashi Memorial Prize, all from the IEICE Japan, the 11th Kenjiro Sakurai Memorial Prize from OEIDA, and the Electronics Letters Premium from the IEE. He is a senior member of the IEEE, a member of the IEICE Japan, the Japan Society of Applied Physics, the Laser Society of Japan, and the Optical Society of America (OSA).

André Scavennec received his Doctor-Engineer degree from the University of Paris-Orsay in 1972 for his work on optical pulse generation in lasers. With France Télécom-CNET since 1968, his main activities have involved the technology and characterization of III–V transistors and photodetectors for optical fibre telecom applications, as researcher, group leader, and, finally, department manager. In 1998, he was appointed deputy manager of OPTO+.

Meint K. Smit studied Electrical Engineering at the Delft University of Technology. In 1974 he started work in a Microwave Remote Sensing group on FM-CW radar development. He switched to optical communication in 1981, where he worked on semiconductor-based Photonic Integrated Circuits.

He invented the phased-array wavelength demultiplexer (PHASAR, AWG or WGR) and worked on multimode interference (MMI) couplers, optical switches, and the measurement and characterization of electro-optical devices and the development of Computer-Aided Design Tools. In 1997 he received an IEEE/LEOS Technical Award for the invention of the phased-array demultiplexer.

Wolfgang Sohler was born in Wangen, Germany, in 1945. He received the Diplom-Physiker and Dr. rer. nat. degrees in physics from the University of Munich, in 1970 and 1974, respectively. From 1975 to 1980 he was with the University of Dortmund, working on integrated optics. In 1980 he joined the Fraunhofer Institut für Physikalische Meßtechnik, Freiburg, as head of the Department of Fibre Optics. Since 1982 he has been with the University of Paderborn, as Professor of Applied Physics. His research interests include integrated optics, fibre optics and laser physics. He is the (co)author of more than 100 journal contributions and of several book chapters. Dr. Sohler has been a member of the program committee of several (international) conferences on integrated optics. He is a member of the German Physical Society, the German Society of Applied Optics, and the Institute of Electrical and Electronics Engineers.

Hiroshi Terui was born in Hokkaido, Japan in 1950. He received his B.Sc. and M.Sc. degrees in physics from Tohoku University, Sendai, in 1972 and 1974, respectively. In 1974, he joined NTT Electrical Communication Laboratories, where he was engaged in research on materials for magneto-optic recording. Since 1976, he has been engaged in research on guided-wave optical devices. He is currently an engineering director in the NTT Electronics Corporation (NEL) and is developing hybrid integrated PLCs.

Réal Vallée has been Associate Professor of Physics at l'Université Laval, in Québec city, Canada, since 1996. He received his B.Sc. and Ph.D. degrees in physics from the same university in 1982 and 1986, respectively. He joined the Centre d'Optique Photonique et Laser (COPL) and the Physics department of the l'Université Laval in 1987 after a postdoctoral year at the Laboratory for Laser Energetics of the University of Rochester. He is a member of the Optical Society of America and of the advisory editorial board of the journal Optics Communications. He has been the director of the COPL since February 2000.

Herbert Venghaus received a diploma and doctoral degree in physics at Hamburg University. Subsequently he investigated the optical properties of semiconductors for several years at the Max-Planck-Institut für Festkörperforschung in Stuttgart. After five years with Siemens Corporate Research he joined the Heinrich-Hertz-Institut für Nachrichtentechnik, Berlin, in 1986, where he is currently heading the Integration Technology Department, focusing on the development of InP-based monolithic Optoelectronic Integrated Circuits. He has been a subproject leader in European projects, served repeatedly as an auditor to the European Commission, and has also been a programme committee member for various international conferences.

Carl Michael Weinert received his Dr. rer. nat. degree in 1980 from the Technische Universität Berlin where he was engaged in theoretical work on deep impurities in semiconductors. In 1987 he joined the Integrated-Optics Division at the Heinrich-Hertz Institut Berlin (HHI) where he worked on the modeling and numerical simulation of integrated-optics devices. At present he is in the Optical Signal Processing department at HHI and focuses on the modeling of photonic systems.

Yasufumi Yamada was born in Tokyo, Japan in 1956. He received his B.Sc., M.Sc. and Ph.D. degrees from Waseda University, Tokyo, in 1980, 1982 and 1990, respectively. In 1982, he joined the NTT Electrical Communication Laboratories, where he has been engaged in the research and development of planar lightwave circuits (PLCs) and optical hybrid integration technology using PLCs. He is currently an engineering director in the NTT Electronics Corporation (NEL) and is developing hybrid integrated PLCs.

Mikhail N. Zervas joined the Optoelectronics Research Centre (ORC), University of Southampton (UoS) as a Research Fellow in 1991. In 1995, he was appointed Research Lecturer at the ORC and Electronics and Computer Science (ECS) Department, UoS. In 1999, he was appointed Professor in Optical Communications at ORC/ECS, UoS. His main research interests are in the areas of fibre amplifiers, fibre and planar waveguide gratings, DFB fibre lasers and optical trapping. In 1996 he shared an award on "Metrology for World Class Manufacturing" for the development of a high-accuracy fibre grating characterisation system. He is the author and co-author of over 110 technical publications and 15 patents. He has served as a member of the programme committees of various international conferences. He was the 1999 General Chair of the OSA Topical Meeting on Optical Amplifiers and Their Applications.

Index